A Handbook of Applied Biopolymer Technology
Synthesis, Degradation and Applications

RSC Green Chemistry

Series Editors:
James H Clark, *Department of Chemistry, University of York, York, UK*
George A Kraus, *Department of Chemistry, Iowa State University, Iowa, USA*

Titles in the Series:
1: The Future of Glycerol: New Uses of a Versatile Raw Material
2: Alternative Solvents for Green Chemistry
3: Eco-Friendly Synthesis of Fine Chemicals
4: Sustainable Solutions for Modern Economies
5: Chemical Reactions and Processes under Flow Conditions
6: Radical Reactions in Aqueous Media
7: Aqueous Microwave Chemistry
8: The Future of Glycerol: 2nd Edition
9: Transportation Biofuels: Novel Pathways for the Production of Ethanol, Biogas and Biodiesel
10: Alternatives to Conventional Food Processing
11: Green Trends in Insect Control
12: A Handbook of Applied Biopolymer Technology: Synthesis, Degradation and Applications

How to obtain future titles on publication:
A standing order plan is available for this series. A standing order will bring delivery of each new volume immediately on publication.

For further information please contact:
Book Sales Department, Royal Society of Chemistry, Thomas Graham House, Science Park, Milton Road, Cambridge, CB4 0WF, UK
Telephone: +44 (0)1223 420066, Fax: +44 (0)1223 420247
Email: books@rsc.org
Visit our website at http://www.rsc.org/Shop/Books/

A Handbook of Applied Biopolymer Technology
Synthesis, Degradation and Applications

Edited by

Sanjay K. Sharma
Jaipur Engineering College & Research Centre, Jaipur, Rajasthan, India

Ackmez Mudhoo
*Department of Chemical and Environmental Engineering,
University of Mauritius, Reduit, Mauritius*

RSCPublishing

RSC Green Chemistry No. 12

ISBN: 978-1-84973-151-5
ISSN: 1757-7039

A catalogue record for this book is available from the British Library

© Royal Society of Chemistry 2011

All rights reserved

Apart from fair dealing for the purposes of research for non-commercial purposes or for private study, criticism or review, as permitted under the Copyright, Designs and Patents Act 1988 and the Copyright and Related Rights Regulations 2003, this publication may not be reproduced, stored or transmitted, in any form or by any means, without the prior permission in writing of The Royal Society of Chemistry or the copyright owner, or in the case of reproduction in accordance with the terms of licences issued by the Copyright Licensing Agency in the UK, or in accordance with the terms of the licences issued by the appropriate Reproduction Rights Organization outside the UK. Enquiries concerning reproduction outside the terms stated here should be sent to The Royal Society of Chemistry at the address printed on this page.

The RSC is not responsible for individual opinions expressed in this work.

Published by The Royal Society of Chemistry,
Thomas Graham House, Science Park, Milton Road,
Cambridge CB4 0WF, UK

Registered Charity Number 207890

For further information see our web site at www.rsc.org

Preface

Researchers are conducting active research in different fields of engineering, science and technology by adopting the 12 Principles of Green Chemistry and the inherent green methodologies to devise new processes with a view to help protect and ultimately save the environment from further anthropogenic interruptions and damage. With this in focus, this book provides an up-to-date, coherently written and objectively presented set of book chapters from eminent international researchers who are actively involved in academic and technological research in the synthesis, degradation, testing and applications of biodegradable polymers and biopolymers. Hence, the overall pool of latest ideas and recent research and technological progress achieved in the synthesis, degradation, testing and applications of biodegradable polymers/biopolymers together with a high level of thinking have been presented in a comprehensive perspective to make progress in the emerging field of biodegradable polymer science and engineering (or bio-based polymers). The element of environmental sustainability as linked to biopolymer technology also constitutes the essence and novelty of this very relevant book in today's era of environmental depredation.

This book consists of book chapters written and contributed by international experts from academia who are world leaders in research and technology regarding sustainability and biopolymer and biodegradable polymer synthesis, characterization, testing and use. The book highlights the following areas: Green polymers; Biopolymers and bionanocomposites; Biodegradable and injectable polymers; Biodegradable polyesters: Synthesis and physical properties; Discovery and characterization of biopolymers; Degradable bioelastomers, Lactic acid-based biodegradable polymers; Biodegradation of biodegradable polymers; Biodegradation of polymers in the composting environment; and Recent research and application development in biodegradable polymers. The book is aimed at technical, research-oriented and marketing people in industry, universities and

RSC Green Chemistry No. 12
A Handbook of Applied Biopolymer Technology: Synthesis, Degradation and Applications
Edited by Sanjay K. Sharma and Ackmez Mudhoo
© Royal Society of Chemistry 2011
Published by the Royal Society of Chemistry, www.rsc.org

institutions. The book will also be of value to the worldwide public interested in sustainability issues and biopolymer development and as well as others interested in the practical means that are being used to reduce the environmental impacts of chemical processes and products, to further eco-efficiency, and to advance the utilization of renewable resources in bio-based production and the supplier chain. The main outcomes of reading this book should be that the reader will have a comprehensive and consolidated overview of the immense potential and ongoing research in bio-based and biodegradable polymer science, engineering and technology, which is earnestly attempting to make the world of tomorrow greener. Hence, this handbook is a reasonably comprehensive and applied treatise of the topic and provides up-to-date information to a very wide audience on the applied research areas of biopolymers.

Sanjay K. Sharma
Ackmez Mudhoo

This book is for Kunal and Kritika, my twin angels.
Sanjay K. Sharma

For you Neelam.
Ackmez Mudhoo

Contents

About the Editors   xvii

Chapter 1   History of Sustainable Bio-based Polymers   1
Tim A. Osswald and Sylvana García-Rodríguez

    1.1   Background   1
    1.2   Silk: From a Royal Stitch to a Wounded Peasant   4
    1.3   Cellulose: The Quintessential Bio-based Plastic   7
    1.4   Casein Plastics: From Food to Plastic   9
    1.5   Soy Protein Plastic: Back to Nature   12
    1.6   Building Scaffolds for Our Bodies: Collagen
         and Chitosan   15
    1.7   Letting Bacteria Make Our Plastics   17
    1.8   Conclusions   19
    References   19

Chapter 2   Synthetic Green Polymers from Renewable Monomers   22
Naozumi Teramoto

    2.1   Introduction   22
    2.2   Triglycerides of Fatty Acids and their Derivatives   24
         2.2.1   Monomers from Triglycerides   24
         2.2.2   Polymers Synthesized from Triglycerides   29
    2.3   Essential Oils, Natural Phenolic Compounds and
         their Derivatives   34
         2.3.1   Terpenoids   35
         2.3.2   Phenylpropanoids   40
         2.3.3   Lignin Digests or Extracts and
                Liquefied Wood   46
         2.3.4   Other and Natural Phenolic Compounds   48

	2.4	Carbohydrates and their Derivatives	51
		2.4.1 Polymers from Popular Carbohydrates	51
		2.4.2 Furan Derivatives	54
	2.5	Monomers Obtained by Fermentation	55
	2.6	Conclusions and Outlook	60
	References		61
Chapter 3	**Polyhydroxyalkanoates: The Emerging New Green Polymers of Choice**		**79**
	Ranjana Rai and Ipsita Roy		
	3.1	Introduction	79
	3.2	History of Polyhydroxyalkanoates	80
	3.3	Chemical Organization of PHAs	80
	3.4	Occurrence and Biosynthesis of PHAs	81
	3.5	Cheap Substrates for Cost-effective PHA Production	86
	3.6	Physical Properties of PHAs	86
	3.7	Biocompatibility of PHAs	88
	3.8	Biodegradation of Polyhydroxyalkanoates	89
		3.8.1 Factors Affecting Biodegradation	89
		3.8.2 Biodegradation in the Environment	89
		3.8.3 Biodegradation and Biocompatibility	90
	3.9	Applications of Polyhydroxyalkanoates	90
		3.9.1 Industrial Application	90
		3.9.2 Medical Applications	91
	3.10	PHAs as Green Biofuels	94
	3.11	Market and Economics of PHAs	95
	3.12	Concluding Remarks	97
	Acknowledgement		97
	References		97
Chapter 4	**Fully Green Bionanocomposites**		**102**
	P. M. Visakh, Sabu Thomas and Laly A. Pothan		
	4.1	Green Composites – Introduction	102
	4.2	Green Materials: Fibres, Whiskers, Crystals and Particles	103
		4.2.1 Cellulose Fibres	103
		4.2.2 Chitin Whiskers	105
		4.2.3 Starch Crystals	107
		4.2.4 Soy Protein Particles	110
		4.2.5 Polylactic Acid	111
		4.2.6 Natural Rubber Uncross-linked Particles	114
	4.3	Green Nanocomposites	114
		4.3.1 Cellulose-based Green Composites	114
		4.3.2 Chitin and Chitosan-based Green Composites	116

Contents   xi

	4.3.3 Starch-based Green Composites	117
	4.3.4 Soy Protein-based Green Composites	117
	4.3.5 PLA-based Green Composites	118
4.4	Applications	121
4.5	Conclusion	122
	References	123

Chapter 5 Biopolymer-based Nanocomposites   129
Kikku Fukushima, Daniela Tabuani and Cristina Abbate

5.1	Introduction	129
5.2	Experimental	131
	5.2.1 Materials and Methods	131
	5.2.2 Biodegradation Conditions and Evaluation Methods	133
5.3	Results and Discussions	133
	5.3.1 Characterization	133
	5.3.2 Biodegradation	139
5.4	Conclusions	145
	Acknowledgements	146
	References	146

Chapter 6 Biodegradable Polyesters: Synthesis and Physical Properties   149
Jasna Djonlagic and Marija S. Nikolic

6.1	Introduction	149
6.2	Poly(α-hydroxy acid)s	153
	6.2.1 Poly(glycolic acid)	153
	6.2.2 Poly(lactic acid)	154
6.3	Poly(ε-caprolactone)	161
	6.3.1 Synthesis of Poly(ε-caprolactone)	161
	6.3.2 Properties and Degradation of Poly(ε-caprolactone)	164
6.4	Poly(hydroxyalkanoate)s	166
	6.4.1 Synthesis of Poly(hydroxyalkanoate)s	166
	6.4.2 Properties and Degradation of Poly(hydroxyalkanoate)s	171
6.5	Poly(alkylene dicarboxylate)s	174
	6.5.1 Synthesis of Poly(alkylene dicarboxylate)s	174
	6.5.2 Properties and Degradation of Poly(alkylene dicarboxylate)s	177
6.6	Application of Biodegradable Polyesters	181
	6.6.1 Ecological Applications	181
	6.6.2 Medical Applications	184
6.7	Future Trends in Biodegradable Polyesters	185
	References	185

Chapter 7	Synthesis and Characterization of Thermoplastic Agro-polymers	197
	C. J. R. Verbeek and J. M. Bier	

	7.1	Introduction	197
		7.1.1 Polysaccharides	198
		7.1.2 Proteins	200
	7.2	Synthesis	200
		7.2.1 General Considerations	201
		7.2.2 The Role of Additives	201
		7.2.3 Starch	202
		7.2.4 Proteins	204
	7.3	Characterization	206
		7.3.1 Overview of Characterization	206
		7.3.2 Mechanical Behaviour	210
		7.3.3 Thermal Properties	218
	7.4	Conclusions	236
	References		237

Chapter 8	Degradable Bioelastomers: Synthesis and Biodegradation	243
	Q. Y. Liu, L. Q. Zhang and R. Shi	

	8.1	Character, Definition and Category of Degradable Bioelastomers	243
	8.2	Requirements of Degradable Bioelastomers	245
		8.2.1 Safety	245
		8.2.2 Biodegradation	246
		8.2.3 Cross-linking	247
	8.3	Synthesis and Biodegradation of Degradable Bioelastomers	247
		8.3.1 Degradable Segmented Polyurethane Bioelastomers	248
		8.3.2 Poly(ε-caprolactone) Related Bioelastomers	254
		8.3.3 Polylactide-related Bioelastomers	259
		8.3.4 Polycarbonate-related Bioelastomers	264
		8.3.5 Poly(glycerol sebacate) Bioelastomer and its Derivatives	267
		8.3.6 Citric Acid-related Polyester Bioelastomers	271
		8.3.7 Poly(ether ester) Bioelastomers	276
		8.3.8 Poly(ester amide) Bioelastomers	278
		8.3.9 Other Novel Degradable Bioelastomers	280
	8.4	Conclusions	284
	Acknowledgements		284
	References		284

Chapter 9 Functionalization of Poly(L-lactide) and Applications of the Functionalized Poly(L-lactide)   291
Xiuli Hu and Xiabin Jing

    9.1 Introduction   291
    9.2 PLA Functionalization   293
        9.2.1 Morpholine Diones   293
        9.2.2 α-Amino acid *N*-Carboxyanhydride (NCA)   295
        9.2.3 Cyclic Carbonates   297
        9.2.4 Lactones   299
        9.2.5 Cyclic Diesters   299
    9.3 Applications of the Functionalized PLAs   299
        9.3.1 Drug Delivery Systems   300
        9.3.2 Artificial Oxygen Carriers   302
        9.3.3 Protein Separation and Purification   304
    9.4 Conclusions   305
    References   305

Chapter 10 Biodegradation of Poly (3-hydroxyalkanoates)   311
Rachana Bhatt, Kamlesh Patel and Ujjval Trivedi

    10.1 Introduction   311
    10.2 Degradation of Plastics   314
        10.2.1 Abiotic Degradation   314
        10.2.2 Biotic Degradation   314
        10.2.3 Standard Methods for Plastic Biodegradation Studies   315
    10.3 Biodegradation of Polyhydroxyalkanoates   316
        10.3.1 Extracellular Degradation of PHA   316
        10.3.2 Intracellular Degradation of PHA   325
    10.4 Conclusions   327
    References   327

Chapter 11 Degradation of Biodegradable and Green Polymers in the Composting Environment   332
Ackmez Mudhoo, Romeela Mohee, Geeta D. Unmar and Sanjay K. Sharma

    11.1 Introduction   332
        11.1.1 Biodegradable Polymers   333
        11.1.2 Degradability through Composting   333
    11.2 Degradation of Biodegradable Polymers   334
        11.2.1 Polymer Biodegradation Mechanisms   334
        11.2.2 Assessment of Biodegradable Polymers Degradability   336
        11.2.3 Biodegradable Polymers Blends   338

	11.3	Composting Process Essentials		339
		11.3.1 Composting Chemistry		339
		11.3.2 Physical Parameters in Composting		344
		11.3.3 Composting Systems		345
		11.3.4 Vermicomposting		345
	11.4	Biopolymer Degradation and Composting		345
		11.4.1 Polyhydroxyalkanoates		346
		11.4.2 Poly(lactic acid)-based Polymers		348
		11.4.3 Polyethylenes		350
		11.4.4 Poly-ε-caprolactones		355
	11.5	Concluding Remarks		357
	Acknowledgements			358
	References			358

Chapter 12 Biodegradable Polymers: Research and Applications 365
X. W. Wei, G. Guo, C. Y. Gong, M. L. Gou and Zhi Yong Qian

12.1	Introduction		365
	12.1.1	Biodegradable Polymers and the Environment	366
	12.1.2	Biodegradable Polymers and Biomedical Uses	367
12.2	Natural Biodegradable Polymers and their Derivatives		368
	12.2.1	Starch and Derivatives	368
	12.2.2	Cellulose and Derivatives	369
	12.2.3	Chitin and Chitosan	370
	12.2.4	Alginic Acid	371
	12.2.5	Collagen	372
	12.2.6	Gelatin	373
	12.2.7	Other Biodegradable Natural Polymers	373
12.3	Synthetic Polymers		374
	12.3.1	Polyesters	374
	12.3.2	Polyurethanes	379
	12.3.3	Polyamides	380
	12.3.4	Polyanhydrides	380
	12.3.5	Polyphosphoesters	381
	12.3.6	Others	382
12.4	Conclusions		383
References			383

Chapter 13   Impacts of Biodegradable Polymers: Towards Biomedical Applications   388
Y. Omidi and S. Davaran

13.1	Introduction		388
13.2	Classification of Biodegradable Polymers		390
13.3	Biodegradable Polyesters		391
	13.3.1	Properties of PLA/PGA Polymers	392
	13.3.2	Pharmaceutical Application of Biodegradable Polyesters	393
	13.3.3	Impacts of Micro and Nano Fabrication of PLGA-based Copolymers	394
	13.3.4	Biocompatible Magnetite-PLGA Composite Nanoparticles	399
	13.3.5	PLGA-based Carriers for Macromolecule Delivery	400
	13.3.6	Application of Polyester Polymers in Tissue Engineering	405
13.4	Functional Polymers: Cellular Toxicity		408
13.5	Genocompatibility and Toxicogenomics of Polymers		409
13.6	Final Remarks		413
Acknowledgements			414
References			414

Chapter 14   Biodegradable Injectable Systems for Bone Tissue Engineering   419
Richard T. Tran, Dipendra Gyawali, Parvathi Nair and Jian Yang

14.1	Introduction		419
14.2	Rationale and Requirements for Injectable Bone Tissue Engineering		421
	14.2.1	Injectability and *In Situ* Cross-linking	422
	14.2.2	Mechanical Properties	422
	14.2.3	Porosity	423
	14.2.4	Biodegradation	423
	14.2.5	Cellular Behavior	424
	14.2.6	Biocompatibility	425
14.3	Network Formation		425
	14.3.1	Free Radical Polymerization (FRP)	426
	14.3.2	Chemical Cross-linking Systems (CCS)	427
	14.3.3	Thermally Induced Gelation Systems (TGS)	428
	14.3.4	Self-assembly Systems (SAS)	428
	14.3.5	Ion-mediated Gelation Systems (IGS)	429

14.4	Injectable Ceramics	429
14.5	Injectable Cell Vehicles	431
	14.5.1 Naturally Derived Hydrogels	432
	14.5.2 Synthetic-based Hydrogels	433
14.6	Injectable Drug Delivery Systems	434
	14.6.1 Antibiotic Delivery	435
	14.6.2 Growth Factor Delivery for Osteogenesis	436
14.7	Citric Acid-based Systems	439
14.8	Future Directions	441
14.9	Conclusions	441
	Acknowledgments	442
	References	442

Chapter 15 Production of Polyhydroxybutyrate (PHB) from Activated Sludge — 452
M. Suresh Kumar and Tapan Chakrabarti

15.1	Introduction	452
15.2	Polymers	453
15.3	Storage Polymers in Microorganisms	454
	15.3.1 PHB Biosynthesis	454
	15.3.2 Enzymes Involved in Biosynthesis and Degradation	455
	15.3.3 Properties of PHB	456
	15.3.4 Potential Applications	457
	15.3.5 Biodegradation of PHB	458
15.4	PHB Production	458
	15.4.1 PHB Production with Pure Substrates	458
	15.4.2 PHB Production with Wastes	459
	15.4.3 PHB Production by Mixed Culture	460
15.5	Factors Affecting PHB Production	462
	15.5.1 Feast/Famine Conditions	462
	15.5.2 Microaerophilic Conditions	463
	15.5.3 Carbon/Nitrogen Limitation Conditions	464
	15.5.4 Phosphate Limitation Conditions	464
15.6	PHB Yields and Recovery Processes	465
15.7	Techno-economic Feasibility	467
15.8	Conclusions	468
	References	468

Subject Index — 473

About the Editors

Prof. (Dr.) Sanjay K. Sharma is a very well known author and editor of many books, research journals and hundreds of articles from last twenty years. Dr. Sharma did his Post Graduation (1995) and Ph.D. (1999) from the University of Rajasthan, Jaipur. His field of work was synthetic organophosphorus chemistry and computational chemistry for his Ph.D. In 1999, he joined the Institute of Engineering & Technology, Alwar and started working additionally in the field of environmental chemistry and established a green chemistry research laboratory. His work in the field of green corrosion inhibitors is very well recognized and praised by international research community. Other than this he is known as a person who is dedicated to educate people about environmental awareness, especially for rain water harvesting. He is member of the American Chemical Society (USA), International Society for Environmental Information Sciences (ISEIS, Canada) and Green Chemistry Network (Royal Society of Chemistry, UK) and is also life member of various international professional societies including International Society of Analytical Scientists, Indian Council of Chemists, International Congress of Chemistry and Environment and Indian Chemical Society. Dr. Sharma has 9 textbooks and over 40 research papers of national and international repute to his credit, which is evidence of his excellent track record as a researcher. Dr. Sharma is also serving as Editor-in-Chief for three international research journals: *RASAYAN Journal of Chemistry*, *International Journal of Chemical, Environmental and Pharmaceutical Research* and *International Journal of Water Treatment & Green Chemistry* and is reviewer of many other international journals including prestigious *Green Chemistry Letters & Reviews*. Recently, he has also published an edited book *Green Chemistry for Environmental Sustainability* in the *Sustainability* series (Publisher: Taylor & Francis Group, LLC, Florida, Boca Raton, USA, CRC Press). He was formerly Professor in the Department of Chemistry in IET and was teaching engineering chemistry and environmental engineering courses to B.Tech. students and biochemistry and environmental science to post-graduate students. Dr. Sharma has delivered many guest lectures on different topics of applied chemistry in various reputed institutions. He is presently Professor of Chemistry at Jaipur Engineering College & Research Centre, JECRC Foundation, Jaipur (Rajasthan) India.

Mr Ackmez Mudhoo, presently Lecturer in the Department of Chemical and Environmental Engineering at the University of Mauritius, obtained his Bachelors degree (B.Eng. (Hons.)) in chemical and environmental engineering from the same university in 2004. He then read a master of philosophy (M.Phil.) degree by research in the same department from 2005 to 2007. His research interests encompass the bioremediation of solid wastes and wastewaters by composting, anaerobic digestion, phytoremediation and biosorption. Ackmez has 48 international journal publications (original research papers, critical reviews and book chapters), 5 conference papers to his credit, and an additional 7 research and review papers in the pipeline in his early career. Ackmez also serves as peer reviewer for *Waste Management, International Journal of Environment and Waste Management, Journal of Hazardous Materials, Journal of Environmental Informatics, Environmental Engineering Science, RASAYAN Journal of Chemistry, Ecological Engineering, Green Chemistry Letters and Reviews, Chemical Engineering Journal* and *Water Research*. He is also the Editor-in-Chief for *International Journal of Process Wastes Treatment* and *International Journal of Wastewater Treatment and Green Chemistry*; and serves as handling editor for *International Journal of Environment and Waste Management* and *International Journal of Environmental Engineering*. Ackmez also has professional experience as consultant chemical process engineer for China International Water & Electric Corp. (CWE, Mauritius) from February 2006 to March 2008. Ackmez is also the co-editor of *Green Chemistry for Environmental Sustainability* (Publisher: Taylor & Francis Group, LLC, Florida, Boca Raton, USA, CRC Press).

CHAPTER 1
History of Sustainable Bio-based Polymers

TIM A. OSSWALD AND SYLVANA
GARCÍA-RODRÍGUEZ

University of Wisconsin-Madison, Department of Mechanical Engineering, 1513 University Avenue, Madison, WI 53706-1572, United States of America

1.1 Background

Hidden in an old Art Nouveau townhouse in Essen, Germany, is the world's largest collection of plastic artifacts. This is the private Kölsch Collection, created by the architects Ulrich and Ursula Kölsch, which houses tens of thousands of plastic items, all numbered and placed on shelves from floor to ceiling in every room of the ground floor flat. With every single item perfectly and painstakingly catalogued by Mr and Mrs Kölsch, the collection is the most complete archive of the developments in the plastics industry over a 150-year span. We not only see the development of plastics, but also how this unique material helped in the development and change of design and aesthetics. The Kölsch's collection shows us how, in the first 80 years of the plastics industry, products were made exclusively of biopolymers, most from renewable resources such as cellulose, casein, shellac and ebonite. Here and there, we stumble upon items made of plastic materials that most of us have never even heard of, such as Bois Durci and kopal resin. However, these materials, long dropped from our collective memories, helped shape what is today the plastics industry.

For example, Bois Durci or *Hardened Wood*, an early plastic that dates back to the 1850s, is a mixture of sawdust (usually of a hardwood such as ebony or

---

RSC Green Chemistry No. 12
A Handbook of Applied Biopolymer Technology: Synthesis, Degradation and Applications
Edited by Sanjay K. Sharma and Ackmez Mudhoo
© Royal Society of Chemistry 2011
Published by the Royal Society of Chemistry, www.rsc.org

rosewood), carbon black, metal particles and blood or egg. The sawdust was mixed with vegetable oils, mineral or metallic fillers, and the blood with a gelatinous substance diluted in water. The dry and wet components were mixed and compressed into finished parts in a steel mold held in a steam-heated screw press. Figure 1.1 shows a French mirror made of this early plastic material by compression molding using a screw press.

Continuing the walk through the Kölsch's art townhouse, we find colorful kopal artifacts brought to us in the Art Deco period between 1921 and 1931 by the EBENA company in Belgium. Also molded in a heated screw press, these items, most of which are beautiful decorative Art Deco bowls, containers, boxes, lamps, radios (Figure 1.2), and clocks, are made of a mixture of wood fiber, pigments and a fossilized tree resin from Congo called *kopal*. Kopal is found in the ground in the jungles of Africa. It can be considered to be the inland counterpart of amber, which is found in the North and Baltic Seas. The EBENA factory in Wijnegem, Belgium, existed between 1921 and 1931, and was the only manufacturing facility to ever produce kopal products. Since EBENA closed in 1931, the expensive fossilized resins such as kopal and amber have been forgotten as molding materials, except for perhaps a few jewellery applications.

In their collection, it is not difficult to stumble upon items made of casein, a milk protein obtained *via* acidification or enzymatic action. The resulting curds can be dried, molded, and treated with a hardening agent to yield commercial plastic. Casein had its heyday as a commercial plastic in the early 1900s; the same time period when many other items that are found in the Kölsch's collection were produced.

Figure 1.1 Bois durci mirror from France, circa 1880 (Kölsch Collection).

History of Sustainable Bio-based Polymers 3

Figure 1.2 Jewellery box molded from kopal resin by EBENA (Kölsch Collection).

The recent revival of interest in these materials makes the story of their discovery and development even more relevant today. The great success of petrochemical-based polymers from the Second World War through the present is testament to the versatility, economy, and durability of such synthetic materials. However, the indestructibility that for decades made petrochemical-based synthetics so desirable has increasingly become a liability. Over 25 million tons of plastic entered the municipal solid waste stream in the USA in 2001. This non-biodegradable plastic waste accounted for over 11% of the municipal solid waste in the USA, up from 1% in 1960.[1] Incinerating plastics can cause toxic air pollution; plastic litter is unsightly. Thus there are health, environmental, and aesthetic problems with continued use of non-biodegradable petrochemical-based polymers. The expanded use of renewable, biodegradable biopolymers would alleviate the problems associated with disposal of non-biodegradable polymers. In addition, the increasing monetary and political costs of American and European dependence on foreign sources of oil make sustainable, domestically grown resources a desirable alternative.

Renewed interest in biopolymers since the mid-1990s has shown itself in new research into the processing and properties of renewable, biodegradable materials like casein, soy protein, polylactic acid (PLA), polyhydroxyalkanoates (PHA), and other novel materials. This work has demonstrated the opportunity for renewable, biodegradable biopolymers to replace their synthetic counterparts in a variety of applications. But the interest, needs, and materials that are re-emerging are a continuation of a story that began centuries earlier; the opening chapter of this ongoing story is told here.

This chapter presents various biopolymers within a historical perspective in relation to today's needs. We will cover silk, casein, and soy, as well as other materials that have recently been found useful in recent applications such as chitin, collagen, PHA, and PLA.

1.2 Silk: From a Royal Stitch to a Wounded Peasant

According to the Chinese legend, the Empress Si Ling-Chi was drinking a cup of tea under a mulberry tree, when something suddenly fell from the tree and into her cup. As she removed it, she slowly unraveled a thread from what she found was a silkworm cocoon. This 5000-year-old legend marks the discovery of silk, a material that was then processed into fine threads and considered the cloth of the gods. Only royalty was allowed to have it, and anyone who tried to trade silk, silkworms, or mulberry trees was punished with death: silk and its production was kept secret for approximately 3000 years. In the first century A.D., this secret slowly leaked to the outside world, and soon silk became a luxurious fabric throughout Asia and Europe, giving rise to a network of trade routes called the Silk Road.[2]

Silk is a protein polymer (Figure 1.3), whose amino acid composition depends on the producing species. Several lepidoptera larvae are capable of producing silk, but silks produced by spiders and silkworms have been the most studied. The silkworm *Bombyx mori* has been the most popular species for silk production. Its silk is characterized by fibroin fibers that are held together by a coat of a glue-like protein called sericin (absent in spider silk). This way, composite fibers are arranged and held together to protect the worm inside the cocoon. Silk provides great toughness and strength, as well as very high elasticity. Its resistance to failure under compression makes it comparable to Kevlar.[3]

Silk could be traced back, as a suturing material, to the second century A.D., when Galen of Pergamon wrote *De Methodo Medendi*. He was well known in the times of the Roman Empire for treating and suturing injured tendons of gladiators. In his work, he stated: "Moreover let ligatures be of a material that does not rot easily like that of those brought from Gaul and sold especially in the Via Sacra ..." (referring to linen or Celtic thread). He continued: "... In many places under Roman rule you can obtain silk, especially in large cities where there are many wealthy women".[5]

In the battlefields of Crécy, northern France, in 1346, cobweb was popular for stopping a wound from bleeding.[5,6] Its styptic properties (for stopping

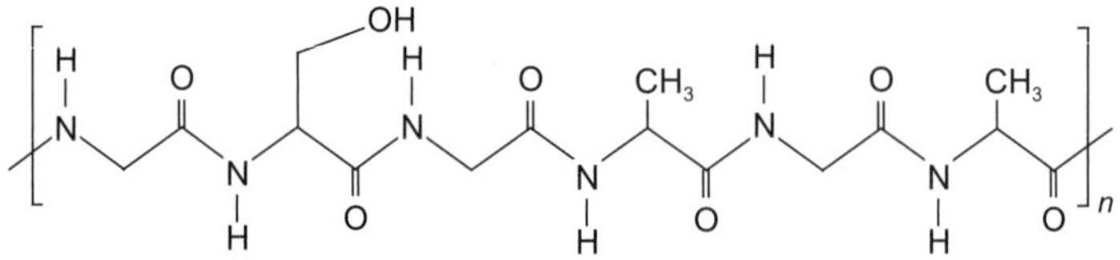

Figure 1.3 Repeating chemical structure of silk fibroin, composed of the amino acid sequence: glycine-serine-glycine-alanine-glycine-alanine.[4]

bleeding) were still popular two centuries later, as reflected in one of Shakespeare's comedies, *A Midsummer Night's Dream*, where he wrote: "I shall desire you of more acquaintance, good master cobweb. If I cut my finger I shall make bold with you ...".[7]

Around the end of the eighteenth century, it was established that bleeding vessels were better treated by ligatures (tying up the ends of the vessels) than by cautery (burning the ends of the vessels). By this time, waxed thread had been replaced by silk as the material of choice. Philip Syng Physick (1768–1837) was an American who became the first professor of surgery at the University of Pennsylvania. Following the teachings of his mentor, the famous John Hunter, a Scot who became the founder of experimental surgery and surgical pathology, Physick used adhesive leather strips to close a wound. He then noticed that these dissolved in contact with fluids from the wound. He thought this characteristic would be of great advantage in the use of ligatures. This idea was historic, since no one had previously thought of a suture that would be absorbed after performing its function.

In 1867, Joseph Lister, among his great contributions in antisepsis, wrote an article "*Observations of ligature of arteries on the antiseptic system*". He believed that a silk ligature could be left in the body if bacteria lying within the interstices of the threads could be killed. At that time, ligatures were left long and protruding through the wound to then be pulled out along with the necrosed or dead tissue at the end of the vessel, increasing the risk of a secondary hemorrhage. In his experiments, he started using antiseptic silk ligatures soaked in an aqueous solution of carbolic acid, where he found the ligature was not absorbed after ten months of implantation on the external iliac artery of a 50-year-old woman. These results lead him to explore, in 1868, the use of ox peritoneum and catgut in a carbolic acid solution, seeking an antiseptic absorbable ligature.

In 1881, arguing that carbolized catgut (referring to Lister's mixture of olive oil and carbolic acid) was not an effective antiseptic, Kocher of Berne started a campaign against catgut and in favor of silk. In his rules of surgery, Halsted, who introduced thin rubber gloves in 1890, recommends: "... gentle handling of tissues, meticulous haemostasis, and interrupted silk sutures".

By 1900, however, the catgut industry was firmly established in Germany, using the intestines of sheep, important in their sausage industry.[5] Nevertheless, both catgut and silk were important base materials for the production of sutures for the following 100 years.[8]

By then, the silk industry was large and of great economic importance, as silk was being used for a large variety of consumer goods, from clothing, weavings and stockings for women. However, most of the silk, the raw material for this growing industry, came from Japan and China, a relatively unstable part of the world in the first part of the twentieth century. This prompted the growing Western industries to concentrate on finding replacements for silk, as had been done with natural rubber, through chemistry. The downfall of silk as an industrial material began in 1927, when the DuPont Company hired the chemist Wallace Hume Carothers to run their "pure research laboratories".

The exit road for silk was paved by 1938, a year after Carothers's death, when nylon was introduced to the world, primarily as a replacement for silk in hose and stockings and as toothbrush bristles. It is certain that the invention of nylon gravely affected the Japanese trade balance, and in consequence, the overall position of the Japanese industry in world markets at the threshold of the Second World War. The influence of this miracle fiber, that could be produced at the fraction of the cost as its natural counterpart, is indisputable. Allied use of nylon in parachutes during the invasion of Normandy may have played a decisive role in the war's military outcome. The most obvious influence may come from its impact on consumer consumption.

As time progressed, the news for silk turned even more dire; in the 1960s, the use of virgin silk was found to produce an adverse biological response in sutured patients. This was later attributed, in the late 1970s and early 1980s, to sericin in the inner fibroin fibers of the silkworm silk, which was found to cause a type I allergic reaction. Virgin silk was then processed to extract sericin from its fibroin fibers, followed by a coating of wax or silicone to improve material properties and reduce fraying, and received the name of black braided silk (*e.g.* Perma-Hand™). Due to the biocompatibility issues, however, between the 1960s and the 1980s, silk decreased in its popularity as a suturing material.[9–11] In addition, these events ran in parallel with the development of synthetic biocompatible polymers, based on polyglycolic acid (PGA) (Dexon™, Maxon™) and polylactic acid (PLA) (Vicryl™). These two, together with catgut (Catgut™), are classified as biodegradable suture materials, according to the definition of an absorbable suture material by the US Pharmacopeia: one that loses most of its tensile strength within 60 days after being placed below the skin surface.[8,12] Silk-based sutures, along with other kinds based on braided polyester (Ethibond™, Mersilene™, Tevdek™), nylon (Ethilon™) and polypropylene (Prolene™, Surgilene™), are classified as non-absorbable suturing materials.[12] Some studies on silk, however, have showed its susceptibility to proteolytic degradation and the loss of the majority of its tensile strength *in vivo* after 1 year of implantation.[13–16] The braided structure of silk-based sutures increases the risk of infection, but these sutures have great handling and tying capabilities, and therefore it is still used today around eyelids and lips, where incidence of infection is low.[8]

Today, silk-based biomaterials are reviving, accompanied with advances in molecular and genetic manipulations. Its great mechanical properties and degradation characteristics have opened doors in the fabrication of tissue engineering scaffolds, which need ample time to interact with the host tissue before degrading. Some studies have found silk scaffolds comparable to those based on collagen for culturing bone and ligament tissue, as well as fibroblasts and bone marrow stromal cells.[12,17–19] The capability of processing silk fibroin into foams, meshes, fibers, and films make silk a promising material for several biomedical applications.[12,20,21] Advancements in genetic manipulation and protein tailoring have included spider silk in this array of opportunities, offering even superior mechanical properties when compared to *B. mori* silk.[20–22]

1.3 Cellulose: The Quintessential Bio-based Plastic

If we step back to the nineteenth century, another natural polymer, cellulose, in addition to rubber, impacted everyday life. The invention of cellulose plastics, also known as Celluloid, Parkesine, Xylonite, or Ivoride, has been attributed to three people: the Swiss professor Christian Schönbein, the English inventor Alexander Parkes, and the American entrepreneur John Wesley Hyatt.

Christian Friedrich Schönbein, a chemistry professor at the University of Basel, loved to perform chemistry experiments in the kitchen of his home, much to his wife's dismay. Early one morning in the spring of 1845, Schönbein spilled a mixture of nitric and sulfuric acids, part of that day's experiment, on the kitchen counter. He quickly took one of his wife's cotton aprons and wiped the mess up, then rinsing it with water before the acid would damage the cloth. As he hung the apron to dry over the hot stove, it exploded in a loud bang and flame in front of his very eyes. After he recovered from the shock, Schönbein's curiosity led him to impregnate wads of cotton with the acid mixture. Every time, he was able to ignite the mass, leading to an enormous, uncontrollable explosion. He called his invention guncotton. He had invented cellulose nitrate. Guncotton was three times as powerful as gunpowder and did not leave a black cloud after the explosion. Schönbein sold his patent to the Austrian Empire's army, but found no buyers in Prussia, Russia, or France. Finally, he sold his patent to John Taylor, his English agent, who immediately began production of guncotton in England. The production ended when his factory exploded, killing 20 workers. Although there were no buyers, several laboratories did spring up across Europe to investigate guncotton; often blowing up faster than they were being built. In addition to its military applications, Schönbein envisioned other uses for the nitrated cotton mass. He added a solvent or plasticizer made of ether and alcohol and found a way to nitrate the cellulose fibers into a less explosive material which he called kollodium, *glue* in Greek. He reported to his friend Michael Faraday that this mass "is capable of being shaped into all sorts of things and forms …". In the spring of 1846, after accidentally cutting himself on the hand, he covered the wound with a thin elastic translucent film made of kollodium. He sold his idea to the English, who for years supplied the world with the first adhesive bandages. In England, there was one person that took particular interest in the Swiss professor's inventions. His name was Alexander Parkes.

Alexander Parkes started playing around with cellulose nitrate in 1847, and spent the next 15 years in the laboratory perfecting the formulas and processes to manufacture cellulose nitrate. His final process took the nitrated cotton and added vegetable oils and organic solvents producing a "plastic mass" that was easily molded into any shape or form after it was softened under heat. He called his plastic mass Parkesine. The new applications for this versatile material, such as combs, knife handles, and decorations, made their debut at the 1862 World Exposition in London. In 1866, Parkes launched the Parkesine Company Ltd. Due to the low quality of its products, Parkesine was not a success and the company was liquidated in 1868. The poor mixing of the additives and solvents caused Parkesine products to significantly warp only a few weeks after manufacture. In 1869, Parkes sold his patents to Daniel Spill, his chief engineer, who founded the Xylonite Company and renamed the compound Xylonite. Parkes continued working on his material until his death in 1890 at the age of 77. Alexander Parkes, the inventor and engineer, can be credited with improving on Schönbein's invention, paving the road for the future of the plastics industry. He is also credited with fathering a total of 20 children. A very busy man, to say the least.

At the same time as the plastics industry seemed to be going under in England, in the United States John Wesley Hyatt was launching an enterprise that finally made cellulose nitrate a success, under the name of celluloid. As the story goes, it all began when in 1865 the billiard ball manufacturer Phelan & Collendar placed an ad that promised $10 000 to the person who would find a replacement for ivory in the manufacture of billiard balls. Elephants were being slaughtered at a rate of 70 000 per year, which would have led to the extinction of this great animal, exorbitant prices for the 'white gold' from Africa, and reduced profits for the billiard ball industry. The $10 000 prize attracted the 28-year-old Hyatt's attention. After returning home from his job as a printer, he worked on this project until eventually he stumbled upon nitrocellulose in 1869. After finding a better way to mix all the components as well as allowing the solvents to completely evaporate from the mass before solidification, he was soon manufacturing high-quality billiard balls. Instead of cashing in on the $10 000 prize, John Hyatt founded the Albany Billiard Ball Company with his brother Isaiah, becoming a direct competitor to Phelan and Collendar. For the next 30 years, until Bakelite replaced celluloid on the billiard table, many guns were pulled in the Wild West when the volatile balls sometimes exploded upon collision.

Another immediate application of celluloid was dentures, which up until then were made of hard rubber. In view of losing a rather profitable business to plastics, the rubber industry started a propaganda campaign against cellulose in all major US newspapers. They falsely claimed that celluloid dentures could easily explode in one's mouth when coming in contact with hot food. This not only cheated people of a much prettier smile, but also started a rivalry between the two industries, which has caused them to remain as completely separate entities to this day. In fact, despite the materials and processing similarities between plastics and rubber, the plastics industry and the rubber industry have completely separate societies and technical journals. A plastics engineer is likely to be found in meetings organized by the Society of Plastics Engineers (SPE) or the Society of the Plastics Industry (SPI), while a member of the rubber industry will attend meetings organized by their own society, the Rubber Division of the American Chemical Society.

With a new and versatile material, Hyatt and his co-workers needed equipment to mass-produce plastic products. Based on experience from metal injection molding, the Hyatt brothers built and patented the first injection molding machine in 1872, to mold cellulose materials,[23] as well as the first blow molding machine, to manufacture hollow products. In the summer of 1869,

History of Sustainable Bio-based Polymers

Figure 1.4 Jewellery box made from celluloid (Kölsch Collection).

Hyatt and Spill, respectively, filed for patents dealing with the manufacture of nitrocellulose materials. This started a lengthy and costly litigation that eventually ruled in Hyatt's favor in 1876. Spill died soon after, at the age of 55, of complications from diabetes. John Wesley Hyatt lived another 44 productive years in which he invented the injection molding and the blow molding machines with which he processed celluloid products. The combination of blow molding and celluloid resulted in many toys and household products such as jewellery boxes (Figure 1.4). Hyatt can certainly be credited for being the first person to successfully mold a plastic mass into a useful, high-quality final product. However, above all, we should credit him for saving the elephant on the road to a $10 000 prize he never claimed.

With the mass production of rubber, gutta-percha, cellulose, and shellac articles during the height of the industrial revolution, the polymer-processing industry after 1870 saw the invention and development of internal kneading and mixing machines for the processing and preparation of raw materials.[24] One of the most notable inventions was the Banbury mixer, developed by Fernley Banbury in 1916. This mixer, with some modifications, is still in use today for rubber compounding.

1.4 Casein Plastics: From Food to Plastic

Humans have utilized the milk protein casein as a food source since the domestication of livestock. The use of casein in non-food applications dates back to two centuries BCE, when the Egyptians used it to bind pigments in paints, and it has been used as an adhesive from the eighteenth century onwards.[25] The first patents for making plastics from casein were granted in 1885 and 1886, in Germany and the USA, respectively.[26] However, as these patents made no mention of any hardening agent, the patented material easily biodegraded and no useful products are known to have resulted. Easy biodegradation was an unwanted effect a century ago, but is a desirable property in some applications today.

In Germany the demand for white "blackboards" circa 1897 led to Adolf Spitteler's serendipitous discovery of casein.[25,26] As the story goes, one night in 1897, Spitteler's cat knocked over a small bottle of formaldehyde. The formaldehyde dripped from the chemist's counter down to the floor into the cat's milk dish. The next morning, when Spitteler returned to his laboratory, he found that the formaldehyde that had dripped into the dish had caused the milk to curdle, turning it into a hard horn-like substance, much like celluloid. In fact, his cat had just invented the first semi-synthetic plastic since cellulose (Figure 1.5). Soon, Spitteler started experimenting with cheese curds and formaldehyde, and he found that the milk protein was rendered water-insoluble by letting it sit in a formaldehyde solution for extended periods of time.[25] Spitteler and his new-found business partner, Ernst Krische, secured several patents in 1899 and coined the trade name "Galalith", from the Greek words for milk (*gala*) and stone (*lithos*).

Casein-formaldehyde soon was being produced and utilized in numerous applications across Europe. By 1915 European manufacturers of casein had found a viable method of making high-quality casein plastic. Dry casein could be slightly moistened with water, then processed into a plastic by

Figure 1.5 Spitteler's cat "inventing" casein plastics (courtesy of Luz M. Daza).

applying pressure and shear, as in a heated extruder.[27,28] Water was added to casein resins as a plasticizer prior to molding. Water enhanced processability, but resulted in greater shrinkage and warpage upon drying.[26] The same gains in processability could be achieved by using a 10% borax solution in place of the water.[28] This reduced shrinkage, but the finished material still easily absorbed large amounts of moisture and was therefore subject to significant warpage. This shortfall was addressed by "hardening" the casein plates and rods in a 40% formaldehyde solution, rendering them water insoluble.[28] Even so, the water absorption of casein was approximately 20–30%.[26]

The development of casein in the USA proceeded more slowly than in Europe. It was not until 1919 that the first casein plastic that was up to European quality standards was made in the USA.[26] Even then, the industry in the USA did not take off for several reasons: climatic differences meant the European processes could not be copied in a straightforward manner in the USA, the long and costly process did not yield easily to faster processing in automatic machines, and the scrap could not be reworked, so 50% waste was not uncommon. All this made competition with established materials more difficult.[26]

In 1929, Christensen added aluminum stearate, a water-soluble aluminum salt, to the resin and found that a non-hardened plastic rigid enough to be worked in automatic machines was produced. This not only allowed the casein plastic to be worked in machines, but aluminum casein scraps were easily re-workable, greatly reducing waste.[26] In 1926, 55% of the world's buttons were made of casein.[26]

Casein production in the 1930s steadily increased, from 10 000 tons produced worldwide in 1930 to 60 000 in 1932 and 70 000 in 1936.[26] However, buttons, belt buckles, jewellery, and ornaments (Figure 1.6) remained the main products in which casein could be used, pending a method of reducing the absorption of water from humid air.[26] Additionally, casein plastic still required hardening in formaldehyde, making its manufacture a long and costly process. The hardening and subsequent drying could take anywhere from two weeks to a year, depending on the thickness of the part.[29]

An event that led to further improvements occurred in 1938 in Utica, New York, when a tannery was ordered to shut down because it was polluting a local stream. William S. Murray, a prominent Utica politician and chemist, obtained a 30-day stay and found a method of solidifying the waste runoff, thereby protecting the stream. He applied the same methods to the skimmed milk being discarded by the nearby powdered milk plant. By changing the natural milk sugar to an aldehyde in the presence of casein, he found a method of casein plastic production that avoided formaldehyde hardening.[30] He secured a patent, but the Second World War prevented any further development of his promising new process.[31] The continued use of phenol formaldehyde and widespread adoption of synthetic plastics during and after the Second World War drastically curtailed the use of casein, which is produced at a minimal level today.[25]

Figure 1.6 Different ornaments produced with casein (Kölsch Collection).

1.5 Soy Protein Plastic: Back to Nature

Soy protein can be easily isolated from soybeans and molded. Soy plastics, which for a short period were used in automotive knobs and even body panels, never saw the widespread adoption predicted by Henry Ford in the 1930s. The Second World War sent his dream of "growing a car like a crop" into oblivion. Yet, as bio-based materials were being phased out by a growing synthetic, petroleum-based plastics industry, soy-protein plastics were being researched at the Ford Motor Company in Dearborn, Michigan, probably propelled by its leader Henry Ford's conscience, in an effort to give back to the American farmer what Ford tractors had taken away.

Actually, soybeans were introduced to the United States from the Far East in 1804. They were grown more as a curiosity than a substantial crop until the First World War, when concerns about vegetable oil shortages made the oil-rich soybean an attractive crop.[32] The success of extruded plastics from casein in the early 1900s spurred research into other agricultural sources of plastics, including plastics made from soy protein.[33] By 1913 patents had been issued for preparing plastics from soy protein in Britain and France.[34] The first US patent for soy protein plastics was issued to Sadakichi Satow in 1917.[35] Satow's process was a "wet process", analogous to that for casein. It involved a long hardening period in formaldehyde, and since drying occurred after shaping, the final products were susceptible to warping, shrinking, porosity, and water absorption;[36] problems that exist to this day, keeping various research groups busy around the world. These drawbacks prevented any commercialization of Satow's patent.

In the 1920s and 1930s there was a significant drive in America to help the nation's farmers by making industrial products from agricultural byproducts, known as the *chemurgy movement*. Chemurgy attracted a lot of intellectual and monetary capital, most notably from Henry Ford and Thomas Edison. These titans brought their financial, research, and manufacturing resources to bear on the potential for making industrial goods from agricultural by-products. Ford spearheaded a strong effort to utilize soy meal and protein in making parts for automobiles. In 1930 Ford laid out his vision to his lieutenants at the Ford Motor Company. A year-long survey of crops that might have potential industrial applications led to soybeans as the most promising candidate. By 1932 the company had planted 8000 acres of soybeans, an acreage that would increase to 12 000 by 1936.[37,38]

Ford developed a process to prepare molded plastics using soybean meal, which contained 48% protein, hardened with formaldehyde. These soy-formaldehyde parts absorbed more than 20% moisture by weight, causing severe warping and cracking. Phenolic resin, known to be water resistant, was therefore mixed in with the soy-formaldehyde resin to give a more water-resistant plastic. These thermoset phenolic-soy-formaldehyde plastic parts were filled with 30% wood flour and were cheaper to produce than straight phenolic resin parts, as well as weighing less and needing no costly surfacing or finishing.[38] In 1936 each of one million Fords had 15 pounds of soy plastic parts in gear shift knobs, window frames, electrical switches, horn buttons, and distributor caps, using over 3 million pounds of oil-free soybean meal.[38] This was in addition to 5 pounds per vehicle of soybean oil used in foundry sand cores and enamel. The photograph in Figure 1.7 shows Henry Ford driving an automobile with soy protein body panels in 1939. In 1942, 150 million pounds of soy meal extruded plastic, made of phenol-formaldehyde, soy and wood flour, were produced.[34] However, the water absorption of even these soy plastics remained high enough that products such as Ford's automotive body panels warped excessively. Still, the utility of soybeans in industrial products was a genie let out of the bottle. A 1936 article noted that the oil and meal of "soybeans are used in making ... paint, enamel, varnish, glue, printing ink, rubber substitutes, linoleum, insecticides, plastics, glycerin, flour, soy sauce, breakfast foods, candies, roasted beans with nut-like flavor [and] livestock feeds".[37]

At the same time that the Ford Company was implementing its process for producing soy protein plastics, research into other techniques to make plastic parts from soybeans was ongoing at the US Regional Soybean Industrial Products Laboratory in Urbana, Illinois. From 1936 through 1942 the lab

Figure 1.7 Henry Ford sits behind the wheel of a car whose body panels are made from a soy protein-based plastic (courtesy of the Henry Ford Museum).

published a steady stream of research articles on their progress in making plastics from soy meal and soy protein. One innovation was the advent of a dry process wherein the soy meal in powder form was pre-hardened with formaldehyde before shaping.[32] This eliminated the post-shaping hardening step, drastically reducing production time as well as cost, since soy proteins needed more highly concentrated formaldehyde than casein to be hardened in a wet process.[39] The pre-hardened material was claimed to be thermoplastic as it could be shaped repeatedly.[36]

However, further experimentation indicated that instead of being truly thermoplastic, the pre-hardened soy-formaldehyde resin displayed thermoset and thermoplastic behavior, perhaps due to a long curing reaction.[32] Work was needed to find a soy meal based material that was either thermoplastic enough to be injection molded or that would thermoset completely so it could be removed from a hot die. A variety of hardening agents were explored, with formaldehyde being the most effective in terms of low water absorption, followed by furfural, and propionaldehyde.[29]

The thermoplastic character of pre-hardened soy resin appeared only with adequate plasticization. Water was the most effective plasticizer, but the high water absorption of the molded parts made finding another plasticizer a priority. Seventy different candidates were screened, with none producing parts with better water resistance than those plasticized with water itself.[36] Ethylene glycol was the top performing non-water plasticizer. Water repellants were investigated as a means to increase the moisture resistance of the final parts. Aluminum stearates[29] and oleanolic acid[36] were found to be the most effective at decreasing water absorption. A 1940 article explored the possibility of making laminates by coating paper with a solution of soy-formaldehyde, then hot pressing multiple sheets together into a single plate.[40] Water absorption was again found to be problematic, but could be reduced by adding outer laminate sheets treated with phenolic resin. Significant research continued in making soy protein plastics less water absorbing and therefore more useful as industrial materials.[41,42]

The Second World War effectively put an end to the chemurgy movement, including Ford's desire to introduce large quantities of soy plastics into automobiles. After the Second World War, cheaper and better performing synthetic petrochemical based resins replaced soy and milk based protein plastics.[43]

From the Second World War through the 1980s there was little research done on protein plastics. However, such research is experiencing a resurgence. Environmental concerns associated with petrochemical plastics have led to industrial and commercial interest in environmentally friendly alternatives. These materials display comparable properties to petrochemical plastics. This, in combination with the increasing environmental concern of consumers, has led these materials to gain some acceptance despite their high cost in comparison with established synthetic materials.[43]

1.6 Building Scaffolds for Our Bodies: Collagen and Chitosan

The history of tissue engineering as a formal field can be traced back to just the past two decades. It was defined as a formal field by the National Science Foundation bioengineering panel meeting in 1987 (Washington DC, USA).[44] Bell *et al.* made a great contribution in 1981 when they cultured fibroblasts and epidermal cells from skin biopsies of donor rats in collagen lattices. This preparation was implanted on skin wounds of the donor rats and found to be a successful alternative to a skin graft.[45] Extraordinary contributions could be elucidated. Apart from the capability of culturing skin grafts of larger sizes, this could decrease or eliminate the issue of an adverse immune response from the donor,[46] since the cells were obtained from the graft host. New doors started to open and the field quickly branched into research areas that focused on the cellular level, the scaffold (or lattice) material, as well as its mechanical properties. Ethical issues involving the type of cells (stem cells and embryonic cells) arose and have since been involved in several controversies. Soon the questions came: "Can a whole organ be grown *in vitro*?" And even more profound questions were raised when referring to the use of embryonic cells in research: "When does life begin?"

The tissue engineering field and the paper by Bell *et al.* cannot stand alone without the previous existing knowledge of cells and proteins. This inevitably takes us back to the findings of Robert Hooke in 1665, when he discovered small holes in cross-sections of cork when viewing them under a microscope. He named these holes "cells" and described them in his book *Micrographia*.[47] About two centuries later, in 1805, the German Lorenz Oken stated that living organisms are composed of cells. This was later formalized within the "Cell Theory", established in 1839 in Germany by Matthias Schleiden and Theodor Schwann. With Rudolph Virchow's foundations on cell regeneration (every cell comes from a pre-existing cell), Loeb reported the growth of cells outside of the body in 1897. Cell culture research was then started, and Harrison was the first to grow ectodermal cells from frogs *in vitro* in 1907. With the later discovery by Roud and Jones at the turn of the nineteenth century, that trypsin could be used to degrade matrix proteins, scientists found that cells could be separated and research in cell differentiation and culture *in vitro* kept expanding into later decades.[48]

For tissue growth, you need the presence of a cell, but also a means for the cell to attach, proliferate, and function. A substrate and support was needed, and Bell *et al.* looked at the natural cell environment to obtain a proper cell support, or what today is termed 'scaffold'. They saw how the process of wound healing involved the interaction of skin cells with surrounding protein fibers, mainly collagen.[49] They extracted collagen from tendon bundles in rat tails and prepared it in a solution in which human foreskin fibroblasts were successfully seeded.[46] Among their conclusions, they saw collagen lattices as ideal scaffolds due to the low antigenic characteristics of collagen and the capability of harvesting it from the graft recipient.[46]

Collagen had been used in the medical field for hundreds of years, contained in catgut obtained from the walls of animal intestines. In the 1930s, Clark *et al.*[50] used X-ray diffraction to analyze the proteins making up catgut, which is mainly composed of collagen. Wycoff *et al.*[51] were also analyzing X-ray diffraction patterns of different crystalline proteins in kangaroo tail tendon. Both Clark and Wycoff recognized a regular structure in collagen. In 1955, Professor Ramachandran, from the University of Madras, India, made great contributions on the detailed structure of collagen and its triple helix configuration.[52] Subsequent reports since then have tried to describe the exact composition, structure, and configuration of collagen fibers: the main component in connective tissue. This highly organized protein polymer (Figure 1.8) plays an important role in providing the extracellular matrix (the substance surrounding cells) with structural integrity,[53] and has been an important scaffold material in the tissue engineering field since the past two decades.

The biocompatibility and biodegradability properties of natural polymers have turned biomedical researchers' attention to other materials, such as chitosan,[55] a crystalline polysaccharide derived from chitin (found in the exoskeletons of arthropods) by partial deacetylation. Chitin was first discovered and isolated by the French pharmacist and chemist Henri Braconnot in 1811 when studying the mushroom. This discovery, 30 years before that of cellulose, makes chitin an ancient polysaccharide.[56] The same substance was found 20 years later in insects and plants. In 1859, C. Rouget found how to process chitin to obtain chitosan. Ledderhose showed in 1894 that chitin was made of glucosamine and acetic acid; this was then reconfirmed in the 1930s. The biocompatibility and degradation characteristics have allowed the use of chitosan in the biomedical world. Applications include wound dressings, drug delivery systems, and space-filling implants. Together with collagen, chitosan has been a key natural polymer for use in tissue engineering scaffolds.[57,58]

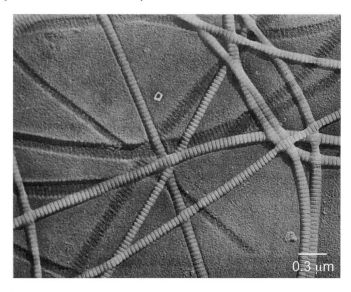

Figure 1.8 Electron micrograph of collagen fibers (reproduced with permission from ref. 54).

The use of natural polymers for the culture of cells is advantageous due to the better capability of cells to attach and interact with particular amino acid sequences. However, there is great variability from batch to batch,[59] and synthetic polymers have found their way into the tissue engineering scaffold world. The properties of synthetic polymer scaffolds are more controllable and reproducible, but a great portion of research has been necessary to account for optimal cell interaction.[59] Several studies have combined natural polymers or critical amino acid sequences with synthetic polymers to find an ideal cell–scaffold relationship. Although synthetic materials are still on the rise for use in tissue engineering scaffolds, much attention is steered towards the use of natural polymers. Growing knowledge about genetic and molecular manipulation allows researchers to tailor some of the properties of these proteins, as well as blend them with other synthetic materials in search of the ideal tissue engineering scaffold.

1.7 Letting Bacteria Make Our Plastics

When experimenting with wine, the well-known Louis Pasteur discovered in 1861 that particular microorganisms, which he termed "butyric vibrios",[60] were able to live without air, and that they were involved in butyric fermentation.[61] These studies marked the discovery of dextran, a polysaccharide made up of several glucose molecules, when the French botanist Van Tieghem identified the bacterium (*Leuconostoc mesenteriodes*) responsible for its formation in 1878. Soon after, A. J. Brown found that cellulose could also be produced by bacteria from acetic fermentation. Further research on microbiology took place, and the

Italian A. Borzi discovered how some bacteria were able to store polymeric substances encapsulated in their cytoplasm or in intracellular polymer reserves. Such phenomenon was observed with the polyamide cyanophycin in cyanobacteria or blue-green algae, as published by Borzi in 1887.[62]

Several years later, in 1926, the French microbiologist Maurice Lemoigne, at the Pasteur Institute, published his work "*Produits de deshydration et de polymerization de l''ácide β-oxybutyrique*",[63,64] in which he described the production of polyhydroxybutyrate in *Bacillus megaterium*. This material belongs to the family of polyhydroxyalkanoate (PHA) polyesters, which are sometimes termed bacterial bioplastics, offering biodegradability and a great array of material properties.[65] After Lemoigne's findings, other bacterial polymers were identified, and research was focused on their metabolic pathways.[66]

In the 1980s, several companies dedicated to the production of PHAs were established, such as Chemie Linz, in Australia, ICI in UK, and TianAn in China, with the idea that the price of petroleum would increase and biodegradable plastics would make their triumphant way into industry.[67] However, the economic prediction on the price of oil did not come true, and petroleum maintained its strong status as a raw material for plastics. Due to the advancements on molecular biology and the ability to genetically manipulate the synthesis of bacterial biopolymers, a growing industry is emerging again.[67] However, limitations today are mainly related to high costs with respect to the petroleum-based synthesis of plastics.

The applications of PHA have included packaging and drug delivery systems, as well as medical bio-implants (produced by Tepha, USA, since the 1990s).[67] The tissue engineering scaffold and drug-release fields have also kept an eye on this versatile material due to its biocompatibility, biodegradability, and thermoprocessibility, in addition to the probability of blending it with other materials, such as hydroxyapatite for bone tissue engineering.[68] PHA has even been found useful as a biofuel,[69] broadening even more its array of applications and motivating the bacterial biopolymer industry. More recently, Mirel™, a new PHA resin from metabolics, has been used to manufacture household goods, such as pens (Figure 1.9), allowing the bacteria to fulfill the

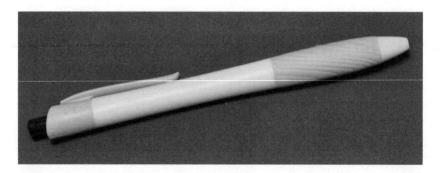

Figure 1.9 Pen produced with Mirel™ material.

consumers' increased needs and leaving a much smaller environmental footprint behind.

1.8 Conclusions

From the Second World War through the 1980s there was little research done on bio-based plastics. However, research in these materials is experiencing a resurgence. Environmental concerns, on the part of both developers and consumers, associated with petrochemical plastics have led to an increased interest in environmentally and biofriendly alternatives, despite their high cost in comparison with established synthetic materials. Additionally, great advances in biomedical research, genetics, and molecular biology have directed much attention to biodegradable and biocompatible natural polymers. These advances have also allowed the development and growth of green industry with bacteria as producers of raw materials that substitute for petroleum-based resources.

Walking down a quiet street in Essen and through the door to the Kölsch's collection is more than just paying a visit to the past. It is also passing through the doorway to a place where time remains frozen, to help us step into the future where renewable, as well as biocompatible, materials will help us free ourselves from fossil fuel dependence and will usher us into a sustainable industrial age, and will bring us opportunities and solutions in the biomedical field.

References

1. United States Environmental Protection Agency (US EPA), *Municipal Solid Waste in the United States: 2001 Facts and Figures*, 2003.
2. http://www.orientations-online.com/articleshow.php?id=3 (last accessed 18 October 2010).
3. P. M. Cunniff, S. A. Fossey, M. A. Auerbach, J. W. Song, D. L. Kaplan, W. W. Adams, R. K. Eby, D. Mahoney and D. L. Vezie, *Polym. Adv. Technol.*, 1994, **5**, 401.
4. P. Munk and T. M. Aminabhavi, *Introduction to Macromolecular Science*, Wiley, 2002.
5. D. Mackenzie, *Medical History*, 1973, **17**, 158.
6. W. Ley, E. Topsell, T. Moffett and C. Gesner, *The History of Four-Footed Beasts and Serpents and Insects: Theater of Insects*, Da Capo Press, 1967.
7. J. Newman and C. Newman, *Int. J. Dermatol*, 1995, **34**, 290.
8. R. L. Moy, A. Lee and A. Zalka, *Am. Fam. Physician*, 1991, **44**, 2123.
9. R. W. Postlethwait, in *Repair and Regeneration: the Scientific Basis for Surgical Practice*, Blakiston Division, McGraw-Hill, New York, 1968, 263.
10. R. W. Postlethwait, *Ann. Surg.*, 1970, **171**, 892.
11. R. W. Postlethwait, M. D. Dillon and J. W. Reeves, *Arch. Surg.-Chicago*, 1962, **84**, 698.
12. G. H. Altman, F. Diaz, C. Jakuba, T. Calabro, R. L. Horan, J. Chen, H. Lu, J. Richmond and D. L. Kaplan, *Biomaterials*, 2003, **24**, 401.

13. E. Rossitch, D. E. Bullard and W. J. Oakes, *Childs. Nerv. Syst.*, 1987, **3**, 375.
14. H. K. Soong and K. R. Kenyon, *Ophthalmology*, 1984, **91**, 479.
15. K. H. Lam, A. J. Nijenhuis, H. Bartels, A. R. Postema, M. F. Jonkman, A. J. Pennings and P. Nieuwenhuis, *J. Appl. Biomater.*, 1995, **6**, 191.
16. T. N. Salthouse, B. F. Matlaga and M. H. Wykoff, *Am. J. Ophthalmol*, 1977, **84**, 224.
17. N. Minoura, S. Aiba, Y. Gotoh, M. Tsukada and Y. Imai, *J. Biomed. Mater. Res.*, 1995, **29**, 1215.
18. K. Inouye, M. Kurokawa, S. Nishikawa and M. Tsukada, *J. Biochem. Bioph. Meth.*, 1998, **37**, 159.
19. S. Sofia, M. B. McCarthy, G. Gronowicz and D. L. Kaplan, *J. Biomed. Mater. Res.*, 2001, **54**, 139.
20. F. G. Omenetto and D. L. Kaplan, *Nat. Photon*, 2008, **2**, 641.
21. F. G. Omenetto and D. L. Kaplan, *Science*, 2010, **329**, 528.
22. J. M. Gosline, P. A. Guerette, C. S. Ortlepp and K. N. Savage, *J. Exp. Biol.*, 1999, **202**, 3295.
23. R. Sonntag, *Kunststoffe*, 1985, **75**, 4.
24. H. Herrmann, *Kunststoffe*, 1985, **75**, 2.
25. T. A. Osswald and G. Menges, *Materials science of polymers for engineers*, Hanser Verlag, 2003.
26. E. Sutermeister, *Casein and Its Industrial Applications*, Reinhold Publishing Corporation, 1939.
27. *Kunststoffe*, 1912, **2**, 225.
28. *Kunststoffe*, 1915, **5**, 106.
29. G. H. Brother and L. L. Mckinney, *Ind. Eng. Chem. Res.*, 1938, **30**, 1236.
30. Anonymous, *Mod. Plast.*, 1938, **15**, 46.
31. W. S. Murray, *Moldable Casein Composition*, Patent Number 2115316, 1938.
32. G. H. Brother and L. L. McKinney, *Mod. Plast.*, 1938, **16**, 41.
33. D. Myers, *Past, Present and Potential Uses of Soy Proteins in Nonfood Industrial Applications*, World Conference on Oilseed Technology and Utilization, 1992.
34. L. A. Johnson, D. J. Myers and D. J. Burden, *Inform*, 1992, **3**, 282.
35. S. Satow, *Process of Making Celluloid-Like Substances*, Patent Number 1245983, 1917.
36. G. H. Brother and L. L. Mckinney, *Ind. Eng. Chem. Res.*, 1939, **31**, 84.
37. E. F. Lougee, *Mod. Plast.*, 1936, **13**, 54.
38. R. L. Taylor, *Chem. Metall. Eng.*, 1936, **43**, 172.
39. A. C. Beckel, G. H. Brother and L. L. Mckinney, *Ind. Eng. Chem. Res.*, 1938, **30**, 436.
40. G. H. Brother, L. L. McKinney and W. C. Suttle, *Ind. Eng. Chem. Res.*, 1940, **32**, 1648.
41. A. C. Beckel, W. C. Bull and T. H. Hopper, *Ind. Eng. Chem. Res.*, 1942, **34**, 973.
42. B. E. Ralston and T. A. Osswald, *Plast. Eng.*, 2008, **64**, 36.

43. D. J. Myers, *Cereal Food World*, 1993, **38**, 355.
44. Y. M. Bello, A. F. Falabella and W. H. Eaglstein, *Am. J. Clin. Dermatol.*, 2001, **2**, 305.
45. E. Bell, H. P. Ehrlich, D. J. Buttle and T. Nakatsuji, *Science*, 1981, **211**, 1052.
46. E. Bell, B. Ivarsson and C. Merrill, *Proc. Natl. Acad. Sci. USA*, 1979, **76**, 1274.
47. R. Hooke, *Micrographia, or Some Physiological Descriptions of Minute Bodies Made by Magnifying Glasses with Observations and Inquiries Thereupon*, Royal Society, London, England, 1665.
48. http://www.woundsresearch.com/article/7895 (last accessed 18 October 2010).
49. H. Cottier, R. Dreher, H. Keller, B. Roos and M. Hess, in *The Ultrastructure of Collagen*, ed. J. J. Longacre, Charles C. Thomas, Springfield, IL, USA, 1976.
50. G. L. Clark, E. A. Parker, J. A. Schaad and W. J. Warren, *J. Am. Chem. Soc.*, 1935, **57**, 1509.
51. R. W. G. Wyckoff, R. B. Corey and J. Biscoe, *Science*, 1935, **82**, 175.
52. G. N. Ramachandran and G. Kartha, *Nature*, 1955, **176**, 593.
53. A. Aszodi, K. R. Legate, I. Nakchbandi and R. Fässler, *Annu. Rev. Cell Dev. Bi.*, 2006, **22**, 591.
54. G. Karp, *Cell and Molecular Biology: Concepts and Experiments*, John Wiley and Sons, 2008.
55. T. Chandy and C. P. Sharma, *Biomater. Artif. Cells Artif. Organs*, 1990, **18**, 1.
56. C. Becq and P. Labrude, *Pharm.*, 2003, **91**, 61.
57. S. V. Madihally and H. W. T. Matthew, *Biomaterials*, 1999, **20**, 1133.
58. D. W. Hutmacher, *Biomaterials*, 2000, **21**, 2529.
59. R. Langer and J. P. Vacanti, *Science*, 1993, **260**, 920.
60. P. Schutzenberger, *The International Scientific Series*, 1876.
61. L. Pasteur, *Bull. Soc. Chim.*, 1861, **11**, 30.
62. A. Borzi, *Malpighia*, 1887, **1**, 28.
63. M. Lemoigne, *Bull. Soc. Chim. Bio.*, 1926, **8**, 770.
64. G. W. Huisman, O. de Leeuw, G. Eggink and B. Witholt, *Appl. Environ. Microbiol.*, 1989, **55**, 1949.
65. B. H. Rehm, *Nat. Rev. Microbiol.*, 2010, **8**, 578.
66. V. Oeding and H. G. Schlegel, *Biochem. J.*, 1973, **134**, 239.
67. G. Chen, *Chem. Soc. Rev.*, 2009, **38**, 2434.
68. G. Chen and Q. Wu, *Biomaterials*, 2005, **26**, 6565.
69. X. Zhang, R. Luo, Z. Wang, Y. Deng and G. Chen, *Biomacromolecules*, 2009, **10**, 707.

CHAPTER 2
Synthetic Green Polymers from Renewable Monomers

NAOZUMI TERAMOTO

Chiba Institute of Technology, Faculty of Engineering, Department of Life and Environmental Sciences, 2-17-1 Tsudanuma, Narashino, Chiba 275-0016, Japan

2.1 Introduction

Synthetic polymer materials made from petroleum chemicals have become major materials over the past 50 years, because of their durability, plasticity, low cost, mechanical flexibility, abundant variety, lightness, and so on.[1] They play very important roles in many industries to produce not only daily commodities but also technological devices and biomedical devices. Behind their convenience, however, the abundant use of synthetic polymer materials has brought gradual spreading of durable polymer wastes into the environment everywhere from farmland to deep sea;[1,2] and these wastes are threatening biodiversity.[3] Moreover, the vast increase of the petroleum consumption relating to the economic growth of developing countries is threatening the future supply of the resource. Recently, global warming has become a pressing international issue; and incineration of the polymer wastes, though it has advantages on waste reduction, leads directly to the increase of carbon dioxide. In these situations, it is reasonable that polymers which undergo biodegradation and/or are produced from renewable resources became the mainstream of environmentally benign polymers.

Nowadays, many names are used to refer to environmentally benign polymers, especially for polymers synthesized from renewable resources,[4] which are the focus of this book. We often hear the terms, 'green polymer', 'biomass polymer', 'bio-based polymer', 'biopolymer', 'renewable polymer', and so on. If you try to search as many recent topics as you can on environmentally benign polymers on your web browser, you should search many times with many words. The situation will get worse because some websites use the term 'plastics' instead of 'polymer'. Among them, 'green polymer', though it is relatively obscure, is the most useful in this chapter, because it includes both biodegradable polymers and polymers prepared from biomass. In addition, the expression 'environmental benign polymer' covers many types of polymers such as easily recyclable polymers, very stable polymers without any eluted substances in water, and polymers synthesized *via* atom-efficient (green) chemical processes, as well as biodegradable polymers and polymers prepared from biomass.

In this chapter we describe 'synthetic green polymers from renewable monomers', including but not limited to biodegradable polymers. In the past, synthetic green polymers attracted attention mainly as biodegradable polymers which would resolve the problems of plastic waste disposal.[5-9] Most of them are aliphatic polyesters, and some of them are synthesized from petroleum chemicals. On the other hand, the involvement of greenhouse gases with the global warming since middle of the twentieth century has been suspected, and the increase of greenhouse gases has been attributed to human activity including the increasing use of fossil resources. On 11 December 1997, the Kyoto Protocol was adopted by Conference of the Parties COP 3 held at Kyoto in Japan under the United Nations Framework Convention on Climate Change (UNFCCC).[10] In the protocol, countries should reduce greenhouse gas emissions at the average emission in 2008–2012 by the reduction rate of 5.2% from the 1990 level of collective emissions of the world. After the enforcement of the protocol, emissions of carbon dioxide became the major part of the environmental issues, and many countries started to act against carbon dioxide emissions. One of the prominent concepts for reducing carbon dioxide emissions is 'carbon neutrality', contemplating the carbon atom's cycle only on the surface of the earth.[11,12]

According to the 'carbon neutrality' concept, resources that have their origin in products of photosynthesis are synthesized from carbon dioxide and their use is not counted as carbon dioxide emission.[13,14] Under this concept, although people should avoid overspending forestry resources without planting, much fossil fuel use can be covered by renewable resources, achieving reduction of carbon dioxide emission. The substitution from fossil fuel resources to renewable fuel resources may be easy, *e.g.* most gasoline can be substituted by ethanol and most fossil diesel can be substituted by biodiesel. The substitution for polymer materials, however, is not so simple because of the necessity of polymers with a variety of physical properties in industrial applications and therefore the necessity of a variety of resources for polymer production. If all the polymers used in the world were limited to polyethylene, polypropylene, and polystyrene, the situation would be similar to that of fuels; but in fact it is not the case.

We should remind the reader that polymers were originally synthesized from natural products. In ancient times, materials were made of substances available without any chemical processing: *e.g.* branches, trunks, and leaves from trees, and skins, hair, and bones from animals were used as they were. Since many chemicals began to be industrially produced in the second industrial revolution (later half of the nineteenth century), natural polymers have been chemically modified to yield 'naturally derived' polymer materials for industrial applications. Though there may be some discussion, vulcanized natural rubber is the first cross-linked polymer that was produced and used industrially, and cellulose nitrate plasticized with camphor was the first thermoplastic polymer.[15] Polymer materials derived from natural polymers, such as regenerated cellulose and cellulose acetate, were developed late in the nineteenth century. While most polymer materials were made from natural polymers in the nineteenth century, synthetic polymers made from petroleum have dominated in the twentieth century with the greatly development of petrochemical industries. Naturally derived monomers (excluding monomers of natural polymers) had not attracted much attention in the polymer industry, perhaps because many of them needed costly extraction processes from plants, animals, and microorganisms, while substitutes obtained from petroleum have been available at a lower cost. However, an air of fragility of sustainable petroleum supply (and the consequent increase of petroleum price) and the recent developments of chemical engineering and biotechnology for effective monomer production and purification have accelerated research on synthetic polymers containing naturally derived monomers (so-called renewable monomers) with the 'carbon neutral' concept.[16,17] Table 2.1 summarizes the classification of renewable monomers. In this chapter, renewable monomers that are used or have the potential to be used for polymer syntheses and examples of their polymerization strategy are reviewed.

2.2 Triglycerides of Fatty Acids and their Derivatives

2.2.1 Monomers from Triglycerides

Photosynthetic products are synthesized by plants from water and carbon dioxide, and therefore renewable monomers obtained from plants are very attractive. Though natural polymers from plants are also very attractive for the same reason, monomers are preferentially focused on in this chapter due to limitations of space (monomers obtained by digestion of natural polymers are included here). Types of renewable monomers obtained from plants without any chemical, enzymatic, and microbial conversion are limited, since plant cells are mainly composed of polymers. The principal monomer compounds that can be directly extracted from plants are oils: triglycerides of fatty acids and essential oils.

Triglycerides of fatty acids are esters composed of glycerol and fatty acids. Many types of fatty acids are found in plant-originated triglycerides, generally called plant oils or vegetable oils (Table 2.2).[18–21] Unsaturated C=C bonds impart characteristic features to triglycerides and fatty acids. The degree of

Table 2.1 Classification of renewable monomers.

	For condensation polymers	For addition polymers	For Diels-Alder reaction
Direct extraction from plants	Castor oil Sucrose Malic acid Tartaric acid Fumaric acid Citric acid Glycolic acid Terpin (p-menthane-1,8-diol) p-Menthane-3,8-diol Cineol Mandelic acid	p-Coumaric acid (4-Hydroxycinnamic acid) Vaniline Vanillic acid Gallic acid Eugenol Cardanol Urushiol Curcumin	Triglycerides containing unsaturated bonds (e.g. tung oil, linseed oil, soybean oil) Limonene Pinene Phellandrene Myrcene Cinnamaldehyde Cinnamic acid Eugenol Cardanol Urushiol	α-Phellandrene α-Terpinene Rosin
Chemical conversion of plant extracts	Epoxidized triglycerides Triglyceride polyol Ricinoleic acid 12-Hydroxystearic acid Sebacic acid 11-Aminoundecanoic acid Glycerol 5-(Hydroxymethyl)-furfural 2,5-Furandicarboxylic acid Furfural	β-Malolactonate esters Levulinic acid γ-Valerolactone Limonene derivatives 1,8-Diamino-p-menthane Rosin derivatives Lignin digests Ferulic acid 8-(3-Hydroxyphenyl)octanoic acid 2[5H]-Furanone Furfuryl alcohol	Fatty acids containing unsaturated bonds (e.g. oleic acid, linoleic acid, linolenic acid) Acrylated triglycerides Terpene-based acrylate Cinnamyl alcohol Tulipalin Pyruvic acid derivatives (e.g. α-acetoxy-acrylate esters)	Furfural Furfuryl alcohol Rosin derivatives Conjugated triglycerides

Table 2.1 (Continued)

	For condensation polymers	For addition polymers	For Diels-Alder reaction
Microbial or enzymatic conversion of renewable resources	Lactic acid Succinic acid 1,3-Propanediol Itaconic acid 2,3-Butanediol 3-Amino-4-hydroxy-benzoic acid Amino acids	Glucose Trehalose Glucaric acid Erythritol β-Hydroxybutyric acid β-Hydroxypropionic acid	Itaconic acid Isoprene
Collection from animals (some *via* chemical conversion)	Cholesterol and its derivatives Lactose	Peptides (protein digests) Amino acids	Animal triglycerides and fatty acids containing unsaturated bonds (*e.g.* fish oil, docosahexaenoic acid)

unsaturation is usually expressed by the iodine values. Plant oils containing many unsaturated bonds are called drying oil, because they are gradually solidified under exposure to oxygen (air) by oxidative cross-linking. Linseed oil and tung oil are classified as drying oils; soybean oil and sunflower oil are as semi-drying oils; while castor oil, palm oil, and coconut oil are non-drying oils. Autooxidation, peroxide formation, and subsequent radical cross-linking

Table 2.2 Composition of triglycerides and chemical structures of fatty acids.

Triglycerides	Fatty acid (x:y)								
	10:0	12:0	14:0	16:0	18:0	18:1	18:2	18:3	20:0
Palm oil	–	–	1–6	32–47	1–6	40–52	2–11	–	–
Soybean oil	–	–	<1	7–11	3–6	22–34	50–60	2–10	<1
Sunflower oil	–	–	–	6	4	19	69	<1	<2
Linseed oil	–	–	<1	5–9	<1	9–29	8–29	45–67	–
Coconut oil	4–10	44–51	13–18	7–10	1–4	5–8	1–3	–	–
Tung oil	–	–	–	–	4	8	4	84*	–
Castor oil	–	–	–	1	1	89–93†	4–5	<1	–

Fatty acid (x:y)
[Chemical structure]

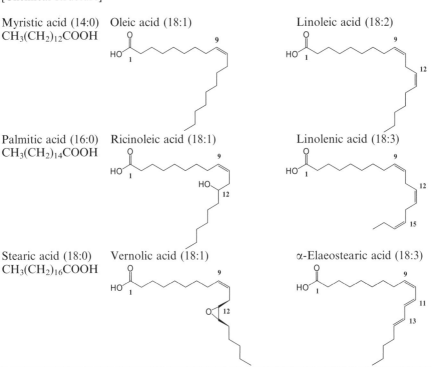

Myristic acid (14:0)
CH$_3$(CH$_2$)$_{12}$COOH

Palmitic acid (16:0)
CH$_3$(CH$_2$)$_{14}$COOH

Stearic acid (18:0)
CH$_3$(CH$_2$)$_{16}$COOH

* Mainly elaeostearic acid.
† Mainly ricinoleic acid.
x is the number of carbon atoms, and y is the number of unsaturated bonds.

reactions is a well-known reaction sequence involving drying oils, accelerated by driers such as cobalt, lead, and zirconium 2-ethylhexanoates. The reaction has been used for a long time to form resins for coatings, paintings, flexible films, binders for composite materials, and so on.[19,21–24] The initial reaction step of autooxidation is the abstraction of the hydrogen radical from methylene groups at the bisallylic position (*e.g.* C11 of linoleic acid).

Triglycerides with some functional groups are used as multifunctional monomers for polymerization or cross-linking agents, though it is difficult to synthesize linear polymers from reactive triglycerides. In many cases, unsaturated C=C bonds plays a role of reactive groups,[21] and they are often transformed into epoxides and polyols, which sometimes undergo the further chemical transformation into multifunctional carboxyl monomers by reaction with dicarboxylic acid and into multifunctional acryl monomers by reaction with acrylic acid (Figure 2.1).[19,25]

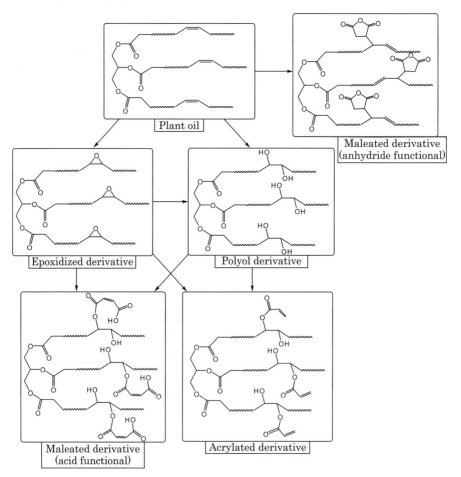

Figure 2.1 Strategies for synthesis of multifunctional reactive monomers derived from triglycerides.

Fatty acids and glycerol are products in hydrolysis reaction of triglycerides. Glycerol is a trifunctional polyol obtained as a by-product of soap production (nowadays it is also known as a by-product of biodiesel production). It has been used in the food industry, the pharmaceutical industry, and the cosmetics industry, and especially used as a monomer for alkyd resin and polyurethane in the polymer industry.[26,27] More recently, hyperbranched polymers are synthesized on the basis of trifunctional reactivity of glycerol.[28–30]

Castor oil is mainly composed of glycerol and ricinoleic acid, which is a characteristic fatty acid compound possessing one hydroxyl group and one unsaturated C=C bond (Table 2.2). Therefore, castor oil is developed for its special usage in the chemical industry and the polymer industry.[31,32] Ricinoleic acid (12-hydroxy-9-*cis*-octadecenoic acid), 12-hydroxystearic acid, sebacic acid,[33] and 11-aminoundecanoic acid[34] are attractive monomers derived from castor oil (Figure 2.2); and these monomers can be used as a component of polyesters and polyamides.

2.2.2 Polymers Synthesized from Triglycerides

Autooxidation of drying oils are mentioned above. This polymerization process has been used for a long time, but it takes a long time for curing and does not meet requirements for usual molded products. Other polymerization reactions involving C=C bonds are ene reaction and Diels–Alder reaction. Polymerization of triglycerides containing C=C bonds by ene reaction or Diels–Alder reaction were studied by several research groups. The advantage of ene reaction

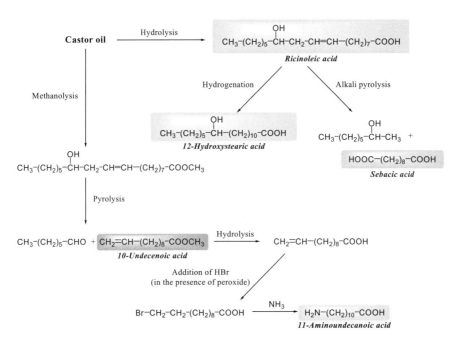

Figure 2.2 Strategies for synthesis of reactive monomers derived from castor oil.

and Diels–Alder reaction is that these reactions occur without a catalyst. Shibata and his co-workers[35,36] used bismaleimide as a cross-linker to yield tung oil-based and dehydrated castor oil-based thermosets. In the polymerization process, ene reaction, Diels–Alder reaction, and radical reaction occurred in parallel. Çaylı and Küsefoğlu[37] reported polymerization of linseed oil with phenolic resin via ene reaction in the presence of a small amount of maleic acid. Ene reaction of soybean oil with p-dinitrosobenzene (DNB) were reported by Mutlu and Küsefoğlu.[38] The polymeric form of DNB was prepared and reacted with soybean oil via ene reaction to give hydroxylamine (–N–O–H) and subsequently azomethine (–C=N–) groups at cross-linking sites. Recently, ultraviolet (UV)-curing of triglycerides via thiol-ene reaction has been reported. Thiol-ene reaction is a free radically initiated reaction of thiol adding to unsaturated C=C bonds. Thiol-ene reaction of fatty acid was investigated by Samuelsson et al.[39] They reported the importance of trans unsaturated C=C bonds in the reaction. Chen et al.[40] prepared a multifunctional thiol monomer and a multifunctional allyl monomer from epoxidized soybean oil (ESBO). The mixture underwent photopolymerization to form coating films. Black and Rawlins[41] synthesized allyl, acrylate, and vinyl ether derivatives of castor oil and photopolymerized them with a multifunctional thiol compound, trimethylolpropane tris(3-mercaptopropionate), in the presence of photoinitiator. Physical properties of cured films increased by aging for a week after UV irradiation. Ene reaction and Diels–Alder reaction are also used for modification of triglycerides with reactive groups. Eren et al.[42] reacted soybean oil with maleic anhydride, and obtained a multifunctional anhydride compound. Reaction of the anhydride compound with polyols gave cross-linked polyesters, and the products were soft solids. Larock and his co-workers[43,44] used Dilulin®, which is a compound derived from linseed oil and cyclopentadiene via Diels–Alder reaction and ene reaction. They polymerized Dilulin® with dicyclopentadiene (DCPD) by cationic polymerization using boron trifluoride diethyl etherate,[43] or by ring-opening metathesis polymerization using Grubbs' second generation catalyst.[44] DCPD content influenced physical properties of the thermosets.

Usual polymerization techniques for vinyl polymers, such as radical polymerization and cationic polymerization, have been often applied to cross-linking of triglycerides and their derivatives. Commercial vinyl monomers are usually added as a diluent to improve properties. In 1940s and 1950s, several researchers reported 'styrenated oils' as copolymers of styrene and triglycerides. Hewitt and Armitage,[45] Schroeder and Terrill,[46] Hoogsteen et al.,[47] Boelhouwer et al.,[48] and Crofts[49] investigated the preparation of copolymers of styrene and triglycerides, mainly using drying oils. This type of copolymers has been revisited since 1990s by two research groups – Yagci and his co-workers and Larock and his co-workers. Yagci and his co-workers[50–52] devised a macroinitiator method and a macromonomer method. In the macroinitiator method, they synthesized a glyceride-based azo initiator by reacting 4,4′-azobis(4-cyanopenanoyl chloride) with partial glycerides which is prepared by glycerolysis of triglycerides.[50] The glyceride-based azo initiator was reacted with styrene to yield a clear solid after monomer distillation in vacuum. In the macromonomer method, they synthesized mixed triglycerides by interesterification of castor oil and linseed oil, and reacted them with acrylic acid.[51] The acrylated triglyceride derivatives were reacted with styrene and dried to yield flexible films. Larock and his co-workers[53] showed a simple method to prepare cross-linked copolymers from tung oil, styrene, and divinylbenzene by thermal polymerization. Light yellow and transparent polymers were obtained in the thermal copolymerization. Physical properties of the polymers varied from those of elastomers to those of rigid plastics. Using the same method, cross-linked polymers from 87% conjugated linseed oil, styrene, and divinylbenzene were prepared.[54] Larock and his co-workers[55–57] also established cationic polymerization system using triglycerides, styrene, and divinylbenzene. Boron trifluoride diethyl etherate was used as an acid catalyst. They obtained cross-linked polymers with good performance comparable to conventional polymers. Composite materials from corn oil, styrene, divinylbenzene, and glass fibers were also prepared.[58]

Modification of triglycerides with acrylate groups is an improved method for polymerization of triglycerides. There are more than two routes to prepare acrylated triglyceride derivatives directly from epoxidized triglycerides or from triglyceride polyols (Figure 2.1): triglyceride polyols are also derived from epoxidized triglyceride, or derived *via* oxidation, hydroxybromination, or hydroxymethylation of triglycerides.[26] Alternatively, castor oil itself is used as a polyol compound for preparation of acrylated triglycerides. The acrylate groups are homopolymerized or copolymerized by conventional polymerization methods such as radical polymerization. Wool and his co-workers[25] achieved well-organized researches on polymer materials from acrylated triglycerides, accompanied with materials from maleated triglycerides and composites of these triglycerides with glass fibers, flax fibers, and hemp fibers. In another report,[59] they reacted hydroxyl groups and residual epoxy groups of acrylated ESBO with maleic anhydride, copolymerized the products with styrene using radical initiator, and investigated the physical properties of the cured polymers. The mechanical properties and the glass transition temperature (T_g) were comparable to those of a cured polymer of a commercial unsaturated polyester resin. Another method to prepare acrylated triglycerides was proposed by de Espinosa et al.[60] They reacted high oleic sunflower oil with oxygen under irradiation by a high-pressure sodium vapor lamp in the presence of *meso*-tetraphenylporphyrin at first, and subsequently reacted with sodium borohydride. The obtained product was hydroxyl-bearing sunflower oil, and it was used for synthesis of acrylated sunflower oil. Free radical polymerization of maleated triglycerides were also reported.[61–63] Wang et al.[63] succeeded in the preparation of foam plastics from maleated castor oil and styrene. The foam rigidity could be tuned depending on the curing initiator concentration and on the feed ratio of maleated castor oil to styrene. Acrylated triglycerides can be used for photopolymerization using radical photoinitiators. Gandini and his co-workers[64] reported the photopolymerization of acrylated soybean oil. The gel content after Soxhlet extraction was up to 80%. Most recently, nanocomposites of acrylated triglycerides and methacrylated triglycerides were prepared by Åkesson et al.[65]

As described above, triglycerides polyol are derived by several methods. Polyurethanes can be obtained when triglyceride polyols are reacted with multifunctional isocyanate compounds. Numerous reports on this type of polyurethane are found, and some of them involve polyurethane foam. In 1959, castor oil-derived polyurethane foams were prepared by Yeadon et al.[66] They investigated physical properties such as density, tensile elongation, compression deflection, and energy absorption. In the same year, three groups, Patton and Metz,[67] Toone and Wooster,[68] Wilson and Stanton,[69] reported urethane coatings based on castor oil and urethane-modified drying oils. Thereafter, many researches on triglyceride-based polyurethanes were conducted and many types of polyurethanes were synthesized. For examples, Petrović and Fajnik[70] synthesized polyurethane elastomers from diphenylmethane diisocyanate and castor oil. Qipeng et al.[71] reported the polyurethanes from 2,4-tolylene diisocyanate and mixture of castor oil and hydroxyether of bisphenol A. Suresh and Thachil[72] synthesized millable polyurethanes from partially acetylated castor oil and tolylene diisocyanate, and showed the possibility of subsequent vulcanization in a Brabender Plasticorder. Hu et al.[73] reported a rigid polyurethane foam from a rape seed oil-based polyol and polymeric diisocyanate diphenylmethane. Güner et al. comprehensively summarized interpenetrating polymer networks of castor oil-based polyurethanes and vinyl/acrylic components in their review paper.[26] Polyisocyanate derivatives of soybean oil were prepared by Çaylı and Küsefoğlu.[74] The soybean oil-derived polyisocyanate was synthesized by allylic bromination of soybean oil with N-bromosuccinimide and subsequently reaction with silver isocyanate (AgNCO). The product can be used as a monomer for polyurethanes and this enables synthesis of polyurethanes from triglycerides as all components.

Epoxidized triglycerides are also important monomers. They can be used similarly to common epoxy resins, although the mechanical properties of the homopolymerization products are not good. Usually, homopolymerization of epoxy resins are carried out via cationic polymerization. Park et al.[75] homopolymerized ESBO and epoxidized castor oil (ECO) using a thermally latent cationic initiator. The T_g of cured ECO was higher than that of cured ESBO. Copolymerization of ESBO with cycloaliphatic epoxy resin, 3,4-epoxycyclohexylmethyl-3,4-epoxycyclohexane carboxylate, were achieved by Raghavachar et al.[76] using a thermally latent cationic initiator. Vernonia oil is a naturally occurring epoxy functional triglyceride, which is composed of vernolic acid (Table 2.2) and glycerol. Trumbo and his co-workers[77] reported copolymerization of vernonia oil with acrylic copolymers of glycidyl methacrylate, n-butylacrylate, and styrene using a cationic initiator. Addition of styrene increased the tear resistance and addition of vernonia oil increased flexibility. Uyama et al.[78] prepared nanocomposites by curing ESBO and epoxidized linseed oil (ELO) via cationic polymerization in the presence of organophilic montmorillonite. The mechanical properties improved with an increase of clay content. Photo-initiated cationic photopolymerization of epoxidized triglycerides has been also investigated. Crivello and Narayan[79] reported photo-initiated cationic polymerization of epoxidized triglycerides using diaryliodonium and triarylsulfonium salt photoinitiator. They investigated the thermal properties and mechanical properties of the photocured films. Chakrapani and Crivello,[80] Thames and Yu,[81] Decker et al.,[82] and Wan Rosli et al.[83] respectively reported cationic photopolymerization of epoxidized triglycerides. Some of them carried out copolymerization with other epoxy resins. Shibata and his co-workers[84] polymerized ESBO using photoinitiator in the presence of supramolecular hydroxystearic acid nanofibers. The supramolecular nanofibers exhibited the reinforcing effects in the photocured ESBO.

Epoxidized triglycerides can be cured with multifunctional amine, acid anhydride, or phenol compounds. Miyagawa et al.[85–88] reported preparation of several types of epoxidized triglyceride-based thermosets. ELO was cured with anhydride, methyltetrahydrophthalic anhydride,[85] and with multifunctional amine, poly(oxypropylene) triamine.[86] They used bisphenol F diglycidyl ether (DGEBF) as an epoxy resin mixed with ELO. The feed ratio of ELO to DGEBF was varied and it was found that the Izod impact strength decreased with ELO contents for anhydride-cured resin,[85] though the Izod impact strength increased with ELO contents for amine-cured resin.[86] When ESBO were used instead of ELO and cured with anhydride, the Izod impact strength and the fracture toughness increased compared with cured resin of DGEBF without epoxidized triglyceride.[87] In their discussion, the improvement of the impact strength and the fracture toughness was thought to be due to a phase separation of the ESBO into rubbery particles. They also reported reinforcement of anhydride-cured resins of DGEBF with ELO or ESBO using alumina-nanowhisker or organophilic montmorillonite.[88] Prior to Miyagawa et al., Rösch and Mülhaupt[89] prepared anhydride-cured resins of ESBO using various dicarboxylic anhydrides. They tried reactive blending of polypropylene with phthalic anhydride-cured ESBO in the presence of maleic anhydride-grafted polypropylene. Terpene-based acid anhydride was used as a curing agent for ESBO by Shibata and his co-workers.[90] They obtained a pale brown thermoset with the T_g of 48.4 °C. The mechanical properties were significantly improved by addition of regenerated cellulose fibers to prepare biocomposites. Munoz et al.[91] prepared thermosets from phenolic resin and epoxy resin containing ELO. Addition of ELO resulted in an increase of flexibility of the thermosets. Tsujimoto et al.[92] prepared thermosets from ESBO and bio-based renewable phenolic polymers such as terpene-modified phenolic resin and lignin digests. This type of thermoset is attractive because it contains renewable resources at a high level. They also reported ESBO-based silica hybrid materials[93] as a novel type of composite material from renewable resources. They obtained transparent flexible films.

As described above, ricinoleic acid, 12-hydroxystearic acid, sebacic acid, and 11-aminoundecanoic acid are products derived from castor oil (Figure 2.2). These resources are also useful for polymer synthesis. Some examples are as follows: Slivniak and Domb[94] reported synthesis of poly(ricinoleic acid) by ring-opening polymerization of ricinoleic acid lactone mixture containing various size of lactones. The molecular weight of the polymer products was relatively low: the weight average molecular weight (M_w) was up to 5700. Alternatively, copolymerization of ricinoleic acid lactone mixture with lactide proceeded to yield copolymers. The molecular weight of copolymers decreased with an increase in the feed ratio of ricinoleic acid lactones. Poly(ricinoleic acid) with high molecular weight (up to $M_w = 98\,000$) was successfully synthesized by Ebata et al.[95] via lipase-catalyzed polymerization of methyl ricinoleate. The obtained poly(ricinoleic acid) was a viscous liquid with the T_g of –74.8 °C. Domb and Nudelman[96] synthesized ricinoleic acid-based polyanhydride from ricinoleic acid esters of maleate or sebacate by melt polycondensation. They obtained film-forming polymers with high molecular weight ($M_w > 50\,000$). The polyanhydride was biodegradable and applicable to controlled release of drugs. Poly(12-hydroxystearic acid) (PHS) can be synthesized by polycondensation of 12-hydroxystearic acid[97] and low molecular weight PHS is commercially available. It is usually used as a lubricant, a plasticizer, and a reagent for emulsifier. Sebacic acid is a useful diacid for synthesis of aliphatic polyesters and polyanhydrides. The T_g of aliphatic polyesters from sebacic acid is not so high and the polyesters can be used as polyols for polyurethane elastomers and foams. Nylon 6,10 is synthesized from sebacic acid and hexamethylenediamine, and the polyamide, whose physical properties are similar to nylon 6, is commercially produced. Nylon 11 is also a castor oil-based commercial product, which is synthesized by polycondensation of 11-aminoundecanoic acid. Nylon 11 has good dimensional stability due to a low moisture uptake, and T_g (42 °C) and T_m (184 °C) of nylon 11[98] are somewhat lower than nylon 6. Polycondensation of 11-aminoundecanoic acid was reported by Coffman et al.[99] in 1948. Commercial polymer products derived from castor oil are beginning to be used in the motor vehicle industry.[100]

2.3 Essential Oils, Natural Phenolic Compounds and their Derivatives

Essential oils are generally used in food and perfumes for their favorable fragrance and flavor.[101] They are extracted from plants by steam distillation, and most recently an extraction technique using supercritical carbon dioxide is under study. Though essential oils are usually composed of many components, some contain one major component up to 80% and other trace components. For example, limonene is the major component of lemon oil (90%); and cinnamaldehyde is the major component of cinnamon oil (70 ~ 80%). Here, we focus on relatively major components found in essential oils, and discuss their availability as monomers.

Components of essential oils are roughly classified in three category: terpenoids, phenylpropanoids, and others (Figure 2.3). Terpenoids are a series of natural-occurring chemicals whose skeleton is formed with isoprene units, including limonene, pinene, menthane, camphor, and citral. Phenylpropanoids are a series of naturally occuring chemicals whose skeleton is formed from a benzene ring and a three-carbon propyl tail, derived from phenylalanine, and including cinnamaldehyde, coniferyl alcohol, eugenol, safrole, coumarins, and flavonoids. Coniferyl alcohol, sinapyl alcohol, and coumaryl alcohol are known as phenylpropanoid monomers used in the synthesis of lignin.[102] Therefore, lignin digests and their

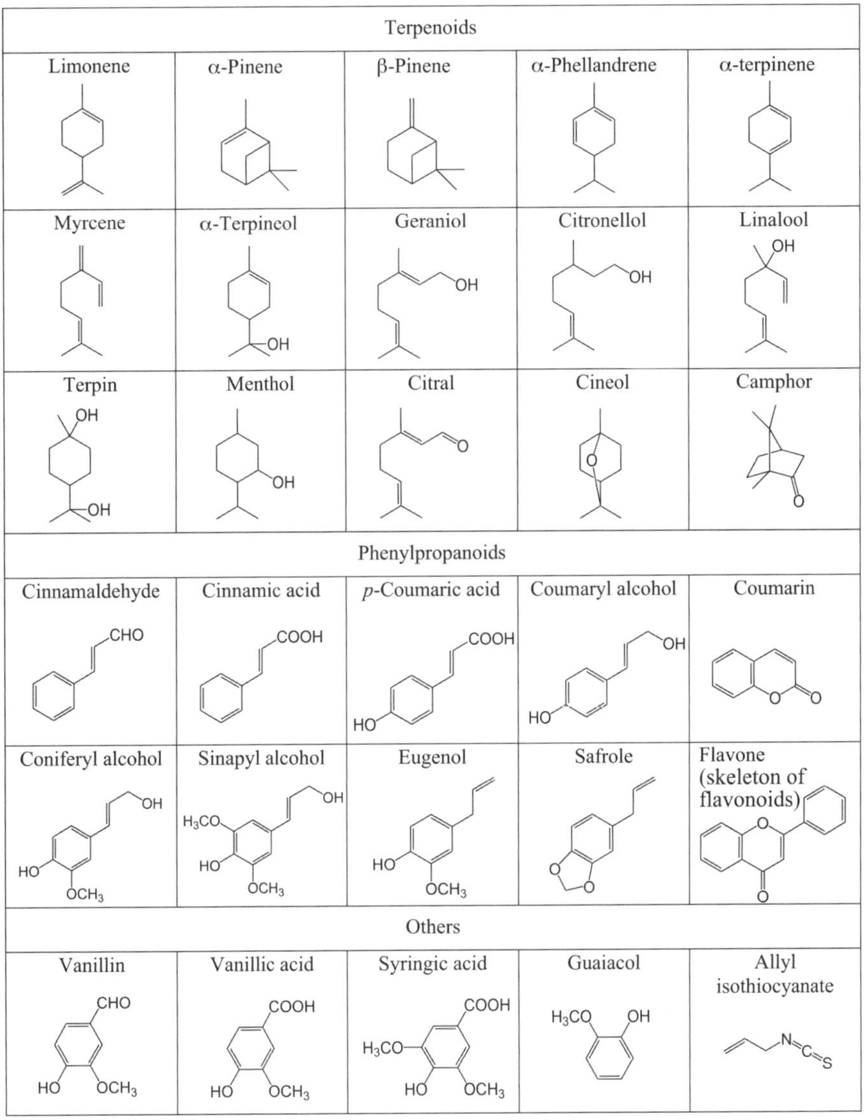

Terpenoids					
Limonene	α-Pinene	β-Pinene	α-Phellandrene	α-terpinene	
Myrcene	α-Terpineol	Geraniol	Citronellol	Linalool	
Terpin	Menthol	Citral	Cineol	Camphor	
Phenylpropanoids					
Cinnamaldehyde	Cinnamic acid	p-Coumaric acid	Coumaryl alcohol	Coumarin	
Coniferyl alcohol	Sinapyl alcohol	Eugenol	Safrole	Flavone (skeleton of flavonoids)	
Others					
Vanillin	Vanillic acid	Syringic acid	Guaiacol	Allyl isothiocyanate	

Figure 2.3 Structure of components obtained from essential oils.

derivatives are also included in this section, though most of them are not volatile (and so are not called essential oils). Some other phenolic compounds, most of which are derived from the shikimic acid pathway, are also extracted from plants.

2.3.1 Terpenoids

Homopolymerization of terpenoids using C=C bonds is not as easy as homopolymerization of industrially used vinyl monomers. The trial study on the polymerization of terpenes was reported by Carmody and Carmody in 1937.[103] They tried to polymerize pinene by aluminum chloride and obtained the mixture of an oily polymer product and a hard polymer product. Roberts and Day[104] also tried to polymerize α-pinene, β-pinene, and limonene by several kinds of Lewis acid catalysts in 1950. The approximate molecular weights of the polymer obtained from α-pinene, β-pinene, and limonene were reported as 700, 3100, and 1200, respectively. Emulsion polymerization of myrcene was studied by Johanson et al.[105] in 1948. The obtained polymers were low in mechanical properties, and the tercopolymer from myrcene, styrene, and butadiene showed good tensile properties. Marvel and Hwa[106] reported polymerization of myrcene with triisobutylaluminum and vanadium trichloride. Inherent viscosities of the products are 2 to 5.5 dL g^{-1}. Cawse et al.[107] synthesized hydroxyl-terminated polymyrcene using hydrogen peroxide-initiated polymerization, and the number-average molecular weight (M_n) of the polymer was up to 4000. In 1992 Keszler and Kennedy[108] reported synthesis of high molecular weight poly(β-pinene) by 'H$_2$O'/EtAlCl$_2$ system at -80 °C (up to $M_n = 39\,900$); and Sawamoto and his co-workers[109] reported cationic polymerization of α-pinene and β-pinene with the AlCl$_3$/SnCl$_3$ binary catalyst in 1996 (up to $M_n = 6400$). Subsequently Sawamoto and his co-workers[110] achieved the living cationic isomerization polymerization of β-pinene using an initiating system with HCl-2-chloroethyl vinyl ether adduct and TiCl$_3$(OiPr) in 1997. Though M_n of the polymer products was less than 5×10^3, the polymerization system opened the synthetic route to block copolymerization and end functionalization.[111–113] Guiné and Castro[114] polymerized β-pinene with C$_2$H$_5$AlCl$_2$ (up to $M_w = 2430$) and determined the Mark-Houwink-Sakurada constants. More recently, the polymerization to obtain high molecular weight poly(β-pinene) (up to $M_n = 25\,100$) was achieved by Satoh et al.[115] in 2006, using EtAlCl$_2$, Et$_3$Al$_2$Cl$_3$, and AlCl$_3$ as a Lewis acid and a 1:1 mixture solvent of CH$_2$Cl$_2$ and methylcyclohexane. After hydrogenation reaction, the T_g of the polymer was found to increase to 130 °C, and the polymer has good transparency and thermal properties comparable to polystyrene and poly(methyl methacrylate). They also achieved the polymerization of α-phellandrene. Figure 2.4 shows the mechanisms of cationic polymerizations of β-pinene and α-phellandrene. A polymer synthesized from a myrcene derivative was reported by Hillmyer and his co-workers,[116] where a cyclic diene 3-methylenecyclopentene prepared by ring-closing metathesis of myrcene was polymerized by living cationic polymerization using i-BuOCH(Cl)Me/ZnCl$_2$ system (up to $M_w = 22\,000$) (Figure 2.5). Radio frequency plasma polymerization of terpinen-4-ol, the major component of tea tree oil, was reported by Bazaka and Jacob.[117] They obtained the smooth thin film on a glass substrate. The structure of the polymer may differ from the repetition of the monomer because of branching and cross-linking.

Research on the copolymerization of terpenoids with industrially used vinyl monomers or other monomers have been often reported. Littmann[118] introduced terpene-maleic anhydride resins for industrial use in alkyd resins in 1936. The resin was synthesized from α-terpinene and maleic anhydride via

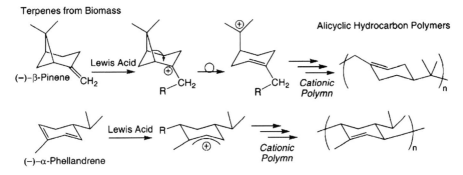

Figure 2.4 Mechanisms of cationic polymerizations of β-pinene and α-phellandrene.[115] Courtesy and reproduced with permission of the authors and the RSC publishing. © RSC, UK.

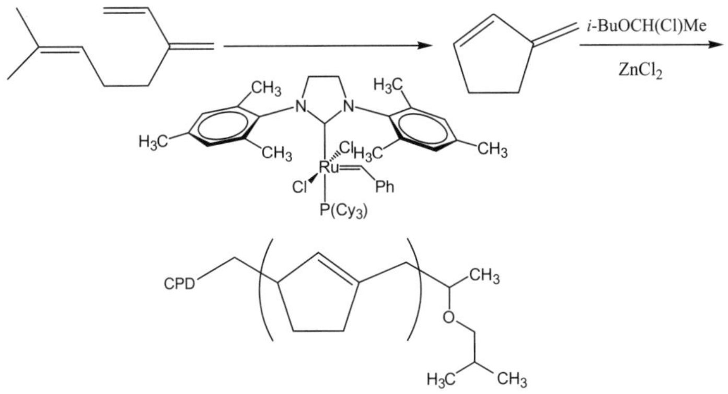

Figure 2.5 Polymer synthesis from myrcene.

Diels–Alder reaction. Marvel et al.[119] reported polymerization of vinyl ester monomers of pinic acid, a terpene-derived acid,[120] and its copolymer with vinyl chloride in 1960. Sheffer and his co-workers[121–123] reported the cationic copolymerization of β-pinene with styrene in 1970 and verified the copolymer synthesis by pyrolysis in 1983. Limonene was copolymerized with maleic anhydride using a radical initiator, α,α'-azobisisobutyronitrile (AIBN), by Doiuchi et al.[124] in 1981. They proposed the cyclocopolymerization mechanism and found that the copolymer was optically active. Copolymers of β-pinene or limonene with maleic anhydride were also synthesized by Maslinska-Solich and Rudnicka[125,126] and found to be optically active polymers. Satoh et al.[127] achieved the living radical chain copolymerization of limonene with maleimide by an end-to-end sequence-regulated AAB copolymer. Khan et al.[128] reported the cationic copolymerization of α-pinene with styrene using $AlCl_3$. The molecular weight of the copolymer was low (2320–3080). Srivastava and his co-workers have vigorously studied the radical copolymerization of terpenes and vinyl monomers. They reported many copolymers: geraniol-co-styrene,[129] α-terpineol-co-methylmethacrylate (MMA),[130] citronellol-co-vinyl acetate,[131] limonene-co-MMA,[132] citronellol-co-acrylonitrile (AN),[133] α-terpineol-co-styrene,[134] α-terpineol-co-butylmethacrylate,[135] linalool-co-vinyl acetate,[136] limonene-co-styrene,[137] and geraniol-co-MMA.[138] They found that the poly-(geraniol-co-MMA) was a liquid crystalline polymer.[138]

Nomura et al.[139,140] achieved the polycondensation between terpene derivatives and malonate esters by the palladium-catalyzed allylic substitution reaction (Figure 2.6). They used geraniol and citronellol for synthesis of monomers and obtained polymers with M_n up to 44 000. Mathers et al.[141] reported the ring-opening metathesis polymerization of dicyclopentadiene in the presence of terpenes to yield functional hyperbranched polymers. They used terpenes as a chain transfer reagent and a solvent, and the terpene units were incorporated only at the ends of the polymer products. Polymerization of limonene monooxide with carbon dioxide was reported by Coates and his co-workers.[142] They reacted the *trans* and *cis* diastereomers with CO_2 using β-diiminate zinc acetate catalysts, but the reaction proceeded selectively on the *trans* monomer to yield a regioregular polycarbonate (Figure 2.7). Terpin (*p*-menthane-1,8-diol) and *p*-menthane-3,8-diol are also interesting diol

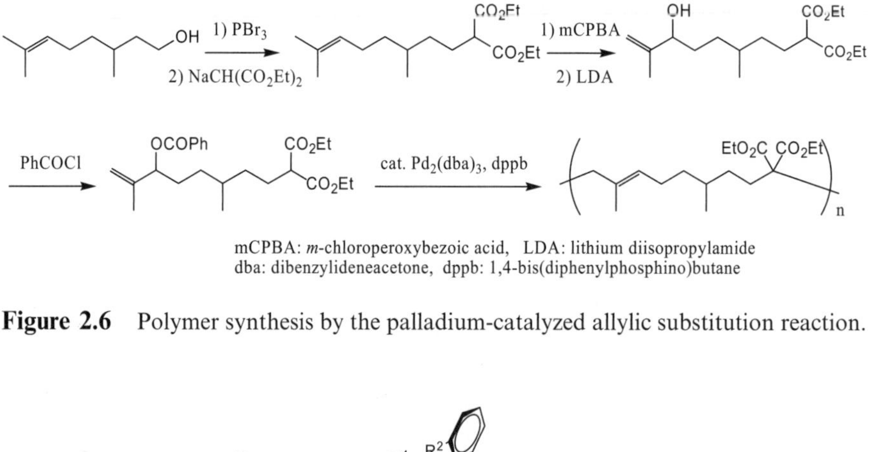

mCPBA: *m*-chloroperoxybezoic acid, LDA: lithium diisopropylamide
dba: dibenzylideneacetone, dppb: 1,4-bis(diphenylphosphino)butane

Figure 2.6 Polymer synthesis by the palladium-catalyzed allylic substitution reaction.

R^1, R^2 = Et or ^{i}Pr
R^3 = H, R^4 = CH_3 or CF_3

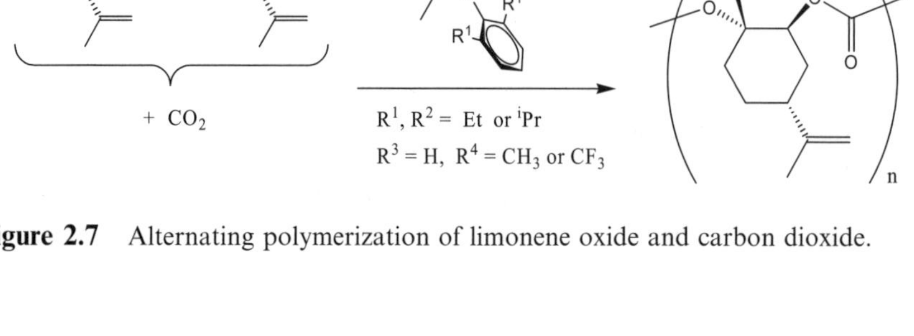

Figure 2.7 Alternating polymerization of limonene oxide and carbon dioxide.

compounds which may be available to use in polyester synthesis. (We could not find literature references about these terpenoids in polymer synthesis.)

Thermosets and cross-linked polymers containing terpenoids have also been reported. Crivello and Yang[143] polymerized γ-terpinene diepoxide and limonene diepoxide using a photolabile cationic initiator to obtain cross-linked polyethers. Trumbo and his co-workers[144,145] reported terpene-anhydride resins which can react with epoxy resins, bisphenol A-based epoxy resin, and epoxidized soybean oil, to give coating films. The films were found to be hard, glossy, and brittle when using the bisphenol A-based epoxy resin. Many types of terpene-based reactive monomers are available from a chemical company, Yasuhara Chemical Co., Ltd. (Fuchu, Hiroshima, Japan). Kimura et al.[146] prepared the terpenediphenol-based benzoxiazine and cured it with epoxy resin. The benzoxiazine ring opens when heated and formed phenolic hydroxyl groups which can react with epoxy groups. The cured resin showed good properties in thermal stability, mechanical properties, electrical insulation, and water resistance, compared with the cured resin from bisphenol A type resin. Xu et al.[147] synthesized an epoxy resin from 1-naphthol, limonene, paraformaldehyde, and epichlorohydrin (Figure 2.8). The epoxy resin was reacted with dicyandiamide or bisphenol A-formaldehyde novolac resin. The cured polymers had a high glass transition temperature, high thermal stability, and good water resistance compared with bisphenol A type epoxy resin. Wu et al.[148] synthesized an terpinene-based epoxy resin from the hydrogenated terpinene-maleic anhydride and epichlorohydrin. Subsequently they prepared polyols from the epoxy resin, and cured the polyols with a 1,6-hexamethylene diisocyanate derivative to give polyurethanes with high impact strength, high flexibility, and high thermal stability.[149] Shibata and his co-workers[150] reported organic-inorganic hybrid nanocomposites synthesized from terpene-based acrylate resin and methacryl-substituted polysilsesquioxane. They prepared the transparent hybrid nanocomposites by photocuring the acrylate and the

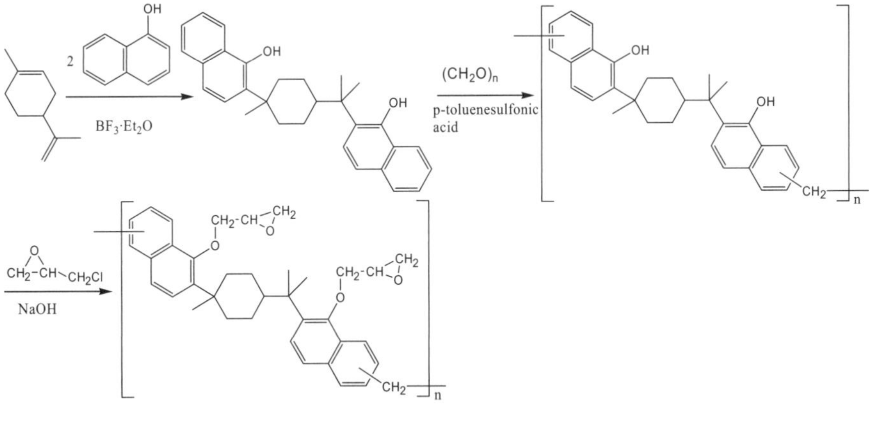

Figure 2.8 Synthesis of an epoxy resin from limonene and 1-naphthol.

methacrylate groups in the presence of a radical photoinitiator. The mechanical properties of cured terpene-based acrylate resins were improved by polysilsesquioxane.

2.3.2 Phenylpropanoids

Phenylpropanoids are biologically synthesized from phenylalanine as described above. Among them, cinnamic acid is synthesized directly from phenylalanine by phenylalanine ammonia-liase (PAL), and p-hydroxycinnamic acid (p-coumaric acid) is synthesized from cinnamic acid by cinnamic acid 4-hydroxylase (C4H, an enzyme in the cytochrome P-450 family).[102] The phenylpropanoid metabolic pathway is important for plants to synthesize lignin, and some phenylpropanoids are seen at junctions of cell wall polysaccharides such as hemicellulose and pectin.[151]

Phenylpropanoids having a cinnamoyl group (Ph–CH=CH–CO–), often found in essential oils, are most usable monomers among phenylpropanoids. Cinnamaldehyde, which is the main component of cinnamon oil, can be easily transformed into cinnamic acid. Though the cinnamoyl group has a C=C double bond, the successful addition homopolymerization has not been reported, as far as we could find. The usual bulk polymerization of ethyl cinnamate using a radical initiator gave a polymer with the molecular weight of 2300.[152] On the other hand, its photodimerization reaction (2+2 cycloaddition, Figure 2.9) by irradiation of UV light ($\lambda \sim 280$ nm) is a well-known reaction and often used for preparation of photocross-linked materials. Recently, homopolymerization of cinnamyl groups (Ph–CH=CH–CH$_2$–), which also had been known as an 'unpolymerizable' group for a long time, was reported by Washburn et al.[153] using the topological control process, in which cinnamyl alcohol was absorbed onto a metallic substrate and subsequently photopolymerized at very low temperature (100 K) to yield very thin film of poly (cinnamyl alcohol) with M_n 5.0×10^5. However, this polymerization process is very particular and will not become popular for industrial production.

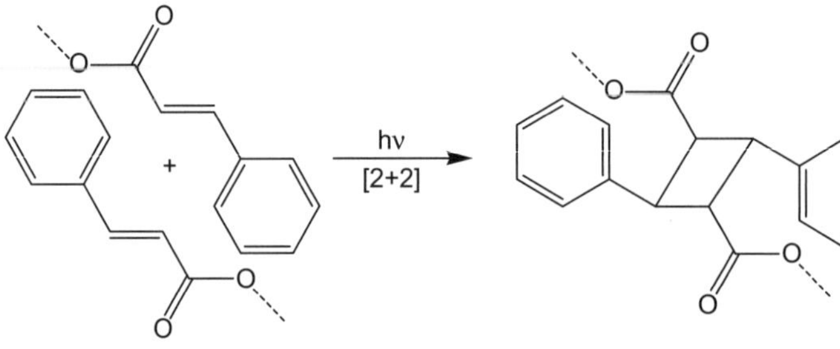

Figure 2.9 Photodimerization reaction of cinnamoyl groups.

The condensation (homo- or co-)polymerization of *p*-hydroxycinnamic acid (*p*-coumaric acid) is achieved by several research groups as described below.

Copolymerization of cinnamoyl esters with other vinyl monomers using unsaturated C=C bonds was reported by several research groups. Barson[154] copolymerized cinnamic acid with styrene, and determined the monomer composition. Homolytic propagation of styrene was found to predominate slightly. Amrani and his co-workers synthesized polystyrene-co-(cinnamic acid) and studied its compatibility with other polymers.[155,156] David and Ioanid[157] reported synthesis of poly(2-methyl-2-oxazoline) macromonomers having a cinnamoyl end group and its copolymerization with styrene. They carried out the dispersion copolymerization of the macromonomers with styrene to obtain monodisperse micrometer size beads. Esen and Küsefoğlu[158] reported synthesis of cinnamate esters of epoxidized soybean oil and its homopolymerization and copolymerization with other vinyl monomers such as styrene and MMA. In the homopolymerization using benzoyl-*t*-butyl peroxide, a soft solid sample insoluble in all common solvents was obtained, though the cinnamate C=C bonds were not completely used up. On the other hand, the copolymerization with other vinyl monomers proceeded successfully to give rigid thermosets. They also carried out the photocross-linking of the cinnamate ester of epoxidized soybean oil to obtain a thin film. Satoh et al.[159] reported synthesis of alternating copolymers of *p*-methoxystyrene and phenylpropanoids having a β-methylstyrene skeleton *via* aqueous-controlled cationic polymerization using BF_3OEt_2.

Preparation of cross-linked polymers *via* the photodimerization of cinnamate groups was reported very often. Minsk et al.[160] synthesized cinnamate esters of poly(vinyl alcohol) (PVA) and cellulose to obtain photosensitive polymers in 1959. Van Paesschen et al.[161] reported the polymerization of vinyl cinnamate in 1960; and Kato[162] reported the polymerization of *p*-vinyl phenyl cinnamate in 1969. Poly(2-vinyloxyethylcinnamate),[163] epoxy-functionalized poly(dimethylsiloxane) modified with cinnamic acid,[164] cinnamoylated polyvinylamine,[165] comb-like poly(methacrylamides) containing cinnamoyl groups in the side chain,[166] functionalized polymers synthesized from hydroxyethylmethacrylate-based monomer modified with various substituted cinnamoyl chloride,[167] copolymer of *N,N*-dimethylacrylamide and hydroxyethylmethacrylate-based monomer modified with cinnamate,[168] cinnamoyl shell-modified poly(amidoamine) dendrimers,[169] and many other cinnamate-based polymers have been reported. In most researches among them, photocross-linking by irradiation of UV light was conducted. Ni and Zheng[170] synthesized a photosensitive octacinnamamidophenyl polyhedral oligomeric silsesquioxane (OcapPOSS) (Figure 2.10a) and prepared the homogeneous nanocomposites of OcapPOSS with poly(vinyl cinnamate). They investigated in situ photocross-linking of the nanocomposites. Teramoto and Shibata[171] synthesized two kinds of trehalose cinnamate with different degrees of substitution on the hydroxyl groups of trehalose (Figure 2.10b), and prepared transparent thin films on a glass substrate by photocross-linking. Langer and his co-workers[172] successfully obtained light-induced shape-memory polymers containing cinnamic or cinnamytidene groups (Figure 2.11). The UV photocross-linked polymers

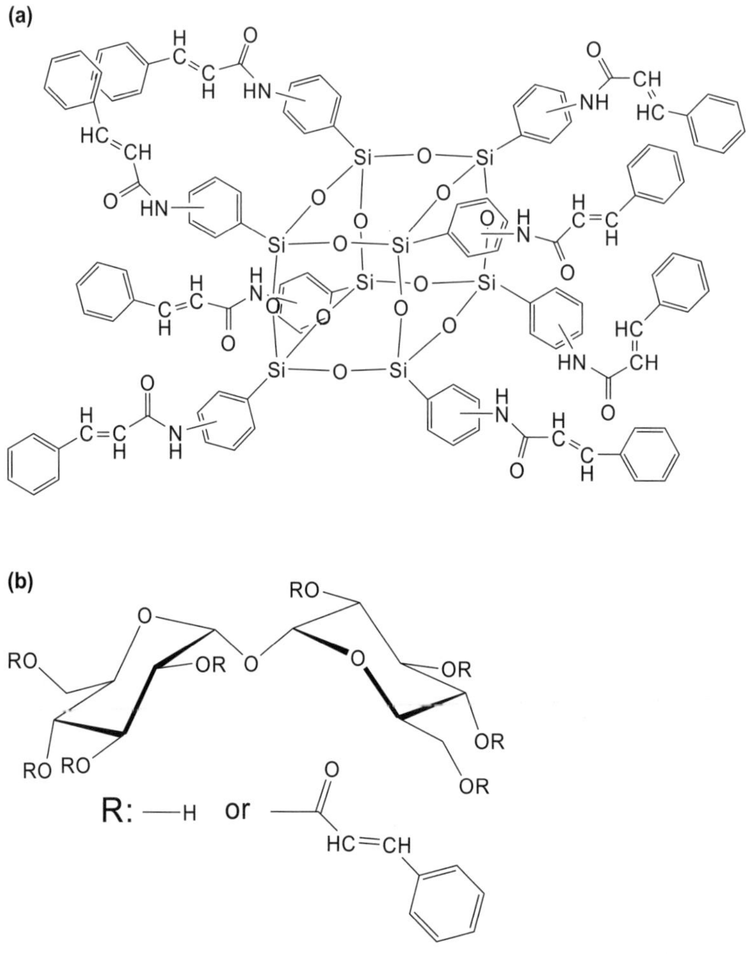

(a)

(b)

Figure 2.10 Example structures of photo-cross-linkable monomers carrying multiple cinnamoyl groups.

deformed when heated at 50 °C, and they can recover their original shapes at ambient temperatures when exposed to UV of a different wavelength.

Since photodimerization of cinnamoyl-carrying polymer using linearly polarized light was reported by Schadt and his co-workers,[173] anisotropically photocross-linked cinnamate polymers have been known as a durable material for liquid crystal alignment[174] and have been studied by many researchers.[175–182] Murase et al.[183] reported the large refractive index change by photodimerization of a cinnamate polymer and Nagata et al.[184] proposed cinnamate polymers as a material for holographic recording.

Coumarin and its derivatives, which are found in many plants, also show photodimerization by irradiation of UV light with longer wavelength ($\lambda = 300 \sim 350$ nm) than a cinnamoyl group. Trenor et al.[185] wrote a detailed

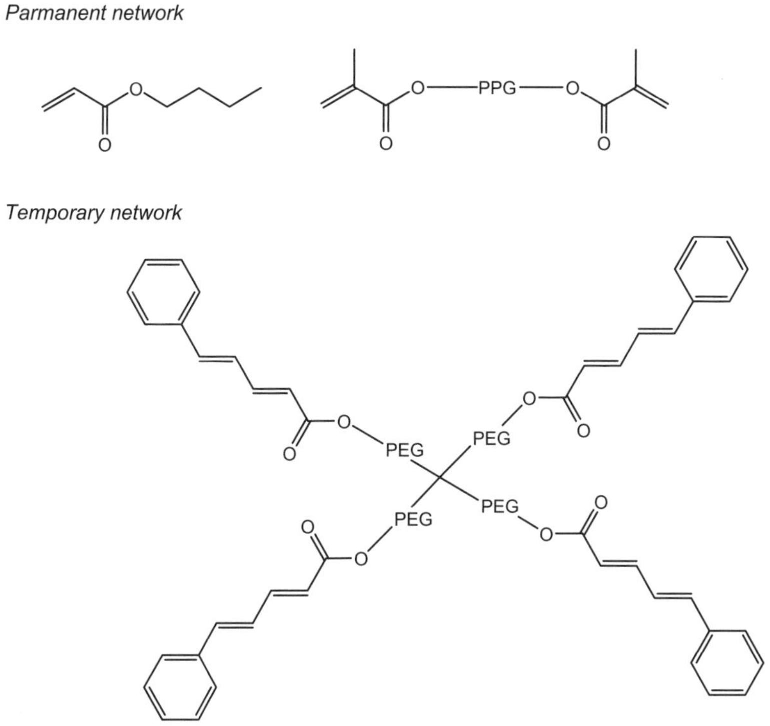

Figure 2.11 Structures of components for light-induced shape-memory polymers (IPN polymer type).

review paper on coumarins in polymers. Coumarin and its derivatives are used for functional groups in polymers.[186–195] Polymers containing coumarin groups are also used for photoalignment of liquid crystals.[190–193] Some coumarin-containing polymers were found to exhibit electroluminescence.[188] Biopolymers carrying coumarin groups were investigated by Yamamoto et al.[194] and Wondraczek et al.[195] to obtain cross-linkable poly(Lys) and polysaccharide sulfates, respectively.

Hydroxyl-functionalized cinnamic acid derivatives such as p-coumaric acid (p-hydroxycinnamic acid), ferulic acid, and sinapinic acid are attractive monomers for syntheses of high-performance polyesters. The obtained polyesters are also expected to be biodegradable in the case of copolymerization with aliphatic hydroxy acids such as lactic acid. Tanaka et al.[196] reported the thermal polycondensation of p-coumaric acid at 550 °C without any catalyst under high pressure up to 80 kbar (in the solid state) in 1975. They obtained red or brownish-red hard solids insoluble in conventional organic solvents. Higashi and his co-workers[197] synthesized copolyesters of p-coumaric acid and 4-hydroxybenzoic acid or their methoxy substitutions (ferulic acid, vanillic acid, or syringic acid) by polycondensation using hexachlorocyclotri(phosphazene) in pyridine in 1981. The obtained polymers that exhibited UV spectra different from those of monomers and inherent viscosity of 0.3 to 1.3 dL g^{-1}. Palacios and his co-workers reported syntheses of poly(p-coumaric acid)[198] and poly(ferulic acid)[199] using thionyl chloride in 1985. The melting temperatures of poly(p-coumaric acid) and poly(p-ferulic acid) were 313 °C and 325 °C, respectively. These polymers showed signs of thermotropic liquid crystallinity in DSC measurements. Some aromatic polyesters are known as liquid crystalline polymers and the cinnamoyl-based main-chain liquid crystalline polymers were explicitly reported by Creed et al.[200] in 1988, followed by Navarro and Serrano[201] in 1989. Creed et al.[200] also investigated the photocross-linking of the liquid crystalline polyesters. Synthesis of cinnamoyl-based polyanhydrides was reported by Ritter and his co-workers.[202] They synthesized polyanhydrides from 4,4'-(octamethylenedioxy)di-*trans*-cinnamic acid and 4,4'-(sebacoyldioxy)di-*trans*-cinnamic acid by using acetic anhydride and subsequent melt polycondensation. The polyanhydrides had M_w of ca. 11 000, and underwent UV irradiation to give photocross-linked films. Jin et al.[203] synthesized ternary copolymers of p-hydroxybenzoic acid, glycolic acid, and p-coumaric acid. They observed a Schlieren-type texture for low molecular weight copolymers or banded textures for high molecular weight copolymers under polarizing optical microscopy (POM), indicating that the polyesters were liquid crystalline polymers. The copolymers degraded *via* simple hydrolysis at amorphous or less aligned regions. Three research groups, Kricheldorf and his co-workers,[204–208] Nagata and his co-workers,[209–213] and Akashi and his co-workers[214–218] studied the liquid crystalline properties, thermal properties, biodegradability, photocross-linking, and mechanical properties of hydroxycinnamate-based or hydroxyphenylpropionate-based polyesters. These polymers sometimes contain p-hydroxybenzoate units,[204,207–209] gallate units (for hyperbranched polymers),[207,208] vanillate units,[204,208,209] lactate units,[213,214] and/or other aliphatic and aromatic ester units. Considering the higher performance of aromatic polyesters, these polymers may overwhelm poly(L-lactide) someday. Other groups studied nanostructures,[219] microspheres,[220,221] thermal decomposition kinetics,[222] biodegradability,[223,224] and mechanical properties.[224] Mizutani and Matsuda[225] synthesized coumarin-endcapped biodegradable copolyestercarbonates and photocured the copolymers by UV irradiation to give cross-linked biodegradable films. After surface erosion by hydrolysis, the copolymer films exhibited no cell adhesion in the cell culturing test. Figure 2.12 summarizes structures of polyesters containing phenylpropanoid units.

Eugenol is another type of phenylpropanoid. It has an allyl group and a phenolic hydroxyl group. Because of the low reactivity on homopolymerization of the allyl group, I could not find a research paper on the successful homopolymerization of eugenol except oxidative polymerization. The two reactive groups on eugenol were effectively used by several researchers to incorporate eugenol into polymers. The allyl group reacts with the hydrosilyl group (Si–H) in the presence of a Pt catalyst, and therefore it is easy to obtain eugenol end-capped siloxane polymers (Figure 2.13). The eugenol end-capped siloxane polymers can be used as a diol or bisphenol compound, and they are

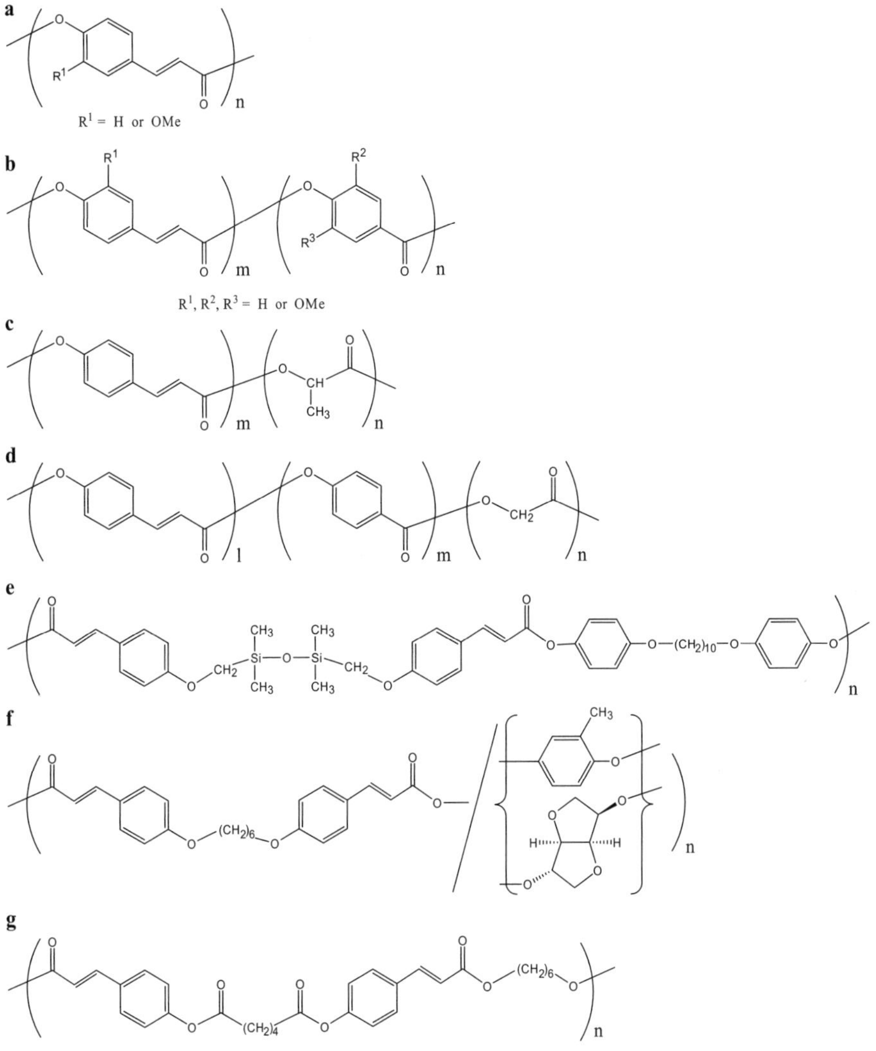

Figure 2.12 Example structures of polyesters containing phenylpropanoid units.

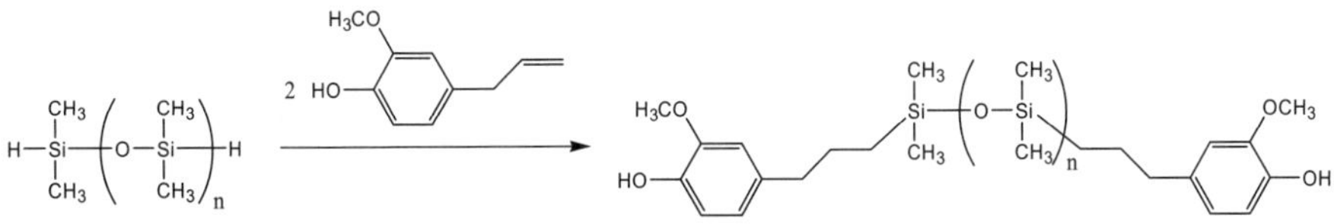

Figure 2.13 Synthesis of eugenol end-capped siloxane oligomers.

available from General Electric Co. Hagenaars et al.[226] reported characterization of polycarbonates containing bisphenol A, eugenol, and siloxane oligomer units. Waghmare et al.[227,228] synthesized polyesters from phenylindane bisphenol (or bisphenol A), diphenyl terephthalate, diphenyl isophthalate, and the eugenol end-capped siloxane polymer. Masuda and his co-workers[229] synthesized 4-allyl-2-methoxy-1-prop-2-ynyloxybenzene, an ether synthesized from eugenol and propargyl chloride. The monomer was polymerized with Rh or Mo catalyst, to yield a novel polyacetylene. Rojo et al.[230] reported the polymerization of eugenol-based methacrylate monomers. They achieved the polymerization *via* radical polymerization and found that some allylic C=C double bonds were involved in the radical polymerization to give cross-linked polymers. The polymers were designed for dental composites and orthopedic bone cements.

Lignin is synthesized from phenylpropanoids *via* oxidative polymerization, perhaps by peroxidase, an H_2O_2-dependent hemoprotein, and/or laccase, an oxygen-dependent oxidase containing four copper atoms.[102] Oxidative polymerization (dehydrogenative polymerization) of phenylpropanoids by enzymes and chemical oxidants has been studied for a long time since 1951, mainly for the purpose of understanding lignin biosynthesis and lignin structures.[231–236] For example, lignin-like polymers were synthesized from eugenol and ferulic acid by peroxidase activity of tissue sections by Stafford;[235] and Higuchi and his co-workers polymerized 3,5-disubstituted *p*-coumaryl alcohols using a peroxidase with H_2O_2 or with $FeCl_3$.[236] The researches on enzymatic oxidative polymerization of phenolic compounds (Figure 2.14) have been developed and applied to the field of polymer materials from phenylpropanoids[237–240] including flavonoid derivatives.[241–244] Horseradish peroxidase (HRP) and soybean peroxidase (SBP) are often used for this purpose; and laccases are sometimes used. The enzymatic oxidative polymerization of other naturally occurring phenolic compounds are described below and the polymerization techniques in detail are reviewed elsewhere.[245–249]

2.3.3 Lignin Digests or Extracts and Liquefied Wood

Lignin digests and extracts are also important renewable materials, because abundant amount of lignin waste is generated in the pulping industry and will be generated in the production of ethanol, an upcoming biofuel from cellulose. Repolymerization of lignin digests or extracts has been researched in several

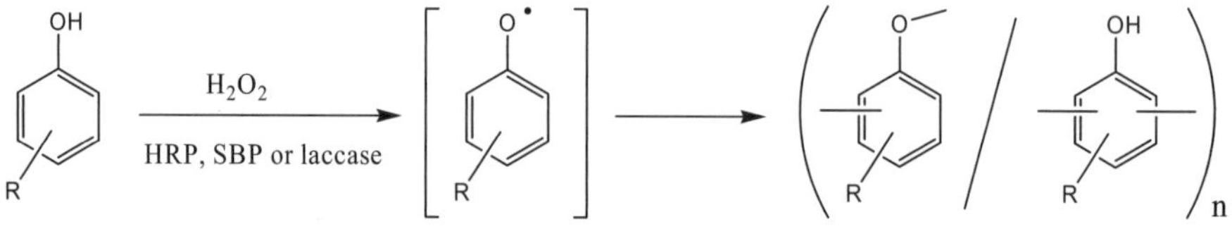

Figure 2.14 Enzymatic oxidative polymerization of phenolic compounds.

methods.[250,251] Oxidative polymerization of lignin digests and extracts is achieved mainly by enzymatic reaction in a similar manner as described above. Allan et al.[252] reported the modification of cellulose surface by enzymatic polymerization of isoeugenol as a lignin analog, and produced a composite with a striking wood-like appearance. Dordick and his co-workers reported enzymatic modification of lignin with some phenolic compounds.[253–255] The copolymers of Kraft lignin with p-cresol or p-phenylphenol showed markedly lower T_g and higher exothermic enthalpy of curing than unreacted lignin.[255] Hüttermann and his co-workers[256] polymerized lignin soluble in organic solvents by laccase and analyzed the structure of products by UV and solid-state ^{13}C-NMR spectroscopy. In other literature,[257,258] they also modified the surface of wood fibers using laccase to give wood composites like medium density fiberboards (MDF). Batog et al.[259] showed that laccase can be used for adhesive bonding of lignocellulosic materials by enzymatic oxidation of lignin.

The synthetic method of phenol-formaldehyde resins have been applied to the polymerization of natural phenolic compounds. Lignin has many phenol units which can react with formaldehyde, and its digests and extracts that are soluble in water or organic solvents are used as raw material.[250] This reaction was sometimes carried out in the presence of phenol to improve the mechanical and molding performance. Phenol-lignin-aldehyde resins were known for a long time[260] and studied by many researchers.[261–271] Recently ligninformaldehyde resins with higher performance were reported;[269–271] and mechanical and thermal properties of the thermosets for plywood or fiberboard are comparable to phenol-formaldehyde resins.

Polyurethanes from lignin or lignin-aldehyde resins were prepared using multifunctional isocyanates.[272–280] It is known that lignin digests and extracts, and its reaction products with formaldehyde, have many hydroxyl groups which can react with isocyanate groups. The mechanical properties of this type of polyurethanes varied with the lignin contents. Generally, the mechanical rigidity increases with an increase of lignin contents. Thermoplastic polyesters synthesized from lignin and sebacoyl chloride was reported by Thanh Binh et al.[281]

Epoxy resins were also synthesized from lignin digests and extracts, mainly by reaction with epichlorohydrin.[250] Because epoxy resins react with multifunctional phenolic compounds to give network polymers, thermosets can be prepared from lignin-derived epoxy resins and lignin. This type of epoxy resin was reported to have good mechanical properties recently.[282] Epoxy resins also react with anhydrides and amines, and research has been carried out on the lignin-derived epoxy resins.[283–288] Printed circuit boards are usually prepared from phenol-formaldehyde resins or epoxy resins, and therefore lignin derivatives can substitute for these resins.[289,290] Figure 2.15 shows the photographs of a printed circuit board manufactured from lignin derivatives developed by Hitachi, Ltd.

Funaoka and his co-workers[291–294] studied on the lignin digests and developed their researches extensively into the large field of preparation of polymers and chemicals. They adapted their original method to degrade lignin through the phase-separated system, and obtained polymeric or oligomeric compounds carrying many hydroxyl groups.

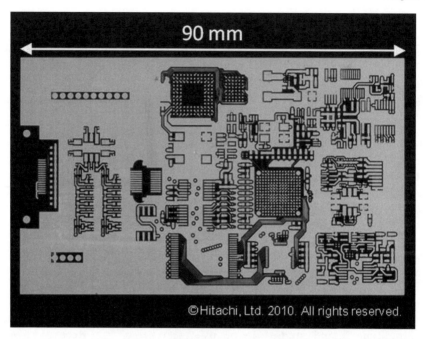

Figure 2.15 Printed circuit board manufactured from lignin derivatives developed by Hitachi, Ltd. Courtesy and reproduced with permission of Hitachi Ltd., Japan. © Hitachi Ltd, Japan.

Liquefied wood is also an interesting raw material for polymer preparation. Wood flour contains lignin, cellulose, and hemicellulose, and can be liquefied with phenol or glycol in the presence of an acid or base catalyst, or at high temperature (>250 °C) or in sub-/super-critical fluid (*e.g.* water, alcohols, and phenols) in the absence of a catalyst.[295–298] The liquefaction technique is very attractive because it can be applied to most biomass other than wood products. The liquefied products mainly contains polyols which are used for synthesis of polyurethanes,[299–303] polyesters,[304,305] phenol-formaldehyde type resins,[306–312] and epoxy resins.[313–316]

2.3.4 Other and Natural Phenolic Compounds

Many phenolic compounds other than phenylpropanoids can be obtained from plants: catechol, guaiacol, syringic acid, syringaldehyde, gallic acid, vanillin, and vanillic acid are obtained from plants. These phenolic compounds have become candidates for monomers used in the polymer industry. They undergo oxidative polymerization and serve as monomers for phenol-formaldehyde type resins, and some compounds having a hydroxyl group and a carboxylic acid serve as monomers for polycondensation.

Kobayashi and his co-workers[317] showed that the laccase-catalyzed oxidative polymerization of syringic acid (3,5-*O*-methyl substituted gallic acid) gave a

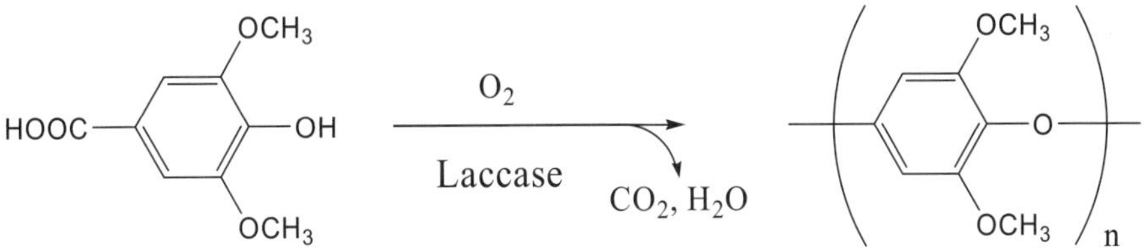

Figure 2.16 Synthesis of poly(phenylene oxide) by laccase-catalyzed oxidative polymerization of syringic acid.

poly(phenylene oxide). The polymerization involves dehydrogenation and decarboxylation (Figure 2.16). The molecular weight of the product was up to 18 000 at the peak top of the GPC profile. Syringaldehyde, 3,5-dimethoxy-4-hydroxybenzaldehyde, was copolymerized with bisphenol A in the presence of peroxidase and H_2O_2 by An et al.[318] The M_w of the product synthesized at the feed ratio of 1:1 (syringaldehyde:bisphenol A as weight) in acetone was 5700.

Catechol (o-hydroxyphenol), guaiacol (o-methoxyphenol), and resorcinol (m-hydroxyphenol) can be extracted from plants or derived from other naturally occurring precursors. Dubey et al.[319] reported the polymerization of catechol using peroxidases, HRP, and SBP. The polymer prepared by the HRP-catalyzed polymerization shows the regular pattern with spongy rock type morphology at SEM observation. Shin et al.[237] polymerized catechol, guaiacol, and other phenolic compounds in situ for wool coloration as a novel dying method. Vachoud et al.[320] reported the graft polymerization of gallate esters onto chitosan to propose a novel modification method. Gallic acid is obtained from tannic acid, whose structure is mainly composed of glucose esters of gallic acid. Tannic acid is biosynthesized in *Schlechtendalia chinensis*, a gall-aphid commonly found on the nutgall sumac tree. It has condensed phenolic hydroxyl groups and can be used as a hardener of epoxy resins. Shibata and his co-workers[321,322] reported the preparation of thermosets and composites from biomass-derived epoxy resins cross-linked with tannic acid. Gallic acid is often used as a branching point of hyperbranched polymers.[207,208,323,324]

Cardanol is a main component of thermally treated cashew nut shell liquid (CNSL), and is a phenolic compound with a long unsaturated hydrocarbon chain substituted in the *meta* position (Figure 2.17a). Urushiol, which is obtained from lacquer tree, poison ivy, poison oak, and poison sumac (*Toxicodendron*), and used for a raw material of a lacquer (urushi) in East Asia, is also a phenolic compound of catechol with a long unsaturated or saturated hydrocarbon chain (Figure 2.17b). Cardanol-based polymers have been reported very often, while there are a few research reports on urushiol-based polymers. Research on polymers synthesized from cardanol or CNSL are reviewed elsewhere.[325] In the late 1980s, cardanol or CNSL-based polymers began to be reported as novel phenol-formaldehyde type resins[326,327] and novel epoxy resins.[328,329] Thereafter, Pillai and his co-workers have vigorously studied synthesis of various type of cardanol-based polymers: polymers obtained

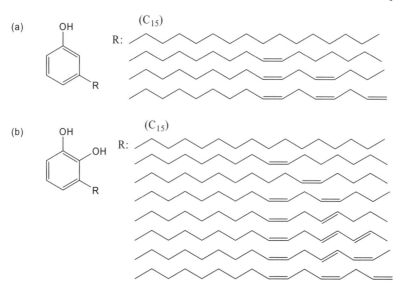

Figure 2.17 Structures of (a) cardanol and (b) urushiol (typical structures).

by cationic polymerization,[330] semi-interpenetrating network polymers of cardanol-formaldehyde resins and MMA polymer,[331,332] flame retardant phenol-formaldehyde type resins from cardanyl phosphoric acid or bromine-modified cardanol,[333] a self-cross-linkable polymer from cardanyl acrylate.[334] Polyesters[335] and polyurethanes[336–338] derived from cardanol, and cardanol-formaldehyde thermoset composites reinforced with natural fibers[339,340] were studied by several researchers. 8-(3-Hydroxyphenyl) octanoic acid (3-HPOA) is a hydroxy acid derivative obtained by the oxidative conversion of cardanol. Thermotropic liquid crystalline copolyesters from 3-HPOA and p-hydroxybenzoic acid, and homopolymers of 3-HPOA were synthesized.[341,342]

The urushi lacquer has been used for more than 5000 years in China[343] and it is known as a highly durable material. Polymerization of urushiol, the major component of the lacquer, involves laccase-catalyzed dimerization and aerobic oxidative polymerization,[343] and the drying process takes a very long time. Several studies on shortening of this time have been carried out: UV curing[344] and hybridizing with other reactive polymers or monomers.[345–347] Cardanol has a similar structure to urushiol, and the enzymatic oxidative polymerization of cardanol were reported by three research groups.[348–350] The development of the polymerization process leads to 'artificial urushi'.[351]

Vanillin is the main component of vanilla oil and it has a phenolic hydroxyl group and an aldehyde group. Dordick and his co-workers[352] reported the oxidative oligomerization of vanillin and other guaiacol derivatives using SBP, and the monomers were not polymerized to high molecular weight products. Vanillic acid and syringic acid are a kind of hydroxybenzoic acid and they have been used to prepare condensation polymers.[197,204,208,209] Because of steric hindrance by methoxy groups adjacent to the hydroxyl group, vanillic acid and

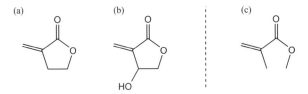

Figure 2.18 Structures of (a) tulipalin A and (b) tulipalin B, in comparison with (c) MMA.

syringic acid are usually copolymerized with *p*-hyrdoxybenzoic acid, *p*-coumaric acid, and ferulic acid. In 1955, Bock and Anderson[353] synthesized dibasic acids by etherifying vanillic acid with alkylene dihalides, and polymerizing the dibasic acids with ethylene glycol to yield linear polyesters. Kricheldorf and Löhden[354] synthesized poly(ester-amide)s from vanillic acid and 4-aminobenzoic acid and studied morphologies of the crystallites. They also synthesized hyperbranched poly(ester-amide)s from vanillic acid, 4-aminobenzoic acid, and gallic acid to reduce the crystallinity and improve the solubility.[324] Other groups, Sazanov *et al.*[355] and Li *et al.*,[356] reported condensation polymers containing vanillic acid as one of monomer components.

Tulipalin is a naturally occurring α-methylene-γ-butyrolactone (Figure 2.18) isolated from tulips:[357] tulipalin A is α-methylene-γ-butyrolactone (MBL), and tulipalin B is β-hydroxy-α-methylene-γ-butyrolactone. Tulipalin has a cyclic structure partially containing a MMA unit, and polymerization of tulipalin A has been often reported. Akkapeddi[358] showed the synthesis of poly(MBL) by both free-radical and anionic polymerization in 1979. The polymer from free-radical polymerization is an amorphous transparent polymer like PMMA with a higher glass transition temperature of 195 °C. Ueda *et al.*[359] investigated the kinetics of radical homopolymerization of MBL and that of copolymerization of MBL with styrene. More recently well-controlled polymerization of MBL is reported by two groups: Mosnácek and Matyjaszewski[360] reported the atom transfer radical polymerization (ATRP) of MBL, and Miyake *et al.*[361] reported the polymerization of MBL using ambiphilic silicon propagating species consisting of both the nucleophilic silyl ketene acetal initiating moiety and the electrophilic silylium catalyst.

2.4 Carbohydrates and their Derivatives

2.4.1 Polymers from Popular Carbohydrates

The most popular carbohydrate obtained from plants is sucrose. Glucose, maltose, and cellobiose are also popular because they are obtained from starch and cellulose by hydrolysis. Despite its abundant production by plants, it is difficult to use them as monomers for synthetic polymers. These carbohydrates have three or four hydroxyl groups per one glucose unit, and they are very hydrophilic. Therefore they are insoluble in most major organic solvents.

Multiple hydroxyl groups restrict usage of these carbohydrates as monomers for linear polymers. Sachinvala et al.[362] reported sucrose-based monomers carrying unsaturated C=C bonds for cross-linked polymers. Preparation of sugar-based linear polymers are more challenging than cross-linked sugar-based polymers. Researchers have found many ways to synthesize linear polymers from sugars, but synthesized polymers have been somewhat biased towards side-chain polymers (called as glycopolymers) mainly pursuing bio-functionality.[363] For example, 3-O-methacryloyl-1,2:5,6-diisopropylidene-D-glucose (MDG) was synthesized from 1,2:5,6-diisopropylidene-D-glucose and methacryloyl chloride, and polymerized using AIBN as radical initiator by Kimura and Imoto.[364] After polymerization, the isopropylidene protecting groups were hydrolyzed to give poly(3-O-methacryloyl-D-glucose). From the point of view of environmental polymeric materials, removing of protecting groups is not necessary, even if it enhances the biocompatibility of the polymer material. Al-Bagoury and Yaacoub[365] investigated thermal, mechanical, and rheological properties of copolymers of MDG and butyl acrylate prepared by emulsion polymerization. The T_g of the copolymers varied from −54 °C to 167 °C depending on the composition of MDG to butyl acrylate. The Young's modulus increased with an increase of MDG contents, but the film became more brittle. Most recently, Barros et al.[366] adapted a selective Mitsunobu reaction to synthesize a sucrose-based methacryl monomer (Figure 2.19). This selective reaction is very attractive for synthesis of sugar-based polymers.

Preparation of cross-linked sucrose-based polymers by photopolymerization was achieved by Acosta Ortiz et al.[367] using thiol-ene reaction. They synthesized sucrose allyl ether, mainly diallyl sucrose, and polymerized the sucrose-based monomers with dithiothreitol by UV irradiation in the absence or presence of a photoinitiator.

Condensation polymers containing sucrose or glucose units in the main chain were also reported. Cross-linked polyurethanes prepared from polyethylene glycol (PEG), diphenylmethane diisocyanate (MDI) and a sugar (glucose, fructose, or sucrose), were reported by Zetterlund et al.[368] The storage modulus, the T_g, and the stress at break increased with an increase of sucrose content; and the strain at break decreased with an increase of sucrose content.

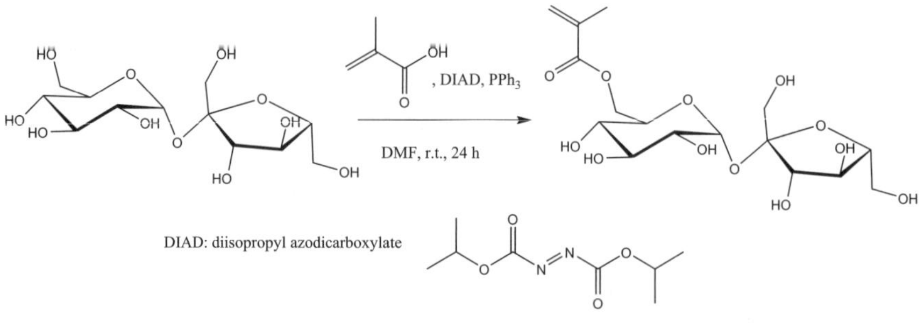

Figure 2.19 Regioselective Mitsunobu reaction of sucrose with methacrylic acid.

Donnely[369] synthesized mono- or bisglucoside-modified polytetrahydrofuran (PTHF) polyols and cured them with MDI. When the glucoside-modified PTHF was cured with MDI in the presence of ethylene glycol as a chain extender, the cured polyurethanes exhibited good mechanical properties comparable to commercial reaction injection molding polyurethanes. Kurita and his co-workers synthesized polyesters[370] and polyurethanes[371] by direct reaction of D-cellobiose with dicarboxylic acid dichlorides and diisocyanates, respectively. Interestingly, they obtained soluble polymers without crosslinking. Most recently, a rigid polyurethane foam reinforced with cellulose whiskers was prepared by Li et al.[372] using sucrose-based polyol and glycerol-based polyol with polymeric diphenylmethane diisocyanate.

For synthesis of linear condensation polymers, 'sugar diols' are the most desirable monomers.[373] Isosorbide, 1,4:3,6-dianhydro-D-sorbitol (Figure 2.20a), and isomannide, 1,4:3,6-dianhydro-D-mannitol (Figure 2.20b), are typical diol compounds derived from D-sorbitol and D-mannitol. Isosorbide is synthesized from D-glucose by hydrogenation (production of D-sorbitol) and subsequent dehydration in the presence of an acid catalyst.[373] Many condensation polymers from isosorbide have been reported. Polyesters,[374–378] polyurethanes,[375,379–381] and polycarbonates[375,382–384] were synthesized. Among these polymers, copolymers with aliphatic compounds are biodegradable. Polymers from 1,4:3,6-dianhydrohexitols are described in detail elsewhere.[385] Other types of sugar diol are methyl 4,6-O-benzylidene-α-D-glucopyranoside (Figure 2.20c) and methyl 2,6-di-O-pivaloyl-α-D-glucopyranoside (Figure 2.20d). These types of monomers are synthesized by the technique of regioselectively protecting the hydroxyl groups. Polyurethanes from these two monomers and 1,6-hexamethylene diisocyanate were reported by Garçon et al.[386] Hasimoto et al.[387] also reported

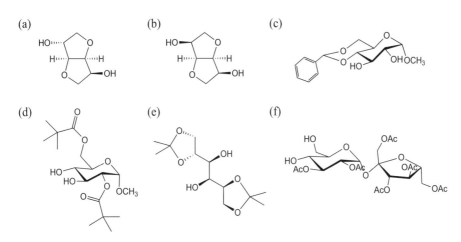

Figure 2.20 Structures of sugar diols: (a) D-isosorbide, (b) isomannide, (c) methyl 4,6-O-benzylidene-α-D-glucopyranoside, (d) methyl 2,6-di-O-pivaloyl-α-D-glucopyranoside, (e) 1,2:5,6-di-O-isopropylidene-D-sorbitol, and (f) 2,3,1′,3′,4′,6′-hexa-O-acetylsucrose.

synthesis of polyurethanes from methyl 4,6-*O*-benzylidene-α-D-glucopyranoside and two diisocyanates, hexamethylene diisocyanate and methyl 2,6-diisocyanatohesanoate (lysine diisocyanate). They also synthesized another sugar diol, 1,2:5,6-di-*O*-isopropylidene-D-sorbitol (Figure 2.20e), as a monomer of polyurethane. Jhurry and Deffieux[388] synthesized a sugar diol from sucrose (Figure 2.20f) by selective ketalization using 2-methoxypropene (production of 4,6-*O*-isopropylidenesucrose), acetylation of residual hydroxyl groups, and subsequent deprotection of the isopropylidene group. Sucrose-based polyurethanes were synthesized by polyaddition of this sugar diol and 1,4-phenylene diisocyanate or hexamethylene diisocyanate.

Enzymatic and chemoenzymatic reactions have been applied to synthesize polymers and polymerizable monomers from sugar.[389] Two-step enzymatic synthesis of sucrose-based polyesters were reported by Dordick and his co-workers.[390] In the first step, sucrose was reacted with an excess amount of divinyl adipate using lipase in an organic solvent to give sucrose 6,6′-*O*-divinyl adipate; and in the second step, the monomer was reacted with equimolar amounts of aliphatic diols using lipase to give linear polyesters with M_w of up to 22 000 (Figure 2.21). Tokiwa and his co-workers adopted the chemoenzymatic technique for synthesis of addition polymers carrying sugar ester units as a pendant group. They prepared polymers carrying glucose units[391] and maltose units.[392]

2.4.2 Furan Derivatives

Furan is a heterocyclic aromatic compound containing one oxygen atom in the five-membered ring. Furan substitutes are easily produced by heating carbohydrates in the presence of an acid catalyst. Liquefied cellulose and liquefied starch are known to contain considerable amount of furan compounds. Generally, 5-hydroxymethylfurfural is produced from hexose carbohydrates, and furfural is produced from pentose carbohydrates (Figure 2.22). Recent technological advances in this conversion process achieved very high yields of

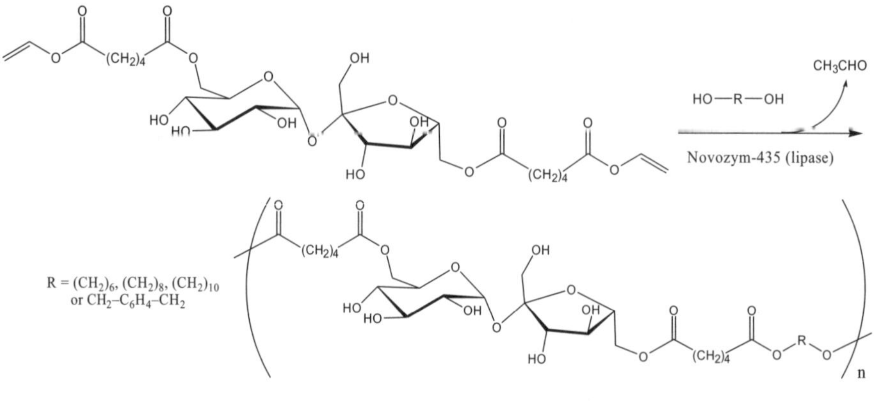

Figure 2.21 Synthesis of polyesters by lipase-catalyzed polymerization.

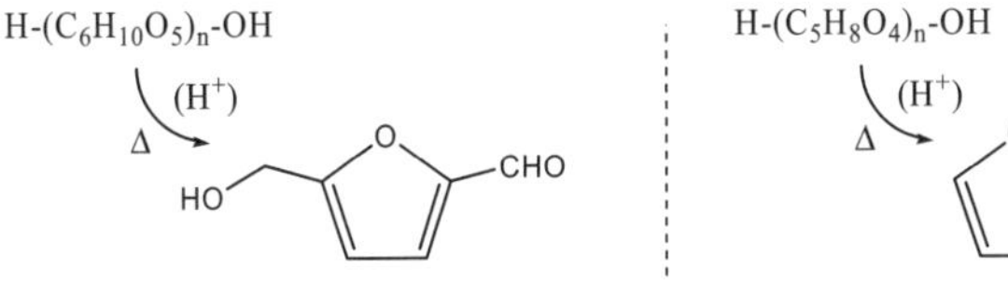

Synthetic Green Polymers from Renewable Monomers 55

H-(C$_6$H$_{10}$O$_5$)$_n$-OH $\xrightarrow[\Delta]{(H^+)}$ HOCH$_2$-furan-CHO

H-(C$_5$H$_8$O$_4$)$_n$-OH $\xrightarrow[\Delta]{(H^+)}$ furan-CHO

Figure 2.22 Production of 5-hydorxymethylfurfural from hexose carbohydrates and fufural from pentose carbohydrate.

up to 80%.[393–396] Moreover, this process can be applied to most biomass-containing carbohydrates such as corn stalk, rice straw, and wood waste.

Furan resin has been used for thermosets (for example, matrix of fiber reinforced composites) and mortar cements for a long time.[397] Gandini and his co-workers vigorously studied and developed many polymers from furan compounds (Figure 2.23). A part of their researches are introduced here. One of the interesting reactions characteristic to furan compounds is Diels–Alder reaction, which involves 4+2 cycloaddition of diene and dienophile. Furan is a very good dienophile for the reaction with maleimide, and furthermore this reaction is reversible.[398] Gandini and his co-workers[399] synthesized polyurethanes bearing furan moieties and cross-linked the polymers by a bismaleimide compound in 1997. Then, they did not succeed in the reversible cross-linking by Diels–Alder and retro-Diels–Alder reactions, but they succeeded in the reversible cross-linking using other flexible-chain polymers.[400] Other polymers synthesized from furan compounds are polyesters,[401–403] polyamides,[404,405] polyurethanes,[399,406] poly(furylene vinylene),[407,408] and poly(furfurylmaleimide),[409] and so on.[397] The furylene vinylene unit was reported to undergo dimerization by UV irradiation.[407] Polyesters from 2,5-furandicarboxylic acid and ethylene glycol showed thermal properties similar to those exhibited by poly(ethylene terephthalate) (PET)[403] (T_m of the furan-derived polyester is somewhat lower than that of PET). Polyamides from 2,5-furandicarbxylic acid and 1,4-phenylenediamine showed thermal properties similar to those of Kevlar.[404]

2.5 Monomers Obtained by Fermentation

Advances in fermentation technology have been remarkable over recent decades, and many chemical compounds are produced in an industrial scale.[410] Ethanol and lactic acid have been produced by fermentation for very long time. These two products are easy to obtain because they are the final products of each metabolic pathway. Fortunately, lactic acid is a monomer available for a biodegradable polyester, and ethanol is available as biofuel. Poly(lactic acid) (PLA) is a well-known polymer, whose physical properties are comparable to those of hard plastics produced from petroleum chemicals. The amount of industrial PLA production is increasing, and the application field of PLA is being extended. Much literature covers the properties, synthetic methods,

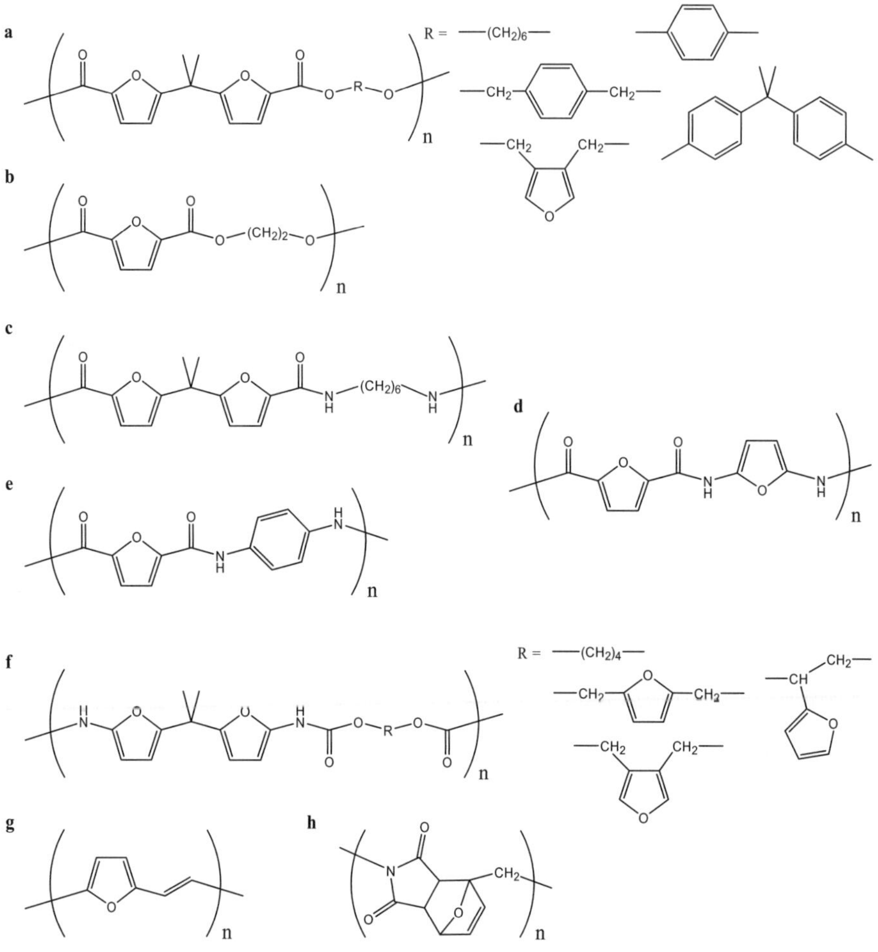

Figure 2.23 Example structures of synthetic polymers containing furan units.

crystallization techniques, and improvement of PLA.[411,412] Here I do not describe PLA due to limitations of space, and ask you to refer to other sections of this book or other literature. 3-Hydroxypropionic acid also can be synthesized by a fermentation process.[413] 3-Hydroxypropionic acid is applicable to synthesis of acrylic acid and acrylamide as well as to synthesis of poly(3-hydroxypropionic acid) (P(3HP)), a biodegradable polyester. Drawn films of P(3HP) have very remarkable tensile strength up to 482 MPa with high Young's modulus of 2.79 GPa at the draw ratio of 14.4.[414] Zhang et al.[415] reported synthesis of P(3HP) via macrocyclic esters from 3-hydroxypropionic acid. The melting point of P(3HP) was reported to be 76 °C.

Recently, Du Pont Co. started the industrial production of 1,3-propanediol from corn via an original fermentation process.[416] Poly(trimethylene terephthalate) (PTT) is produced from 1,3-propanediol and therefore became one of green polymers that are manufactured using renewable resources. PTT has unique properties compared to PET; its fibers are stretchable and its textile is soft and flexible. This stretchable property is due to its unique crystalline structure.[417] Biosynthesis of 1,3-propanediol by native organisms is achieved by fermentation of glycerol under anaerobic conditions. However, it is desirable to obtain 1,3-propandiol from D-glucose by fermentation using one organism strain. Therefore a strain that produces 1,3-propanediol from D-glucose under aerobic conditions was developed based on *Escherichia coli* (*E. coli*) K12 strain.[418] Currently, succinic acid is becoming a renewable resource synthesized by fermentation.[410,419,420] Succinic acid is frequently used for synthesis of biodegradable aliphatic polyesters, as represented by Bionolle®, poly(butylene succinate), produced by Showa Denko K.K. Considering succinic acid has a C_4 skeleton, many other important chemicals, such as butanediol, maleic acid and γ-butyrolactone, can be derived from succinic acid. Though many organisms are known to excrete succinic acid, only several organisms are suitable for the industrial production. Generally, four organisms, *Anaerobiospirillum succiniciproducens*, *Actinobacillus succinogenes*, *Corynebacterium glutamicum*, and *Manheimia succiniciproducens* are used as native strain, and genetically modified strains of these four organisms and *E. coli* are also used for production of succinic acid.[419] As for butanediol, 2,3-butanediol is a fermentation product of *Bacillus polymyxa* and *Klebsiella pneumoniae*.[421,422] The monomer is applicable not only to synthesis of aliphatic polyesters but also to synthesis of 1,3-butadiene, a monomer for synthetic rubber. The advantage of the fermentation production of 2,3-butanediol is the possible usage of wood hydrolysates and hemicellulose as a resource.[421,422] Itaconic acid is an unsaturated dicarboxylic acid, mainly obtained by fermentation using *Aspergillus terreus*. Itaconic acid is an interesting monomer because it can be polymerized with two different mechanisms – addition and condensation. Polyesters synthesized from itaconic acid with aliphatic monomers are expected to be biodegradable polymers, and they are cross-linkable.[423,424] After cross-linking, their biodegradability will be retained at a low content of itaconic acid. Still, the problem is that the polycondensation reaction is very technical because of high reactivity of the C=C bond of itaconate[425] and high isomerization activity to citraconic acid or mesaconic acid. If itaconic acid is polymerized firstly using the C=C bond, we can obtain a polyanion, poly(itaconic acid), like poly(acrylic acid).[426] Polyion complex from poly(itaconic acid) and chitosan was prepared by Lárez et al.[427] By esterification of the carboxyl groups, the polymer becomes hydrophobic and thermoplastic.[426,428]

Several carbohydrates, sugar alcohol and their derivatives, such as erythritol, trehalose, and glucaric acid, are also synthesized by fermentation or an enzymatic process. Erythritol is a C_4 compound carrying four hydroxyl groups. Erythritol is produced from glucose, sucrose, or *n*-alkanes by the ermentation method using osmophilic yeasts and many yeast-like organisms such as *Pichia*, *Zygopichia*, *Candida*, *Torulopsis*, *Trigonopsis*, *Moniliella*, *Trichosporonoides*, and *Aureobasidium*.[429–433] Trehalose is a non-reducing disaccharide, D-glucopyranosyl-D-glucopyranoside, and it has two glucose units bound at each C1 position. Trehalose is found in many organisms, such as bacteria, fungi, plants, and invertebrates including insects. Trehalose is known to play an important role in cryptobiosis of many organisms. Recently, an enzymatic process for the industrial synthesis of trehalose was established by Hayashibara Co. Ltd,[434] and trehalose became available at a low cost (<$3 per 1 kg). Both erythritol and trehalose have two primary hydroxyl groups and several secondary hydroxyl groups. The difference in reactivity between primary hydroxyl groups and secondary hydroxyl groups is applicable to regioselective reaction which enable facile synthesis of linear polymers from erythritol or trehalose. Examples of regioselective reactions are acetalization with benzaldehyde derivatives, iodine substitution, and enzymatic esterification. Polymers synthesized from trehalose are reviewed in detail elsewhere.[435] Research reports on polymers from erythritol are very few, and most of them are related to polyester. Uyama et al.[436] and Hu et al.[437] reported polymerization of sugar alcohols with divinyl sebacate and adipic acid, respectively, using lipase. Barrett et al.[438] reported the thermal polycondensation of erythritol with dicarboxylic acids to obtain cross-linked elastomeric materials. The regioselective esterification using rare-earth trifluoromethanesulfonate, reported by Takasu et al.,[439] will be a convenient method for preparation of linear polymer from erythritol or glucaric acid. 1,4-Anhydroerythritol is a cyclic bifunctional compound derived from erythritol. This will be used as a monomer for polycondensation. Imai et al.[440] reported the ring-opening polymerization of 1,4-anhydroerythritol to obtain hyperbranched carbohydrate polymers, and Wachenfeld and Burchard[441] synthesized epoxy resins from 1,4-anhydroerythritol and bisphenol A diglycidyl ether. Bachmann and Thiem[442] transformed 1,4-anhydroerythritol into 2,3-diamino-2,3-dideoxy-1,4-anhydroerythritol, and polymerized the resulting diamine with aromatic and aliphatic dicarboxylic acid dichlorides to obtain polyamides (Figure 2.24).

Glucaric acid is a compound having two carboxyl groups and four secondary hydroxyl groups. It can be obtained by oxidation of glucose, and recently Moon et al.[443] constructed a recombinant E. coli that produces glucaric acid. Polyamides based on glucaric acid were synthesized by Kiely et al.[444] via simple condensation reactions of esterified glucaric acid with diamines (Figure 2.25).

Interestingly, even an aromatic compound with a benzene ring is also found to be synthesized by fermentation. Horinouchi and his co-workers[445] reported that two genes, griI and griH, conferred the ability of in vivo production of 3-amino-4-hydroxybenzoic acid (3,4-AHBA) on E. coli (Figure 2.26). This pathway is independent of the shikimate pathway, which is a well-known pathway for production of aromatic amino acids in bacteria, fungi, algae, and higher plants. Homopolymerization of 3,4-AHBA yields polybenzoxazole (ABPBO) (Figure 2.27), which shows the liquid-crystalline phase during polymerization.[446–449] In the near future, high-performance industrial plastics from renewable monomers may appear.

Monomers that can be obtained by fermentation are shown in Figure 2.28.

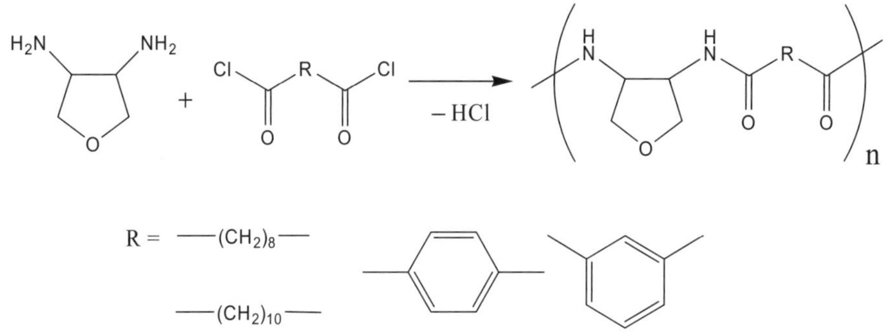

Figure 2.24 Synthesis of polyamide from erythritol-derived diamine and dicarboxylic acid dichloride.

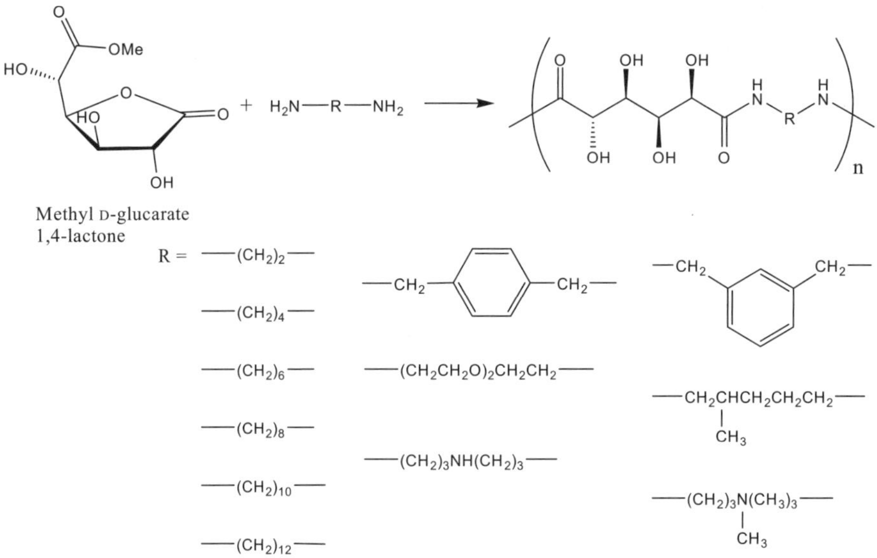

Figure 2.25 Synthesis of polyamides from esterified glucaric acid diamine and dicarboxylic acid dichloride.

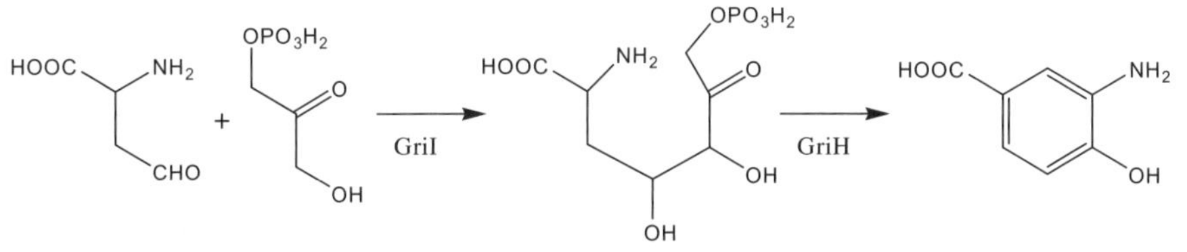

Figure 2.26 Biosynthesis of 3-amino-4-hydroxybenzoic acid by *E. coli* transformed with *GriI* and *GriH* genes.

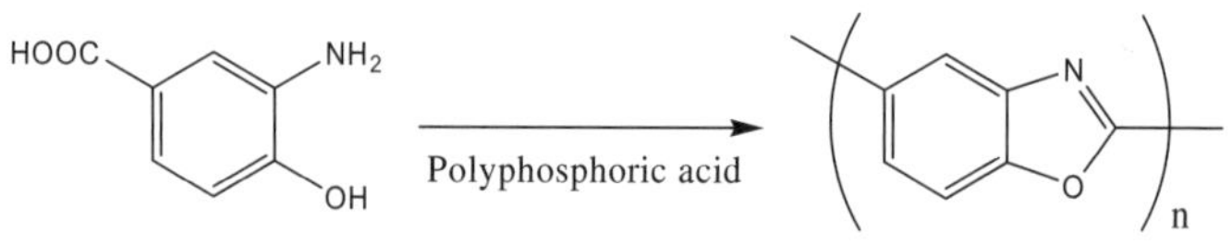

Figure 2.27 Polymerization of 3-amino-4-hydroxybenzoic acid to give poly(2,5-benzoxazole) (ABPBO).

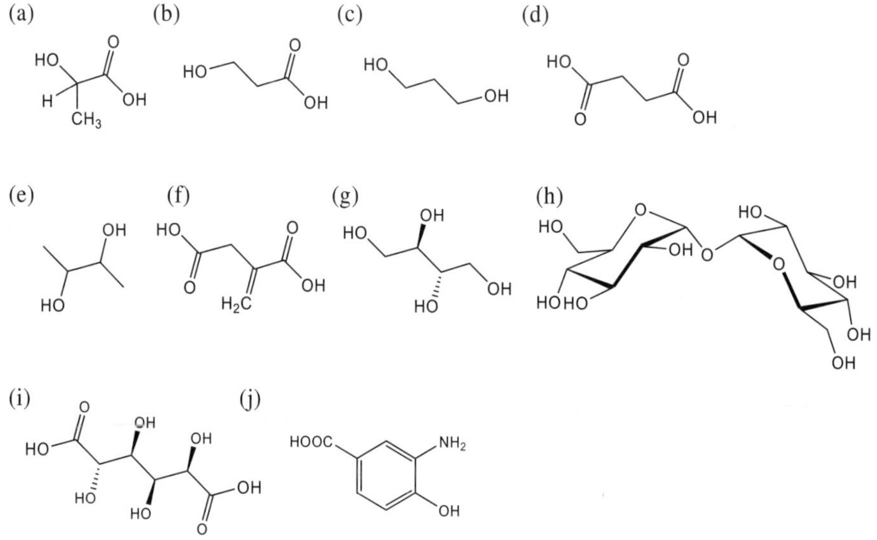

Figure 2.28 Monomers that can be obtained by fermentation.

2.6 Conclusions and Outlook

Insofar as space is available, renewable monomers usable for polymer synthesis are reviewed. This review, even though quite long, is not perfectly comprehensive, because large numbers of renewable monomers have been found while I have been seeking references. For examples, rosin derivatives, amino acids, and many intermediate products from metabolic pathways of sugar, fatty acid, and proteins are not included. For the same reason, detailed descriptions about the synthesis and properties of each polymer is not included.

Now we know how the many polymers are made, the next issue is 'how do we cast them'? The answer is not simple, and the limitations in opening a huge market are firstly cost performance and secondly physical performance. In other words, the largest reason for suppressing the availability of renewable polymers is the generally higher cost compared with polymers from petroleum chemicals. However, we should remind ourselves that, even if a renewable monomer we are now planning to use for polymerization is expensive, it may become inexpensive in the decade ahead. The answer to another question, 'What is really most environmentally friendly?' is not simple, either. The developments in biotechnology and green chemistry are seamless and the situation is changing from day to day. Even if a renewable monomer we are planning to use requires many processes for its production, it may become improved in the decade ahead. Alternatively, even if a monomer is now derived only from petroleum, it may become a renewable compound in the decade ahead.

Most current technology is beginning to change the view of renewable polymers; and nowadays ethylene and propylene are becoming renewable monomers, because ethylene and propylene can be derived from ethanol, which is a renewable compound.[450–452] Furthermore, Bio-PET, which is synthesized using ethylene glycol produced from sugar cane, will start to be used in motor vehicles from the year 2011.[453] In addition, synthesis of terephthalic acid from p-cymene (an essential oil) or limonene has been reported.[454] Possibly, these technologies would greatly change the present concept of green polymers, because current commodity polymers become green polymers. However, there is no assurance that the day will come soon. At least now, we should look towards naturally occurring compounds and struggle to find many green polymers that covers our livelihood. Hundreds varieties of renewable monomers will offer the solutions.

This review was written without an emphasis on biodegradability. Green polymers from renewable resources are not necessarily biodegradable. From the view of carbon neutrality, it is often said that carbon dioxide is released by biodegradation of polymers as well as by incineration of polymers. However, I do not wish to downplay the importance of the biodegradability of green polymers. I would rather emphasize the importance of biodegradability, by considering differences in the releasing rate of carbon dioxide: biodegradation is a slow reaction, while incineration is a rapid reaction. Though it might be a misleading expression, we can fix carbon dioxide as a form of biodegradable polymers. Of course, the amount should be limited to avoid overspending resources. At least biodegradable polymers can prevent the environmental pollution with nondegradable polymers which are unconsciously discarded.

References

1. G. Scott, *Polymers and the Environment*, The Royal Society of Chemistry, Cambridge, 1999.
2. C. J. Moore, *Environ. Res.*, 2008, **108**, 131–139.
3. D. K. A. Barnes, *Nature*, 2002, **416**, 808–809.
4. K. G. Satyanarayana, G. G. C. Arizaga and F. Wypych, *Prog. Polym. Sci.*, 2009, **34**, 982–1021.
5. P. P. Klemchuk, *Polym. Degrad. Stability*, 1990, **27**, 183–202.
6. T. M. Aminabhavi, R. H. Balundgi and P. E. Cassidy, *Polym.-Plastics Technol. Eng.*, 1990, **29**, 235–262.

7. A. Wendy, A. Allan and T. Brian, *Polym. Int.*, 1998, **47**, 89–144.
8. R. A. Gross and B. Kalra, *Science*, 2002, **297**, 803–807.
9. M. Okada, *Prog. Polym. Sci.*, 2002, **27**, 87–133.
10. http://unfccc.int/kyoto_protocol/items/2830.php (last accessed October 2010).
11. F. S. Zeman and D. W. Keith, *Philos. Trans. Royal Soc. A*, 2008, **366**, 3901–3918.
12. T. Abbasi and S. A. Abbasi, *Renewable Sustainable Energy Rev.*, 2010, **14**, 919–937.
13. D. L. Klass, in *Encyclopedia of Energy*, Elsevier, New York, 2004, pp. 193–212.
14. A. J. Ragauskas, C. K. Williams, B. H. Davison, G. Britovsek, J. Cairney, C. A. Eckert, W. J. Frederick, Jr., J. P. Hallett, D. J. Leak, C. L. Liotta, J. R. Mielenz, R. Murphy, R. Templer and T. Tschaplinski, *Science*, 2006, **311**, 484–489.
15. D. Feldman, *Design. Monom. Polym.*, 2008, **11**, 1–15.
16. J. B. Van Beilen and Y. Poirier, *Plant J.*, 2008, **54**, 684–701.
17. A. Gandini, *Macromolecules*, 2008, **41**, 9491–9504.
18. M. S. Graboski and R. L. McCormick, *Prog. Energy Combust. Sci.*, 1998, **24**, 125–164.
19. M. A. Meier, J. O. Metzger and U. S. Schubert, *Chem. Soc. Rev.*, 2007, **36**, 1788–1802.
20. R. Binder, T. Applewhite, G. Kohler and L. Goldblatt, *J. Am. Oil Chem. Soc.*, 1962, **39**, 513–517.
21. V. Sharma and P. P. Kundu, *Prog. Polym. Sci.*, 2006, **31**, 983–1008.
22. M. Lazzari and O. Chiantore, *Polym. Degrad. Stability*, 1999, **65**, 303–313.
23. J. Mallégol, J. Lemaire and J.-L. Gardette, *Prog. Org. Coatings*, 2000, **39**, 107–113.
24. D. Deffar and M. Soucek, *J. Coatings Technol.*, 2001, **73**, 95–104.
25. S. N. Khot, J. J. Lascala, E. Can, S. S. Morye, G. I. Williams, G. R. Palmese, S. H. Kusefoglu and R. P. Wool, *J. Appl. Polym. Sci.*, 2001, **82**, 703–723.
26. F. S. Güner, Y. Yağcı and A. T. Erciyes, *Prog. Polym. Sci.*, 2006, **31**, 633–670.
27. P. Deligny and N. Tuck, *Resins for Surface Coatings*, Vol. II: *Alkyds and Polyesters*, ed. P. K. T. Oldring, Wiley, New York, 2000.
28. R. Haag, A. Sunder and J.-F. Stumbé, *J. Am. Chem. Soc.*, 2000, **122**, 2954–2955.
29. J.-F. Stumbé and B. Bernd, *Macromol. Rapid Commun.*, 2004, **25**, 921–924.
30. P. G. Parzuchowski, M. Grabowska, M. Tryznowski and G. Rokicki, *Macromolecules*, 2006, **39**, 7181–7186.
31. F. Naughton, *J. Am. Oil Chem. Soc.*, 1974, **51**, 65–71.
32. D. S. Ogunniyi, *Bioresour. Technol.*, 2006, **97**, 1086–1091.

33. A. Vasishtha, R. Trivedi and G. Das, *J. Am. Oil Chem. Soc.*, 1990, **67**, 333–337.
34. S. B. Gudadhe, D. H. S. Ramkumar and A. P. Kudchadker, *Ind. Eng. Chem. Proc. Des. Dev.*, 1986, **25**, 354–357.
35. K. I. Hirayama, T. Irie, N. Teramoto and M. Shibata, *J. Appl. Polym. Sci.*, 2009, **114**, 1033–1039.
36. M. Shibata, N. Teramoto and Y. Nakamura, *J. Appl. Polym. Sci.*, 2011, **119**, 896–901.
37. G. Çaylı and S. Küsefoğlu, *J. Appl. Polym. Sci.*, 2010, **118**, 849–856.
38. H. Mutlu and S. H. Kusefoglu, *J. Appl. Polym. Sci.*, 2009, **113**, 1925–1934.
39. J. Samuelsson, M. Jonsson, T. Brinck and M. Johansson, *J. Polym. Sci. Part A: Polym. Chem.*, 2004, **42**, 6346–6352.
40. Z. Chen, B. Chisholm, R. Patani, J. Wu, S. Fernando, K. Jogodzinski and D. Webster, *J. Coatings Technol. Res.*, 2010, **7**, 603–613.
41. M. Black and J. W. Rawlins, *Eur. Polym. J.*, 2009, **45**, 1433–1441.
42. T. Eren, S. H. Küsefoğlu and R. Wool, *J. Appl. Polym. Sci.*, 2003, **90**, 197–202.
43. Y. Xia, P. H. Henna and R. C. Larock, *Macromol. Mater. Eng.*, 2009, **294**, 590–598.
44. P. Henna and R. C. Larock, *J. Appl. Polym. Sci.*, 2009, **112**, 1788–1797.
45. D. H. Hewitt and F. Armitage, *J. Oil & Colour Chem. Assoc.*, 1946, **29**, 109.
46. H. Schroeder and R. Terrill, *J. Am. Oil Chem. Soc.*, 1949, **26**, 153–157.
47. H. M. Hoogsteen, A. E. Young and M. K. Smith, *Ind. Eng. Chem.*, 1950, **42**, 1587–1591.
48. C. Boelhouwer, F. A. de Roos and H. I. Waterman, *Chem. Ind.*, 1953, 1287–1289.
49. J. B. Crofts, *J. Appl. Chem.*, 1955, **5**, 88–100.
50. F. S. Erkal, A. T. Erciyes and Y. Yagci, *J. Coatings Technol.*, 1993, **65**, 37–43.
51. M. Gultekin, U. Beker, F. S. Güner, A. T. Erciyes and Y. Yagci, *Macromol. Mater. Eng.*, 2000, **283**, 15–20.
52. F. Güner, S. Usta, A. Erciyes and Y. Yagci, *J. Coatings Technol.*, 2000, **72**, 107–110.
53. F. Li and R. C. Larock, *Biomacromolecules*, 2003, **4**, 1018–1025.
54. P. P. Kundu and R. C. Larock, *Biomacromolecules*, 2005, **6**, 797–806.
55. F. Li, M. V. Hanson and R. C. Larock, *Polymer*, 2001, **42**, 1567–1579.
56. F. Li and R. C. Larock, *J. Appl. Polym. Sci.*, 2001, **80**, 658–670.
57. D. D. Andjelkovic, M. Valverde, P. Henna, F. Li and R. C. Larock, *Polymer*, 2005, **46**, 9674–9685.
58. Y. Lu and R. C. Larock, *J. Appl. Polym. Sci.*, 2006, **102**, 3345–3353.
59. J. Lu, S. Khot and R. P. Wool, *Polymer*, 2005, **46**, 71–80.
60. L. M. de Espinosa, J. C. Ronda, M. Galià and V. Cádiz, *J. Polym. Sci. Part A: Polym. Chem.*, 2009, **47**, 1159–1167.

61. E. Can, S. Küsefoğlu and R. P. Wool, *J. Appl. Polym. Sci.*, 2001, **81**, 69–77.
62. E. Can, R. P. Wool and S. Küsefoğlu, *J. Appl. Polym. Sci.*, 2006, **102**, 1497–1504.
63. H. J. Wang, M. Z. Rong, M. Q. Zhang, J. Hu, H. W. Chen and T. Czigány, *Biomacromolecules*, 2008, **9**, 615–623.
64. H. Pelletier, N. Belgacem and A. Gandini, *J. Appl. Polym. Sci.*, 2006, **99**, 3218–3221.
65. D. Åkesson, M. Skrifvars, S. Lv, W. Shi, K. Adekunle, J. Seppälä and M. Turunen, *Prog. Org. Coatings*, 2010, **67**, 281–286.
66. D. Yeadon, W. McSherry and L. Goldblatt, *J. Am. Oil Chem. Soc.*, 1959, **36**, 16–20.
67. T. C. Patton and H. M. Metz, *Ind. Eng. Chem.*, 1959, **51**, 1383–1384.
68. G. C. Toone and G. S. Wooster, *Ind. Eng. Chem.*, 1959, **51**, 1384–1385.
69. G. Wilson and J. M. Stanton, *Ind. Eng. Chem.*, 1959, **51**, 1385–1385.
70. Z. S. Petrović and D. Fajnik, *J. Appl. Polym. Sci.*, 1984, **29**, 1031–1040.
71. G. Qipeng, F. Shixia and Z. Qingyu, *Eur. Polym. J.*, 1990, **26**, 1177–1180.
72. K. I. Suresh and E. T. Thachil, *Die Angewandte Makromolekulare Chemie*, 1994, **218**, 127–135.
73. Y. H. Hu, Y. Gao, D. N. Wang, C. P. Hu, S. Zu, L. Vanoverloop and D. Randall, *J. Appl. Polym. Sci.*, 2002, **84**, 591–597.
74. G. Çaylı and S. Küsefoğlu, *J. Appl. Polym. Sci.*, 2008, **109**, 2948–2955.
75. S. J. Park, F. L. Jin and J. R. Lee, *Macromol. Rapid Commun.*, 2004, **25**, 724–727.
76. R. Raghavachar, G. Sarnecki, J. Baghdachi and J. Massingill, *J. Coatings Technol.*, 2000, **72**, 125–133.
77. E. D. Casebolt, B. E. Mote and D. L. Trumbo, *Prog. Org. Coatings*, 2002, **44**, 147–151.
78. H. Uyama, M. Kuwabara, T. Tsujimoto, M. Nakano, A. Usuki and S. Kobayashi, *Macromol. Biosci.*, 2004, **4**, 354–360.
79. J. V. Crivello and R. Narayan, *Chem. Mater.*, 1992, **4**, 692–699.
80. S. Chakrapani and J. V. Crivello, *J. Macromol. Sci., Part A: Pure Appl. Chem.*, 1998, **35**, 691–710.
81. S. F. Thames and H. Yu, *Surf. Coatings Technol.*, 1999, **115**, 208–214.
82. C. Decker, T. Nguyen Thi Viet and H. Pham Thi, *Polym. Int.*, 2001, **50**, 986–997.
83. W. D. Wan Rosli, R. N. Kumar, S. Mek Zah and M. M. Hilmi, *Eur. Polym. J.*, 2003, **39**, 593–600.
84. M. Shibata, N. Teramoto, Y. Someya and S. Suzuki, *J. Polym. Sci. Part B: Polym. Phys.*, 2009, **47**, 669–673.
85. H. Miyagawa, A. K. Mohanty, M. Misra and L. T. Drzal, *Macromol. Mater. Eng.*, 2004, **289**, 629–635.
86. H. Miyagawa, A. K. Mohanty, M. Misra and L. T. Drzal, *Macromol. Mater. Eng.*, 2004, **289**, 636–641.
87. H. Miyagawa, M. Misra, L. T. Drzal and A. K. Mohanty, *Polym. Eng. Sci.*, 2005, **45**, 487–495.

88. H. Miyagawa, A. Mohanty, L. T. Drzal and M. Misra, *Ind. Eng. Chem. Res.*, 2004, **43**, 7001–7009.
89. J. Rösch and R. Mülhaupt, *Polym. Bull.*, 1993, **31**, 679–685.
90. T. Takahashi, K. Hirayama, N. Teramoto and M. Shibata, *J. Appl. Polym. Sci.*, 2008, **108**, 1596–1602.
91. J. C. Munoz, H. Ku, F. Cardona and D. Rogers, *J. Mater. Proc. Technol.*, 2008, **202**, 486–492.
92. K. Tsujimoto, N. Imai, H. Kageyama, H. Uyama and M. Funaoka, *J. Network Polym., Jap.*, 2008, **29**, 192–197.
93. T. Tsujimoto, H. Uyama and S. Kobayashi, *Macromol. Rapid Commun.*, 2003, **24**, 711–714.
94. R. Slivniak and A. J. Domb, *Biomacromolecules*, 2005, **6**, 1679–1688.
95. H. Ebata, K. Toshima and S. Matsumura, *Macromol. Biosci.*, 2007, **7**, 798–803.
96. A. J. Domb and R. Nudelman, *J. Polym. Sci. Part A: Polym. Chem.*, 1995, **33**, 717–725.
97. C. E. H. Bawn and M. B. Huglin, *Polymer*, 1962, **3**, 257–262.
98. S. Liu, Y. Yu, Y. Cui, H. Zhang and Z. Mo, *J. Appl. Polym. Sci.*, 1998, **70**, 2371–2380.
99. D. D. Coffman, N. L. Cox, E. L. Martin, W. E. Mochel and F. J. Van Natta, *J. Polym. Sci.*, 1948, **3**, 85–95.
100. http://www.toyota.co.jp/en/news/08/1217.html (last accessed October 2010).
101. S. Burt, *Int. J. Food Microbiol.*, 2004, **94**, 223–253.
102. R. Whetten and R. Sederoff, *Plant Cell*, 1995, **7**, 1001–1013.
103. M. O. Carmody and W. H. Carmody, *J. Am. Chem. Soc.*, 1937, **59**, 1312–1312.
104. W. J. Roberts and A. R. Day, *J. Am. Chem. Soc.*, 1950, **72**, 1226–1230.
105. A. J. Johanson, F. L. McKennon and L. A. Goldblatt, *Ind. Eng. Chem.*, 1948, **40**, 500–502.
106. C. S. Marvel and C. C. L. Hwa, *J. Polym. Sci.*, 1960, **45**, 25–34.
107. J. L. Cawse, J. L. Stanford and R. H. Still, *J. Appl. Polym. Sci.*, 1987, **33**, 2217–2229.
108. B. Keszler and J. Kennedy, *Adv. Polym. Sci.*, 1992, **100**, 1–9.
109. J. Lu, M. Kamigaito, M. Sawamoto, T. Higashimura and Y. X. Deng, *J. Appl. Polym. Sci.*, 1996, **61**, 1011–1016.
110. J. Lu, M. Kamigaito, M. Sawamoto, T. Higashimura and Y.-X. Deng, *Macromolecules*, 1997, **30**, 22–26.
111. J. Lu, M. Kamigaito, M. Sawamoto, T. Higashimura and Y.-X. Deng, *Macromolecules*, 1997, **30**, 27–31.
112. J. Lu, M. Kamigaito, M. Sawamoto, T. Higashimura and Y. X. Deng, *J. Polym. Sci. Part A: Polym. Chem.*, 1997, **35**, 1423–1430.
113. J. Lu, H. Liang, A. Li and Q. Cheng, *Eur. Polym. J.*, 2004, **40**, 397–402.
114. R. P. F. Guiné and J. A. A. M. Castro, *J. Appl. Polym. Sci.*, 2001, **82**, 2558–2565.

115. K. Satoh, H. Sugiyama and M. Kamigaito, *Green Chem.*, 2006, **8**, 878–882.
116. S. Kobayashi, C. Lu, T. R. Hoye and M. A. Hillmyer, *J. Am. Chem. Soc.*, 2009, **131**, 7960–7961.
117. K. Bazaka and M. V. Jacob, *Mater. Lett.*, 2009, **63**, 1594–1597.
118. E. R. Littmann, *Ind. Eng. Chem.*, 1936, **28**, 1150–1152.
119. C. S. Marvel, Y. Shimura and F. C. Magne, *J. Polym. Sci.*, 1960, **45**, 13–24.
120. Y. Ma, T. R. Willcox, A. T. Russell and G. Marston, *Chem. Commun.*, 2007, 1328–1330.
121. H. Pietila, A. Sivola and H. Sheffer, *J. Polym. Sci. Part A-1: Polym. Chem.*, 1970, **8**, 727–737.
122. C. Snyder, W. McIver and H. Sheffer, *J. Appl. Polym. Sci.*, 1977, **21**, 131–139.
123. H. Sheffer, G. Greco and G. Paik, *J. Appl. Polym. Sci.*, 1983, **28**, 1701–1705.
124. T. Doiuchi, H. Yamaguchi and Y. Minoura, *Eur. Polym. J.*, 1981, **17**, 961–968.
125. J. Maslinska-Solich and I. Rudnicka, *Eur. Polym. J.*, 1988, **24**, 453–456.
126. J. Maślińska-Solich, T. Kupka, M. Kluczka and A. Solich, *Macromol. Chem. Phys.*, 1994, **195**, 1843–1850.
127. K. Satoh, M. Matsuda, K. Nagai and M. Kamigaito, *J. Am. Chem. Soc.*, 2010, **132**, 10003–10005.
128. A. R. Khan, A. H. K. Yousufzai, H. A. Jeelani and T. Akhter, 1985, **22**, 1673–1678.
129. A. K. Srivastava and P. Pandey, *Eur. Polym. J.*, 2002, **38**, 1709–1712.
130. S. Yadav and A. K. Srivastava, *J. Polym. Res.* 2002, **9**, 265–270.
131. P. Pandey and A. K. Srivastava, *J. Polym. Sci. Part A: Polym. Chem.*, 2002, **40**, 1243–1252.
132. S. Sharma and A. K. Srivastava, *J. Macromol. Sci., Part A: Pure Appl. Chem.*, 2003, **40**, 593–603.
133. P. Pandey and A. K. Srivastava, *Des. Monomers Polym.*, 2003, **6**, 197–209.
134. S. Yadav and A. K. Srivastava, *J. Polym. Sci. Part A: Polym. Chem.*, 2003, **41**, 1700–1707.
135. S. Yadav and A. K. Srivastava, *Polym.-Plast. Technol. Eng.*, 2004, **43**, 1229–1243.
136. A. Shukla and A. K. Srivastava, *J. Appl. Polym. Sci.*, 2004, **92**, 1134–1143.
137. S. Sharma and A. K. Srivastava, *Eur. Polym. J.*, 2004, **40**, 2235–2240.
138. G. Misra and A. Srivastava, *Colloid Polym. Sci.*, 2008, **286**, 445–451.
139. N. Nomura, K. Tsurugi and M. Okada, *J. Am. Chem. Soc.*, 1999, **121**, 7268–7269.
140. N. Nomura, N. Yoshida, K. Tsurugi and K. Aoi, *Macromolecules*, 2003, **36**, 3007–3009.
141. R. T. Mathers, K. Damodaran, M. G. Rendos and M. S. Lavrich, *Macromolecules*, 2009, **42**, 1512–1518.

142. C. M. Byrne, S. D. Allen, E. B. Lobkovsky and G. W. Coates, *J. Am. Chem. Soc.*, 2004, **126**, 11404–11405.
143. J. V. Crivello and B. Yang, *J. Polym. Sci. Part A: Polym. Chem.*, 1995, **33**, 1881–1890.
144. D. L. Trumbo, C. L. Giddings and L. R. A. Wilson, *J. Appl. Polym. Sci.*, 1995, **58**, 69–76.
145. C. L. Giddings and D. L. Trumbo, *Prog. Org. Coatings*, 1997, **30**, 219–224.
146. H. Kimura, Y. Murata, A. Matsumoto, K. Hasegawa, K. Ohtsuka and A. Fukuda, *J. Appl. Polym. Sci.*, 1999, **74**, 2266–2273.
147. K. Xu, M. Chen, K. Zhang and J. Hu, *Polymer*, 2004, **45**, 1133–1140.
148. G.-M. Wu, Z.-W. Kong and F.-X. Chu, *Chem. Ind. Forest Prod.*, 2007, **27**, 57–62.
149. G. M. Wu, Z. W. Kong, H. Huang, J. Chen and F. X. Chu, *J. Appl. Polym. Sci.*, 2009, **113**, 2894–2901.
150. S. Ando, Y. Someya, T. Takahashi and M. Shibata, *J. Appl. Polym. Sci.*, 2010, **115**, 3326–3331.
151. C. Lapierre, B. Pollet, M.-C. Ralet and L. Saulnier, *Phytochemistry*, 2001, **57**, 765–772.
152. C. S. Marvel and G. H. McCain, *J. Am. Chem. Soc.*, 1953, **75**, 3272–3273.
153. S. Washburn, J. Lauterbach and C. M. Snively, *Macromolecules*, 2006, **39**, 8210–8212.
154. C. A. Barson, *J. Polym. Sci.*, 1962, **62**, S128–S130.
155. F. Amrani and N. Rahmoun-Haddadine, *Polym. Preprints*, 1996, **37**, 360–361.
156. N. Bouslah, R. Hammachin and F. Amrani, *Macromol. Chem. Phys.*, 1999, **200**, 678–682.
157. G. David and A. Ioanid, *J. Appl. Polym. Sci.*, 2001, **80**, 2191–2199.
158. H. Esen and S. H. Küsefoğlu, *J. Appl. Polym. Sci.*, 2003, **89**, 3882–3888.
159. K. Satoh, S. Saitoh and M. Kamigaito, *J. Am. Chem. Soc.*, 2007, **129**, 9586–9587.
160. L. M. Minsk, J. G. Smith, W. P. van Deusen and J. F. Wright, *J. Appl. Polym. Sci.*, 1959, **2**, 302–307.
161. G. Van Paesschen, R. Janssen and R. Hart, *Die Makromolekulare Chemie*, 1960, **37**, 46–52.
162. M. Kato, *J. Polym. Sci. Part B: Polym. Lett.*, 1969, **7**, 605–608.
163. S. Watanabe, M. Kato and S. Kosakai, *J. Polym. Sci.: Polym. Chem. Ed.*, 1984, **22**, 2801–2808.
164. A. Hajaiej, X. Coqueret, A. Lablache-Combier and C. Loucheux, *Die Makromolekulare Chemie*, 1989, **190**, 327–340.
165. A. El Achari and X. Coqueret, *J. Polym. Sci. Part A: Polym. Chem.*, 1997, **35**, 2513–2520.
166. J. Stumpe, O. Zaplo, D. Kreysig, M. Niemann and H. Ritter, *Die Makromolekulare Chemie*, 1992, **193**, 1567–1578.
167. A. Hyder Ali and K. S. V. Srinivasan, *Polym. Int.*, 1997, **43**, 310–316.
168. Y. Nakayama and T. Matsuda, *J. Polym. Sci. Part A: Polym. Chem.*, 1992, **30**, 2451–2457.

169. J. Wang, X. Jia, H. Zhong, H. Wu, Y. Li, X. Xu, M. Li and Y. Wei, *J. Polym. Sci. Part A: Polym. Chem.*, 2000, **38**, 4147–4153.
170. Y. Ni and S. Zheng, *Chem. Mater.*, 2004, **16**, 5141–5148.
171. N. Teramoto and M. Shibata, *Polym. Adv. Technol.*, 2007, **18**, 971–977.
172. A. Lendlein, H. Jiang, O. Junger and R. Langer, *Nature*, 2005, **434**, 879–882.
173. M. Schadt, K. Schmitt, V. Kozinkov and V. Chigrinov, *Jpn. J. Appl. Phys. Vol*, 1992, **31**, 2155–2164.
174. K. Ichimura, *Chem. Rev.*, 2000, **100**, 1847–1874.
175. X. T. Li, H. Saitoh, H. Nakamura, S. Kobayashi and Y. Iimura, *J. Photopolym. Sci. Technol.*, 1997, **10**, 13–17.
176. N. Kawatsuki, H. Ono, H. Takatsuka, T. Yamamoto and O. Sangen, *Macromolecules*, 1997, **30**, 6680–6682.
177. N. Kawatsuki, K. Goto, T. Kawakami and T. Yamamoto, *Macromolecules*, 2002, **35**, 706–713.
178. K. Ichimura, Y. Akita, H. Akiyama, K. Kudo and Y. Hayashi, *Macromolecules*, 1997, **30**, 903–911.
179. K. Rajesh, M. K. Ram, S. C. Jain, S. B. Samanta and A. V. Narliker, *Thin Solid Films*, 1998, **325**, 251–253.
180. S.-J. Sung, K.-Y. Cho, J.-H. Yoo, W. S. Kim, H.-S. Chang, I. Cho and J.-K. Park, *Chem. Phys. Lett.*, 2004, **394**, 238–243.
181. S. J. Sung, K. Y. Cho, H. Hah, S. Lee and J. K. Park, *Jap. J. Appl. Phys.*, 2005, **44**, L412–L415.
182. H. Ono, T. Shinmachi, A. Emoto, T. Shioda and N. Kawatsuki, *Appl. Opt.*, 2009, **48**, 309–315.
183. S. Murase, K. Kinoshita, K. Horie and S. y. Morino, *Macromolecules*, 1997, **30**, 8088–8090.
184. A. Nagata, T. Sakaguchi, T. Ichihashi, M. Miya and K. Ohta, *Macromol. Rapid Commun.*, 1997, **18**, 191–196.
185. S. R. Trenor, A. R. Shultz, B. J. Love and T. E. Long, *Chem. Rev.*, 2004, **104**, 3059–3078.
186. Y. Chujo, K. Sada and T. Saegusa, *Macromolecules*, 1990, **23**, 2693–2697.
187. M. A. Tlenkopatchev, S. Fomine, L. Fomina, R. Gavino and T. Ogawa, *Polym. J.*, 1997, **29**, 622–625.
188. S. Fomine, E. Rivera, L. Fomina, A. Ortiz and T. Ogawa, *Polymer*, 1998, **39**, 3551–3558.
189. T. Ngai and C. Wu, *Macromolecules*, 2003, **36**, 848–854.
190. M. Obi, S. Morino and K. Ichimura, *Macromol. Rapid Commun.*, 1998, **19**, 643–646.
191. P. O. Jackson, M. O'Neill, W. L. Duffy, P. Hindmarsh, S. M. Kelly and G. J. Owen, *Chem. Mater.*, 2001, **13**, 694–703.
192. N. Kawatsuki, K. Goto and T. Yamamoto, *Liquid Cryst.*, 2001, **28**, 1171–1176.
193. C. Kim, A. Trajkovska, J. U. Wallace and S. H. Chen, *Macromolecules*, 2006, **39**, 3817–3823.

194. H. Yamamoto, T. Kitsuki, A. Nishida, K. Asada and K. Ohkawa, *Macromolecules*, 1999, **32**, 1055–1061.
195. H. Wondraczek, A. Pfeifer and T. Heinze, *Eur. Polym. J.*, 2010, **46**, 1688–1695.
196. Y. Tanaka, T. Tanabe, Y. Shimura, A. Okada, Y. Kurihara and Y. Sakakibara, *J. Polym. Sci.: Polym. Lett. Ed.*, 1975, **13**, 235–242.
197. F. Higashi, Y. Ito and K. Kubota, *Die Makromolekulare Chemie, Rapid Commun.*, 1981, **2**, 29–33.
198. H. G. Elias, J. H. Tsao and J. A. Palacios, *Die Makromolekulare Chemie*, 1985, **186**, 893–905.
199. H. G. Elias and J. A. Palacios, *Die Makromolekulare Chemie*, 1985, **186**, 1027–1045.
200. D. Creed, A. C. Griffin, J. R. D. Gross, C. E. Hoyle and K. Venkataram, *Molecular Cryst. Liquid Cryst. Incorporating Nonlinear Optics*, 1988, **155**, 57–71.
201. F. Navarro and J. L. Serrano, *J. Polym. Sci. Part A: Polym. Chem.*, 1989, **27**, 3691–3701.
202. P. Pinther, M. Hartmann, K. Wermann, W. Günther and H. Ritter, *Die Makromolekulare Chemie*, 1992, **193**, 2669–2675.
203. X. Jin, C. Carfagna, L. Nicolais and R. Lanzetta, *Macromolecules*, 1995, **28**, 4785–4794.
204. H. R. Kricheldorf and T. Stukenbrock, *Macromol. Chem. Phys.*, 1997, **198**, 3753–3767.
205. B. Sapich, J. Stumpe, T. Krawinkel and H. R. Kricheldorf, *Macromolecules*, 1998, **31**, 1016–1023.
206. B. Sapich, J. Stumpe, H. R. Kricheldorf, A. Fritz and A. Schönhals, *Macromolecules*, 2001, **34**, 5694–5701.
207. H. R. Kricheldorf and T. Stukenbrock, *J. Polym. Sci. Part A: Polym. Chem.*, 1998, **36**, 2347–2357.
208. A. Reina, A. Gerken, U. Zemann and H. R. Kricheldorf, *Macromol. Chem. Phys.*, 1999, **200**, 1784–1791.
209. M. Nagata, *J. Appl. Polym. Sci.*, 2000, **78**, 2474–2481.
210. M. Nagata and S. Hizakae, *Macromol. Biosci.*, 2003, **3**, 412–419.
211. M. Nagata and S. Hizakae, *J. Polym. Sci. Part A: Polym. Chem.*, 2003, **41**, 2930–2938.
212. M. Nagata and Y. Sato, *Polymers*, 2004, **45**, 87–93.
213. M. Nagata and Y. Sato, *Polym. Int.*, 2005, **54**, 386–391.
214. M. Matsusaki, A. Kishida, N. Stainton, C. W. G. Ansell and M. Akashi, *J. Appl. Polym. Sci.*, 2001, **82**, 2357–2364.
215. T. Kaneko, M. Matsusaki, T. T. Hang and M. Akashi, *Macromol. Rapid Commun.*, 2004, **25**, 673–677.
216. M. Matsusaki, T. Hang Thi, T. Kaneko and M. Akashi, *Biomaterials*, 2005, **26**, 6263–6270.
217. T. Kaneko, T. H. Thi, D. J. Shi and M. Akashi, *Nat. Mater.*, 2006, **5**, 966–970.

218. T. Kaneko, H. T. Tran, M. Matsusaki and M. Akashi, *Chem. Mater.*, 2006, **18**, 6220–6226.
219. T. Fujiwara, T. Iwata and Y. Kimura, *J. Polym. Sci. Part A: Polym. Chem.*, 2001, **39**, 4249–4254.
220. K. Kimura, S.-i. Kohama, Y. Yamashita, T. Uchida and Y. Sakaguchi, *Polymer*, 2003, **44**, 7383–7387.
221. K. Kimura, H. Inoue, S.-i. Kohama, Y. Yamashita and Y. Sakaguchi, *Macromolecules*, 2003, **36**, 7721–7729.
222. W. Tang, X. G. Li and D. Yan, *J. Appl. Polym. Sci.*, 2004, **91**, 445–454.
223. J. Du, Y. Fang and Y. Zheng, *Polymer*, 2007, **48**, 5541–5547.
224. J. Du, Y. Fang and Y. Zheng, *Polym. Degrad. Stability*, 2008, **93**, 838–845.
225. M. Mizutani and T. Matsuda, *Biomacromolecules*, 2002, **3**, 249–255.
226. A. C. Hagenaars, C. Bailly, A. Schneider and B. A. Wolf, *Polymer*, 2002, **43**, 2663–2669.
227. P. B. Waghmare, S. A. Deshmukh, S. B. Idage and B. B. Idage, *J. Appl. Polym. Sci.*, 2006, **101**, 2668–2674.
228. P. B. Waghmare, S. B. Idage, S. K. Menon and B. B. Idage, *J. Appl. Polym. Sci.*, 2006, **100**, 3222–3228.
229. E. A. Rahim, F. Sanda and T. Masuda, *J. Macromol. Sci., Part A: Pure Appl. Chem.*, 2004, **41**, 133–141.
230. L. Rojo, B. Vazquez, J. Parra, A. López Bravo, S. Deb and J. San Roman, *Biomacromolecules*, 2006, **7**, 2751–2761.
231. K. Freudenberg, R. Kraft and W. Heimberger, *Chem. Ber.*, 1951, **84**, 472–476.
232. S. M. Siegel, *Physiologia Plantarum*, 1955, **8**, 20–32.
233. T. Higuchi, *Physiologia Plantarum*, 1957, **10**, 356–372.
234. T. Higuchi, *Wood Sci. Technol.*, 1990, **24**, 23–63.
235. H. A. Stafford, *Plant Physiol.*, 1960, **35**, 108–114.
236. M. Tanahashi, H. Takeuchi and T. Higuchi, *Wood Res.: Bull. Wood Res. Institute Kyoto Univ.*, 1976, **61**, 44–53.
237. H. Shin, G. Guebitz and A. Cavaco-Paulo, *Macromol. Mater. Eng.*, 2001, **286**, 691–694.
238. H. Uyama, H. Kurioka, J. Sugihara, I. Komatsu and S. Kobayashi, *J. Polymer Sci. Part A: Polym. Chem.*, 1997, **35**, 1453–1459.
239. M. Micic, K. Radotic, M. Jeremic and R. M. Leblanc, *Macromol. Biosci.*, 2003, **3**, 100–106.
240. H. Yoshizawa, J. Shan, S. Tanigawa and Y. Kitamura, *Colloid Polym. Sci.*, 2004, **282**, 583–588.
241. L. Mejias, M. H. Reihmann, S. Sepulveda-Boza and H. Ritter, *Macromol. Biosci.*, 2002, **2**, 24–32.
242. H. Uyama, *Macromol. Biosci.*, 2007, **7**, 410–422.
243. M. Kurisawa, J. E. Chung, Y. J. Kim, H. Uyama and S. Kobayashi, *Biomacromolecules*, 2003, **4**, 469–471.
244. M. Kurisawa, J. E. Chung, H. Uyama and S. Kobayashi, *Macromol. Biosci.*, 2003, **3**, 758–764.

245. S. Kobayashi, S. Shoda and H. Uyama, *Adv. Polym. Sci.*, 1995, **121**, 1–30.
246. M. Ayyagari, J. A. Akkara and D. L. Kaplan, *Acta Polym.*, 1996, **47**, 193–203.
247. M. Reihmann and H. Ritter, *Adv. Polym. Sci.*, 2006, **194**, 1–49.
248. H. Uyama and S. Kobayashi, *Adv. Polym. Sci.*, 2006, **194**, 51–67.
249. S. Kobayashi and H. Higashimura, *Prog. Polym. Sci.*, 2003, **28**, 1015–1048.
250. D. Stewart, *Ind. Crops Prod.*, 2008, **27**, 202–207.
251. J. H. Lora and W. G. Glasser, *J. Polym. Environ.*, 2002, **10**, 39–48.
252. G. G. Allan, G. Bullock and A. N. Neogi, *J. Polym. Sci.: Polym. Chem. Ed.*, 1973, **11**, 1759–1764.
253. J. L. Popp and J. S. Dordick, *Enzyme Microb. Technol.*, 1991, **13**, 522.
254. J. L. Popp, T. K. Kirk and J. S. Dordick, *Enzyme Microb. Technol.*, 1991, **13**, 964–968.
255. A. M. Blinkovsky and J. S. Dordick, *J. Polym. Sci. Part A: Polym. Chem.*, 1993, **31**, 1839–1846.
256. O. Milstein, A. Hüttermann, A. Majcherczyk, K. Schulze, R. Fründ and H.-D. Lüdemann, *J. Biotechnol.*, 1993, **30**, 37–48.
257. A. Kharazipour, A. Huettermann and H. D. Luedemann, *J. Adhesion Sci. Technol.*, 1997, **11**, 419–427.
258. A. Hüttermann, C. Mai and A. Kharazipour, *Appl. Microbiol. Biotechnol.*, 2001, **55**, 387–394.
259. J. Batog, R. Kozlowski and A. Przepiera, *Molecular Cryst. Liquid Cryst.*, 2008, **484**, 35–42.
260. Carroll A. Hochwalt and Mark Plunguian, US Patent 2,282,518, 1942.
261. A. M. A. Nada, H. El-Saied, A. A. Ibrahem and M. A. Yousef, *J. Appl. Polym. Sci.*, 1987, **33**, 2915–2924.
262. T. Sellers, J. H. Lora and M. Okuma, *Mokuzai Gakaishi*, 1994, **40**, 1073–1078.
263. G. Vázquez, J. González, S. Freire and G. Antorrena, *Bioresour. Technol.*, 1997, **60**, 191–198.
264. R. S. J. Piccolo, F. Santos and E. Frollini, *J. Macromol. Sci., Part A: Pure Appl. Chem.*, 1997, **34**, 153–164.
265. N. S. Çetin and N. Özmen, *Int. J. Adhesion Adhesives*, 2002, **22**, 477–480.
266. N.-e. El Mansouri, A. Pizzi and J. Salvadó, *Eur. J. Wood Wood Prod.*, 2007, **65**, 65–70.
267. W. Doherty, P. Halley, L. Edye, D. Rogers, F. Cardona, Y. Park and T. Woo, *Polym. Adv. Technol.*, 2007, **18**, 673–678.
268. M. Wang, M. Leitch and C. Xu, *Eur. Polym. J.*, 2009, **45**, 3380–3388.
269. B. Danielson and R. Simonson, *J. Adhesion Sci. Technol.*, 1998, **12**, 923–939.
270. A. Tejado, G. Kortaberria, C. Peña, M. Blanco, J. Labidi, J. M. Echeverría and I. Mondragon, *J. Appl. Polym. Sci.*, 2008, **107**, 159–165.
271. W. Hoareau, F. B. Oliveira, S. Grelier, B. Siegmund, E. Frollini and A. Castellan, *Macromol. Mater. Eng.*, 2006, **291**, 829–839.
272. V. P. Saraf and W. G. Glasser, *J. Appl. Polym. Sci.*, 1984, **29**, 1831–1841.

273. S. S. Kelley, W. G. Glasser and T. C. Ward, *J. Appl. Polym. Sci.*, 1988, **36**, 759–772.
274. H. Hatakeyama, S. Hirose, T. Hatakeyama, K. Nakamura, K. Kobashigawa and N. Morohoshi, *J. Macromol. Sci., Part A: Pure Appl. Chem.*, 1995, **32**, 743–750.
275. R. W. Thring, M. N. Vanderlaan and S. L. Griffin, *Biomass Bioenergy*, 1997, **13**, 125–132.
276. D. V. Evtuguin, J. P. Andreolety and A. Gandini, *Eur. Polym. J.*, 1998, **34**, 1163–1169.
277. S. Sarkar and B. Adhikari, *Polym. Degrad. Stability*, 2001, **73**, 169–175.
278. N.-E. El Mansouri, A. Pizzi and J. Salvado, *J. Appl. Polym. Sci.*, 2007, **103**, 1690–1699.
279. H. Lei, A. Pizzi and G. Du, *J. Appl. Polym. Sci.*, 2008, **107**, 203–209.
280. A. Tejado, G. Kortaberria, C. Peña, J. Labidi, J. M. Echeverria and I. Mondragon, *Ind. Crops Prod.*, 2008, **27**, 208–213.
281. N. T. Thanh Binh, N. D. Luong, D. O. Kim, S. H. Lee, B. J. Kim, Y. S. Lee and J.-D. Nam, *Composite Interfaces*, 2009, **16**, 923–935.
282. H. Kagawa, Y. Okabe and Y. Takezawa, *Polym. Preprints Jap.*, 2009, **58**, 5431–5432.
283. C. I. Simionescu, V. Rusan, M. M. Macoveanu, G. Cazacu, R. Lipsa, C. Vasile, A. Stoleriu and A. Ioanid, *Composites Sci. Technol.*, 1993, **48**, 317–323.
284. K. Hofmann and W. G. Glasser, *J. Wood Chem. Technol.*, 1993, **13**, 73–95.
285. K. Hofmann and W. Glasser, *Macromol. Chem. Phys.*, 1994, **195**, 65–80.
286. B. Zhao, G. Chen, Y. Liu, K. Hu and R. Wu, *J. Mater. Sci. Lett.*, 2001, **20**, 859–862.
287. G. Sun, H. Sun, Y. Liu, B. Zhao, N. Zhu and K. Hu, *Polymer*, 2007, **48**, 330–337.
288. H. Ito and N. Shiraishi, *Mokuzai Gakaishi*, 1987, **33**, 393–399.
289. L. L. Kosbar, J. D. Gelorme, R. M. Japp and W. T. Fotorny, *J. Ind. Ecol.*, 2000, **4**, 93–105.
290. Y. Okabe, H. Kagawa, C. Sasaki and Y. Nakamura, *Polym. Preprints Jap.*, 2009, **58**, 5433–5434.
291. M. Funaoka, M. Matsubara, N. Seki and S. Fukatsu, *Biotechnol. Bioeng.*, 1995, **46**, 545–552.
292. M. Funaoka, *Polym. Int.*, 1998, **47**, 277–290.
293. Z. Xia, T. Yoshida and M. Funaoka, *Biotechnol. Lett.*, 2003, **25**, 9–12.
294. Y. Nagamatsu and M. Funaoka, *Green Chem.*, 2003, **5**, 595–601.
295. N. Shiraishi, S. Onodera, M. Ohtani and T. Masumoto, *Mokuzai Gakaishi*, 1985, **31**, 418–420.
296. M. Alma, M. Yoshioka, Y. Yao and N. Shiraishi, *Wood Sci. Technol.*, 1995, **30**, 39–47.
297. S. P. Mun, C. S. Ku and S. B. Park, *J. Ind. Eng. Chem.*, 2007, **13**, 127–132.
298. S. H. Lee and T. Ohkita, *Wood Sci. Technol.*, 2003, **37**, 29–38.
299. D. Maldas and N. Shiraishi, *Int. J. Polym. Mater.*, 1996, **33**, 61–71.

300. S. H. Lee, M. Yoshioka and N. Shiraishi, *J. Appl. Polym. Sci.*, 2000, **78**, 319–325.
301. S. H. Lee, Y. Teramoto and N. Shiraishi, *J. Appl. Polym. Sci.*, 2002, **83**, 1482–1489.
302. Y. Wei, F. Cheng, H. Li and J. Yu, *J. Appl. Polym. Sci.*, 2004, **92**, 351–356.
303. Y. Yan, H. Pang, X. Yang, R. Zhang and B. Liao, *J. Appl. Polym. Sci.*, 2008, **110**, 1099–1111.
304. F. Yu, Y. Liu, X. Pan, X. Lin, C. Liu, P. Chen and R. Ruan, *Appl. Biochem. Biotechnol.*, 2006, **130**, 574–585.
305. M. Kunaver, E. Jasiukaityte, N. Čuk and J. T. Guthrie, *J. Appl. Polym. Sci.*, 2010, **115**, 1265–1271.
306. N. Shiraishi and H. Kishi, *J. Appl. Polym. Sci.*, 1986, **32**, 3189–3209.
307. A. Trosa and A. Pizzi, *Eur. J. Wood Wood Prod.*, 1998, **56**, 229–233.
308. M. H. Alma and S. S. Kelley, *J. Polym. Eng.*, 2000, **20**, 365–380.
309. M. H. Alma and M. A. Basturk, *Ind. Crops Prod.*, 2006, **24**, 171–176.
310. H. Pan, T. F. Shupe and C. Y. Hse, *J. Appl. Polym. Sci.*, 2008, **108**, 1837–1844.
311. E. B. Hassan, M. Kim and H. Wan, *J. Appl. Polym. Sci.*, 2009, **112**, 1436–1443.
312. M. Wang, M. Leitch and C. C. Xu, *J. Ind. Eng. Chem.*, 2009, **15**, 870–875.
313. T. Xie and F. Chen, *J. Appl. Polym. Sci.*, 2005, **98**, 1961–1968.
314. H. Kishi, A. Fujita, H. Miyazaki, S. Matsuda and A. Murakami, *J. Appl. Polym. Sci.*, 2006, **102**, 2285–2292.
315. C.-C. Wu and W.-J. Lee, *Wood Science and Technology*, 2010, in print (Abstract published online in July 2010).
316. H. Kishi, A. Fujita, H. Miyazaki, S. Matsuda and A. Murakami, *J. Adhesion Soc. Jap.*, 2005, **41**, 344–352.
317. R. Ikeda, H. Uyama and S. Kobayashi, *Macromolecules*, 1996, **29**, 3053–3054.
318. E. S. An, D. H. Cho, J. W. Choi, Y. H. Kim and B. K. Song, *Enzyme Microb. Technol.*, **46**, 287–291.
319. S. Dubey, D. Singh and R. A. Misra, *Enzyme Microb. Technol.*, 1998, **23**, 432–437.
320. L. Vachoud, T. Chen, G. F. Payne and R. Vazquez-Duhalt, *Enzyme Microb. Technol.*, 2001, **29**, 380–385.
321. M. Shibata and K. Nakai, *J. Polym. Sci. Part B: Polym. Phys.*, 2010, **48**, 425–433.
322. M. Shibata, N. Teramoto, Y. Takada and S. Yoshihara, *J. Appl. Polym. Sci.*, 2010, **118**, 2998–3004.
323. X. Li, Y. Su, Q. Chen, Y. Lin, Y. Tong and Y. Li, *Biomacromolecules*, 2005, **6**, 3181–3188.
324. H. R. Kricheldorf and O. Bolender, *J. Macromol. Sci., Part A: Pure Appl. Chem.*, 1998, **35**, 903–918.
325. M. C. Lubi and E. T. Thachil, *Des. Monomers Polym.*, 2000, **3**, 123–153.
326. A. K. Misra and G. N. Pandey, *J. Appl. Polym. Sci.*, 1984, **29**, 361–372.

327. D. O'Connor and F. D. Blum, *J. Appl. Polym. Sci.*, 1987, **33**, 1933–1941.
328. M. B. Patel, R. G. Patel and V. S. Patel, *Thermochim. Acta*, 1988, **129**, 277–284.
329. D. O'Connor, *Polym. Mater. Sci. Eng.*, 1990, **63**, 821–825.
330. R. Antony, C. K. S. Pillai and K. J. Scariah, *J. Appl. Polym. Sci.*, 1990, **41**, 1765–1775.
331. S. Manjula, C. K. S. Pillai and V. G. Kumar, *Thermochim. Acta*, 1990, **159**, 255–266.
332. S. Manjula, C. Pavithran, C. Pillai and V. Kumar, *J. Mater. Sci.*, 1991, **26**, 4001–4007.
333. R. Antony and C. K. S. Pillai, *J. Appl. Polym. Sci.*, 1993, **49**, 2129–2135.
334. G. John and C. K. S. Pillai, *J. Polym. Sci. Part A: Polym. Chem.*, 1993, **31**, 1069–1073.
335. H. P. Bhunia, A. Basak, T. K. Chaki and G. B. Nando, *Eur. Polym. J.*, 2000, **36**, 1157–1165.
336. H. P. Bhunia, R. N. Jana, A. Basak, S. Lenka and G. B. Nando, *J. Polym. Sci. Part A: Polym. Chem.*, 1998, **36**, 391–400.
337. T. T. M. Tan, *Polym. Int.*, 1996, **41**, 13–16.
338. K. I. Suresh and V. S. Kishanprasad, *Ind. Eng. Chem. Res.*, 2005, **44**, 4504–4512.
339. A. Maffezzoli, E. Calò, S. Zurlo, G. Mele, A. Tarzia and C. Stifani, *Composites Sci. Technol.*, 2004, **64**, 839–845.
340. R. da Silva Santos, A. A. de Souza, M.-A. De Paoli and C. M. L. de Souza, *Composites Part A: Appl. Sci. Manuf.*, 2010, **41**, 1123–1129.
341. C. K. S. Pillai, D. C. Sherrington and A. Sneddon, *Polymer*, 1992, **33**, 3968–3970.
342. S. Rajalekshmi, M. Saminathan, C. K. S. Pillai and C. P. Prabhakaran, *J. Polym. Sci. Part A: Polym. Chem.*, 1996, **34**, 2851–2856.
343. J. Kumanotani, *Prog. Org. Coatings*, 1995, **26**, 163–195.
344. J. Xia, Y. Xu, J. Lin and B. Hu, *Prog. Org. Coatings*, 2008, **61**, 7–10.
345. K. Nagase, R. Lu and T. Miyakoshi, *Chem. Lett.*, 2004, **33**, 90–91.
346. K. Taguchi, S. Hirose and Y. Abe, *Prog. Org. Coatings*, 2007, **58**, 290–295.
347. H. S. Kim, J. H. Yeum, S. W. Choi, J. Y. Lee and I. W. Cheong, *Prog. Org. Coatings*, 2009, **65**, 341–347.
348. K. S. Alva, P. L. Nayak, J. Kumar and S. K. Tripathy, *J. Macromol. Sci., Part A: Pure Appl. Chem.*, 1997, **34**, 665–674.
349. R. Ikeda, H. Tanaka, H. Uyama and S. Kobayashi, *Polym. J.*, 2000, **32**, 589–593.
350. Y. H. Kim, E. S. An, B. K. Song, D. S. Kim and R. Chelikani, *Biotechnol. Lett.*, 2003, **25**, 1521–1524.
351. S. Kobayashi, H. Uyama and R. Ikeda, *Chem. – Eur. J.*, 2001, **7**, 4754–4760.
352. S. Antoniotti, L. Santhanam, D. Ahuja, M. G. Hogg and J. S. Dordick, *Org. Lett.*, 2004, **6**, 1975–1978.
353. L. H. Bock and J. K. Anderson, *J. Polym. Sci.*, 1955, **17**, 553–558.

354. H. R. Kricheldorf and G. Löden, *Polymer*, 1995, **36**, 1697–1705.
355. Y. N. Sazanov, I. V. Podeshvo, G. M. Mikhailov, G. N. Fedorova, M. Y. Goikhman, M. F. Lebedeva and V. V. Kudryavtsev, *Russ. J. Appl. Chem.*, 2002, **75**, 777–780.
356. X. G. Li, M. R. Huang, G. H. Guan and T. Sun, *J. Appl. Polym. Sci.*, 1996, **59**, 1–8.
357. M. W. P. C. van Rossum, M. Alberda and L. H. W. van der Plas, *Phytochemistry*, 1998, **49**, 723–729.
358. M. K. Akkapeddi, *Macromolecules*, 1979, **12**, 546–551.
359. M. Ueda, M. Takahashi, Y. Imai and C. U. Pittman, *J. Polym. Sci.: Polym. Chem. Ed.*, 1982, **20**, 2819–2828.
360. J. Mosnácek and K. Matyjaszewski, *Macromolecules*, 2008, **41**, 5509–5511.
361. G. M. Miyake, Y. Zhang and E. Y. X. Chen, *Macromolecules*, 2010, **43**, 4902–4908.
362. N. D. Sachinvala, W. P. Niemczura and M. H. Litt, *Carbohydrate Res.*, 1991, **218**, 237–245.
363. V. Ladmiral, E. Melia and D. M. Haddleton, *Eur. Polym. J.*, 2004, **40**, 431–449.
364. S. Kimura and M. Imoto, *Die Makromolekulare Chemie*, 1961, **50**, 155–160.
365. M. Al-Bagoury and E.-J. Yaacoub, *Eur. Polym. J.*, 2004, **40**, 2617–2627.
366. M. T. Barros, K. T. Petrova and R. P. Singh, *Eur. Polym. J.*, 2010, **46**, 1151–1157.
367. R. Acosta Ortiz, A. Y. R. Martinez, A. E. García Valdez and M. L. Berlanga Duarte, *Carbohydrate Polym.*, 2010, **82**, 822–828.
368. P. Zetterlund, S. Hirose, T. Hatakeyama, H. Hatakeyama and A. C. Albertsson, *Polym. Int.*, 1997, **42**, 1–8.
369. M. J. Donnelly, *Polym. Int.*, 1995, **37**, 297–314.
370. K. Kurita, N. Hirakawa and Y. Iwakura, *J. Polym. Sci.: Polym. Chem. Ed.*, 1980, **18**, 365–370.
371. K. Kurita, N. Hirakawa and Y. Iwakura, *Die Makromolekulare Chemie*, 1979, **180**, 855–858.
372. Y. Li, H. Ren and A. J. Ragauskas, *Nano-Micro Lett.*, 2010, 89.
373. H. R. Kricheldorf, *J. Macromol. Sci. – Rev. Macromol. Chem.Phys.*, 1997, **37**, 599–631.
374. J. Thiem and H. Lüders, *Polym. Bull.*, 1984, **11**, 365–369.
375. D. Braun and M. Bergmann, *J. Prakt. Chem.*, 1992, **334**, 298–310.
376. M. Okada, Y. Okada and K. Aoi, *J. Polym. Sci. Part A: Polym. Chem.*, 1995, **33**, 2813–2820.
377. G. Schwarz and H. R. Kricheldorf, *J. Polym. Sci. Part A: Polym. Chem.*, 1996, **34**, 603–611.
378. S. Chatti and H. R. Kricheldorf, *J. Macromol. Sci., Part A: Pure Appl. Chem.*, 2006, **43**, 967–975.
379. J. Thiem and H. Lüders, *Die Makromolekulare Chemie*, 1986, **187**, 2775–2785.

380. E. Cognet-Georjon, F. Méchin and J. P. Pascault, *Macromol. Chem. Phys.*, 1995, **196**, 3733–3751.
381. M. Beldi, R. Medimagh, S. Chatti, S. Marque, D. Prim, A. Loupy and F. Delolme, *Eur. Polym. J.*, 2007, **43**, 3415–3433.
382. D. Braun and M. Bergmann, *Die Angewandte Makromolekulare Chemie*, 1992, **199**, 191–205.
383. H. R. Kricheldorf, S.-J. Sun, A. Gerken and T.-C. Chang, *Macromolecules*, 1996, **29**, 8077–8082.
384. M. Okada, M. Yokoe and K. Aoi, *J. Appl. Polym. Sci.*, 2002, **86**, 872–880.
385. F. Fenouillot, A. Rousseau, G. Colomines, R. Saint-Loup and J. P. Pascault, *Prog. Polym. Sci.*, 2010, **35**, 578–622.
386. R. Garçon, C. Clerk, J. P. Gesson, J. Bordado, T. Nunes, S. Caroço, P. T. Gomes, M. E. Minas da Piedade and A. P. Rauter, *Carbohydrate Polym.*, 2001, **45**, 123–127.
387. K. Hashimoto, K. Yaginuma, S.-i. Nara and H. Okawa, *Polym. J*, 2005, **37**, 384–390.
388. D. Jhurry and A. Deffieux, *Eur. Polym. J.*, 1997, **33**, 1577–1582.
389. Y. Tokiwa and M. Kitagawa, *ACS Symp. Ser.*, 2008, **999**, 379–410.
390. O. J. Park, D. Y. Kim and J. S. Dordick, *Biotechnol. Bioeng.*, 2000, **70**, 208–216.
391. M. Kitagawa and Y. Tokiwa, *Carbohydrate Lett.*, 1997, **2**, 343–358.
392. M. Kitagawa, P. Chalermisrachai, H. Fan and Y. Tokiwa, *Macromol. Symposia*, 1999, **144**, 247–256.
393. J. B. Binder and R. T. Raines, *J. Am. Chem. Soc.*, 2009, **131**, 1979–1985.
394. M. Mascal and E. Nikitin, *Ang. Chem. Int. Ed.*, 2008, **47**, 7924–7926.
395. H. Zhao, J. E. Holladay, H. Brown and Z. C. Zhang, *Science*, 2007, **316**, 1597–1600.
396. Z. Zhang and Z. K. Zhao, *Bioresour. Technol.*, 2010, **101**, 1111–1114.
397. A. Gandini and M. N. Belgacem, *Prog. Polym. Sci.*, 1997, **22**, 1203–1379.
398. A. Gandini and M. N. Belgacem, *ACS Symposium Series*, 2007, **954**, 280–295.
399. H. Laita, S. Boufi and A. Gandini, *Eur. Polym. J.*, 1997, **33**, 1203–1211.
400. R. Gheneim, C. Perez-Berumen and A. Gandini, *Macromolecules*, 2002, **35**, 7246–7253.
401. A. Khrouf, M. Abid, S. Boufi, R. E. Gharbi and A. Gandini, *Macromol. Chem. Phys.*, 1998, **199**, 2755–2765.
402. S. Gharbi, J.-P. Andreolety and A. Gandini, *Eur. Polym. J.*, 2000, **36**, 463–472.
403. A. Gandini, A. J. D. Silvestre, C. P. Neto, A. F. Sousa and M. Gomes, *J. Polym. Sci. Part A: Polym. Chem.*, 2009, **47**, 295–298.
404. A. Mitiakoudis and A. Gandini, *Macromolecules*, 1991, **24**, 830–835.
405. S. Gharbi and A. Gandini, *Acta Polym.*, 1999, **50**, 293–297.
406. S. Boufi, A. Gandini and M. N. Belgacem, *Polymer*, 1995, **36**, 1689–1696.
407. G. Koßmehl and A. Yaridjanian, *Die Makromolekulare Chemie*, 1981, **182**, 3419–3426.
408. C. Méalares, Z. Hui and A. Gandini, *Polymer*, 1996, **37**, 2273–2279.

409. C. Goussé and A. Gandini, *Polym. Bull.*, 1998, **40**, 389–394.
410. T. Carole, J. Pellegrino and M. Paster, *Appl. Biochem. Biotechnol.*, 2004, **115**, 871–885.
411. J. Lunt, *Polym. Degrad. Stability*, 1998, **59**, 145–152.
412. L. T. Lim, R. Auras and M. Rubino, *Prog. Polym. Sci.*, 2008, **33**, 820–852.
413. A. J. A. v. Maris, W. N. Konings, J. P. v. Dijken and J. T. Pronk, *Metab. Eng.*, 2004, **6**, 245–255.
414. M. Yamashita, N. Hattori and H. Nishida, *Polym. Prepr. Jpn.*, 1994, **43**, 3980–3981.
415. D. Zhang, M. A. Hillmyer and W. B. Tolman, *Macromolecules*, 2004, **37**, 8198–8200.
416. J. V. Kurian, *J. Polym. Environ.*, 2005, **13**, 159–167.
417. H. H. Chuah, *Macromolecules*, 2001, **34**, 6985–6993.
418. C. E. Nakamura and G. M. Whited, *Curr. Opin. Biotechnol.*, 2003, **14**, 454–459.
419. J. McKinlay, C. Vieille and J. Zeikus, *Appl. Microbiol. Biotechnol.*, 2007, **76**, 727–740.
420. P. Zheng, L. Fang, Y. Xu, J.-J. Dong, Y. Ni and Z.-H. Sun, *Bioresour. Technol.*, 2010, **101**, 7889–7894.
421. S. K. Garg and A. Jain, *Bioresour. Technol.*, 1995, **51**, 103–109.
422. M. J. Syu, *Appl. Microbiol. Biotechnol.*, 2001, **55**, 10–18.
423. N. Teramoto, M. Ozeki, I. Fujiwara and M. Shibata, *J. Appl. Polym. Sci.*, 2005, **95**, 1473–1480.
424. M. Yasuda, H. Ebata and S. Matsumura, *ACS Symp. Ser.*, 2007, **954**, 280–295.
425. S. Takenouchi, A. Takasu, Y. Inai and T. Hirabayashi, *Polym. J.*, 2001, **33**, 746–753.
426. B. E. Tate, *Adv. Polym. Sci.*, 1967, **5**, 214–232.
427. C. V. Lárez, F. Canelón, E. Millán and I. Katime, *Polym. Bull.*, 2002, **48**, 361–366.
428. J. M. G. Cowie, M. Yazdani Pedram and R. Ferguson, *Eur. Polym. J.*, 1985, **21**, 227–232.
429. H. Onishi, *Bull. Agr. Chem. Soc. Jpn.*, 1960, **24**, 131–140.
430. K. Hattori and T. Suzuki, *Agric. Biol. Chem.*, 1974, **38**, 581–586.
431. M. A. Y. Aoki, G. M. Pastore and Y. K. Park, *Biotechnol. Lett.*, 1993, **15**, 383–388.
432. S.-J. Lin, C.-Y. Wen, J.-C. Liau and W.-S. Chu, *Process Biochem.*, 2001, **36**, 1249–1258.
433. H. Ishizuka, K. Wako, T. Kasumi and T. Sasaki, *J. Ferment. Bioeng.*, 1989, **68**, 310–314.
434. T. Higashiyama, *Pure Appl. Chem.*, 2002, **74**, 1263–1269.
435. N. Teramoto, N. Sachinvala and M. Shibata, *Molecules*, 2008, **13**, 1773–1816.
436. H. Uyama, E. Klegraf, S. Wada and S. Kobayashi, *Chem. Lett*, 2000, **29**, 800–801.
437. J. Hu, W. Gao, A. Kulshrestha and R. A. Gross, *Macromolecules*, 2006, **39**, 6789–6792.

438. D. G. Barrett, W. Luo and M. N. Yousaf, *Polym. Chem.*, 2010, **1**, 296–302.
439. A. Takasu, Y. Shibata, Y. Narukawa and T. Hirabayashi, *Macromolecules*, 2006, **40**, 151–153.
440. T. Imai, T. Satoh, H. Kaga, N. Kaneko and T. Kakuchi, *Macromolecules*, 2004, **37**, 3113–3119.
441. E. Wachenfeld and W. Burchard, *Polymer*, 1987, **28**, 817–824.
442. J. Thiem and F. Bachmann, *Die Makromolekulare Chemie*, 1991, **192**, 2163–2182.
443. T. S. Moon, S.-H. Yoon, A. M. Lanza, J. D. Roy-Mayhew and K. L. J. Prather, *Appl. Environ. Microbiol.*, 2009, **75**, 589–595.
444. D. E. Kiely, L. Chen and T. H. Lin, *J. Am. Chem. Soc.*, 1994, **116**, 571–578.
445. H. Suzuki, Y. Ohnishi, Y. Furusho, S. Sakuda and S. Horinouchi, *J. Biol. Chem.*, 2006, **281**, 36944–36951.
446. W. W. Moyer, C. Cole and T. Anyos, *J. Polym. Sci. Part A: General Papers*, 1965, **3**, 2107–2121.
447. Y. Imai, K. Uno and Y. Iwakura, *Die Makromolekulare Chemie*, 1965, **83**, 179–187.
448. A. W. Chow, S. P. Bitler, P. E. Penwell, D. J. Osborne and J. F. Wolfe, *Macromolecules*, 1989, **22**, 3514–3520.
449. S.-M. Eo, S.-J. Oh, L.-S. Tan and J.-B. Baek, *Eur. Polym. J.*, 2008, **44**, 1603–1612.
450. T. Zaki, *J. Colloid Interface Sci.*, 2005, **284**, 606–613.
451. K. Murata, M. Inaba and I. Takahara, *J. Jap. Petroleum Institute*, 2008, **51**, 234–239.
452. H. Oikawa, Y. Shibata, K. Inazu, Y. Iwase, K. Murai, S. Hyodo, G. Kobayashi and T. Baba, *Appl. Catal. A: General*, 2006, **312**, 181–185.
453. http://pressroom.toyota.com/pr/tms/tmc-to-use-bio-pet-ecological-173305.aspx (last accessed October 2010).
454. C. Berti, E. Binassi, M. Colonna, M. Fiorini, G. Kannan, S. Karanam, M. Mazzacurati and I. Odeh, *US Patent*, 168,461, 2010.

CHAPTER 3
Polyhydroxyalkanoates: The Emerging New Green Polymers of Choice

RANJANA RAI AND IPSITA ROY

University of Westminster, Department of Molecular and Applied Biosciences, 115 New Cavendish Street, London, W1W 6UW, UK

3.1 Introduction

Synthetic plastics are imposing serious threats to our environment. Since their introduction in the 1950s plastics have become an absolute necessity of our life. However these plastics are not degradable. Non-biodegradable plastics accumulate at the rate of 25 million tons per year and have therefore become one of the main environmental hazards. This problem is compounded with the fact that the resources for crude oil is also decreasing. Therefore, in the present scenario when global warming, climate change and dwindling fossil carbon resources have taken centre stage, scientists from all over the world are looking for greener eco-friendly alternatives to petrochemical-derived plastics. This alternative has come in the form of biodegradable plastics and one such family attracting considerable interest is the polyhydroxyalkanoates, PHAs. Polyhydroxyalkanoates are linear polyesters of 3, 4, 5 and 6-hydroxyalkanoic acids accumulated by numerous Gram positive and Gram negative bacteria through the fermentation of sugars, lipids, alkanes, alkenes and alkanoic acids (Figure 3.1). These are, therefore, bio-based polymers and once extracted exhibit thermoplastic and elastomeric properties closely resembling synthetic plastics.

In fact, the properties of these polymers can be tailored by controlling the carbon feed. They are recyclable and degrade into CO_2 and water. In addition, these polymers are also biocompatible. It is because of these reasons that PHAs are becoming increasingly popular as environmentally friendly materials with industrial, agricultural and medical applications.[1-3]

3.2 History of Polyhydroxyalkanoates

PHAs were first discovered in 1923, by a French scientist Maurice Lemoigne at the Institute Pasteur. He demonstrated that aerobic spore-forming *Bacillus megaterium* accumulated 3-hydroxybutyric acid under anaerobic conditions.[4] The next step in understanding PHAs was when, in 1958, Macrae and Wilkinson established the role of nutrient limitations in the production of PHAs. They observed that an asporogenous strain of *B. megaterium* accumulated more PHAs as the carbon to nitrogen ratio increased. Their results therefore suggested that like polyphosphate and carbohydrate reserves PHA accumulation occurred in response to an imbalance in growth brought about by nutrient limitations.[5] Commercial exploration of P(3HB) occurred only in the early 1960s. Werber at W.R. Grace & Co. (USA) carried out many pioneering pieces of work which earned them several patents for P(3HB) production and isolation. They were also the first to explore P(3HB) for fabricating articles like sutures and prosthetic devices.[6] Discovery of polyhydroxyalkanoates other than P(3HB) by Wallen and Rohwedder in 1974 and further discovery of medium-chain length PHAs (mcl-PHAs) by Witholt and co-workers in 1983 were important turning points in the research and development of PHAs.[7,8] This is because unlike P(3HB) which is brittle and stiff, mcl-PHAs were more flexible and elastomeric. By the end of 1980 numerous studies had taken place to produce and characterize novel monomers of PHAs. By this time scientists were also studying the molecular aspects of PHA production and had already cloned genes for PHA production from *Cupriavidus necator* (formerly known as *Alcaligenes eutrophus*) into *E. coli*.[9-11] This successful cloning of PHA biosynthetic genes paved another era for creating transgenic plants and recombinant organisms for PHA production. Since the 1990s scientists all over the world have been working on using PHAs for various industrial, agricultural and medical applications. Another avenue for research has been in making PHA production more cost effective and green. Cheap renewable and agricultural wastes are being studied as carbon feed for PHA production.

3.3 Chemical Organization of PHAs

Polyhydroxyalkanoates (PHAs) are polyesters of hydroxyalkanoates (HAs) which have the general structure shown in Figure 3.1.[12] In these polymers, the carboxyl group of one monomer forms an ester bond with the hydroxyl group of the neighbouring monomer.[13] Here, 'R' refers to the type of side chain whereas 'X' refers to the number of 'CH_2' groups. Both 'R' and 'X' determine the type of HA unit constituting the polymer chain.

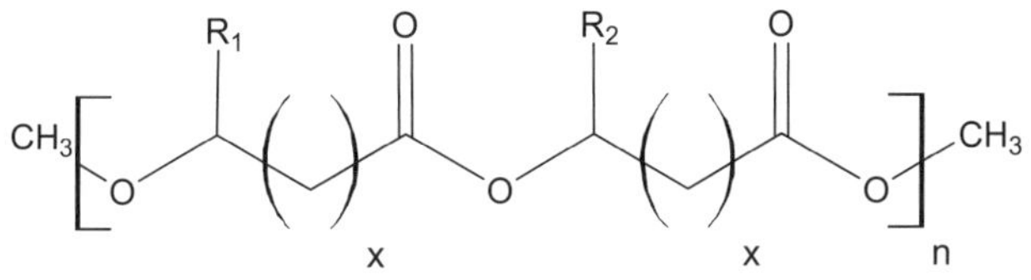

Figure 3.1 The general structure of polyhydroxyalkanoates. $R_1/R_2 = C_1-C_{13}$ alkyl groups, $x = 1-4$, $n = 100-30\,000$.

Depending on the number of carbon atoms in the monomeric unit, PHAs are classified as short-chain length PHAs, scl-PHAs, that contain 3–5 carbon atoms, examples are poly(3-hydroxybutyrate), P(3HB), poly(4-hydroxybutyrate), P(4HB), and medium-chain length PHAs, mcl-PHAs, that contain 6–14 carbon atoms, examples are poly(3-hydroxyhexanoate), P(3HHx), and poly(3-hydroxyoctanoate), P(3HO). Also depending on the kind of monomer present, PHAs can be a homopolymer containing only one type of hydroxyalkanoate as the monomer unit, e.g. P(3HB), P(3HHx), or a heteropolymer containing more then one kind of hydroxyalkanoate as monomer units, e.g. poly(3-hydroxybutyrate-co-3-hydroxyvalerate), P(3HB-co-3HV), and poly-3-hydroxybutyrate-co-3-hydroxyhexanoate, P(3HB-co-3HHx).[14,15] Mcl-PHAs are more structurally diverse than scl-PHAs. Here, the 'R' group can vary from propyl to tridecyl (e.g. 3-hydroxyhexanoate), may contain an aromatic group (e.g. mcl-PHA with para-methylphenoxy and meta-methylphenoxy) and the alkyl side-chain can be saturated (e.g. 3-hydroxyoctanoate) and unsaturated, e.g. 4- hexenoic, 3-hydroxy-8-nonynoate and 3-hydroxy-10-undecynoate.[16,18] Some mcl-PHAs have also been found with the hydroxyl group on the C-2, C-4, C-5 and C-6 carbon atoms. In mcl-PHAs, the presence of functional groups like halogen, carboxyl, hydroxyl, epoxy, phenoxy, cyanophenoxy and nitrophenoxy are particularly important, as they allow further chemical modifications of these PHAs, leading to the production of novel biomaterials with tailorable properties. At the time of writing, more then 150 units of PHA monomers have been reported (Table 3.1 and Figure 3.2).

3.4 Occurrence and Biosynthesis of PHAs

PHAs are synthesized by numerous Gram-positive bacteria, aerobic (cyanobacteria) and anaerobic (non-sulfur and sulfur purple bacteria) photosynthetic bacteria, Gram negative bacteria as well as some Archaebacteria. They serve as intracellular carbon and energy storage compounds.[1,29] PHAs are accumulated by bacteria as water-insoluble cytoplasmic inclusions, the number per cell and size of which varies among different species (Figure 3.3).[1,2,30]

Usually, bacteria produce these granules when subjected to an unbalanced growth condition with excess carbon and simultaneous limitation of nutrient(s) such as oxygen, nitrogen, sulfur, magnesium and phosphorous.[1,12,30] Some bacteria, however, produce them without being subjected to any kind of

Table 3.1 The broad spectrum of monomers found in PHAs (adapted from Zinn et al.[28]).

3-Hydroxy acid	3-Hydroxy acid (unsaturated)	3-Hydroxy acid (branched)	3-Hydroxy acid (substituted side chain)	Other than 3-hydroxy acid	Aromatic side chain	Other functional groups
Butyanoic	4-hexenoic	2,6-dimethyl-5-heptenoic	7-fluoroheptanoic	4-hydroxybutanoic	Dimethyl esters of 3,-6-epoxy-7-nonenoic acid	3-hydroxy-7-oxooctanoate
Pentanoic	5-hexenoic	7-cyanoheptanoic	9-fluorononanoic	4-hydroxyhexanoic	3-hydroxyphenylhexanoic	3-hydroxy-5-oxohexanoate
Hexanoic	6-heptenoic	5-methylhexanoic	6-chlorohexanoic	4-hydroxyoctanoic	3-hydroxyphenylheptanoic	8-acetoxy-3-hydroxyoctanoate
Heptanoic	6-octenoic	4-methyloctanoic	8-chlorooctanoic	5-hydroxyheptanoic	3-hydroxyphenyloctanoic	6-acetoxy-3-hydroxyhexanoate
Octanoic	7-octenoic	5-methyloctanoic	6-bromohexanoic	5-hydroxyhexanoic	3-hydroxy-6-p-methylphenoxyhexanoate	
Nonanoic	8-nonenoic	6-methyloctanoic	8-bromooctanoic	4-hydroxyhexanoic		
Decanoic	9-decenoic	6-methylnonanoic	11-bromoundecanoic	2-hydroxydodecanoic		
Undecanoic	10-undecenoic	7-methylnonanoic	7-cyanoheptanoic			
Dodecanoic	6-dodecenoic	8-methylnonanoic	9-cyanononanoic			
7-cyanoheptanoic	5-tetradecenoic	7-methyldecanoic	12-hydroxydodecanoic			
	5,8-tetradecadienoic		Succinic methylester acid			
	5,8,11-tetradecatrienoic		Adipic acid methylester			
	4-hexadecenoic		Suberic acid methylester			
	4,7-hexadecadienoic		Suberic acid ethyl ester			
			Pimelic acid propylester			

References for the table data: 16, 19–27.

nutritional constraints, for example, *Alcaligenes latus*.[1,30] PHAs make an ideal carbon-energy storage material due to their low solubility and high molecular weight, which exerts negligible osmotic pressure on the bacterial cell. Thus, loss of this energy reserve out of the bacterial cell as leakage is prevented and

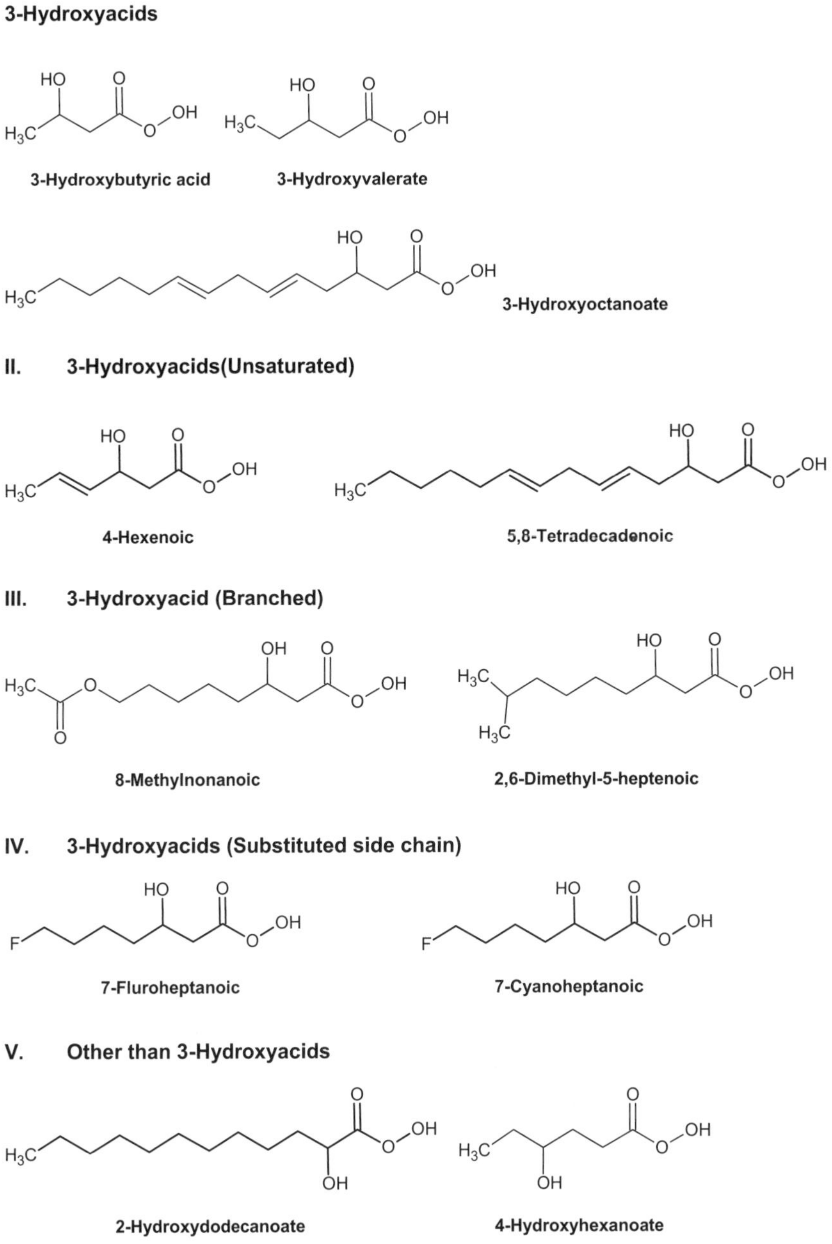

3-Hydroxyacids

3-Hydroxybutyric acid

3-Hydroxyvalerate

3-Hydroxyoctanoate

II. **3-Hydroxyacids(Unsaturated)**

4-Hexenoic

5,8-Tetradecadenoic

III. **3-Hydroxyacid (Branched)**

8-Methylnonanoic

2,6-Dimethyl-5-heptenoic

IV. **3-Hydroxyacids (Substituted side chain)**

7-Fluroheptanoic

7-Cyanoheptanoic

V. **Other than 3-Hydroxyacids**

2-Hydroxydodecanoate

4-Hydroxyhexanoate

Figure 3.2 Compilation of some PHA structures from Table 3.1.

VI. Aromatic side group

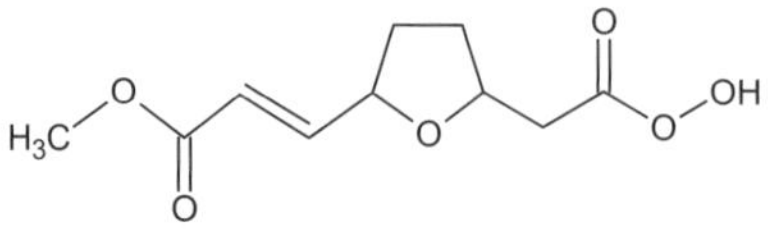

3,6-Epoxy-7-nonenoic acid

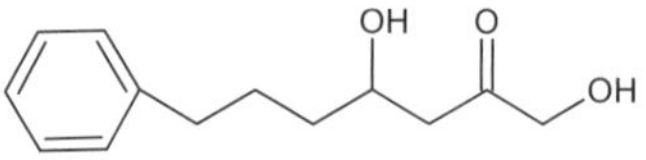

3-Hydroxyphenylhexanoic

VII. Other functional groups

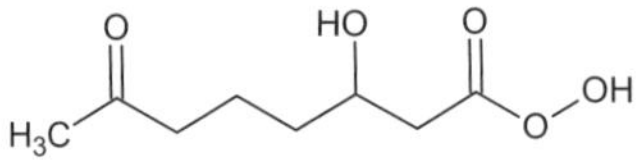

3-Hydroxy-7-oxooctanoate

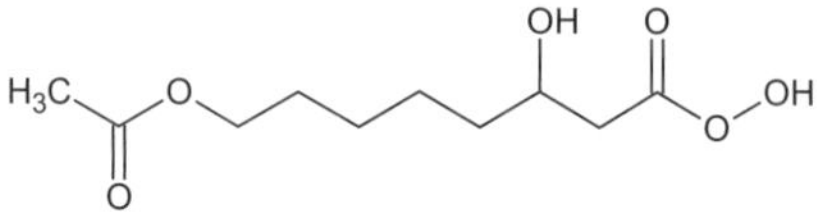

8-Acetoxy-3-hydroxyoctanoate

Figure 3.2 Continued

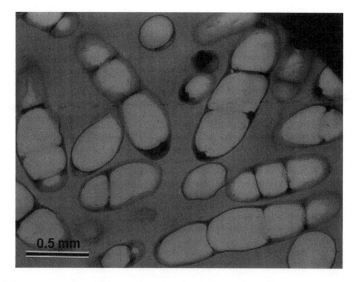

Figure 3.3 Transmission electron micrograph of thin sections of recombinant *R. eutropha* PHB⁻4 cells.[29]

therefore it is securely available to the organism for future use.[29,31] The biosynthesis of PHAs involves two major steps – the first step leads to the generation of hydroxyacyl-CoA substrates and the second leads to the polymerization of these substrates into PHAs. This polymerization of PHAs is carried out by the enzyme called PHA synthase encoded by the *phaC* gene (Figure 3.4).[32]

Three metabolic pathways (Pathway I: Chain elongation reaction; Pathway II: Fatty acid β-oxidation; Pathway III: Fatty acid *de novo* biosynthesis) are used by organisms to produce the hydroxyacyl-CoA substrates (HACoA), which are then polymerized into PHAs (Figure 3.4). Pathway I is the best known among the PHA

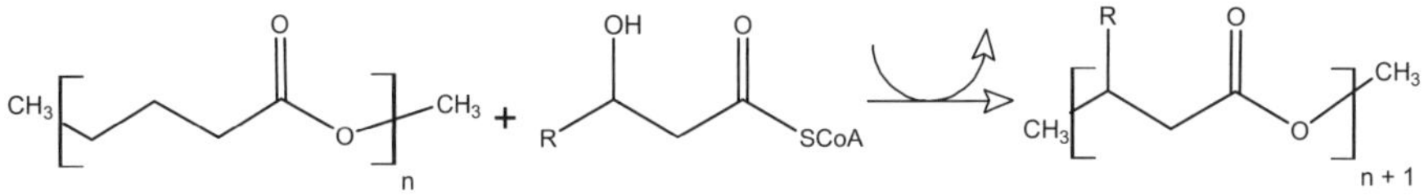

Polyhydroxyalkanoates: The Emerging New Green Polymers of Choice

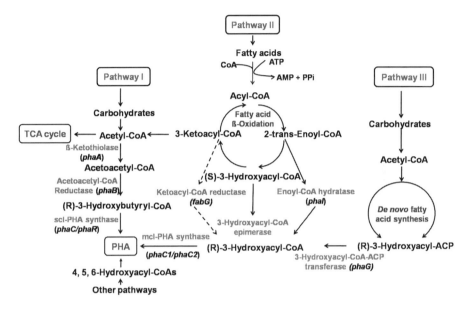

Figure 3.4 Reaction catalysed by the PHA synthase enzyme.[32]

Figure 3.5 Metabolic pathways for the production of PHAs (adapted from Kim et al.[38]).

biosynthetic pathways and is found in *Cupriavidus necator*.[29,33] Here, condensation of two acetyl-CoA molecules from the tricarboxylic acid (TCA) cycle takes place to form acetoacetyl-CoA.[34,36] Acetoacetyl-CoA is then converted to (R)-3-hydroxybutyryl-CoA by the (R)-specific acetoacetyl-CoA reductase (*phaB*). Finally, the PHA synthase enzyme catalyses the polymerization *via* esterification of 3-hydroxybutyryl-CoA into poly(3-hydroxybutyrate), P(3HB).[36] Pathway II, the fatty acid β-oxidation-involving pathway, was deciphered when fluorescent *Pseudomonads*, such as *Pseudomonas oleovorans*, *Pseudomonas putida* and *Pseudomonas aeruginosa*, were found to accumulate PHA consisting of mcl-(R)-3HA units from alkanes, alkanols, alkanoates and oils.[33,37] Here, the fatty acids are first converted to the corresponding acyl-CoA thioesters which are then oxidized by fatty acid β-oxidation *via trans*-2-enoyl-CoA and (S)-3-hydroxyacyl-CoA to form 3-ketoacyl-CoA. 3-ketoacyl-CoA is then cleaved by a β-ketothiolase to form acetyl-CoA and an acyl-CoA comprising of two fewer carbon atoms compared to the acyl-CoA that entered the first cycle. The *trans*-2-enoyl-CoA, (S)-3-hydroxyacyl-CoA and 3-ketoacyl-CoA intermediates formed are utilized for the generation of (R)-3-hydroxyacyl-CoA derivatives which are then polymerized to form mcl-PHAs. Fatty acid *de novo* biosynthesis is involved in Pathway III and is of significant interest because it helps generate PHA monomers from structurally unrelated, simple, inexpensive carbon sources such as glucose, sucrose and fructose (Figure 3.5). In this pathway, the enzyme acyl-CoA-ACP transferase (encoded by *phaG*) transfers the hydroxyacyl moiety from (*R*)-3-hydroxy-acyl carrier protein to coenzyme A, thus forming (*R*)-3-hydroxyacyl-CoA, which acts as the substrate for the PHA synthase enzyme.[36,38]

3.5 Cheap Substrates for Cost-effective PHA Production

Raw materials account for 50% of the total production cost of PHAs.[39] Therefore, new approaches in utilizing waste and by-products as raw materials for cost-effective PHA production has been undertaken. Such approaches have an added advantage of addressing the problem of solid waste management of various agricultural wastes and production of value added products from the waste. Raw materials such as malt waste from the beer brewery industry, soya waste from the soya milk industry, lignocellulosic waste from the tequila brewery industry,[40] whey and maize steep liquor and cassava have all been successfully studied for PHA production. Other agricultural wastes studied are sunflower cake, soya bran and a solid residue of biodiesel production. Molasses either from sugar cane or sugar beet,[41,42] or sugar cane liquor[43] have also been studied. However, production of PHAs from these crude inexpensive substrates is low and therefore efficient processes must be developed to increase PHA productivity based on these agro-industry by-products.[44] Table 3.2 represents a compilation of the cheap carbon substrates used for PHA production.

3.6 Physical Properties of PHAs

The physical and material properties of PHAs are greatly influenced by their monomer composition and chemical structure, *i.e.* the length of the pendant groups which extend from the polymer backbone, the chemical nature of this pendant group and the distance between the ester linkages in the polymer backbone. Scl-PHAs for example P(3HB), with a melting temperature (T_m) of 180 °C and a glass transition temperature (T_g) of 4 °C, is highly crystalline, brittle and stiff.[54] These are typically thermoplastic polymers which become fluid and mouldable above their T_m. The brittleness of P(3HB) is largely due to the presence of large crystalline domains in the form of spherulites which form upon cooling of the melt.[55,56] P(3HB), with a Young's modulus of 3.5 GPa and tensile strength of 40 MPa, has mechanical properties similar to that of polypropylene. However, the elongation to break is about 5%, which is significantly lower than that of polypropylene (400%).[12] Another scl-PHA, P(4HB), has a Young's modulus value of 149 GPa, tensile strength of 104 MPa and a high elongation value of 1000%. The weight-average molecular weights (M_w) of P(3HB) have been found to range between 530 000 and 1 100 000 and polydispersity has been found to be around 1.75.[57]

Table 3.2 Compilation of examples of PHA production using cheap carbon sources.

Organism	Carbon source	PHA produced	Reference
Alcaligene latus DSM 1124	Malt waste	P(3HB)	45
	Soya waste	P(3HB-co-3HV)	
Saccharophagus degradans (*ATCC43961*)	Lignocellulosic waste	PHA	40
Azotobacter chroococcum	Soluble starch	P(3HB)	46
Bacillus cereus	Soluble starch	P(3HB)	47
Cupriavidus necator	Wheat hydrolysate	P(3HB)	48
	Bagasse hydrolysate	Not indicated	
Burkholderia sacchari IPT 101	Bagasse hydrolysate	P(3HB)	49
Pseudomonas hydrogenovora	Whey	P(3HB)	50
P. aeruginosa MTCC 7925	Whey	3HB, 3HV, 3HHD, 3HOD	51
	Wheat bran	3HB, 3HV, 3HHD, 3HOD	
Azotobacter vinelandii UWD	Beet molasses	P(3HB)	52
Pseudomonas fluorescens	Sugarcane liquor	P(3HB)	43
Pseudomonas aeruginosa	Cassava wastewater	3HD (major monomer)	53
	Cassava wastewater + waste cooking oil	3HD and 3HO major monomer	

Mcl-PHAs have melting temperature (T_m) values ranging between 40 °C and 60 °C and glass transition temperature (T_g) values ranging between −50 °C and −25 °C. Mcl-PHAs also have low crystallinity, possibly due to the presence of large and irregular pendant side groups which inhibit close packing of the polymer chains in a regular three-dimensional fashion to form a crystalline array.[58] This combination of T_g values below room temperature and a low degree of crystallinity imparts elastomeric behaviour to these polymers. In fact, mcl-PHAs are the only microbially produced biopolymers which exhibit properties of thermoplastic elastomers and resemble natural rubbers produced by *H. brasiliensis*.[59] However, mcl-PHAs act as true elastomers within a very narrow temperature range. At temperatures above or close to its T_m the polymer is completely amorphous and sticky.[59] In mcl-PHAs the crystalline parts act as physical cross-links and therefore they have mechanical properties such as tensile strength and elongation to break that are very different from that of scl-PHAs shown in Table 3.3.[60] In mcl-PHAs, M_w for polymers containing both saturated and unsaturated pendant groups lie in the range of 60 000 and 412 000 and number-average molecular weight, M_n, is between 40 000 and 231 000.[61] These values are relatively low, compared to that of scl-PHAs. The polydispersities of mcl-PHA copolymers are in the range of 1.6 to 4.4 with higher values for mcl-PHAs with unsaturated monomers than those with saturated monomers.[61]

Introduction of a co-monomer into the polymer backbone, as in the case of heteropolymers, greatly affects the polymer properties by increasing its flexibility, toughness and decreasing its stiffness.[62] For example, a copolymer such

Table 3.3 Mechanical properties of various PHAs and conventional plastics (adapted from Castilho et al.[67]).

Polymer	Tensile strength	Modulus	Elongation to break (%)	Reference
P(3HB)	40 MPa	3.5 GPa	6	12
P(4HB)	104 MPa	0.149 GPa	1000	29
P(3HB-co-17%3HHx)	20 MPa	0.173 MPa	850	64
P(3HO-co-12%3HHx)	9 MPa	0.008 MPa	380	65
P(HO)		17 MPa	250–350	56
P(3HO-co-12%3HHx-co-2%3HD)	9.3 MPa	7.6 MPa	380	66
P(3HO-co-4.6%3HHx)	22.9 MPa	599.9 MPa	6.5	15
P(3HO-co-5.4%3HHx)	23.9 MPa	493.7 MPa	17.6	
P(3HO-co-7%HHx)	17.3 MPa	288.9 MPa	23.6	
P(3HO-co-8.5%HHx)	15.6 MPa	232.3 MPa	34.3	
P(3HO-co-9.5%HHx)	8.8 MPa	155.3 MPa	43.0	
High density polyethylene	17.9–33.1 MPa	0.4–1.0 GPa	12–700	67
Low density polyethylene	15.2–78.6	0.05–0.1	150–600	
Polystyrene	50	3.0–3.1	3–4	
Nylon-6,6	83	2.8	60	

as poly(3-hydroxybutyrate-co-3-hydroxyhexanoate), P(3HB-co-3HHx), has a lower melting temperature and crystallinity and is more malleable than P(3HB). In fact P(3HB-co-3HHx) has similar mechanical properties to one of the representative commercial polymers, low-density polyethylene (LDPE).[63] The properties of PHAs vary considerably depending on their monomer content and hence can be tailored by controlling their composition.[64] PHAs are also biocompatible, exhibit piezoelectricity which stimulates bone growth, aid in wound healing and exhibit wide ranging physical and mechanical properties that arise from the diversity in their chemical structures. The physical properties of PHAs and other commercial plastics are given in Table 3.3.

3.7 Biocompatibility of PHAs

The biocompatibility of PHAs originates from the fact that some monomers incorporated into the polymer chain occur naturally in the human body. The monomer (R)-3-hydroxybutyric acid is a normal metabolite found in human blood. This hydroxy acid is present at concentrations of 3–10 mg per 100 ml blood in healthy adults.[6,68] Also low molecular weight PHAs are found complexed to other cellular macromolecules – hence they are called complexed PHAs (cPHAs). For example, cPHAs have been found in human tissues complexed with low-density lipoproteins, carrier protein albumin and in the potassium channel (KcsA) of *Streptomyces lividans*.[69,70] Biocompatibility of PHAs, like any other biomaterial, is dependent on factors such as shape, surface porosity, surface hydrophilicity, surface energy, chemistry of the material and its degradation product.[71–73] In tissue engineering, it is important that the cellular behaviour affected by the degradation products be considered for a comprehensive biocompatibility evaluation of the implant polymers. To this end studies carried out by Sun and his group on the cellular responses of mouse fibroblast cell line L929 to the PHA degradation products, oligo-hydroxyalkanoates (OHAs), showed that mcl-PHAs are more biocompatible than scl-PHAs.[74] Biocompatibilities of PHA scaffolds have also been enhanced by: (1) increasing the hydrophilicity of the polymer, for example by grafting acrylamide and carboxyl ions onto the P(3HO), P(3HB) and P(3HB-co-3HHx) films using plasma treatment;[75] (2) surface modifications of PHAs using NaOH and enzyme treatment;[7,76] (3) coating the polymer surface using a biocompatible compound, for example the surface of both porous and dense P(3HB-co-3HHx) matrices was coated with a biocompatible protein, silk fibroin which is a natural protein generated from silk worm silk fibre.[77] Biomaterials intended for long-term contact with blood must not induce thrombosis, antigenic responses or destruction of blood components and plasma protein. *In vitro* tests showed haemocompatibility of P(3HB-co-3HHx) as blood contact graft material that reduced thrombogenicity and adhesiveness of blood platelets.

3.8 Biodegradation of Polyhydroxyalkanoates

3.8.1 Factors Affecting Biodegradation

PHAs are biodegradable polymers that can degrade under both aerobic and anaerobic conditions. They can also be subjected to thermal degradation and enzymatic hydrolysis. In biological systems, however, PHAs can be degraded using microbial depolymerases as well as by non-enzymatic and enzymatic hydrolysis in animal tissue.[36] Numerous factors affect the biodegradability of PHAs, such as stereoregularity, molecular mass, monomeric composition and crystallinity of the polymer. Studies carried out by Mochizuki[78] and Tokiwa *et al.*[79] showed that biodegradation of PHAs is influenced by its chemical structure, such as the presence of functional groups in the polymer chain, the hydrophilicity/hydrophobicity balance and the presence of ordered structure, like crystallinity, orientation and morphological properties.[78,79] Usually the degradation of the polymer decreases with the increase of highly ordered structure, that is increasing crystallinity. Since more crystalline structures also have higher melting temperatures for the crystalline phase of the polymer, the degradation rate for PHAs also decreases with increasing T_m. Thus mcl-PHAs with low crystallinity and low T_m are more degradable than scl-PHAs which have comparatively higher crystallinity and higher T_m. Studies have also shown that the rate of hydrolysis of PHAs depends on the surface area of the polymer exposed. Hydrolysis starts on the surface and at physical lesions on the polymer and proceeds to the inner part of the material.[80]

3.8.2 Biodegradation in the Environment

In nature, the microbial population present in a given environment and temperature also contributes to the biodegradability of the polymer. Microorganisms from the families *Pseudonocardiaceae*, *Micromonosporaceae*, *Thermomonosporaceae*, *Streptosporangiaceae* and *Streptomycetaceae* predominantly degrade P(3HB) in the environment. These microbes secrete extracellular enzymes that solubilize the polymer and these soluble products are then absorbed through their cell walls and utilized. Some PHA-producing bacteria are able to degrade the polymer intracellularly. During intracellular degradation, the polymer is ultimately broken down to acetyl-CoA which under the aerobic conditions enters the citric acid cycle and is oxidized to CO_2.[12,36] The enzyme involved in the degradation of the PHAs is PHA depolymerase encoded by *phaZ*.[81]

3.8.3 Biodegradation and Biocompatibility

It is of paramount importance that the rate of degradation of the PHA scaffold should equal that of the regenerative rate of the tissues. The *in vivo* and *in vitro* degradation of PHAs has been studied by a number of research groups and various biodegradation rates of PHAs observed. Williams *et al.*[68] observed that P(3HO-co-3HHx) degrades slowly *in vivo*. The subcutaneous implants of P(3HO-co-3HHx) in mice decrease in M_w from 137 000 on implantation to around 65 000 over 40 weeks, and there was no significant differences between the molecular weights of samples taken from the surface and interior of the implants. The latter finding suggests that slow, homogenous hydrolytic breakdown of the polymer occurs.[68] Since the degradability of PHAs decreases with the overall increase in the crystallinity, when Wang *et al.*[80] blended gelatin with P(3HO-co-3HHx) they found that blending of gelatin accelerated the degradation of P(3HO-co-3HHx). They concluded that the weight loss observed was first due to the reduction in the crystallinity of the blended polymer, as confirmed by the weakening of its crystalline peak from X-ray diffraction (XRD) analysis. Secondly, blending with gelatin created a more porous polymer surface, which was exposed for hydrolytic attack, as observed by scanning electron microscopy (SEM) analysis.[80]

3.9 Applications of Polyhydroxyalkanoates

3.9.1 Industrial Application

PHAs have successfully been used for industrial applications. A blend of P(3HB) and P(3HO) is marketed by a US-based company, Metabolix. This polymer has an FDA approval for production of food additives.[82] Tsinghua University (China) in collaboration with Guangdong Jiangmen Center for Biotech Development (China), KAIST (Korea) and Procter & Gamble (USA) have carried out industrial production of P(3HB-3HHx) using *Aeromonas hydrophila*. The polymer produced is used to make flushables, non-wovens, binders, flexible packaging, thermoformed articles, synthetic paper and medical devices.[83] A German company Biomer produces P(3HB) industrially from *Alcaligene latus*. The polymer is used for making articles such as combs, pens, bullets and for use in classical transformation processes.[36,84] Metabolix (Cambridge. MA, USA) also manufactures BIOPOL®, a copolymer of P(3HB-co-3HV). BIOPOL® is used to coat paper and paperboards, blow moulding and film production. It has antistax properties that can be exploited for electric and electronic packaging.[85] Nodax™, developed by Procter and Gamble, is a copolymer of P(3HB) with small quantity of medium chain length monomers with side groups of at least three carbon units or more. This polymer can be used to make flushables that can degrade in septic tank systems and this would include hygienic wipes and tampon applicators. They can also be used to manufacture medical surgical garments, upholstery, carpet, packaging, compostable bags and lids or tubs for thermoformed articles. PHA-based latex can be used for making water-resistant surfaces to cover paper and cardboard.[36,86]

3.9.2 Medical Applications

PHAs have been studied as a biomaterial for scaffolds in tissue engineering of both hard and soft tissues. Encapsulation of drugs in controlled drug delivery using PHAs as a matrix material has also been carried out. In addition, PHAs like P(3HB), P(4HB), P(3HB-co-3HHx) have also been used for making medical devices.

3.9.2.1 Tissue Engineering

PHAs have been used for vascular grafting to repair or replace malfunctioning blood vessels in the arterial or venous systems due to damage or disease. A synthetic graft made of P(3HB-co-4HB) as a graft coating in dogs was investigated for up to 10 weeks. The polymer showed signs of degradation after 2 weeks of implantation. When P(3HHx-co-3HO) was studied as an impregnation substrate in rat models, the scaffold showed slow degradation, of only about 30% reduction in molecular weight after 6 months of implantation. PHAs have also been used for heart valve development. One of the early studies using an elastomeric P(3HO) (Tepha Inc) for the fabrication of a 3-leaflet heart valve scaffold was carried out by Sodian and his group in 2000. Vascular cells were harvested from ovine carotid arteries, expanded *in vitro* and seeded onto the heart valve scaffold. The study concluded that tissue engineered P(3HO) fabricated heart valve can be used for implantation in the pulmonary position with an appropriate function for 120 days in lambs.[87,88] In the same year, Stock *et al.*[89] evaluated the feasibility of creating 3-leaflet, valved, pulmonary conduits from autologous ovine vascular cells and thermoplastic P(3HO), (PHO 3836; TEPHA Inc., Cambridge, MA) in lambs. Composite scaffolds of polyglycolic acid (PGA) and P(3HO) were formed into a conduit and three leaflets consisting of a monolayer of porous P(3HO) were sewn into the conduit as shown in Figure 3.6(a). Figures 3.6(b) and 3.6(c) show the appearance of tissue engineered seeded conduit after 24 weeks *in vivo*. These results indicated that the remodelling of the tissue engineered structure continues for at least 24 weeks.[89] Stereolithography has also been used to construct P(4HB) and P(3HO) (Tepha, Inc., Cambridge, MA) heart valve scaffolds derived from X-ray computed tomography and specific software (CP, Aachen, Germany).[90]

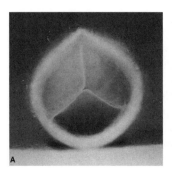
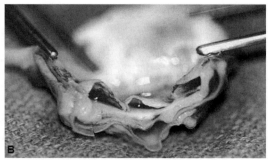
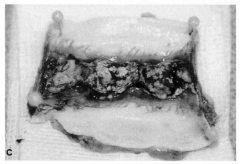

Figure 3.6 (a) P(3HO) scaffold frontal view, 18 mm in diameter and 20 mm in length. Three leaflets of porous P(3HO) were sutured into the conduit proximal view. (b) Gross appearance of tissue-engineered seeded conduit 24 weeks *in vivo* (distal view). Clear separation of all three leaflets from the conduit wall is shown. (c) Gross appearance of tissue-engineered unseeded conduit 4 weeks *in vivo* (proximal view).[89]

Using this technique, P(3HO) and P(4HB) could be moulded into a complete 3-leaflet valve scaffold without the need for suturing.[88,90] Novel hybrid valves which were fabricated from decellularized porcine aortic valves and coated with P(3HB-co-3HHx) were also developed. The results *in vivo* indicated that the P(3HB-co-3HHx) coating reduced calcification and promoted the repopulation of the hybrid valve with the recipient's cells, resembling native valve tissue.[91]

PHAs like P(3HB), P(3HB-co-3HHx) and blends of PHA with PDLLA have also been used as biomaterials for nerve regeneration. One of the early studies on P(3HB-co-3HHx) as conduit material for nerve regeneration was carried out by Yang *et al.*[92] in 2002. This study showed that the foetal mouse cerebral cortex cells were able to grow well when seeded on P(3HB-co-3HHx) films.[92] Similarly PHAs have also been extensively explored for hard tissue engineering. P(3HB) and composites of P(3HB) with bioactive nanobioglass (n-BG) 45S5 have been extensively studied for bone tissue engineering. The composite P(3HB)/n-BG showed good biocompatibility with the seeded MG-63 human osteoblast cell line. Incorporation of vitamin E in the composite further enabled better attachment, proliferation and differentiation of the osteoblast cells.[93–95] Composites of mcl-PHA, P(3HO)/n-BG two-dimensional (2D) scaffold or film have also been developed for the first time by Rai et al.[96] to produce a potential multifunctional wound dressing. The films would act as both a scaffold for skin regeneration and also accelerate wound haemostasis owing to the haemostatic nature of the incorporated bioactive n-BG, thus potentially reducing blood loss following tissue injury.[96] Studies carried out by Wang and his group have shown that P(3HB-co-3HHx) is a more superior biomaterial for osteoblast attachment, proliferation and differentiation for bone marrow cells when compared to poly(lactic acid) (PLA) and P(3HB)[97] (Figure 3.7). Wang's and other studies have shown that P(3HB-co-3HHx) with tailor-made HHx content can be designed to meet the growth requirements of specific cell and tissues for bone tissue engineering.[97–99]

Several investigations have been carried out on three-dimensional polymer scaffold systems consisting of a blend of P(3HB) and P(3HB-co-3HHx) for possible application as a matrix for cartilage tissue engineering.[72,100,101] Biochemical analysis and RT-PCR confirmed that the blend polymer of P(3HB)/P(3HB-co-3HHx) was capable of initiating a redifferentiation process, which

Figure 3.7 Scanning electron micrographs of rabbit bone marrow cells seeded on PHBHHx scaffold after 10 days of incubation (1000×). (a) Cell clumps; (b) round cells with fibrillar collagen (F,C) attached by filapodia; (c) cells with extracellular matrix (M) and calcified globuli (G).[97]

allowed chondrocytes to express and produce type II collagen more than the P(3HB) only scaffold (control). The P(3HB-co-3HHx) component in the blend P(3HB)/P(3HB-co-3HHX) scaffold provided better surface properties for anchoring type II collagen filaments and their penetration into internal layers of the scaffolds. These results suggested that the cells underwent chondrogenic differentiation on P(3HB-co-3HHx)-containing scaffolds and that the presence of the right proportion of P(3HB-co-3HHx) in the blend system of P(3HB)/P(3HB-co-3HHx) highly favoured the production of extracellular matrix of articular cartilage chondrocytes.[101]

3.9.9.2 Drug Delivery

PHAs have been studied as drug carrier scaffolds for controlled drug delivery. The use of P(3HB-co-4HB) rods as antibiotic carriers for the treatment of osteomyelitis using Sulperazone™ or Ducoid was evaluated and compared with P(3HB-co-3HV). P(3HB-co-4HB) was preferred as it was less rigid and easier to handle as opposed to P(3HB-co-3HV). Hasirci and Keskin also studied P(4HB) and P(3HB-co-4HB) as matrices for tetracycline release, because the physical properties like strength, modulus and elongation of the scaffolds were comparable to that of other drug delivery systems.[62,102] Transdermal drug delivery (TDS) is an ideal method for drug administration. However, in this method the hydrophobic stratum corneum represents a major barrier against hydrophilic ionizable drugs. Studies were therefore carried out by Wang et al.[103] to use mcl-PHAs for TDS. Tamsulosin was used as the model drug in this study and polyamidoamine dendrimer, which acted as an enhancer, was added to the polymer matrix. The dendrimer-containing PHA matrix achieved the clinically required amount of tamsulosin permeating through the skin model.[103] Graft copolymerized monoacrylate-poly(ethylene glycol), PEGMA and P(3HO), *i.e.* PEGMA-g-P(3HO), has been used to develop a swelling controlled release delivery system for Ibuprofen as a model drug.[104] In another study P(3HB-co-3HHx) was studied as a matrix for the controlled release of triamcinolone acetonide for possible treatment of cystoid macular oedema (CMO) and acute posterior segment inflammation associated with uveitis.[105]

3.10 PHAs as Green Biofuels

An interesting new area of development has been the research on PHAs as possible biofuels. Zhang et al.[106] carried out this study for the first time. Chemically PHAs are 3-hydroxyalkanoate esters which are similar to biofuels, particularly biodiesel which is methyl esters of long-chain fatty acids. In this study investigations and comparisons of the combustion heats of 3-hydroxybutyrate methyl ester, 3HBME, medium-chain length hydroxyalkanoate methyl ester, 3HAME, ethanol, n-propanol, n-butanol, 0 diesel, 90 gasoline, and 3HBME-based and 3HAME-based blended fuels were carried out. It was found that 3HBME and 3HAME had energy of combustion of 20 and 30 kJ g^{-1}, respectively. Ethanol has an energy of combustion of 27 kJ g^{-1}, while addition of 10% 3HBME or 3HAME enhanced the energy of combustion of ethanol to 30 and 35 kJ g^{-1}, respectively. Combustion heats of blended fuels 3HBME/diesel or 3HBME/gasoline and of 3HAME/diesel or 3HAME/gasoline were lower than that of the pure diesel or gasoline. It was roughly estimated that the production cost of PHA-based biofuels in China should be around (US$1200 per ton) which is higher than the cost of the gasoline products in the Chinese market (US$800 per ton).[106] Carbon dioxide emitted from burning fossil fuel is one of the main contributors to greenhouse gases, GHG. In fact in the United States carbon dioxide released from fossil fuel combustion represents 98% of the total GHG emission. Studies were carried out by Yu et al.[107] on the GHG emissions and fossil energy requirement per kg of bioplastics produced. Three PHA polymers from different carbon sources PHA-G (glucose), PHA-O (vegetable oil) and PHA-BS(black syrup) were studied. The analysis indicates that PHA bioplastics can reduce GHG emissions with only 0.49 kg CO_2-e being emitted from production of 1 kg of resin. Compared with 2–3kg CO_2-e of petrochemical counterparts, PHAs have the potential of about 80% reduction in the CO_2-e. Interestingly, other bio-based polyesters such as polylactide, PLA, showed high GHG emission, 1.8 kg CO_2-e. The total fossil energy requirement per kg of PHAs was between 40 and 59 MJ kg^{-1}, again lower than those of the petrochemical counterparts (78–88 MJ kg^{-1} resin). Figure 3.8 compares the GHG emissions and fossil energy requirements of representative petroleum and bio-based polymers.[107] Although the energy of combustion of the PHA biofuels was lower and the estimated cost of production higher, it has however opened a new avenue for PHA application. The significance of PHA-based biofuels is also enhanced by the fact that these are biodegradable and emit low GHG as oppose to petroleum-based products. Other biofuels are produced from biomass; however, PHAs could be produced from agricultural waste and sewage sludge, thus providing a new insight into the 'food versus fuel' controversy. Further work could be carried out to make PHA biofuels more efficient and cost effective.

3.11 Market and Economics of PHAs

The bioplastic market is growing at the rate of 8–10% per year and covers approximately 10–15% of the total plastic market. A graphical representation of the rising production capacity for bioplastics in the world is shown in Figure 3.9. Its market share is further projected to increase 25–30% by 2020. In 2007 the bioplastic market had reached 1 billion US$ and is expected to be over 10 billion by 2020.[108,109]

Thus, PHAs can be a key player in the bioplastic market; however, to achieve this, production of PHAs must be made more cost effective, so that its price is comparable or cheaper than petroleum-based plastics. The cost for PHA production has been reduced from US$16 per kilogram to US$2 (approx. €1.55 or £1.35) per kilogram.[110] Metabolix in partnership with Archer Daniels Midland in 2009 started to market PHAs under the trade name Mirel at US$ 1.02 per kilogram of Mirel pellets.[111] Therefore, though efforts have been made in reducing the cost of PHA production, it is still three times more expensive than polypropylene, as shown in

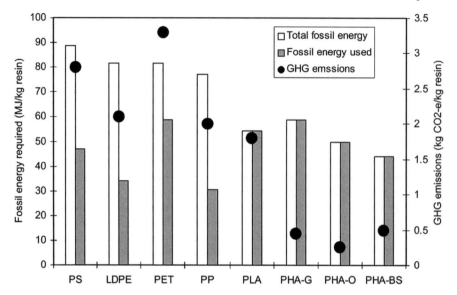

Figure 3.8 Comparison of GHG emissions and fossil energy requirements of representative petroleum and biobased polymers based on 1 kg of resin produced. Symbols: polystyrene (PS), low-density polyethylene (LDPE), polyethylene terephthalate (PET), polypropylene (PP), polylactide (PLA), polyhydroxyalkanoates based on glucose (PHA-G), PHA based on oil (PHA-O), and PHA based on black syrup (PHA-BS).[107]

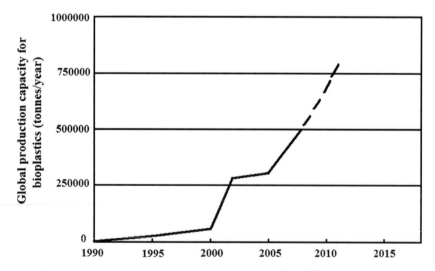

Figure 3.9 The global production capacity for bioplastics.[109]

Table 3.4. One area of application where the high cost of PHA can be tolerated in lieu of its amenable properties is in medical applications. However, if PHAs have to play a dominant role in the bioplastics market and act as a serious contender for

Table 3.4 Market prices for PHAs and petroleum-based polymers (adapted from Castilho et al.[67]).

Polymer market	Price	Reference
PHA (Mirel) Metabolix and ADM	US$ 1.02/kg	108
P(3HB) from Biomer (Germany)	€12/kg	112
P(3HB-co-3HV) from Metabolix (USA)	€10–12/kg	113
Modified starch polymers from Novamont (Italy)	€2.5–3.0/kg	
Polylactic acid from Cargill Dow (USA)	€2.2–3.4/kg	114
Polypropylene (PP)	€0.74/kg	
High-density polyethylene (HDPE)	€0.78/	
Low-density polyethylene (LDPE)	€0.74/kg	
Polyvinyl chloride (PVC)	€0.72/kg	
Polystyrene (PS)	€0.70/kg	

eco-friendly green alternatives to petroleum-derived plastics, then the cost of PHAs must be brought down at par with conventional plastics.

3.12 Concluding Remarks

The properties of biocompatibility, biodegradability and tailorability make PHAs a key player amongst the various bioplastics that are available in the market. However, the relatively high cost of PHA production as opposed to conventional petroleum-derived plastics is proving to be a real bottle neck. This problem is being addressed with different approaches such as using cheap agricultural and waste products as carbon feed, by constructing recombinant strains of organisms and plants that can accumulate higher amounts of PHAs and by developing efficient and cheap downstream processing. Not only do such approaches increase the efficiency of PHA production but it also helps in addressing the problems associated with solid waste management of agricultural and industrial waste. An interesting area of PHA application that has recently emerged is its possible use as biofuels. Thus, with new areas for PHA application emerging, many companies are now trading in PHAs. These companies produce wide ranging products catering to industrial, agricultural and medical sectors. Therefore, PHAs certainly are worthy alternatives for petroleum-derived plastics with a huge potential to be used in a large number of different applications in the near future.

Acknowledgement

The author R. Rai greatly acknowledges the financial support from the Quintin Hogg Foundation of the University of Westminster.

References

1. A. J. Anderson and E. A. Dawes, *Microbiol. Rev.*, 1990, **54**, 450.
2. E. Y. Lee, J. D. Jendrossek, A. Schirmer, C. Y. Choi and A. Steinbüchel, *Appl. Microbiol. Biotechnol.*, 1995, **42**, 901.

3. C. Y. Loo and K. Sudesh, *MPJ*, 2007, **2**, 31.
4. M. Lemoigne, *Bull. Soc. Chem. Biol.*, 1926, **8**, 770.
5. R. M. Macrae and J. R. Wilkinson, *J. Gen. Microbiol.*, 1958, **19**, 210.
6. P. J. Hocking and R. H. Marchessault, in *Chemistry and Technology of BioDegradable Polymers*, ed. G. J. L. Griffin, Blackie Academic & Professional, London, 1994, pp. 48–96.
7. L. L. Wallen and W. K. Rohwedder, *Environ. Sci. Technol.*, 1974, **8**, 576.
8. M. J. de Smet, G. Eggink, B. Witholt, J. Kingma and H. Wynberg, *J. Bacteriol.*, 1983, **154**, 870.
9. S. C. Slater, W. H. Voige and D. E. Dennis, *J. Bacteriol.*, 1988, **170**, 4431.
10. P. Schubert, A. Steinbuchel and H. G. Schlegel, *J. Bacteriol.*, 1988, **170**, 5837.
11. O. P. Peoples and A. J. Sinskey, *J. Biol. Chem.*, 1989, **264**, 15293.
12. S. Y. Lee, *Biotechnol. Bioeng.*, 1995, **49**, 1.
13. L. L. Madison and G. W. Huisman, *Microbiol. Mol. Biol.*, 1999, **63**, 21.
14. D. Byrom, *Trends. Biotechnol.*, 1987, **5**, 246.
15. J. Asrar, H. E. Valentin, P. A. Berger, M. Tran, S. R. Padgette and J. R. Garbow, *Biomacromolecules*, 2002, **3**, 1006.
16. D. Y. Kim, Y. B. Kim and Y. H. Rhee, *Macromolecules*, 1998, **31**, 4760.
17. Y. B. Kim, D. Y. Kim and Y. H. Rhee, *Macromolecules*, 1999, **32**, 6058.
18. G. A. Abraham, A. Gallardo, J. S. Roman, E. R. Olivera and R. Jodra, *Biomacromolecules*, 2001, **2**, 562.
19. A. Timm and A. Steinbüchei, *Appl. Environ. Microbiol.*, 1990, **56**, 3360.
20. R. G. Lageveen, G. W. Huisman, H. Preusting, P. Ketelaar, G. Eggink and B. Witholt, *Appl. Environ. Microbiol.*, 1988, **54**, 2924.
21. H. Matsusaki, S. Manjil, K. Taguchi, M. Kato, T. Fukui and Y. Doi, *J. Bacteriol.*, 1998, **180**, 6459.
22. G. Schmack, V. Goreriflo and A. Steinbüchel, *Macromolecules.*, 1998, **31**, 644.
23. K. Fritzsche, R. W. Lenz and R. C. Fuller, *Macromol. Chem.*, 1990, **191**, 1957.
24. M. Kato, T. Fukyi and Y. Doi, *Bull. Chem. Soc. Jpn.*, 1996, **69**, 515.
25. K. Jung, R. Hany, D. Rentsch, T. Storni, T. Egli and B. Witholt, *Macromolecules*, 2000, **33**, 8571.
26. S. N. Kim, S. C. Shum, W. J. Hammer and R. A. Newmark, *Macromolecules*, 1996, **29**, 4572.
27. Y. B. Kim, R. W. Lenz and R. C. Fuller, *Macromolecules*, 1992, **25**, 1852.
28. M. Zinn, B. Witholt and T. Egli, *Adv. Drug Delivery Rev.*, 2001, **53**, 5.
29. K. Sudesh, H. Abe and Y. Doi, *Prog. Polym. Sci.*, 2000, **25**, 1503.
30. Y. Doi, in *Microbial Polyesters*, VCH Publishers, New York, 1990, pp. 11–16.
31. V. Peters and B. H. A. Rehm, *FEMS Microbiol. Lett.*, 2005, **248**, 93.
32. B. H. A. Rehm, *Biochem. J.*, 2003, **376**, 15.
33. K. Taguchi, T. Tsuge, K. Matsumoto, S. Nakae, S. Taguchi and Y. Doi, *Riken Rev.*, 2001, **42**, 71.
34. P. J. Senior and E. A. Dawes, *Biochem. J.*, 1973, **134**, 225.
35. P. J. Senior and E. A. Dawes, *Biochem. J.*, 1971, **125**, 55.
36. S. Philip, T. Keshavarz and I. Roy, *J. Chem. Technol. Biotechnol.*, 2007, **82**, 233.

37. G. W. Huisman, O. D. Leeuw, G. Eggin and B. Witholt, *Appl. Environ. Microbiol.*, 1989, **55**, 1949.
38. D. Y. Kim, H. W. Kim, M. G. Chung and Y. H. Rhee, *J. Microbiol.*, 2007, **45**, 87.
39. R. S. Makkar and S. S. Cameotra, *J. Surfactants. Deterg.*, 1999, **2**, 237.
40. L. E. A. Munoz and M. R. Riley, *Biotechnol. Bioeng.*, 2008, **100**, 882.
41. M. Yilmaz and Y. Beyatli, *Zuckerindustrie*, 2005, **130**, 109.
42. F. Liu, W. Li, D. Ridgway, T. Gu and Z. Shen, *Biotechnol. Lett.*, 1998, **20**, 345.
43. Y. Jiang, X. Song, L. Gong, P. Li, C. Dai and W. Shao, *Enzyme Microb. Technol.*, 2008, **42**, 167.
44. J. Choi and S. Y. Lee, *Appl. Microbiol. Biotechnol.*, 1999, **51**, 13.
45. H. P. Yu, H. Chua, A. L. Huang and K. P. Ho, *Appl. Biochem. Microbiol.*, 1999, 77.
46. B. S. Kim, *Enzyme Microb. Technol.*, 2000, **27**, 774.
47. P. M. Halami, *World. J. Microbiol. Biotechnol.*, 2008, **24**, 805.
48. A. A. Koutinas, Y. Xu, R. Wang and C. Webb, *Enzyme Microb. Technol.*, 2007, **40**, 1035.
49. L. F. Silva, M. K. Taciro, M. E. M. Ramos, J. M. Carter, J. G. C. Pradella and J. G. C. Gomez, *J. Ind. Microbiol Biotechnol.*, 2004, **31**, 245.
50. M. Koller, R. Bona, E. Chiellini, E. G. Fernandes, P. Horvat, C. Kutschera, P. Hesse and G. Braunegg, *Bioresour. Technol.*, 2008, **99**, 4854.
51. A. K. Singh and N. Mallick, *J. Ind. Microbiol. Biotechnol.*, 2009, **36**, 347.
52. W. J. Page, J. Manchak and B. Rudy, *Appl. Environ. Microbiol.*, 1992, **58**, 2866.
53. S. G. V. A. O. Costa, F. Lépine, S. Milot, E. Déziel, M. Nitschke and J. Contiero, *J. Ind. Microbiol Biotechnol.*, 2009, **36**, 1063.
54. D. P. Martin and S. F. Williams, *Biochem. Eng. J.*, 2003, **16**, 97.
55. P. J. Barham and A. Keller, *J. Polym. Sci. Phys. Ed.*, 1986, **24**, 69.
56. R. H. Marchessault, C. J. Monasterios, F. G. Morin and P. R. Sundarajan, *Int. J. Biol. Macromol.*, 1990, **12**, 158.
57. S. P. Valappil, S. K. Misra, A. R. Boccaccini, T. Keshavarz, C. Bucke and I. Roy, *J. Biotechnol.*, 2007, **132**, 251.
58. R. Sánchez, J. Schripsema, L. F. da Silva, M. K. Taciro, G. C. Pradella and G. C. Gomez, *Eur. Polym. J.*, 2003, **39**, 1385.
59. A. Steinbüchel and T. L. Eversloh, *Biochem. Eng. J.*, 2003, **16**, 81.
60. P. A. Holmes, *Development in Crystalline Polymers*, Elsevier, London, 1988, 1–65.
61. R. Rai, T. Keshavarz, J. A. Roether, A. R. Boccaccini and I. Roy, *Mat. Sci. Eng. R.*, 2011: doi:10.1016/j.mser.2010.11.002.
62. S. P. Valappil, S. K. Misra, A. R. Boccaccini and I. Roy, *Expert Rev. Med. Devices*, 2006, **3**, 853.
63. Y. Doi, S. Kitamura and H. Abe, *Macromolecules*, 1995, **28**, 4822.
64. R. Hartmann, R. Hany, E. Pletscher, A. Ritter, B. Witholt and M. Zinn, *Biotechnol. Bioeng.*, 2006, **93**, 737.
65. H. Preusting, A. Nijenhuis and B. Witholt, *Macromolecules*, 1990, **23**, 4220.
66. K. D. Gagnon, R. W. Lenz and R. J. Farris, *Rubber. Chem. Tech.*, 1992, **65**, 761.

67. L. R. Castilho, D. A. Mitchell and D. M. G. Freire, *Bioresour. Technol.*, 2009, **100**, 5996.
68. S. F. Williams, D. P. Martin, D. M. Horowitz and O. P. Peoples, *Int. J. Biol. Macromol.*, 1999, **25**, 111.
69. T. Nelson, E. Kaufman, J. Kline and L. Sokoloff, *J. Neurochem.*, 1981, **37**, 1345.
70. R. N. Reusch, *Non-storage poly-(R)-3-hydroxyalkanoates (complexed PHAs) in prokaryotes and eukaryotes*, Wiley-VCH, Weinheim Germany, 2002.
71. X. S. Yang, K. Zhao and G. Q. Chen, *Biomaterials*, 2002, **23**, 1391.
72. K. Zhao, Y. Deng, J. C. Chen and G. Q. Chen, *Biomaterials*, 2003, 1041.
73. Z. Zheng, F. F. Bei, Y. Deng, H. L. Tian and G. Q. Chen, *Biomaterials*, 2005, **26**, 2537.
74. J. Sun, Z. Dai, Y. Zhao and G. Q. Chen, *Biomaterials*, 2007, **28**, 3896.
75. H. W. Kim, C. W. Chung, S. S. Kim, Y. B. Kim and Y. H. Rhee, *Int. J. Biol. Macromol.*, 2002, **30**, 129.
76. K. Zhao, X. Yang, G. Q. Chen and J. C. Chen, *J. Mater. Sci. Mater. Med.*, 2002, **13**, 849.
77. N. Mei, P. Zhou, L. F. Pan, G. Chen, C. G. Wu, X. Chen, Z. Z. Shao and G. Q. Chen, *J. Mater. Sci. – Mater. Med.*, 2006, **17**, 749.
78. M. Mochizuki and M. Hirami, *Polymer. Adv. Tech.*, 1997, **8**, 203.
79. Y. Tokiwa and B. P. Calabia, *Biotechnol. Lett.*, 2004, **26**, 1181.
80. Y. W. Wang, Q. Wu and G. Q. Chen, *Biomacromolecules*, 2005, **6**, 566.
81. M. Knoll, T. M. Hamm, F. Wagner, V. Martinez and J. Pleiss, *BMC Bioinf.*, 2009, **10**, 89.
82. A. M. Clarinval and J. Halleux, *Classification of Biodegradable Polymers, in Biodegradable Polymers for Industrial Applications*, CRC Florida, USA, 2005.
83. G. Q. Chen, G. Zhang, S. J. Park and S. Y. Lee, *Appl. Microbiol. Biotechnol.*, 2011, **57**, 50.
84. G. Q. Chen, *Polyhydroxyalkanoates, in Biodegradable Polymers for Industrial Applications*, CRC, Florida, USA, 2005.
85. J. Asrar and K. J. Gruys, in *Biopolymers*. ed. Y. Doi, A. Steinbüchel, Wiley-VCH, Weinheim 2002, pp. IIIa/53 ff.
86. I. Noda, P. R. Green, M. M. Salkowski and L. A. Schectman, *Biomacromolecules*, 2005, **6**, 580.
87. R. Sodian, S. P. Hoerstrup, J. S. Sperling, S. Daebritz, D. P. Martin, A. M. Moran, B. S. Kim, F. J. Schoen, J. P. Vacanti and J. E. Mayer, *Circulation*, 2000, **102**, 22.
88. G. Q. Chen and W. Qiong, *Biomaterials*, 2005, **26**, 6565.
89. U. A. Stock, M. Nagashima, P. N. Kahalil, G. D. Nollert, T. Herden, J. S. Sperling, A. M. Moran, B. Lien, D. P. Martin, F. J. Schoen, J. P. Vacanti and J. E. Mayer, *J. Thorac. Cardiovasc. Surg.*, 2000, **119**, 732.
90. R. Sodian, M. Loebe, A. Hein, D. P. Martin, S. P. Hoerstrup, E. V. Potapov, H. Hausmann, T. Leuth and R. Hetzer, *ASAIO J.*, 2002, **48**, 12.
91. S. Wu, Y. L. Liu, B. Cui, X. H. Qu and G. Q. Chen, *Artif. Organs*, 2007, **31**, 689.
92. F. Yang, X. Li, Li, G., N. Zhao and X. Zhang, *J. Biomed. Eng.*, 2002, **19**, 25.

93. S. K. Misra, S. N. Nazhat, S. P. Valappil, M. M. Torbati, R. J. K. Wood, I. Roy and A. R. Boccaccini, *Biomacromolecules*, 2007, **8**, 2112.
94. S. K. Misra, D. Mohn, T. J. Brunner, W. J. Stark, S. E. Philip, I. Roy, V. Salih, J. C. Knowles and A. R. Boccaccini, *Biomaterials*, 2008, **29**, 1750.
95. S. K. Misra, S. E. Philip, W. Chrzanowski, S. N. Nazhat, I. Roy, J. C. KNowles, V. Salih and A. R. Boccaccini, *J. R. Soc. Interface*, 2009, **6**, 401.
96. R. Rai, A. R. Boccaccini, J. C. Knowlesd, I. C. Lockee, M. P. Gordgee, A. M. Cormick, V. Salihd, T. Keshavarz and I. Roy. *AIP Conf. Proc.*, 2010, **125**, 126.
97. Y. W. Wang, Q. O. Wu and G. Q. Chen, *Biomaterials*, 2004, **25**, 669.
98. Y. W. Wang, W. Qiong, J. Chen and G. Q. Chen, *Biomaterials*, 2005, **26**, 899.
99. Y. W. Wang, F. yang, Q. Wu, Y. C. Cheng, P. H. F. Yu, J. Chen and G. Q. Chen, *Biomaterials*, 2005, **26**, 755.
100. Y. Deng, K. Zhao, X. F. Zhang, P. Hu and G. Q. Chen, *Biomaterials*, 2002, **23**, 4049.
101. Y. Deng, X. J. Lin, Z. Zheng, J. G. Deng, J. C. Chen, H. Ma and G. Q. Chen, *Biomaterials*, 2003, **24**, 4273.
102. F. Türesin, I. Gürsel and V. Hasirci, *J. Biomater. Sci. Polymer. Ed.*, 2001, **12**, 195.
103. Z. Wang, Y. Itoh, Y. Hosaka, I. Kobayashi, Y. Nakano, I. Maeda, F. Umeda, J. Yamakawa, M. Kawase and K. Yagi, *J. Biosci. Bioeng.*, 2003, **95**, 541.
104. H. W. Kim, C. W. Chung, S. J. Hwang and Y. H. Rhee, *Int. J. Biol. Macromol.*, 2005, **36**, 84.
105. C. Bayram and E. B. Denbas, *J. Bioact. Compat. Polym.*, 2008, **23**, 334.
106. X. Zhang, R. Luo, Z. Wang, Y. Deng and G. Q. Chen, *Biomacromolecules*, 2009, **10**, 707.
107. J. Yu and L. X. L. Chen, *Environ. Sci. Technol.*, 2008, **42**, 6961.
108. H. Kaiser, *Bioplastic Market Worldwide Report 2007-2025*, 2008; Available from: http://www.scribd.com/doc/4899724/The-global-market-for-bio-plastic-.[cited Access: 2008/9/03].
109. *Bioplastic Market an Overview*, 2010. Available from: http://www.bioplastics24.com/content/view/89/28/lang,en/. [cited Access: 2010/ 18/03].
110. B. P. Mooney, *Biochem. J.*, 2009, **418**, 219.
111. B. E. DiGregorio, *Chem. Biol.*, 2009, **16**, 1.
112. V. J. Hänggi, 2004. Economics and environmental advantages of biodegradable polymers for food packages. In: *Proceedings of Packtech Seminar*, Gent, Belgium.
113. M. Crank, M. Patel, F. Marscheider-Weidemann, J. Schleich, B. Hüsing and G. Angerer. *Techno-economic Feasibility of Large scale Production of Bio-based Polymers in Europe (PRO-BIP)*. European Commission's Institute for Prospective Technological Studies (IPTS) report., 2004.
114. CMAI. Global, Global Plastics and Polymers Market Report. European Update., 2009, issue no. 002. http://www.cmaiglobal.com/MarketReports/samples/GPPREU002.pdf. [accessed 2009 10/03].

CHAPTER 4
Fully Green Bionanocomposites

P. M. VISAKH,[1] SABU THOMAS[1] AND
LALY A. POTHAN[2]

[1] Centre for Nanoscience and Nanotechnology, Mahatma Gandhi University, P. D. Hills P.O., Kottayam 686560, Kerala, India; [2] Department of Chemistry, Bishop Moore College, Mavelikare, Kerala, India

4.1 Green Composites – Introduction

Green composites are a new class of fully biodegradable composites prepared by combining (natural/bio) fibres with biodegradable resins. The major attractions of green composites are that they are environmentally friendly, fully degradable and sustainable, that is, they are truly 'green'. After use they can be easily disposed off or composted without harming the environment.

Green composites may be used effectively in many applications such as in mass-produced consumer products with short lifecycles or products intended for one-time or short-term use before disposal. Green composites may also be used for indoor applications with a useful life of several years. In addition, they are used for drug/gene delivery, tissue engineering applications and cosmetic orthodontics.

Fibres can be combined with traditional resins or newer plant based resins. The result is a plant-based alternative for many traditional steel and fibreglass products. Green composites have specific advantages over traditional composites like reduced weight, increased flexibility and greater mouldability. They are also less expensive. These composites have sound insulation capability and are sustainable. In this chapter, we discuss the preparation, structure and

---

RSC Green Chemistry No. 12
A Handbook of Applied Biopolymer Technology: Synthesis, Degradation and Applications
Edited by Sanjay K. Sharma and Ackmez Mudhoo
© Royal Society of Chemistry 2011
Published by the Royal Society of Chemistry, www.rsc.org

properties of biofibres and the properties of their composites produced in combination with various polymers.

Composites of bio-nano materials are a relatively new field in nanotechnology. These materials have attracted significant attention during the last decade. The major challenges of bio nanocomposite research are the efficient separation of nano-reinforcements from natural resources, compatibilization of the nano-reinforcements within the matrix polymer and development of suitable methods for processing these novel biomaterials. The energy consumption factor and the cost factor involved in the process are also challenging factors. One interesting way to improve the property of biopolymers is to introduce a reinforcing agent in nanoscale to the polymer. Though this technique has been known for some time it is still in a development phase. Nanoparticles have the advantage of relatively high surface to volume ratio compared to their macroscopic counterparts. Nanofillers that are commercially available today are, for example, different nanoclay minerals, carbon nanotubes and other fullerenes. The drawback with the mentioned nanofillers is that they are not renewable where as bio-nano fibres are renewable. They are also biodegradable where as traditional nanofillers are not biodegradable.

4.2 Green Materials: Fibres, Whiskers, Crystals and Particles

4.2.1 Cellulose Fibres

Cellulose was discovered in 1838 by the French Chemist Anselme Payen, who isolated it from plant matter and determined its chemical formula.[1,2] Cellulose was used to produce the first successful thermoplastic polymer, celluloid, by Hyatt Manufacturing Company in 1870. Hermann Staudinger determined the polymer structure of cellulose in 1920. Cellulose is the most abundant natural biopolymer and is readily available from renewable resources. Wood, cotton and hemp rope are all made of fibrous cellulose. Because cellulose is built out of a sugar monomer, it is called a polysaccharide. Cellulose is an organic compound with the formula $(C_6H_{10}O_5)_n$, a polysaccharide consisting of a linear chain of several hundred to over ten thousand $\beta(1\rightarrow 4)$ linked D-glucose units.[1,4] Some species of bacteria secrete it to form biofilms. About 33% of all plant matter is cellulose (the cellulose content of cotton is 90% and that of wood is 50%.[5] In addition to cellulose, plant fibres contain different natural substances, mainly hemicelluloses, lignin, pectins and waxes.[6–8] Cellulose is considered as a nearly inexhaustible raw material with fascinating structure and properties for the remarkable demand for environmentally friendly and biocompatible products. As is well known, cellulose is insoluble in many solvents, which leads to a limitation in its reactivity and processability for utilization. Preparation of microcrystalline cellulose from materials other than wood and cotton such as water hyacinth, coconut shells, sugar cane bagasse, ramie, wheat and rice straws, jute, flax fibres and flax straw and soybean husk is receiving a lot of interest. Natural cellulose fibres are gaining attention as a reinforcing phase in thermoplastic matrices.[9–11] Its low density, highly reduced wear of the processing machinery, and a relatively reactive surface may be mentioned as attractive properties, together with their abundance and low price. Moreover, the recycling by combustion of cellulose-filled composites is easier in comparison with inorganic filler systems. Nevertheless, such fibres are used only to a limited extent in industrial practice, which may be explained by difficulties in achieving acceptable dispersion levels.

From the molecular structure of cellulose given in Figure 4.1, it can be seen that cellulose is a homopolymer of D-anhydroglucopyranose monomeric units connected through β (1-4) glycosidic linkages. In general, cellulose can be seen as a long chain polymer with D-glucose, a sugar, as its repeating units. Since the glucose units are six-membered rings within a cellulose chain, they are known as pyranoses. These pyranose rings are joined by single oxygen atoms, acetal linkages, between the C-1 of one pyranose ring and the C-4 of the next ring. The glucose units in cellulose polymer are referred to as anhydroglucose units. Often, in nature, cellulose is associated and mixed with other substances such as lignin, pectins, hemicelluloses, fats and proteins. Cellulose that is produced by plants is referred to as native cellulose, which is found in two crystalline forms, cellulose I and cellulose II.[12] Cellulose II, generally occurring in marine algae, is a crystalline form that is formed when cellulose I is treated with aqueous sodium hydroxide.[13–15] Among the four different crystalline polymorphs, cellulose I, II, III and IV, cellulose I is thermodynamically less stable while cellulose II is the most stable structure.

Liquid ammonia treatment of cellulose I and cellulose II gives crystalline cellulose III form[16–18] and the heating of cellulose III generates the cellulose IV crystalline form.[18] Recently a non-crystalline form known as nematic ordered cellulose has been described.[19] The hydroxyl groups on the cellulose chain that are in the equatorial position protrude laterally along the extended molecule. Within the crystalline regions, the extensive and strong inter-chain hydrogen bonds give the resultant fibres good strength and insolubility in most solvents. Cellulose has no taste, is odourless, hydrophilic, chiral and biodegradable. It is

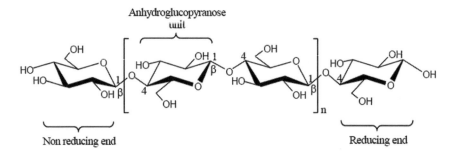

Figure 4.1 Molecular structure of cellulose.

insoluble in water and in most organic solvents. It can be broken down chemically into its glucose units by treating it with concentrated acids at high temperature. The multiple hydroxyl groups on the glucose residues from one chain form hydrogen bonds with oxygen molecules on the same or on a neighbouring chain, holding the chains firmly together side-by-side and forming microfibrils with high tensile strength. This strength is important in cell walls, where the microfibrils are meshed into a carbohydrate matrix, conferring rigidity to plant cells. It has interesting sound insulation properties and low thermal conductivity.

4.2.2 Chitin Whiskers

Chitin is a polymer that can be found in anything from the shells of beetles to webs of spiders. It is the second most abundant natural polymer after cellulose.[20] It is present all around us, in plants and in animal creatures. Chitin and cellulose are molecularly similar polysaccharide compounds. Cellulose contains a hydroxyl group and chitin contains an acetamide group. Crabs, beetles, worms and mushrooms contain large amount of chitin. Chitin is a very stable material, and it helps to protect an insect against harm and pressure. Depending on its thickness, chitin can be rigid or yielding. Often, insect coats contain thick, stiff layers of chitin. Shellfish-containing industry waste (shrimp or crab shells) in which chitin content range between 8% and 33% constitutes the main source of this polymer. Chitin is a high molecular weight linear polysaccharide, specifically β-1,4-N-acetyl-D-glucosamine. Chitin is natural, non-toxic, non-allergenic, anti-microbial and biodegradable. It is insoluble in water and resistant to acid, alkalis and many organic solvents.[21] Chitin has found applications in many areas other than food such as in biosensors.[22] The main uses of chitin film and fibre are in medical and pharmaceutical applications as wound-dressing material[23,24] and controlled drug release.[25,26] Chitin has versatile biological activity, excellent biocompatibility and complete biodegradability in combination with low toxicity. α-chitin is more abundant than β-chitin and χ-chitin.[27] Chitin has been known to form microfibrillar arrangements embedded in a protein matrix and these microfibrils have diameters ranging from 2.5 to 2.8 nm.[28] Crustacean cuticles possess chitin microfibrils with diameters as large as 25 nm.[29,30] The chitin-protein fibres are arranged in horizontal planes forming a typical twisted plywood structure or Bouligand pattern. Chitin can easily be isolated from crab shell, where it is found to be highly thixotropic and liquid crystalline.[31,32] Dufresne and co-workers have successfully isolated the crystalline regions of chitin whiskers from the crab shells and squid pens by hydrochloric acid hydrolysis.[33,34] It has been reported that the reinforcing effect strongly depends on the aspect ratio of the chitin whiskers.[35–37]

The molecular structure of chitin is shown in Figure 4.2. The crystallography of chitin has been investigated for a long time.[38–41] Marguerite[27] has reported in his reviews on chitin and chitosan; at first glance the powder X-ray diagrams of chitins from shrimp shell (α-chitin) and anhydrous squid pen (β-chitin) appeared nearly the same. Further information on the crystalline structure of

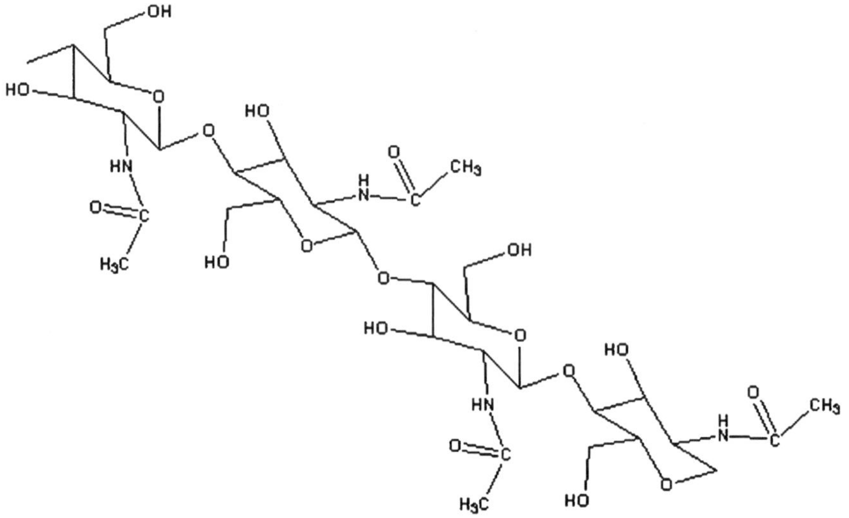

Figure 4.2 Molecular structure of chitin.

α- and β-chitin is obtained by analysis of electron diffraction patterns of highly crystalline samples. The crystallographic parameters of α- and β-chitin reveal that there are two antiparallel molecules per unit cell in α-chitin, whereas only one is present in β-chitin, which consists therefore of a parallel arrangement. Despite this difference, it appears that the N-acetyl glycosyl moiety is the independent crystallographic unit in both allomorphs. The observation of diffraction patterns of various α-chitin samples indicates some discrepancy in their diffraction patterns. In particular, the X-ray pattern of lobster tendon chitin presents a marked 001 diffraction spot[41] which is absent in the more crystalline sagitta chitin.[42-44] Therefore, it appears that more work is required to resolve these ambiguities about the crystal structure of α-chitin. In contrast, the structure of anhydrous β-chitin appears to be well established. However, the crystal structure of the β-chitin hydrate remains to be refined, as some uncertainty exists, even as to its unit cell parameters.[45,46]

Chitin is an example of highly basic polysaccharide. Their unique properties include polyoxy salt formation, optical structural characteristics, ability to form films, chelate metal ions and optical structural characteristics. Neither random nor block orientation is meant to be implied for chitin. Like cellulose, it naturally functions as a structural polysaccharide, but differs from cellulose in its properties. Chitin is highly hydrophobic and is insoluble in water and most organic solvents. It is soluble in hexafluroisopropanol, hexafluoroacetone and chloro alcohols in conjugation with aqueous solutions of mineral acids and dimethylacetamide containing 5% lithium chloride. Chitin is a highly crystalline, intractable material and only a limited number of solvents are known which are applicable as reaction solvents. Chitin degrades before melting, which is typical for polysaccharides with extensive hydrogen bonding.

4.2.3 Starch Crystals

Starch is second to cellulose in terms of its availability in nature. Starch serves as a food reserve for plants and serves as a medium for utilizing the Sun's energy for non-photosynthesizing organisms like man. The most important industrial sources of starch are corn, wheat, potato, tapioca and rice. Today, starch is inexpensive and is available annually from such crops, in excess of current market needs in the United States and Europe. The low price and the availability of starch associated with its very favourable environmental profile in the last 15 years aroused a renewed interest in starch-based polymers as an attractive alternative to polymers based on petrochemicals. Starch is totally biodegradable in a wide variety of environments and permits the development of totally degradable products for specific market demands. Degradation or incineration of starch products recycles atmospheric carbon dioxide trapped by starch-producing plants and does not increase potential global warming.

Starch consists of two major components: amylose, a mostly linear ⟨-D(1,4)-glucan and amylopectin, an ⟨-D-(1,4) glucan which has ⟨-D(1,6) linkages at the branch point. The linear amylose molecules of starch have a molecular weight of 0.2–2 million, while the branched amylopectin molecules have molecular weights as high as 100–400 million.

Starch is a well-known polymer naturally produced by plants in the form of granules (mainly from potatoes, corn and rice). This is because the short branched amylopectin chains are able to form helical structures which crystallize. Starch granules vary from plant to plant, but in general are composed of a linear polymer, amylose (in most cases about 20% of the granule), and a branched polymer amylopectin.[47,48] Amylose is a semi-crystalline biopolymer and is soluble in hot water, while amylopectin is insoluble in hot water during its biodegradation. The production of starch polymers begins with the extraction of starch. Taking as an example corn: starch is extracted from the kernel by wet milling. The kernel is first softened by steeping it in a dilute acid solution, then ground coarsely to split the kernel and remove the oil-containing germ. Finer milling separates the fibre from the endosperm which is then centrifuged to separate the less dense protein from the more dense starch. The starch slurry is then washed in a centrifuge, dewatered and dried. Either prior or subsequent to the drying step, the starch may be processed in a number of ways to improve its properties. Starch undergoes enzyme-catalysed acetal hydrolysis, the α-1,4 link in amylopectin is attacked by glucosidases.[47] Starch or amylum is a polysaccharide carbohydrate consisting of a large number of glucose units joined together by glycosidic bonds. Depending on the plant, starch generally contains 20 to 25% amylose and 75 to 80% amylopectin.[49] Glycogen, the glucose store of animals, is a more branched version of amylopectin. Starch can be used as a thickening, stiffening or gluing agent when dissolved in warm water, giving wheat paste. In nature, starch is found as crystalline beads of about 15–100 μm in diameter, in three crystalline modifications designated A (cereal), B (tuber) and C (smooth pea and various beans), all characterized by double helices: almost perfect left-handed, six-fold structures, as elucidated by X-ray.

Starch molecules arrange themselves in plants in semi-crystalline granules. Each plant species has a unique starch granular size: rice starch is relatively small (about 2 µm) while potato starches have larger granules (up to 100 µm). Although in absolute mass only about one quarter of the starch granules in plants consist of amylose, there are about 150 times more amylose molecules than amylopectin molecules. Amylose is a much smaller molecule than amylopectin. A typical feature of starch is that it becomes soluble in water when heated. The granules swell and burst, the semi-crystalline structure is lost and the smaller amylose molecules start leaching out of the granules. This process is called starch gelatinization. During cooking, starch becomes a paste and becomes viscous. During cooling or prolonged storage of the paste, the semi-crystalline structure partially recovers and starch paste thickens. This is mainly caused by the retrogradation of the amylose. This process is also responsible for staling, hardening of bread and water layer on top of a starch gel. Some cultivated plant varieties have pure amylopectin starch without amylose, known as waxy starch. Waxy starch has less retrogradation; the viscosity of the paste will be more stable. Also high amylose starch, amylomaize, is cultivated for the use of its gel strength. The amorphous and partially crystalline rings start from the centre, called the hilum, and follow each other alternatively. The amorphous rings are amylose, while the partially crystalline ones are amylopectin. The macromolecules are oriented in the radial direction. The amorphous region of partially crystalline rings are formed by those parts of the amylopectin macromolecule where the chain branches. The crystalline part consists of the oriented, double helix molecular chain parts of amylopectin. The structure of amylose and amylopectin are given in Figures 4.3 and 4.4, respectively.

Starch has received considerable attention during the past two decades as a biodegradable thermoplastic polymer and as biodegradable particulate filler. Indeed, products from agricultural sources such as starch, offers an attractive and cheap alternative in developing degradable materials. Starch is not truly thermoplastic as most synthetic polymers. However, it can be melted and made to flow at high temperatures under pressure. If the mechanical shears become too high, then starch will degrade to form products with low molecular weight. Addition of water or other plasticizers enables the starch to flow under milder conditions and reduces the degradation considerably. However, the thermochemical stability is strongly due to the addition of plasticizers.[50] By itself, starch is a poor choice as a replacement for any plastic. It is mostly water insoluble, difficult to process and brittle. In principle some of the properties of starch can be significantly improved by blending it with synthetic polymers. Physical incorporation of granular starch or starch derivatives as a functional additive and filler into synthetic polymers during processing has been largely used, since the first announcement of using starch in combination with synthetic polymer either as starch gel blends with ethylene acrylic acid copolymer by Westhoff *et al.*[51] or as a particulate starch dispersion in polyolefin by Griffin and Priority.[52]

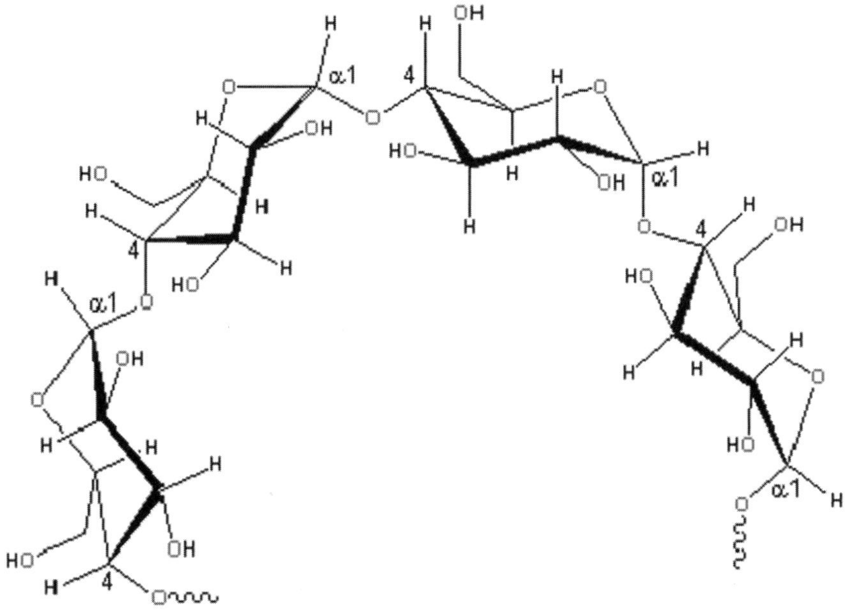

Figure 4.3 Structure of amylose.

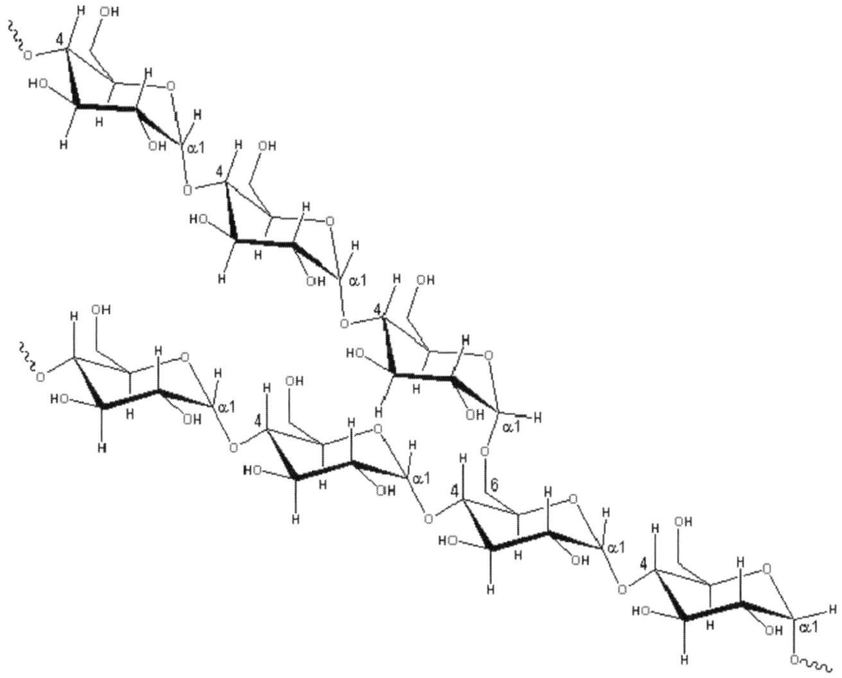

Figure 4.4 Structure of amylopectin.

4.2.4 Soy Protein Particles

Soy protein isolate (SPI), the major component of soybeans, is readily available from renewable resources and agricultural processing by-products. Soy proteins are of two types: soy protein concentrates (SPC) (70% protein minimum dry weight basis) and soy protein isolates (SPI) (90% protein minimum dry weight basis). They are increasingly used as ingredients in various types of prepared foods, meat analogues, dairy and bakery products. Their high nutritional value, essential amino acid content and excellent functional properties makes them a very valuable ingredient. Also, health benefits of soy proteins have recently been recognized by the US Food and Drug Administration, approving a health claim in 1999 indicating that foods containing 6.25 g of soy proteins per serving may reduce the risk of coronary heart disease.[53] Traditionally SPC and SPI are obtained from defatted soy flakes or flour. SPC is obtained by removing soluble sugar and minor constituents from the defatted flakes using aqueous alcohol or a dilute acid solution in the pH range of 4.0–4.8. SPI are produced by extracting the soy flakes/flour using a dilute alkali (pH 8–9) with subsequent centrifugation for the removal of the insoluble materials producing a soy protein extract (SPE) containing soluble protein, oligosaccharides and minerals. Acidification of the SPE to pH 4.5 using a food grade acid (sulfuric acid, phosphoric acid or hydrochloric acid) causes the selective recovery of the proteins due to their precipitation and concentration into a curd. Subsequent washing of the curd for the removal of non-protein solubles, neutralization (pH 7) and spray drying produces SPI.[2] SPC yields for the conventional processes have been reported to vary between 60 and 70% based on the protein in the flakes/flour.[54] SPI yields are between 30 and 40% based on the defatted soy flakes/flour weight or about 60% of the protein in the flakes.[54] Approximately one-third is insoluble residue after the extraction and the remaining one-third are soy whey proteins that are normally lost or are very difficult to recover because of low solids concentration.[55] SPC containing various oil contents can also be produced using full fat soy flakes.[56] The utilization of SPI in the preparation of biodegradable materials, such as adhesives, plastics and various binders, has received more attention in recent years. Plastics from SPI have very high strength and good biodegradability; however, they are also brittle and water sensitive, which limits their applications. Accordingly, the properties of the SPI have commonly been modified by physical, chemical or enzymatic treatments. Such treatments mainly promote cross-linking within SPI or modify the side chains of SPI, for example, acetylation and esterification, denaturation, incorporating fillers and blending with other polymers. Soy protein is generally regarded as the storage protein held in discrete particles called protein bodies, which are estimated to contain at least 60–70% of the total soybean protein. Upon germination of the soybean, the protein will be digested, and the released amino acids will be transported to locations of seedling growth. Legume proteins, such as soy and pulses, belong to the globulin family of seed storage proteins called leguminins (11S) and vicilins (7S), or in the case of soybeans, glycinin and beta-conglycinin. Grains contain a third type of storage protein called gluten or 'prolamines'. Soybeans also contain biologically active or metabolic proteins such as enzymes, trypsin inhibitors, haemagglutinins and cysteine proteases very similar to papain. The soy cotyledon storage proteins, important for human nutrition, can be extracted most efficiently by water, water plus dilute alkali (pH 7–9) or aqueous solutions of sodium chloride (0.5–2 M) from dehulled and defatted soybeans that have undergone only a minimal heat treatment so that the protein is close to being native or undenatured. Soybeans are processed into three kinds of modern protein-rich products: soy flour, soy concentrate and soy isolate. Soy protein has been available since 1936 for its functional properties. In 1936, American organic chemist Percy Lavon Julian designed the world's first plant for the isolation of industrial-grade soy protein. The largest use of industrial grade protein was, and still is, for paper coatings, in which it serves as a pigment binder. However, Dr. Julian's plant must have also been the source of the 'soy protein isolate' which Ford's Robert Boyer and Frank Calvert spun into an artificial silk that was then tailored into that now famous 'silk is soy' suit that Henry Ford wore on special occasions. The plant's eventual daily output of 40 tons of soy protein isolate made the Soya Products Division into Glidden's most profitable division. At the start of World War 2, Glidden sent a sample of Julian's isolated soy protein to National Foam System Inc. (today a unit of Kidde Fire Fighting) of Philadelphia, PA, which used it to develop Aero-Foam, the US Navy's beloved fire-fighting 'bean soup'; while not exactly the brainchild of Percy Lavon Julian, it was the meticulous care given to the preparation of the soy protein that made the fire fighting foam possible. When a hydrolysate of isolated soy protein was fed into a water stream, the mixture was converted into a foam by means of an aerating nozzle. The soy protein foam was used to smother oil and gasoline fires aboard ships, and was particularly useful on aircraft carriers. It saved the lives of thousands of sailors. In 1958, Central Soya of Fort Wayne, Indiana, acquired Julian's Soy Products Division (Chemurgy) of the Glidden Paint Company, Chicago. Recently, Central Soya's (Bunge) Protein Division, in January 2003, joined/merged with DuPont's soy protein business (Solae), which in 1997 had acquired Ralston Purina's soy division, Protein Technologies International (PTI) in St. Louis. Eighth Continent, an 'ersatz' soy milk, is a combined 'venture' product of DuPont and General Mills. Food-grade soy protein isolate first became available on October 2, 1959 with the dedication of Central Soya's edible soy isolate, Promine D, production facility on the Glidden Company industrial site in Chicago. An edible soy isolate and edible spun soy fibre have also been available since 1960 from the Ralston Purina Company in St. Louis, who had hired Boyer and Calvert. In 1987, PTI became the world's leading maker of isolated soy protein.

4.2.5 Polylactic Acid

Figure 4.5 shows that the chemical structure of poly(lactic acid) (PLA). Poly(lactic acid) or polylactide (PLA) is a biodegradable, thermoplastic,

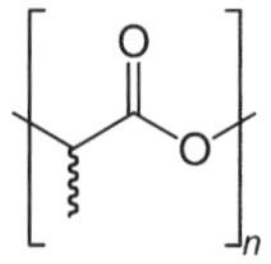

Figure 4.5 Chemical structure of poly (lactic acid).

aliphatic polyester derived from renewable resources, such as corn starch (in the USA) or sugarcanes (the rest of the world). Although PLA has been known for more than a century, it has only been of commercial interest in recent years, in light of its biodegradability. PLA is a sustainable alternative to petrochemical-derived products, since the lactides from which it is ultimately produced can be derived from the fermentation of agricultural by-products such as corn starch or other carbohydrate-rich substances like maize, sugar or wheat.

Poly(lactic acid) is a versatile polymer made from renewable agricultural raw materials, which are fermented into lactic acid.[1,57] Due to its good biocompatibility, biodegradability, mechanical properties and light weight, PLA has been widely used in many aspects, such as medical applications[2,3,58,59] and automotive parts.[4,60] The commercial market for PLA has increased substantially in recent years. PLA offers advantages of relatively high strength and ability to be processed in most equipment, but reinforcement is usually needed for practical applications due to its brittleness.[5,6,62] One way to improve the mechanical and thermal properties of PLA is the addition of fibres or filler materials.[5-7,61-63] Traditional fibres (*e.g.* glass fibre and recycled newspaper fibre) and natural fibres (*e.g.* bamboo and silk fibres) have been used as reinforcements to enhance the mechanical properties of PLA.[8-10,64-66] PLA is currently used in a number of biomedical applications, such as sutures, stents, dialysis media and drug delivery devices. It is also being evaluated as a material for tissue engineering. Because it is biodegradable, it can also be employed in the preparation of bioplastic, useful for producing loose-fill packaging, compost bags, food packaging and disposable tableware. In the form of fibres and non-woven textiles, PLA also has many potential uses, for example as upholstery, disposable garments, awnings, feminine hygiene products and nappies (diapers). PLA has been used as the hydrophobic block of amphiphilic synthetic block copolymers used to form the vesicle membrane of polymersomes.

The basic monomer of PLA is lactic acid, which is derived from starch by fermentation. Lactic acid is then polymerized to poly(lactic acid), either by gradual polycondensation or by ring-opening polymerization.[67-69] PLA is and has been frequently used for biodegradable packing materials.[70,71] However, numerous tests have shown that PLA is also suitable as a matrix for the embedding of fibres in composites. Some products of natural fibre-reinforced PLA are already established at the market: Jakob Winter (Satzung, Germany) produces biodegradable urns from flax and PLA by compression moulding.[72] NEC Corporation and UNITIKA LTD have announced joint development of bioplastic composites for mobile phone shells consisting of PLA and 15–20%

kenaf fibres.[73,74] An example for the use of natural fibre-reinforced bioplastics in the automobile industry is the Toyota RAUM, which is equipped with a spare tire cover made of kenaf fibre-reinforced PLA.[75] Beside plant fibres, artificial fibres based on renewable raw materials – *e.g.* viscose, rayon (Cordenka) or Lyocell – can be used to reinforce PLA.

PLA is more expensive than many petroleum-derived commodity plastics, but its price has been falling as production increases. The demand for corn is growing, both due to the use of corn for bioethanol and for corn-dependent commodities, including PLA. PLA has been developed in the United Kingdom to serve as sandwich packaging. PLA has also been used in France to serve as the binder in Isonat Nat'isol, a hemp fibre building insulation. PLA is used for biodegradable and compostable disposable cups for cold beverages, the lining in cups for hot beverages, deli containers and clamshells for food packaging. PLA has become a significant commercial bioplastic. Its clarity makes it useful for recyclable and biodegradable packaging, such as bottles, yoghurt pots and candy wrappers. It has also been used for food service ware, lawn and food waste bags, coatings for paper and cardboard, and fibres for clothing, carpets, sheets and towels, and wall coverings. In biomedical applications, it is used for sutures, prosthetic materials and materials for drug delivery.

There are several promising markets for biodegradable polymers such as polylactide. Plastic bags for household bio waste, barriers for sanitary products and nappies (diapers), planting cups, disposable cups and plates are some typical applications. To date no commercial large-scale production of polylactide exists, but this is likely to change in the near future.

The starting material, lactic acid, will also need new capacity. Commercial markets for biodegradable polymers are expected to increase substantially in the coming years.

Due to high production costs, in the early stages of its development PLA was used in limited areas, such as preparation of medical devices (bone surgery, suture and chemotherapy, *etc.*). Since its production cost has been lowered by new technologies and large-scale production, the application of PLA has been extended to other commodity areas such as packaging, textiles and composite materials.[76,77] However, brittleness and other properties such as low thermal stability, medium gas barrier properties and low solvent resistance (*e.g.* against water) of the pure polymer are often insufficient for food packaging applications.[78]

PLA-based composites are attractive for the manufacture of electrical/electronic devices or automotive parts requiring high level of mechanical, thermal and flame retardant properties. Unfortunately, because of their flammability and dripping combustion, PLAeAII composites cannot be recommended for applications where advanced flame retardant properties are required. As for traditional engineering polymers, different approaches can be investigated to tailor the flame retardant nature of PLA products. Among them has to be mentioned the addition of selected nanofillers such as organo-modified clays,[79–81] melamine derivatives,[82] metal hydroxides[83] and phosphorous derivatives such as ammonium polyphosphate[84] or phosphinates.[85] Unfortunately, most of the products that can provide PLA with flame retardancy sometimes trigger problems such as loss of mechanical and thermal properties or degradation of the polyester matrix, aspects that need to be considered when targeting a potential application.

Polylactide has still not found any meaningful market acceptance as an engineering resin, because of its non-satisfying impact resistance and low heat distortion temperature. Therefore, manufacturing of PLA lightweight parts with a high impact resistance would lead to new application fields, *e.g.* automotive or electrical industry. PLA already has found applications in textile and medical fields and also in the packaging industry. Besides, PLA is in principle compostable under certain conditions, that is, in the presence of the right triggers the material can be degraded into harmless natural compounds. Nonetheless, the inherent brittleness of PLA has been the main obstacle to expand its commercial use. Many approaches like plasticization, block copolymerization, blending with tougher polymers or rubber have been proposed but all led to significant decreases in modulus and strength of the toughened PLA.[86]

4.2.6 Natural Rubber Uncross-linked Particles

Natural rubber (NR) is a bio-based polymer and renewable resource, extracted from the tree *Hevea Brasiliensis*. The global production of natural rubber in 2003 was 8 million tonnes (Source: United Nations) with the major producers being Thailand (2.8 million tonnes), Indonesia (1.8 million tonnes), Malaysia (0.9 million tonnes) and India (0.6 million tonnes). Kerala accounts for 90% of India's rubber production. NR has excellent extrudability and calanderability and has a high rate of cure. Apart from its application in tire industry, NR is used in the production of thin walled soft products with high strength.

4.3 Green Nanocomposites

4.3.1 Cellulose-based Green Composites

Cellulose nanowhiskers have been used as reinforcement in many polymer matrixes, such as plastics and rubbers using different processing methods. Most of the work has been done to understand the influence of processing conditions and the effect of whisker content on the morphology and properties. Kulpinski *et al.*[87] reported on the cellulose nanofibres obtained by the electrospinning process from spinning drops containing cellulose dissolved in an *N*-methylmorpholine-*N*-oxide/water system. Under different electrospinning process conditions, a non-woven fibre network and a cellulose membrane were obtained. Torres *et al.*[88] prepared a nanocomposite material formed by bacterial cellulose (BC) networks and calcium-deficient hydroxyapatite (HAp) powders. Hajji *et al.*[89] prepared nanocomposites of cellulose whiskers using styrene copolymers latex as a matrix. Three different processing methods were used for the preparation of composites: (1) water evaporation method; (2) freeze drying and hot pressing; and (3) freeze drying, extruding (cylindrical extrudates were randomly dispersed in the mould) and then hot pressing. They used DSC, DMA and UTM for the characterization of the nanocomposites. The thermo-mechanical behaviour has been enhanced by increasing the filler content. Differential scanning calorimetry was used to determine T_g (glass transition temperature). The thermo-mechanical properties of these nanocomposites have been investigated, and the influence of processing conditions and the effect of whisker content have been considered. The T_g of the composites systems had been found to be relatively insensitive to different cellulose contents (average value of $T_g = -2$ °C). They have concluded that T_g is nearly independent of both filler content and processing conditions. Grunert and Winter[90] were successful in working using cellulose acetate butyrate by solution casting method. Samples were prepared containing 0, 2.5, 5.0, 7.5 and 10.0 wt% cellulose crystals. The resulting films were transparent. Favier et al.[91] prepared the nanocomposites of cellulose whiskers and copolymer of styrene and butyl acrylate and a small amount of acrylic acid. The suspensions were poured into poly(tetrafluoroethylene) moulds and allowed to dry slowly for 1 month at room temperature. Oksman et al.[92] prepared nanocomposites using three different types of cellulose reinforcements, microcrystalline cellulose (MCC), cellulose fibres (CFs) and wood flour (WF) with polylactic acid (PLA) as a matrix using a twin-screw extruder and injection-moulding techniques. They in fact treated micro cellulose with N,N-dimethylacetamide (DMAc) containing lithium chloride (LiCl) in order to swell the micro cellulose and partly separate the cellulose whiskers. The suspension of whisker was pumped into the polymer melt during the extrusion process. Samir et al.[93] prepared cellulose nanocrystals extracted from tunicate as the reinforcing phase for nanocomposite and poly(oxyethylene) (POE) as the matrix. They used SEM, polarized optical micrographs for morphological studies of the nanocomposites and TGA, DMA and DSC for thermal, mechanical and crystallization behaviour respectively. The glass–rubber transition temperature of POE was not influenced by the cellulosic filler. The melting temperature and degree of crystallinity which were found to decrease for highly filled (10 wt% and above) materials. This restricted crystallinity was confirmed by dynamic cooling crystallization experiments and polarized optical microscopic observations. It was ascribed to both strong interactions between the POE chains and cellulosic surface and increased viscosity of the melt composite. Cellulose/POE interactions were quantified using heat flow microcalorimetry measurements. The mechanical behaviour of cellulose whiskers/POE nanocomposites was evaluated in the linear range over a broad temperature range from dynamic mechanical analysis. The main effect of the filler was a thermal stabilization of the storage modulus for the composites above the melting temperature of the POE matrix. Oksman and Bondenson[94] characterized polylactic acid/PVOH cellulose whisker nanocomposites by using SEM, TEM and DMA. TEM analysis showed that the whiskers were better dispersed in the nanocomposite produced with liquid feeding. Analysis of microtomed and fractured samples in

FE-SEM showed that PLA and PVOH formed two immiscible phases with a continuous PLA phase and a discontinuous PVOH phase. The thermal stability of the nanocomposites was not improved compared to its unreinforced counterpart, probably because the majority of the whiskers were located in the PVOH phase and only a negligible amount was located in the PLA phase. The small improvements for the nanocomposites in tensile modulus, tensile strength and elongation to break were noted compared to its unreinforced counterpart.

4.3.2 Chitin and Chitosan-based Green Composites

Gopalan and Dufresne[95] prepared nanocomposite materials from a colloidal suspension of high aspect ratio chitin whiskers as the reinforcing phase and poly(caprolactone) as the matrix. The chitin whiskers, prepared by acid hydrolysis of Riftia tubes, consisted of slender parallelepiped rods with an aspect ratio close to 120. Films were obtained by both freeze-drying and hot-pressing or casting and evaporating the preparations. Amorphous poly-(styrene-co-butyl acrylate) latex was also used as a model matrix. Sriupayo and co-workers[96] prepared α-chitin whisker-reinforced poly(vinyl alcohol) (PVA) nanocomposite films by solution-casting techniques and they studied the thermal stability of the chitin nanocomposites by TGA. The presence of the whiskers did not much affect the thermal stability and the apparent degree of crystallinity of the chitosan matrix. The tensile strength of α-chitin whisker-reinforced chitosan films increased from that of the pure chitosan film with initial increase in the whisker content to reach a maximum at the whisker content of 2.96 wt% and decreased gradually with further increase in the whisker content, while the percentage of elongation at break decreased from that of the pure chitosan with initial increase in the whisker content and levelled off when the whisker content was greater than or equal to 2.96 wt%. They also studied crystallinity of nanocomposites; the presence of the whiskers did not have any effect on the crystallinity of the PVA matrix. They suggested that the cast PVA film was essentially amorphous for the α-chitin whiskers, their WAXD pattern exhibits two major scattering peaks at 2θ angles of about 9° and 19°, respectively for the resulting α-chitin whisker-reinforced PVA films. The WAXD patterns were intermediate to those of the pure components with the strong scattering peaks of α-chitin whiskers (*i.e.* at about 9° and 19°) being more pronounced with increasing whisker content. To verify whether or not incorporation of α-chitin whiskers into PVA resulted in an increase in the crystallinity of the PVA matrix, FT-IR spectra were considered. The peak at 1144 cm^{-1} (C–O of doubly H-bonded OH in crystalline regions) was useful for indication of the crystallinity of PVA. Apparently, the relative intensity of this peak was not found to increase with increasing whisker content, indicating that incorporation of α-chitin whiskers did not have an effect on the crystallinity of the PVA matrix. The properties of high performance chitin filled natural rubber nanocomposites were analysed by Gopalan and Dufresne.[97] It was concluded that the whiskers form a rigid network in the NR matrix which is assumed to be governed by a percolation mechanism. A percolated filler–filler network is formed by hydrogen bonding interactions between chitin particles above the percolation threshold. The values of diffusion coefficient, bound rubber content and relative weight loss also supported the presence of a three-dimensional chitin network within the NR matrix. The mechanical behaviour of the composites gives additional insight and evidence for this fact.

4.3.3 Starch-based Green Composites

Starch-based nanocomposites were made by Angellier et al.[98] using the water evaporation method with NR latex as matrix and they were characterized using scanning electron microscopy, water and toluene absorption experiments, differential scanning calorimetry and wide-angle X-ray diffraction. In their work the nanocomposite films NR/starch nanocrystals were characterized by X-ray diffraction. The diffraction patterns recorded for a film of pure waxy maize starch nanocrystals obtained by pressing freeze-dried nanocrystals displayed typical peaks. By adding starch nanocrystals into NR, the peaks corresponding to the amylose allomorph become stronger and stronger, as expected. This shows that an increase of the starch content results in an increase of the global crystallinity of the composite material. The diffraction patterns of the various NR/starch nanocrystals do not exactly correspond to a simple mixing rule of the diffractograms of the two pure parent components.

The barrier properties of the nanocomposites to water vapour and oxygen were also investigated, and the effect of surface chemical modification of starch nanocrystals was studied. By introduction of starch nanocrystals in NR, the swelling by toluene decreased and the swelling by water increased. It was assumed that these phenomena were due to the formation of a starch nanocrystals network through hydrogen linkages between starch nanoparticles clusters and also to favourable interactions between the matrix and the filler. As explained earlier, the formation of the network of starch nanocrystals is governed by a percolation mechanism. According to the authors, the critical volume fraction of starch nanocrystals at the percolation should be around 6.7 vol% (*i.e.* 10 wt%). The platelet-like morphology of starch nanocrystals seems to be responsible for the decrease of both the permeability to water vapour and oxygen of natural rubber filled films.

4.3.4 Soy Protein-based Green Composites

Lu and co-workers[99] developed environmentally friendly thermoplastic nanocomposites using soy protein isolate (SPI) plastics as reinforcement with chitin whiskers. They studied the morphology and properties on the chitin whiskers content in the range from 0 to 30 wt% for the glycerol plasticized SPI nanocomposites using dynamic mechanical thermal analysis, scanning electron microscopy, swelling experiments and tensile testing. The results indicate that the strong interactions between fillers and between the filler and SPI matrix play an important role in reinforcing the composites without interfering with their biodegradability. The SPI/chitin whisker nanocomposites at 43% relative humidity increased in both tensile strength and Young's modulus from 3.3 MPa for the SPI sheet to 8.4 MPa and from 26 MPa for the SPI sheet to 158 MPa.

4.3.5 PLA-based Green Composites

To exploit the enhanced properties of such nanocomposites, various studies have been performed on the preparation of PLA-based nanocomposites.[100–102] Although there are many publications on PLA-based nanocomposites, few works on PLA as an antimicrobial carrier for food and non-food packaging materials have been published yet. Rhim et al.[103] and co-workers prepared PLA/nanoclay composite films using a solvent casting method and investigated the effect of the type of nanoclay and its concentration on the properties as well as the antimicrobial activity of the prepared PLA-based composite films. Ochi[104] prepared the unidirectional biodegradable composite materials from kenaf fibres and an emulsion-type PLA resin. Thermal analysis of kenaf fibres revealed that tensile strength of kenaf fibres decreased when kept at 180 °C for 60 min. The unidirectional fibre-reinforced composites showed tensile and flexural strengths of 223 MPa and 254 MPa, respectively. Moreover, tensile and flexural strength and elastic moduli of the kenaf fibre-reinforced composites increased linearly up to a fibre content of 50%. The biodegradability of kenaf/PLA composites was examined for 4 weeks using a garbage-processing machine. Experimental results showed that the weight of composites decreased 38% after four weeks of composting. To what extent the fibre load determines the PLA composite characteristics were investigated by Kimura et al.[105] They examined compression-moulded PLA composites reinforced with ramie fibres, using non-twisted commingled yarn made of ramie and PLA fibres as raw material. The best mechanical values, i.e. tensile strength, bending strength and stiffness, were obtained at ramie fibre volume contents between 45% and 65%. Large proportions of ramie also increased the notched impact strength considerably. Ochi[106] investigated kenaf/PLA composites with different fibre proportions. Tensile and bending strength as well as Young's modulus increased linearly up to a fibre content of 50%. Pan et al.[107] produced kenaf/PLA composites by melt-mixing and injection moulding with fibre mass contents ranging between 0% and 30%. At 30% the tensile strength improved by 30%. A study of injection-moulded flax and artificial Cordenka fibre-reinforced PLA was carried out by Bax and Müssig.[108] The composites' mechanical qualities improved with a rising fibre mass content of 10% up to 30%. Flax/PLA and Cordenka/PLA composites clearly differed in their impact strength characteristics: While the impact strength of pure PLA could be multiplied by adding Cordenka, the values of flax/PLA composites were inferior to the pure matrix. Ganster and Fink[109] investigated injection-moulded Cordenka fibre-reinforced PLA with a fibre mass content of 25%. Stiffness and strength of the composites could be approximately doubled compared to the pure matrix. The impact strength could be tripled. Shibata et al.[110] constructed Lyocell fabric and PLA by compression moulding. The tensile modulus and strength of Lyocell/PLA composites improved with increasing fibre content. Impact strength was considerably higher than that of pure PLA. Some research papers deal with different fibre modification methods to optimize the composites' mechanical characteristics. Tokoro et al.[111] examined three kinds of injection-moulded bamboo fibre-reinforced PLA (short fibre bundles, alkali-treated fibre bundles and steam-exploded fibre bundles). The highest bending strength was obtained with steam-exploded fibres. To improve clearly notched impact strength, samples were alternatively fabricated by hot pressing using medium length bamboo fibre bundles to counter the decrease in length at fabrication. Cho et al.[112] made jute, kenaf and henequen fibre-reinforced PLA composites by compression moulding. Fibres were surface-treated with tap water by static soaking and dynamic ultrasound. Interfacial shear strength, flexural properties and dynamic mechanical properties were investigated. The characteristics of the polymeric resins were clearly improved by incorporating the natural fibres into the resin matrix. Hu and Lim[113] studied the mechanical properties of compression-moulded composites with different volume fractions and the effects of alkali treatment on the fibre surface morphology of hemp. Best results were obtained with 40% volume fraction of alkali-treated fibre bundles. Huda et al.[114] examined compression-moulded PLA laminated composites reinforced with kenaf produced by the film stacking method to see how far alkalization and silane-treatment of the kenaf fibre bundles influence the composites' mechanical and thermal properties. Both treatments improved the mechanical properties. Scanning electron microscopy (SEM) showed better adhesion between treated fibres and matrix and an improvement of notched impact strength for a bast fibre PLA composite, which is an interesting result. We examined unnotched impact strength of kenaf/PLA composites to find out if we see similar trends.

Plasticizing PLA composites to optimize mechanical characteristics is also an important field of research. Oksman et al.[115] studied the impact properties of compression-moulded flax/PLA composites. Masirek et al.[116] prepared composites with hemp fibres, poly(ethylene glycol) (PEG) and PLA by compression moulding. Mechanical tests showed that the composites' Young's modulus markedly increased with the hemp content, in the case of crystallized PLA reinforced with 20 wt% hemp, whereas the elongation and stress at break decreased with an increasing amount of fibres. Plasticization with PEG did not improve the tensile properties of the composites. Serizawa et al.[117] examined kenaf/PLA composites and the effects of adding a flexibilizer (copolymer of lactic acid and aliphatic polyester). The results showed that the flexibilizer improved the strength of the composites. Okubo et al.[118] experimented with compression-moulded bamboo fibre-reinforced PLA, adding micro-fibrillated cellulose (MFC) as an enhancer for the bending characteristics. Three point bending strength was clearly improved. They also found that tangled MFC fibres prevented micro cracks along the interface between bamboo fibre and matrix.

Cheng and co-workers[119] have prepared chicken feather fibre (CFF)/reinforced poly(lactic acid) (PLA) composites using a twinscrew extruder and an injection moulder. The tensile moduli of CFF/PLA composites with different CFF content (2, 5, 8 and 10 wt%) were found to be higher than that of pure PLA, and a maximum value of 4.2 GPa 16% was attained with 5 wt% of CFF without causing any substantial weight increment. The morphology, evaluated by scanning electron microscopy (SEM), indicated that a uniform dispersion of CFF in the PLA matrix existed. The mechanical and thermal properties of pure PLA and CFF/PLA composites were compared using dynamic mechanical analysis (DMA), thermomechanical analysis (TMA) and thermogravimetric analysis (TGA). DMA results revealed that the storage modulus of the composites increased with respect to the pure polymer, whereas the mechanical loss factor (tan δ) decreased. The results of TGA experiments indicated that the addition of CFF enhanced the thermal stability of the composites as compared to pure PLA. The outcome obtained from this study is believed to assist the development of environmentally friendly composites from biodegradable polymers, especially for converting agricultural waste (chicken feathers) into useful products.

Pluta et al.[120] investigated PLA/montmorillonite micro- and nanocomposites, and showed that the microcomposites form a phase separation between the matrix and reinforcement. Furthermore, nanocomposites can be very easily processed by using nanofillers; however, the biodegradability of the composite was affected. Moreover, the thermal stability of PLA/montmorillonite in an oxidative atmosphere could be improved. The influence of chopped glass and recycled newspaper in PLA composites was assayed by Huda et al.[121] With addition of 30 wt% of both fibres the stiffness rose from 3.3 GPa (the original polymer) to 5.4 GPa with recycled newspaper fibres and to 6.7 GPa with glass fibres. The tensile strength was improved from 62.9 MPa (the original polymer) to 67.9 and 80.2, respectively.[122] Also wood polymer composites (WPC) on based on PLA were studied.[123] With addition of different contents of wood flour (WF) the flexural strength improvement from 98.8 to 114.3 MPa could be observed. The modulus of elasticity increases proportionately as the amount of WF increases. However, the thermal stability decreases with the increase of WF. A study on PLA composites with microcrystalline cellulose (MCC)[124] showed that both PLA/WF and PLA/wood pulp (WP) have better mechanical properties than PLA/MCC. This is due to the poor interfacial adhesion of PLA and MCC, which was observed via scanning electron microscopy. Also, a higher level of matrix crystallinity of PLA/WF and PLA/WP compared to PLA/MMC could be seen. It was considered as a reason for the improved stiffness, when compared to PLA/MCC. They showed shown that MCC has better potential as nucleating agent and causes higher degree of crystallinity, like WF or cellulose fibres which were investigated. Biocomposites of PLA/abaca were investigated by Shibata et al.[125] In this research three different kinds of chemical treatment were used: esterification, mercerization and cyanoethylation. It has been shown that for reinforced PLA the flexural strength could not be improved significantly, even if the fibres were treated. In contrast, the stiffness increases continually from approx. 3.6 to 6.0 GPa, depending on the fibre content and treatment method. The biodegradation rate of PLA/untreated abaca fibre (90/10) was much higher, as the neat PLA and PLA/esterificated abaca fibre (90/10). Other natural fibres used as reinforcement for PLA composites were jute,[126] flax[127] and kenaf.[128] The reinforcing effect of most of them brought an increase in stiffness and sometimes in strength. The ductility of the composites was overall worst, which resulted in lower impact strength. To improve this drawback, some other fibrous reinforcement must be used. However, to keep the concept of 'biocomposites', the reinforcement should be obtained from renewable raw materials, for example a viscose fibre made of native cellulose (artificial cellulose).[129] Fink and Ganster[130] published results of PLA/artificial cellulose in comparison to PP composites. They reported that with addition of 30 wt% of regenerated cellulose an improvement of most mechanical characteristics can be achieved. The unnotched and notched Charpy impact strength can be improved by factors of 3.8 and 2.0, respectively. The modulus and strength increase to 150% of that with original PLA. Composites of natural polymers with PLA have been reported in the literature in order to develop biodegradable/biocompatible products for both medical and commodity applications. All PLA resins are manufactured using renewable agricultural resources, such as corn or sugar beets. Cargill Dow prepared a biodegradable product whose applications include compostable food and lawn waste bags, yoghurt cartons, seeding mats and non-woven mulch to prevent weed growth. DuPont's biodegradable Biomax copolyester resin, a modified form of PET, was launched in 1997. Its properties, according to DuPont, are diverse and customizable, but they are generally formulated to mimic PE or PP. Because it is based on PET technology, and can be produced on commercial lines, DuPont believes that Biomax is only marginally more expensive to produce than PET itself, and significantly cheaper to produce than other biodegradable polymers.

4.4 Applications

All green nanocomposites are low-cost materials. These low-cost green composites were found to have mechanical strength and properties suitable for applications in housing construction materials, furniture and automotive parts. The influence of fibre treatment on the properties of biocomposites derived from grass fibre. Studies carried out in the past decades have demonstrated that green composites materials should combine high mechanical and other essential operational and technological properties (*e.g.* stability, low gas permeability, environmental safety, easy moulding) with biodegradability.. Recent work on biocomposites reveals that in the most cases the specific mechanical properties of biocomposites are comparable to widely used glass fibre reinforced plastics. In order to be competitive, ecofriendly composites must have the same desirable properties as obtained in conventional plastics. The most important factors to the formation of a successful green composites material industry include cost reduction as well as public and political acceptance. Existing green and environmental friendly composites materials are mainly blended with different materials with an aim to reduce cost and to tailor the product for some specific applications.

Application of green composites in natural fibre-reinforced composites will broaden their uses. The demand for green and renewable materials continues to rise. Polymeric composites from renewable resources have occupied major applications in green packaging. Although these are emerging as alternatives to existing petroleum derived plastics; the present low level production and high costs restrict their wide spread applications. The barrier properties of such degradable polymers can be improved through nano reinforcements. The incorporation of nanoparticles in a polymer matrix reduces the permeability of penetrant molecules and thus develops high barrier composites. Nanocomposites especially green nanocomposites or nanocomposites obtained from renewable resources are an emerging new class of materials, total environmental/economical impacts of which need to be studied to prove the industrial and environmental potential of targeted nanocomposites for automotive applications. Biopolymers combined with suitably organically modified clays will result in environmental friendly nanocomposites. The need of green and biodegradable plastics has been increased during the past decades not only due to the increasing environmental concerns but also for its biomedical applications. It is now well evident that polymer / plastics waste management through biodegradation or bio-conversion is the most suitable solution for '*Plastic Waste Management*' among the other traditional methods like incineration, pyrolysis and landfilling. Several means have been used to achieve the biodegradability/biodeteriobility /Photo-biodegradability in polymers.

4.5 Conclusion

As a result of environmental awareness and the international demand for green technology, green composites have the potential to replace present petrochemical-based materials. They represent an important element of future waste disposal strategies. In true green nanocomposites, both the reinforcing material (such as a natural fibre) and the matrix are biodegradable. Cellulose, chitin and starch are the most abundant organic compounds in nature; they are also green, inexpensive, biodegradable and renewable. They obviously receive a great attention for non-food applications. The use of natural fibres instead of traditional reinforcement materials, such as glass fibres, carbon and talc, provides several advantages including low density, low cost, good specific mechanical properties, reduced tool wear and biodegradability. Important applications include packaging, wide-ranging uses from environmentally friendly biodegradable composites to biomedical composites for drug/gene delivery, tissue engineering applications and cosmetic orthodontics. They often mimic the structures of the living materials involved in the process in addition to the strengthening properties of the matrix that was used but still providing green and biocompatibility, *e.g.* in creating scaffolds in bone tissue engineering. Bionanocomposites combine plant and animal nanofibres (derived from waste and biomass) with resins and other polymers, such as plastics and rubbers, to create natural-based composite materials. A variety of plant fibres with high tensile strength can be used including kenaf, industrial hemp, flax, jute, sisal, coir, *etc.* Green fibres can be combined with traditional resins or newer plant-based resins. The result is a plant-based alternative for many traditional steel and fibreglass applications. Advantages of green nanocomposites over traditional composites are reduced weight, increased flexibility, greater mouldability, reduced cost, better sound insulation and their renewable nature.

References

1. R. L. Crawford, *Lignin Biodegradation and Transformation*. Wiley, New York (1981). ISBN 0-471-05743-6.
2. R. Young, *Cellulose Structure Modification and Hydrolysis*. Wiley, New York (1986). ISBN 0471827614.
3. D. Klemm, H. Brigitte, F. Hans-Peter and B. Andreas, *J. Chem.-Inform.*, 2005, **36**, 36.
4. D. M. Updegraff, *J. Anal. Biochem.*, 1969, **32**, 420.
5. Cellulose *Encyclopedia Britannica*, Encyclopedia Britannica Online, Retrieved January 11, 2008 (2008).
6. A. K. Bledzki, S. Reihmane and J. Gassan, *J. Appl. Polym. Sci.*, 1996, **59**, 1329.
7. K. G. Satyanarayana, K. Sukumaran, P. S. Mukherjee and C. P. Pavitharan, *J. Compos.*, 1990, **12**, 117.
8. A. Bismarck, S. Mishra and T. Lampke, *Plant Fibres as Reinforcement for Green Composites*, CRC Press, Boca Raton, FL, (2005), pp. 37–108.
9. C. Klason, J. Kubat and H. E. Stromvall, *J. Polym. Mater.*, 1985, **11**, 9.
10. P. Zadorecki and A. J Michell, *J. Polym. Compos*, 1989, **10**, 69.
11. D. Maldas, B. V. Kokta, R. Raj and G. C. Daneault, *J. Polym.*, 1988, **29**, 1255.
12. S. Kuga and R. M. Brown, *J. Carbohydr. Res.*, 1988, **180**, 345.
13. K. R. Z. Andress, *J. Phys. Chem. Abt. B*, 1929, **4**, 190.
14. J. Blackwell and F. Kolpak, *J. Macromol.*, 1976, **9**, 273.
15. H. Chanzy, Y. Nishiyama and P. Langan, *J. Am. Chem. Soc.*, 1999, **121**, 9940.
16. S. Watanabe, J. Ohkita, J. Hayashi and A. Sufoka, *J. Polym. Lett*, 1975, **13**, 23.
17. A. J. Sarko, J. Southwick and J. Hayashi, *J. Macromol.*, 1976, **9**, 857.
18. H. Chanzy and A. Buleon, *J. Polym. Sci. Polym. Phys. Ed*, 1980, **18**, 1209.
19. E. Togawa, R. M. Brown and T. J. Kondo, *J. Biomacromol.*, 2001, **2**, 1324.
20. J. Li, J. F. Revol and R. H. Marchessault, *J. Appl. Polym. Sci.*, 1997, **65**, 373.

21. Y. Yamaguchi, T. T. Nge, A. Takemura, N. Hori and H. Ono, *J. Biomacromol.*, 2005, **6**, 1941.
22. B. Krajewska, *J. Enzyme Microbiol. Technol.*, 2004, **35**, 126.
23. N. L. Yusof, A. Wee, L. Y. Lim and E. Khor, *J. Biomed. Mater. Res. A*, 2003, **66**, 224.
24. S. M. Hudson, A. Domard, G. A. F. Roberts and K. M. Varum (eds.), *Advances in Chitin Science,* vol. 2, pp. 590–599, Jacques Andre Publishers, Lyon, France (1998).
25. M. Kanke, H. Katayama, S. Tsuzuki and H. Kuramoto, *J. Chem. Pharm. Bull.*, 1989, **37**, 523.
26. Y. Kato, H. Onishi and Y. J. Machida, *J. Curr. Pharm. Biotechnol.*, 2003, **4**, 303.
27. R. Marguerite, *J. Prog. Polym. Sci.*, 2006, **31**, 603.
28. J.-F. Rovel and R. H. Marchessaultf, *J. Biomacromol.*, 1993, **15**, 329.
29. R. A. Muzzarelli, Chitin microfibrils, In *Chitin*, pp. 51–55, Pergamon Press, New York (1977).
30. C. J. Brine and P. R. Austin, *ACS Symp. Ser.*, 1975, **18**, 505.
31. S. B. Murry and A. C. Neville, *J. Int. Boil. Macromol.*, 1997, **20**, 123.
32. S. B. Murry and A. C. Neville, *J. Int. Boil. Macromol.*, 1998, **22**, 137.
33. N. K. Gopalan and A. Dufresne, *J. Biomacromol.*, 2003, **4**, 657.
34. A. Morin and A. Dufresne, *J. Macromol.*, 2002, **35**, 2190.
35. N. K. Gopalan and A. Dufresne, *J. Biomacromol.*, 2003, **4**, 666.
36. N. K. Gopalan and A. Dufresne, *J. Biomacromol.*, 2003, **4**, 1835.
37. M. A. S. Samir, A. F. Alloin, J. Y. Sanche, N. E. Kissi and A. Dufresne, *J. Macromol.*, 2004, **37**, 1386.
38. H. W. Gonell, *J. Z. Physiol. Chem.*, 1926, **152**, 18.
39. G. L. Clark and A. F. Smith, *J. Phys. Chem.*, 1936, **40**, 863.
40. K. H. Gardner and J. Blackwell, *J. Biopolym.*, 1975, **14**, 1581.
41. R. Minke and J. Blackwell, *J. Mol. Biol.*, 1978, **120**, 167.
42. Y. Saito, T. Okano, H. Chanzy and J. Sugiyama, *J. Struct. Biol.*, 1995, **114**, 218.
43. H. Chanzy, Chitin crystals, In A. Domard, G. A. F. Roberts, K. M. Varum (eds.) *Advances in Chitin Science*, pp. 11–21. Jacques Andre, Lyon, France (1998).
44. M.-J. Chretiennot-Dinet, M.-M. Giraud-Guille, D. Vaulot, J.-L. Putaux and H. Chanzy, *J. Phycol.*, 1997, **33**, 666.
45. F. Gaill, J. Persson, P. Sugiyama, R. Vuong and H. Chanzy, *J. Struct. Biol*, 1992, **109**, 116.
46. J. Blackwell, *J. Biopolym.*, 1969, **7**, 281.
47. L. Averous, *J. Macromol. Sci. C Polym. Rev.*, 2004, **44**, 231.
48. S. S. Ray and M. Bousmia, *J. Prog. Mater. Sci.*, 2005, **50**, 962.
49. W. H. Brown and T. Poon, *Introduction to Organic Chemistry (3rd edn)*, Wiley, Hoboken, NJ (2005). ISBN 0-471-44451-0.
50. M. Battacharya and U. R. Vaidya, *J. Appl. Polym. Sci*, 1994, **52**, 617.
51. R. P. Westoff, F. H. Oety, C. L. Mehlttter and C. R. Russell, *J. Ind. Eng. Chem., Prod. Res. Dev.*, 1974, **13**, 123.

52. G. J. Griffin and L. Priority, U. K. Patent 1,485,833 (1972).
53. K. Wrick, *Nutraceut. World*, 2003, **6**, 32.
54. T. L. Mounts, W. J. Wolf and W. H. Martinez, Processing and utilization. In: *Soybeans: Improvement, Production, and Uses*, J .R. Wilcox (ed.), The American Society of Agronomy, Madison, Wisconsin, 1987, pp. 819.
55. E. W. Meyer and R. A. Lawrie (ed.), Butterworth's Whitefriars Press, London, England, 1970.
56. N. Sugarman, Process for simultaneously extracting oil and protein from oleaginous materials, US Pat. 4,307,014.
57. K. Oksmana, M. Skrifvars and J.-F. Selin, *J. Comp. Sci. Tech.*, 2003, **63**, 1317.
58. S. Vainionpaa, P. Rokkanen and P. Torrmala, *J. Prog. Polym. Sci.*, 1989, **14**, 679.
59. M. J. Manninen, U. Päivärinta, H. Pätiälä, P. Rokkanen, R. Taurio and M. Tamminmäki, *J. Mater. Sci.*, 1992, **3**, 245.
60. I. Takashi, K. Yuji, K. Junji and M. Taktayasu, *Proc. JSAE Annu. Congr.*, 2003, **60**, 11.
61. K. Oksman and J. F. Selin. Plastics and composites from polylactic acid, In: *Natural Fibers, Plastics and Composites*, F. T. Wallenberger, N. Weston (eds.), Kluwer Academic Publishers, Boston, 2004, p. 149.
62. M. S. Huda, L. T. Drzal, A. K. Mohanty and M. Misra, *J. Comp. Sci. Tech.*, 2006, **6**, 1813.
63. A. K. Mohanty, M. Misra and L. T. Drzal, *J. Polym. Environ.*, 2002, **10**, 19.
64. M. S. Huda, L. T. Drzal, M. Misra, A. K. Mohanty, K. Williams and D. F. Mielewski, *J. Ind. Eng. Chem. Res.*, 2005, **44**, 5593.
65. S. H. Lee and S. Wang, *J. Compos. A*, 2006, **37**, 80.
66. H.-Y. Cheung, K.-T. Lau, X.-M. Tao and D. Hui, *J. Compos. B*, 2008, **39**, 1026.
67. D. Garlotta, *J. Polym. Environ.*, 2001, **9**, 63.
68. D. W. Farrington, J. L. Davies and R. S. Blackburn, Poly(lactic acid) fibers, In: *Biodegradable and Sustainable Fibers*, R. S. Blackburn (ed.), Woodhead Publishing, Cambridge, England, 2005, p. 191.
69. B. Gupta, N. Revagade and J. Hilborn, *J. Prog. Polym. Sci.*, 2007, **32**, 455.
70. M. Leimser, *Werkstoff der Zukunft. Kunststoff Trends*, 2006, **1**, 22 [in German].
71. B. Kucera, *Bioplastics Mag.*, 2007, **2**, 20.
72. C. Grashorn, Erstes Serienprodukt aus naturfaserverstärktem PLA im Spritzguss. In: *Nova Institut, Hürth*, Germany; 5. N-FibreBase Kongress. Hürth 21st–22nd Mai, 2007 [in German].
73. Anonymous, *Bioplastics Mag.*, 2006, **1**, 18.
74. M. Iji, Highly functional bioplastics used for durable products, *The Netherlands Science and Technology, Innovative Technologies in Bio-Based Economy*, Wageningen, The Netherlands, 8th April, 2008 [http://www.twanetwork.nl].
75. Anonymous, *Bioplastics Mag.*, 2007, **2**, 14.

76. D. Garlotta, *J. Polym. Environ.*, 2001, **9**, 63.
77. R. E. Drumright, P. R. Gruber and D. E. Henton, *J. Adv. Mater.*, 2000, **12**, 1841.
78. L. Cabedo, J. L. Feijoo, M. P. Villanueva, J. M. Lagaro and E. Gimenez, *Macromol. Symp.*, 2006, **233**, 191.
79. S. Solarski, M. Ferreira, E. Devaux, G. Fontaine, P. Bachelet and S. Bourbigot, *J. Appl. Polym. Sci.*, 2008, **109**, 841.
80. S. Solarski, F. Mahjoubi, M. Ferreira, E. Devaux, P. Bachelet and S. Bourbigot, *J. Mater. Sci.*, 2007, **42**, 5105.
81. S. Bourbigot, G. Fontaine, S. Duquesne and R. Delobel, *J. Nanotechnol.*, 2008, **5**, 683.
82. J. Zhan, L. Song, S. Nie and Y. Hua, *J. Pol. Deg. Stab.*, 2009, **94**, 291.
83. K. Kimura and Y. Horikoshi, *J. Sci. Technol.*, 2005, **41**, 173.
84. C. Reti, M. Casetta, S. Duquesne, S. Bourbigot and R. Delobel, *J. Polym. Adv. Technol.*, 2008, **19**, 628.
85. S. Bourbigot, S. Duquesne, G. Fontaine, S. Bellayer, T. Turf and F. Samyn, *J. Mol. Cryst. Liq. Cryst.*, 2008, **486**, 325.
86. R. Bhardwaj and A. K. Mohanty, *J. Biomacromol.*, 2007, **8**, 2476.
87. P. Kulpinski, *J. Appl. Polym. Sci.*, 2005, **98**, 1855.
88. G. F. G. Torres, M. G. Clara and M. C. B. Cristian, *J. Biomater.*, 2009, **5**, 1605.
89. P. Hajji, J. Y. Cavaille, V. Favier, C. Gauthier and G. Vigier, *J. Polym. Compos.*, 1996, **17**, 4.
90. M. Grunert and W. T. Winter, *J. Polym. Environ.*, 2002, **10**, 27.
91. V. Favier, H. Chanzy and J. Y. Cavaille, *J. Macromol.*, 1996, **28**, 6365.
92. K. Oksman, A. P. Mathew, D. Bondeson and I. Kvien, *J. Compos. Sci. Technol.*, 2006, **66**, 2776.
93. S. M. A. S. Azizi, F. Alloin, S. Jean-Yves and A. Dufresne, *J. Polym.*, 2004, **45**, 4149.
94. K. Oksman and D. Bondeson, *J. Compos. Part A*, 2007, **38**, 2486.
95. N. K. Gopalan and A. Dufresne, *J. Biomacromol.*, 2003, **4**, 1835.
96. J. Sriupayo, P. Supaphol, J. Blackwell and R. Rujiravanit, *J. Polym*, 2005, **46**, 5637.
97. N. K. Gopalan and A. Dufresne, *J. Biomacromol.*, 2003, **4**, 657.
98. H. Angellier, S. M. Boisseau, L. Lebrun and A. Dufresne, *J. Macromol.*, 2005, **38**, 3783.
99. Y. Lu, L. Weng and L. Zhang, *J. Biomacromol.*, 2004, **5**, 1046.
100. L. S. Liu, V. L. Finkenstadt, C. K. Liu, T. Jin, M. L. Fishman and K. B Hicks, *J. Appl. Polym. Sci.*, 2007, **106**, 801.
101. P. Maiti, K. Yamada, M. Okamoto, K. Ueda and K. Okamoto, *J. Chem. Mater.*, 2002, **14**, 4654.
102. N. Ogata, G. Jimenez, H. Kawai and T. Ogihara, *J. Polym. Sci. Part B: Polym. Phys.*, 1997, **35**, 389.

103. J. W. Rhim, S. I. Hong and C. S. Ha, *J. Food Sci. Technol.*, 2009, **42**, 612.
104. S. Ochi, *J. Mech. Mater.*, 2008, **40**, 446.
105. T. Kimura, M. Kurata, T. Matsuo, H. Matsubara and T. Sakobe, Compression moulding of biodegradable composite using Ramie/PLA non-twisted commingled yarn. In: *5th Global Wood and Natural Fibre Composites Symposium*, Kassel, Germany, April, 2004, p. 27.
106. S. Ochi, *J. Mech. Mater.*, 2008, **40**, 446.
107. P. Pan, B. Zhu, W. Kai, S. Serizawa, M. Iji and Y. Inoue, *J. Appl. Polym. Sci.*, 2007, **105**, 1511.
108. B. Bax and J. Müssig, *J. Compos. Sci. Technol.*, 2008, **68**, 1601.
109. J. Ganster and H. P. Fink, *J. Cellulose*, 2006, **13**, 271.
110. M. Shibata, S. Oyamada, S. Kobayashi and D. Yaginuma, *J. Appl. Polym. Sci.*, 2004, **92**, 3857.
111. R. Tokoro, D. M. Vu, K. Okubo, T. Tanaka, T. Fujii and T. Fujiura, *J. Mater. Sci.*, 2008, **43**, 775.
112. D. Cho, J. M. Seo, H. S. Lee, C. W. Cho, S. O. Han and W. H. Park, *J. Adv. Compos. Mater.*, 2007, **16**, 299.
113. R. Hu and J. K. Lim, *J. Compos. Mater.*, 2007, **41**, 1655.
114. M. S. Huda, L. T. Drzal, A. K. Mohanty and M. Misra, *J. Compos. Sci. Technol.*, 2008, **68**, 424.
115. K. Oksman, M. Skrifvars and J. F. Selin, *J. Compos. Sci. Technol.*, 2003, **63**, 1317.
116. R. Masirek, Z. Kulinski, D. Chionna, E. Piorkowska and M. Pracella, *J. Appl. Polym. Sci.*, 2007, **105**, 255.
117. S. Serizawa, K. Inoue and M. Iji, *J. Appl. Polym. Sci.*, 2006, **100**, 618.
118. K. Okubo, T. Fujii and N. Yamashita, *J. Ser. A*, 2005, **48**, 199.
119. C. Sha, L. Kin-tak, L. Tao, Z. Yongqing, L. Pou-Man and Y. Yansheng, *J. Compos. Part B*, 2009, **40**, 650.
120. M. Pluta, A. Galeski, M. Alexandre, M. A. Paul and P. Dubois, *J. Appl. Polym. Sci.*, 2002, **86**, 1497.
121. H. S. Huda, L. T. Drzal, A. K. Mohanty and M. Misra, *J. Comput. Sci. Technol.*, 2006, **66**, 1813.
122. M. S. Huda, L. T. Drzal, M. Misra and A. K. Mohanty, *J. Appl. Polym. Sci.*, 2006, **102**, 4856.
123. A. P. Mathew, K. Oksman and M. Sain, *J. Appl. Polym. Sci.*, 2005, **97**, 2014.
124. A. P. Mathew, K. Oksman and M. Sain, *J. Appl. Polym. Sci.*, 2006, **101**, 300.
125. M. Shibata, K. Ozawa, N. Teramoto, R. Yosomiya and H. Takeishi, *J. Macromol. Mat. Eng.*, 2003, **288**, 35.
126. D. Plackett, T. L. Andersen, W. B. Pedersen and L. Nielsen, *J. Comput. Sci. Technol.*, 2003, **63**, 1287.
127. K. Oksman, M. Skrifvars and J. F. Selin, *J. Comput. Sci. Technol.*, 2003, **63**, 1317.

128. T. Nishino, K. Hirao, M. Kotera, K. Nakamae and H. Inagaki, *J. Comput. Sci. Technol.*, 2003, **63**, 1281.
129. A. K. Bledzki, A. Jaszkiewicz and V. E. Sperber, Abaca and cellulose fibre reinforced polypropylene, In: *Proceedings of Fourth International Workshop on Green Composites*, Tokyo, September 2006, pp. 1–5.
130. H. P. Fink and J. Ganster, Rayon reinforced thermoplastics for injection moulding, In: *Proceedings of Fifth International Symposium: Materials Made of Renewable Resources*, Erfurt, September 2005, pp. 1–9.

CHAPTER 5
Biopolymer-based Nanocomposites

KIKKU FUKUSHIMA,[1] DANIELA TABUANI[2] AND CRISTINA ABBATE[3]

[1] Politecnico di Torino sede di Alessandria, Viale Teresa Michel 5, 15121 Alessandria, Italy; [2] Consorzio PROPLAST, Strada Savonesa 9, 15057 Rivalta Scrivia, Alessandria, Italy; [3] Dipartimento di Scienze Agronomiche, Agrochimiche e delle Produzioni animali (DACPA), University of Catania, Via S. Sofia 98, 95123 Catania, Italy

5.1 Introduction

Synthetic polymers are used for various purposes, especially in the packaging industrial sector; however, the majority of these materials constitutes at present a serious problem for waste management because more than 80% of plastic production is based on polyolefins which are mostly produced from fossil fuels and after consumption can be discarded into the environment, ending up as practically eternal undegradable wastes.[1] In order to solve this problem, biodegradable polymers have attracted a special attention as the plastics of the 21st century, since they can be biologically degraded and, therefore, can be considered as environmental friendly materials.[2] Among all the studied biodegradable polymers, polyesters play a key role due to the presence of highly hydrolysable ester bonds in their chemical structure,[3] making these polymers highly subject to degradation in humid environments as well as in natural conditions considering that enzymes able to degrade esters (esterases) are ubiquitous in nature.

Among these biodegradable polyesters, poly(lactic acid) (PLA) and poly(ε-caprolactone) (PCL) show a great potential for different applications, such as in agriculture and in everyday life as biodegradable packaging material,[4]

because of their facile availability, good biodegradability and good mechanical properties. PLA (from L-lactic acid, D-lactic acid or mixtures of both) can be obtained by means of a fermentation process using glucose from corn. The proportion of the L- and D- isomeric forms will determine the properties of the polymer, *i.e.* if the material is amorphous or semi-crystalline.[5,6] Conversely, PCL is a linear and partially crystalline polymer, synthesized by ring-opening polymerization of cyclic ε-caprolactone in presence of a catalyst.[1]

As far as polymer hydrolytic degradation is concerned, most of the studies on PLA have concentrated on abiotic hydrolysis, which has been reported to take place mainly in the bulk of the material rather than its surface.[7-10] This process has been described as an autocatalytic hydrolysis of PLA occurring homogeneously along sample cross-section: the formation of polylactic acid oligomers, which follows the chain scission, increases the carboxylic acid end groups concentration in the degradation medium, thus rendering the hydrolytic degradation of PLA a self-catalysed and self-maintaining process.[8,9,11]

Concerning PLA biodegradability, it has been shown that PLA is naturally degraded in soil or compost,[12-14] as the resulting products of its hydrolytic degradation in these media can be totally assimilated by microorganisms such as fungi or bacteria.[2,15-17] Even more, the presence of compost microorganisms in the degradation medium seems to accelerate the hydrolytic degradation of PLA compared to the degradation in the corresponding sterile medium.[18]

In the case of PCL, it has been reported that this is a relatively stable material against abiotic hydrolysis;[19] however, it has been found that this polymer can be easily degraded by microorganisms widely distributed in different environments such as compost and soil;[19,20] indeed several authors have reported on highly active bacteria and some fungal phytopathogens towards PCL degradation.[19,21-25] Contrary to PLA, it has been shown that the biodegradation of PCL proceeds by a rapid weight loss through surface erosion with a minor reduction of molecular weight in contrast to its abiotic hydrolysis, which proceeds by a strong reduction in molecular weight combined with minor weight losses.[19]

The main limitations of these biodegradable polymers towards their wider application are their relatively low thermal and mechanical resistance and limited gas barrier properties, which limit their access to certain industrial sectors, such as food packaging, in which their use would be justified when biodegradability is required.[26] Nevertheless, the above drawbacks could be overcome by enhancing their properties through the use of filler and/or additives. In the last two decades, the addition of nanofillers to polymers has attracted great attention for the potentiality of these materials to improve a high number of polymer properties; for example, polymer layered silicate nanocomposites, because of the nanometer size of the silicate sheets, exhibit, even at low filler content (1–5 wt%), markedly improved mechanical, thermal, barrier and flame retardant properties, in comparison to the unfilled matrix and to the more conventional microcomposites.[3,27-29]

Several authors have recently reported on the preparation and characterization of biodegradable polymer-based nanocomposites. Paul *et al.*[7] reported the preparation of PLA/montmorillonite systems by melt intercalation using a montmorillonite modified with bis-(2-hydroxyethyl) methyl (hydrogenated tallow alkyl) ammonium cations. Chang et al.[29] reported the preparation of PLA-based nanocomposites with three different kinds of layered silicates, via solution intercalation method in N-dimethylacetamide, obtaining intercalated nanocomposites independently of the clay nature. Pantoustier et al.[30] used the in situ intercalative polymerization method to obtain PCL-based nanocomposites with high dispersion level. Di et al.[31] reported the preparation of PCL/layered silicate nanocomposites using a twin-screw extruder, determining a dependence of clay dispersion on the processing conditions.

Concerning the biodegradation of polymer nanocomposites, recently most of researchers report that nanofillers are able to accelerate polymer biodegradation. In particular, Ray et al.[32] conducted a respirometric test to study the degradation of PLA and its organically modified layered silicate nanocomposites in a compost environment, observing a catalytic effect of some of these clays towards PLA biodegradation, whereas other nanocomposites presented the same degradation level of neat PLA. They concluded that the observed differences can be related to the different types of clays used which could influence the hydrolysis of the esters bonds.

In other studies, Lee et al.[33] and Paul et al.[34] have found an increased PLA biodegradation rate in compost due to the presence of nanoparticles. This behaviour was attributed to the high relative hydrophilicity of the clays, allowing an easier permeability of water into the bulk material, thus possibly accelerating the hydrolytic degradation process of the polymer matrix.

Biodegradation of PCL-based nanocomposites has been scarcely studied until now. Tetto et al.[35] reported that organically modified layered silicates accelerate PCL biodegradation. Contrarily, Maiti et al.[36] reported that the biodegradation of PCL-based nanocomposites can be depressed by the presence of nanoparticles, probably due to an improvement of the barrier properties of the polymer matrix.

In summary, despite the addition of nano-sized particles that would potentially confer multifunctional enabling properties to these biodegradable polymers for several industrial applications, until now very few works have dealt with the investigation of the real impact of nanoparticles on biopolymer degradation rate and mechanism.

On this background, the general aim of this work is to study the addition of two different nanoparticles on mechanical properties and biodegradation trends of poly(lactic acid) (PLA) and poly(ε-caprolactone) (PCL), playing particular attention on the influence of polymer filler affinity.

5.2 Experimental

5.2.1 Materials and Methods

Poly(lactic acid) (PLA, 4042D) and poly(ε-caprolactone) (PCL, CAPA®6500) of commercial grade were supplied by NatureWorks and SOLVAY S.A. respectively. Commercial modified montmorillonites were supplied by Southern Clay (USA) (CLOISITE 30B) and Süd-Chemie (Germany) (NANOFIL 804). The characteristics of the clays are listed in Table 5.1.

Table 5.1 Characteristics of clays.[37]

Commercial name	Modifier structure	Code
CLOISITE 30B	H_3C, HT, N$^+$, HO, OH	CLO30B
NANOFIL 804	HT, R, N$^+$, $[O]_n$H, $[O]_n$H	NAN804 (n ~ 1)

HT = hydrogenated linear alkyl chains: $C_{8 \sim 18}$.
R = H or CH_3 (conflicting structures reported in literature [37])

Prior to the mixing step, PLA and PCL were dried at 50 °C under vacuum for 8 h and the clays at 90 °C under vacuum for 6 h. Nanocomposites were obtained at 5% clay loading by melt blending using an internal mixer (Rheomix-Brabender OHG 47055) with a mixing time of 5 min at 75 °C and 165 °C for PCL and PLA respectively. The mixing was performed at two different rotor speeds: 30 rpm in the loading step and 60 rpm during mixing. The batch was extracted from the mixing chamber manually, allowed to cool to room temperature in air and ground in a rotatory mill. Sheets were obtained by compression moulding in a hot-plate hydraulic press at 120 °C (PCL) or at 210 °C (PLA) and allowed to cool to room temperature under pressure. All characterizations were made on 0.6 mm films, except the WAXS analysis which was performed on 3 mm films. Biodegradation tests were performed on 0.13 mm films.

The characterization of the materials was performed in order to evaluate their morphology (X-ray diffraction and electron microscopy) and mechanical properties through dynamic mechanical analysis.

Wide angle X-ray spectra (WAXS) were recorded using a Thermo ARL diffractometer X-tra 48, at room temperature, in the range 1–30° (2θ) (step size = 0.02°, scanning rate = 2 s per step) by using filtered Cu K$_\alpha$ radiation (λ = 1.54 Å).

Scanning electron microscopy (SEM) was carried out on the cryogenic fracture surfaces of the specimens coated by sputtering with gold, using a Leo 14050 VP SEM apparatus equipped with energy dispersive spectroscopy (EDS).

Transmission electron microscopy (TEM) was performed using a high-resolution transmission electron microscope (JEOL 2010). Ultrathin sections about 100 nm thick were cut with a Power TEOMEX microtome equipped with a diamond knife and placed on a 200-mesh copper grid.

Dynamic mechanical thermal analysis (DMTA) was performed on compression moulded 6 × 20 × 0.6 mm^3 films, using a DMTA TA Q800 in tension film clamp. The analysis temperature range was from +25 to +90 °C (in the case of PLA) and from –150 to +50 °C (in the case of PCL), at a heating rate of

2 °C min^{-1}, 1 Hz frequency, preload of 0.4 N (for PCL) and 0.01 N (for PLA), in strain controlled mode and 15 micron of amplitude. All tests were carried out according to the International Standard UNI EN ISO 6721.

5.2.2 Biodegradation Conditions and Evaluation Methods

Biodegradation in compost was performed on compression moulded 25 × 25 × 0.13 mm^3 films at 40 °C by putting them into contact with a compost supplied by Società Metropolitana Acque Torino (SMAT) made of sludges from wastewater treatment plants, wood chips, green clippings, dried leaves and straw, keeping a relative humidity of approximately 50–70%. Around 35–40 samples of each material were vertically buried at 4–6 cm depth to guarantee aerobic degradation conditions at a horizontal distance of 5–6 cm between samples. At selected times, 3 or 4 samples of each material were washed with water and dried at room temperature to constant weight.

The surface of the degraded films extracted from the compost was observed using a LEICA MS5 optical microscope.

Average molecular weights (M_n and M_w) and polydispersity index (M_w/M_n) of degraded samples were determined by means of size exclusion chromatography (SEC). The measurements were made at 20 °C in $CHCl_3$ with a Waters instrument, model HPLC Pump 515, provided with four Waters Styragel high resolution columns, using a differential refraction index detector Waters R401. 200 μL of polymer solution with a concentration of 5 mg mL^{-1} were injected at 1 mL min^{-1}. The number and weight average molecular weights (M_n and M_w, respectively) were calculated from the SEC curve through the Polymer Lab Caliber® software, supplied by Polymer Laboratories, using a calibration curve from standard polystyrene (Labservice Analytica).

5.3 Results and Discussions

5.3.1 Characterization

5.3.1.1 Morphology

WAXS pattern of the polymer matrices (not shown here) are not influenced by the presence of the nanoparticles, neither in the case of PLA nor in that of PCL.

Figure 5.1 represents the WAXS patterns of PLA- and PCL-based nanocomposites, as well as of pristine clays. CLO30B is characterized by a single diffraction peak at $2\theta = 5.0°$ corresponding to the basal reflection (001) and accounting for a 1.8 nm interlayer distance. On the other hand, NAN804 is characterized by two diffraction peaks at $2\theta = 5.0°$ and 1.9°, the former attributed, as in the case of CLO30B, to the (001) basal reflection, the latter to the insertion of two molecular layers of organic modifier between clay layers or to a bulkier spatial arrangement of the modifier.[38]

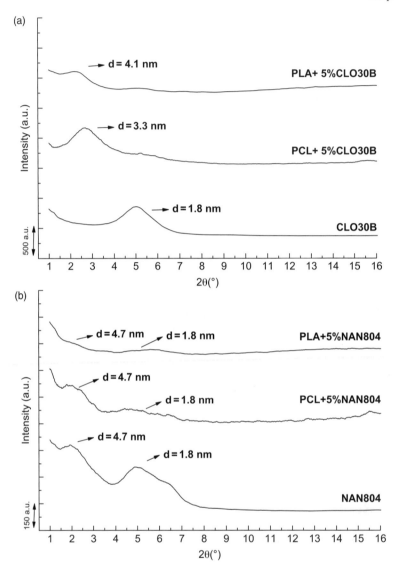

Figure 5.1 WAXS patterns of CLO30B, NAN804 and nanocomposites of PLA and PCL with (a) 5% CLO30B and (b) 5% NAN804.[38] Reproduced from K. Fukushima *et al.* by permission of Elsevier Science Ltd., UK.

CLO30B presents a good interaction with both polymers, especially with PLA (Figure 5.1a), as a shift of the clay diffraction peak to lower angles can be observed, corresponding to an increase of the interlayer distance of 1.5 and 2.3 nm for PCL and PLA, respectively. Moreover, in the case of PLA, a significant decrease of the WAXS peak intensity can be observed and attributed to the formation of a disordered structure. In the case of NAN804-based materials

(Figure 5.1b), an almost complete disappearance of WAXS signals is observed for PLA suggesting extensive dispersion of the clay in the polymer matrix. In PCL, the diffraction peak around 5.0° is reduced in intensity and that at 1.9° is broadened indicating reduced clay dispersion as compared to PLA.

Scanning electron microscopy (SEM) reveals a good degree of distribution of CLO30B in PLA (Figure 5.2a), as no aggregates can be observed. On the other hand, in the case of NAN804/PLA composites, aggregates of about 5–10 μm are detected by the analysis (see circles in Figure 5.2b); taking into account the absence of X-ray diffraction peak in Figure 5.1b, we can conclude that part of NAN804 layers are strongly disordered but not individually separated in the polymer matrix.

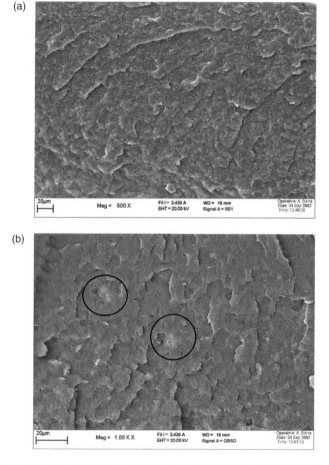

Figure 5.2 Scanning electron micrographs of nanocomposites of PLA with (a) 5% CLO30B and (b) 5% NAN804.[38] Reproduced from K. Fukushima *et al.* by permission of Elsevier Science Ltd., UK.

A similar clay dispersion is observed in the case of PCL-based materials (Figure 5.3), with a good distribution and dispersion level of CLO30B in the matrix (Figure 5.3a), whereas PCL/NAN804 system shows residual aggregates of about 10–20 μm along all the polymer matrix (see circle in Figure 5.3b).

In order to better understand PLA/nanoclay interactions, PLA-based materials were further analysed by means of TEM. PLA nanocomposites based on CLO30B show a high level of intercalation and exfoliation of the silicate layers (Figure 5.4a), as small stacks and single dispersed layers can be observed in the TEM micrograph, results that are in accordance with WAXS analysis. The incorporation of NAN804 into the PLA matrix shows a certain level of intercalation as well as the occurrence of micro-aggregates (Figure 5.4b).

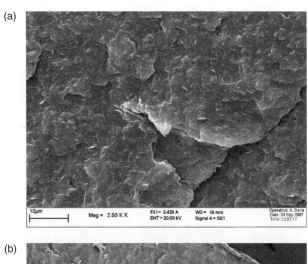

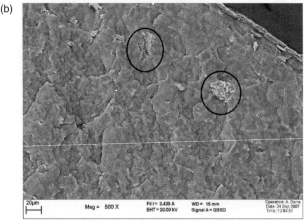

Figure 5.3 Scanning electron micrographs of nanocomposites of PCL with (a) 5% CLO30B and (b) 5% NAN804.

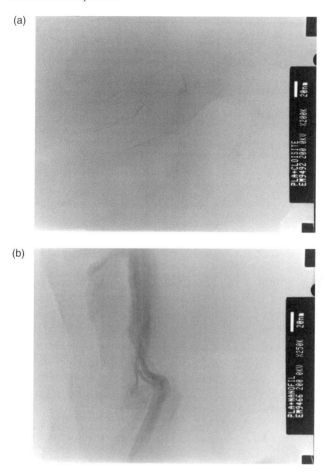

Figure 5.4 TEM patterns of (a) PLA + 5% CLO30B and (b) PLA + 5% NAN804.[38] Reproduced from K. Fukushima *et al.* by permission of Elsevier Science Ltd., UK.

Associating these observations with WAXS and SEM results, we can conclude that the significant decrease of the peak intensity of NAN804 in the composite spectra should be due to disordering by intercalation rather than by exfoliation of the silicate layers.

Evaluating WAXS, SEM and TEM results, it is possible to say that the dispersion and affinity of CLO30B in both polymers is higher than what obtained with NAN804 and this is attributed to higher interactions between the polymers and CLO30B, originated from the hydrogen bonding between the carbonyl groups of the polymers and the hydroxyl groups of CLO30B organic modifier.[1,33,38] Although both CLO30B and NAN804 are characterized by the presence of hydroxyl groups in the organic modifier, it can be assumed that the shorter alkyl chains of CLO30B organic modifier make hydroxyl groups more available for interactions with the polymers than the olygomeric structure of NAN804, thus allowing for a higher dispersion degree.[38]

5.3.1.2 Dynamic-Mechanical Thermal Analysis (DMTA)

The dynamic-mechanical experiments of PLA, PCL and their nanocomposites are reported in Tables 5.2 and 5.3, which show the temperature dependence values of storage modulus (E') for the above materials. These results show that above glass transition temperature of the polymers ($T_{g\ PLA} \sim +50$ to $+60\,°C$ and $T_{g\ PCL} \sim -50$ to $-60\,°C$) considerable increases in E' were found by the addition of both clays, whereas below these respective temperatures the influence of the clays is almost negligible. Such increases indicate that the addition of clays induces reinforcement effects related to enhancements of the thermomechanical stability of PLA and PCL matrices,[39] especially for PLA nanocomposites possibly due to higher polymer/filler interactions as compared to PCL ones.

The more pronounced increases of E' with temperature with respect to that of the matrices are attributed to a higher reinforcement effect of the clays due to restricted movements of the polymer chains above T_g. This phenomenon leads to an increase of E' of 145% and 25% for PLA/CLO30B and PLA/NAN804 at 80 °C (Table 5.2), and of 52% and 19% for PCL/CLO30B and PCL/NAN804 at 30 °C (Table 5.3).

At the same time, the highest increment in E' brought by the addition of CLO30B can be accounted for by its higher dispersion level into the polymers as compared to NAN804.

Table 5.2 E' value of PLA based nanocomposites at different temperatures. Data extracted from K. Fukushima et al.[38] by permission of Elsevier Science Ltd., UK.

Sample	E' at 30 °C (MPa)	E' at 60 °C (MPa)	E' at 80 °C (MPa)
PLA	3058	1544	5
PLA + CLO30B	3573	2761	12
PLA + NAN804	3109	2462	6

Table 5.3 E' value of PCL based nanocomposites at different temperatures. Data extracted from K. Fukushima et al.[38] by permission of Elsevier Science Ltd., UK.

Sample	E' at -125 °C (MPa)	E' at -75 °C (MPa)	E' at 30 °C (MPa)
PCL	4715	3451	503
PCL + CLO30B	5071	3882	762
PCL + NAN804	4864	3663	599

5.3.2 Biodegradation

5.3.2.1 *Physical Alterations*

The preliminary phase of this study was developed in order to determine, by visual observation, the degradation of the materials after contact with compost. Figure 5.5 shows the PLA-based materials before and after different times in compost: all specimens show a considerable surface deformation and whitening, this being more evident for the nanocomposites after only 3 weeks of degradation. The appearance of these degradation signs has been associated to the effective starting of a hydrolytic degradation process of the polymer matrix,[5,15,37,40] thus inducing a refractive index change of the specimen as consequence of a high water absorption level and/or the presence of degradation products into the bulk material formed by the hydrolytic bond scission.

It is worth noticing that the degradation of PLA in compost has been reported to occur in two main phases: during the initial phases of degradation, the high molecular weight PLA chains are hydrolysed to form lower molecular weight chains;[14] only when a certain molecular weight is reached through this process, the compost microorganisms are able to catalyse the degradation by further hydrolytic scission of ester groups into acid and alcohol and finally to convert the lower molecular chains to CO_2, water and humus. Therefore, any factor which increases the hydrolysis tendency of PLA matrix can in principle accelerate the degradation of PLA in compost.[37]

The faster appearance of visual signs of degradation in nanocomposites as compared to neat PLA (Figure 5.5) can be associated to the high presence of hydroxyl groups belonging to the silicate layers of the clays used, which, being highly dispersed in the polymer matrix, can have a catalytic effect on hydrolysis of the polymer ester groups in this medium.[37,41]

However, despite the degradation observed for all samples, these do not present considerable weight decreases within experimental error ($\pm 5\%$) up to 39 weeks. Since surface degradation generally leads to release of degradation products into the medium of degradation,[14] these results indicate either that these products at this stage are adherent to the material surface and/or that the compost microorganisms are not able to noticeably assimilate them. Simultaneously, it is also possible that, in the studied period of time, the degradation mainly proceeds in the bulk material and the release rate of degradation products into the compost is relatively slow.[37] In the case of PCL-based materials, after only 3 weeks in compost, these show a significant visual degradation with roughing and formation of holes along their surface (as shown in Figure 5.6), indicating a faster degradation process as compared to PLA-based materials.

Optical microscopy (Figure 5.7), confirms the considerable level of surface degradation for all PCL-based systems, especially in the case of neat PCL. It is important to highlight that this surface degradation seems to proceed through an inhomogeneous mechanism, taking place at different extents on the sample surface. Note that some surface areas are more degraded (arrows in Figure 5.7) than others (circles in Figure 5.7) – probably due to a different contact level of

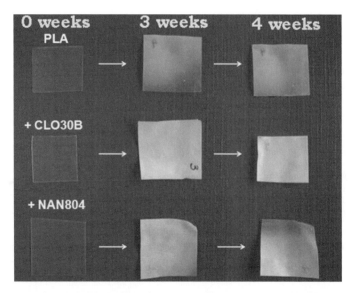

Figure 5.5 PLA and nanocomposites based on CLO30B (+CLO30B) and NAN804 (+NAN804) before degradation (0 weeks) and after 3 and 4 weeks of degradation in compost at 40 °C.[37] Reproduced from K. Fukushima *et al.* by permission of Elsevier Science Ltd., UK.

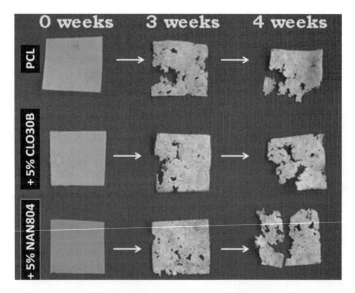

Figure 5.6 PCL and nanocomposites based on CLO30B (+5% CLO30B) and NAN804 (+5% NAN804) before degradation (0 weeks) and after 3 and 4 weeks of degradation in compost.[42] Reproduced from K. Fukushima *et al.* by permission of Elsevier Science Ltd., UK.

Biopolymer-based Nanocomposites 141

the compost along all the specimen surface and/or to an inhomogeneous growth of the microorganisms on the sample surface, leading to fragmentation of the specimen as well as loss of degradation products by migration into the compost.[42]

The significant and fast degradation, previously observed by visual observation and through optical micrographs, can be confirmed by the important weight decreases obtained by all PCL-based materials in short times in compost (Figure 5.8), observing an almost complete specimen weight loss after 8 weeks and indicating the considerable fragmentation level of PCL matrix in this medium.[42] This trend is not influenced by the presence of the clays.

It is worth noticing the different rate and mechanism of degradation observed for PLA- and PCL-based materials in the same compost. These differences could be associated to the different polymer chemical structure of these

Figure 5.7 Optical micrographs of (a) PCL, (b) PCL + CLO30B and (c) PCL + NAN804 after 6 weeks in compost.[42] Reproduced from K. Fukushima *et al.* by permission of Elsevier Science Ltd., UK.

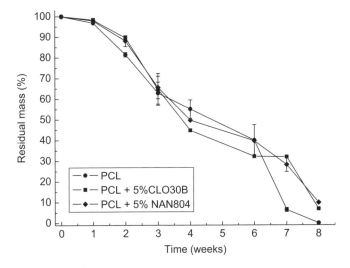

Figure 5.8 Residual mass of PCL and nanocomposites as a function of degradation time in compost.[42] Reproduced from K. Fukushima *et al.* by permission of Elsevier Science Ltd., UK.

polymers and/or to the respective lower glass transition temperature of PCL ($T_g \sim$ –50 to –60 °C) as compared to PLA ($T_g \sim$ + 50 to + 60 °C), as during process of degradation at 40 °C in compost the larger mobility of PCL chains may favour either a faster attack by microorganisms and a preferentially surface degradation mechanism.

5.3.2.2 Size Exclusion Chromatography (SEC)

Size exclusion chromatography (SEC) results of PLA-based materials, before (week 0) and after composting (weeks 4 and 17) are reported in Table 5.4. A considerable decrease of number and weight average molecular weights (M_n and M_w, respectively) after degradation is shown, indicating the effective degradation level of PLA and its nanocomposites in this medium.

The highest decreases are observed for the number average molecular weight (M_n). Indeed after 17 weeks, M_n of the unfilled PLA decreases by ca. 55% with respect to its initial value (from 72 743 g mol^{-1} to 32 775 g mol^{-1}, Table 5.4 and Figure 5.9). In parallel, the losses of M_n for PLA-based nanocomposites are 79% and 41% in the case of CLO30B and NAN804, respectively.

Pure PLA presents M_w decreases from 149 593 g mol^{-1} to 113 096 g mol^{-1}, losing ca. 24% of its M_w after 17 weeks (Table 5.4); PLA/NAN804 shows a decrease of 34% and the highest M_w decrease is again observed for the system PLA/CLO30B (ca. 52%). The above results, considering that the samples do not considerably lose weight during degradation, are a consequence of chain scission occurring during degradation, thus increasing the fraction of the lower molecular weight molecules. Simultaneously, the observed differences in the degradation extent of the polymer matrix between nanocomposites could be related to the different clay dispersion level obtained for these systems, taking into account that bulk hydrolytic degradation of PLA has been reported to start from the interface between polymer and fillers.[8]

Indeed, the larger decrease of both M_w and M_n in PLA/CLO30B as compared to PLA, shows that the presence of CLO30B in PLA accelerates the hydrolytic degradation process of PLA, likely because of a catalysed degradation due to the presence of terminal hydroxylated edge groups of the silicate layers.[8,37,41,43] On the other hand, the higher decrease of M_w and lower decrease

Table 5.4 SEC data on PLA and its nanocomposites at different times of degradation in compost. Data extracted from K. Fukushima et al.[37] by permission of Elsevier Science Ltd., UK.

	Week 0		Week 4		Week 17	
Sample	Mn (g/mol)	Mw (g/mol)	Mn (g/mol)	Mw (g/mol)	Mn (g/mol)	Mw (g/mol)
PLA	72 743	149 593	67 300	138 181	32 775	113 096
PLA + CLO30B	64 270	135 768	40 412	120 642	13 461	64 817
PLA + NAN804	47 528	103 217	34 559	74 128	27 911	67 341

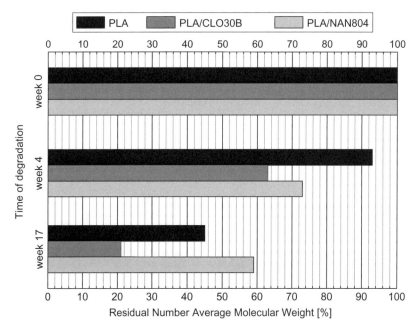

Figure 5.9 Residual number average molecular weight of PLA and its nanocomposites before and after 4 and 17 weeks of degradation in compost.

of M_n in PLA/NAN804 as compared to PLA indicates that, apparently, higher molecular weight PLA molecules are preferentially hydrolysed as compared to neat PLA or PLA/CLO30B, possibly because of a coarser dispersion of the clay in PLA as compared to CLO30B as previously observed through WAXS, SEM and TEM analysis.[42]

As far as SEC characterizations of PCL-based materials is concerned, only PCL-based samples up to week 6 were taken into consideration in this work. We think that the study up to week 6 is significant enough for the determination of the degradation mechanism of the PCL-based materials.

Size exclusion chromatography (SEC) results of the PCL-based materials before and after degradation are reported in Table 5.5. All specimens show decreases of the M_n and M_w with degradation time, indicating the effective degradation of the materials in short times.

Similar to PLA-based materials, the highest decreases upon degradation can be observed for M_n. Indeed, after only 6 weeks, in the case of neat PCL, it decreases by *ca.* 71% with respect to its initial value. In parallel, the losses of M_n for the nanocomposites are 48% and 66% in the case of CLO30B and NAN804, respectively (Figure 5.10). It is also important to underline that in the case of CLO30B, an almost steady state situation is observed until 4 weeks of degradation, after this period of time a significant decrease in M_n can be detected. Conversely, the PCL/NAN804 material follows a similar trend as pristine PCL.

Table 5.5 SEC data of PCL and its nanocomposites before and after degradation in compost. Data extracted from K. Fukushima et al.[42] by permission of Elsevier Science Ltd., UK.

Sample	Week 0		Week 4		Week 6	
	M_n (g/mol)	M_w (g/mol)	M_n (g/mol)	M_w (g/mol)	M_n (g/mol)	M_w (g/mol)
PCL	73 751	115 555	28 989	109 405	21 464	100 288
PCL + CLO30B	72 074	123 179	71 753	123 938	37 737	109 344
PCL + NAN804	77 569	107 656	43 577	115 082	26 034	105 224

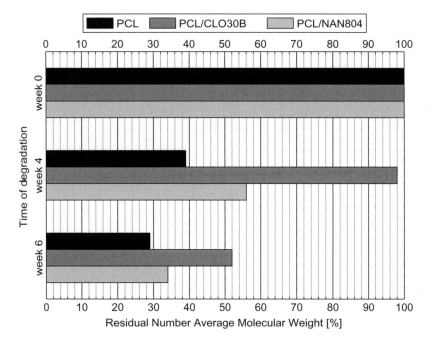

Figure 5.10 Residual number average molecular weight of PCL and its nanocomposites before and after 4 and 6 weeks of degradation in compost.

M_w decreases in unfilled PCL from 115 555 g mol^{-1} to 100 288 g mol^{-1}, losing ca. 13% after 6 weeks (see Table 5.5) and the losses for nanocomposites were 11% and 2% in the case of CLO30B and NAN804, respectively.

The larger decrease of M_n as compared to M_w indicates that in the degraded PCL samples, the quantity of polymer chains shortened by hydrolysis reaction is larger than that of low molecular weight present in the original sample. This is noticeable by observing that M_w is only slightly lower in the degraded specimens.[42] This behaviour was previously obtained in the case of PLA-based materials; however, the mechanistic explanation for the higher decreases of M_n as compared to M_w upon degradation is different in the two situations because, in the studied time of degradation, weight losses of PLA upon degradation were insignificant, whereas those of PCL were extensive. Thus, preferential release of low molecular weight polymer chains into the compost must be taken into account in PCL.

Interestingly enough and contrarily to what obtained for PLA-based materials, the higher decreases of M_n and M_w were found for neat PCL as compared to nanocomposites, indicating that, in this case, the presence of the nanoparticles tends to delay the polymer degradation process. Considering that PCL degradation seems to mainly proceed from the surface to the interior of the sample, it is possible that the presence of the silicate layers acts as a barrier towards microbial attack on PCL ester groups, slowing down the diffusion of enzymes into the polymer matrix. We can assume that adhesion of PCL and/or of enzymes macromolecules from compost microorganisms to clay layers could partially delay the polymer degradation due to less suitable macromolecular conformations for hydrolysis reactions.[42] Indeed, the higher delaying effect is observed upon the addition of CLO30B and possibly because of its higher dispersion level in the PCL matrix as compared to NAN804.

In summary, the effect of nanoparticles on polymer biodegradation was found to be highly dependent on the polymer matrix type and in particular on the type of degradation mechanism to which the polymer is subjected.

5.4 Conclusions

This work addresses the study of the effect of two different organically modified montmorillonites (CLO30B, NAN804) on several properties and biodegradation trends of polylactic acid (PLA) and poly(ε-caprolactone) (PCL). As far as mechanical properties are concerned, the highest increases in storage modulus were obtained for PLA nanocomposites compared to PCL ones, due to higher polymer/filler interactions.

For both polymers, the most considerable enhancements were found upon the addition of CLO30B because of its high dispersion and attributed to the increased number of organic modifier hydroxyl groups available to promote polymer/fillers interactions.

The degradation of PLA, PCL and their nanocomposites in compost under controlled conditions was also studied in this work. According to most literature, the addition of clays in biopolymers could accelerate their biodegradability level in compost; however, in this work we could observe that this is not an iron rule as nanoparticles can have different effects on the polymer degradation trend depending on the degradation mechanism of neat polymer and on the level of clay dispersion into the polymer.

Two different mechanisms of degradation in compost were assumed for neat PLA and PCL, predominant bulk degradation for PLA and a preferentially surface degradation for PCL. The type of dominating degradation mechanism seems to depend on both the structure of the polymer, physical properties (glass transition temperature) and on the environment it is subjected to. In summary, nanoparticles seem to affect polymer degradation as follows:

(1) When the degradation mainly occurs through a bulk mechanism (PLA), the addition of nanoclays was found to increase the polymer degradation rate because of the presence of hydroxyl groups belonging to the silicate layers; this phenomenon can be particularly evident for the highest dispersed clay in the polymer (PLA/CLO30B).
(2) When the degradation occurs through a mainly surface mechanism (PCL), highly dispersed layered clays can partially delay the polymer degradation rate probably due to a more difficult route for microorganisms in order to attack the polymer ester groups; this last phenomenon can be more evident at higher clay dispersion levels in the polymer (PCL/CLO30B).

Acknowledgements

The authors would like to thank gratefully Prof. Giovanni Camino and Prof. Mara Gennari for their great support and guidance throughout this work. The authors would like to acknowledge Dr Orietta Monticelli from Genoa University for TEM characterizations, Dr Paola Rizzarelli and Loredana Ferreri from CNR ICTP (University of Catania) for SEC characterizations, Dr Michele Negre from Università degli Studi di Torino for degradation tests in compost, as well as EU NoE 'NANOFUN POLY' for financial support.

References

1. S. Sinha Ray and M. Bousmina, *Progr. Mater. Sci.*, 2005, **50**, 962.
2. A. Hoshino, M. Tsuji, M. Ito, M. Momochi, A. Mizutani, K. Takakuwa, S. Higo, H. Sawada and S. Uematsu, in *Biodegradable Polymers and Plastics*, ed. E. Chiellini and R. Solaro, Kluwer Academic/Plenum Publishers, New York, 2003, p. 47.
3. E. Pollet, M. A. Paul and P. Dubois, in *Biodegradable Polymers and Plastics*, ed. E. Chiellini and R. Solaro, Kluwer Academic/Plenum Publishers, New York, 2003, p. 535.
4. H. Tsuji and T. Ishizaka, *Macromol. Biosci.*, 2001, **1**, 59.
5. S. Li, A. Girard, H. Garreau and M. Vert, *Polym. Degrad. Stab.*, 2001, **71**, 61.
6. L. Cabedo, J. L. Feijoo, M. P. Villanueva, J. M Lagarón, J. J. Saura and L. Jiménez, *Revista de Plásticos Modernos*, 2005, **89**, 177.
7. M. A. Paul, M. Alexandre, P. Degée, C. Henrist, A. Rulmont and P. Dubois, *Polymer*, 2003, **44**, 443.
8. Q. Zhou and M. Xanthos, *Polym. Degrad. Stab.*, 2008, **93**, 1450.

9. H. Tsuji, in *Biopolymers for Medical and Pharmaceutical Applications*, ed. A. Steinbüchel and H. Robert, Wiley-VCH Weinheim, Germany, 2005, p. 183.
10. I. Grizzi, H. Garreau, S. Li and M. Vert, *Biomaterials*, 1995, **16**, 305.
11. H. Tsuji and Y. Ikada, *Polym. Degrad. Stab.*, 2000, **67**, 179.
12. H. Tsuji, A. Mizuno and Y. Ikada, *J. Appl. Polym. Sci.*, 1998, **70**, 2259.
13. V. M. Ghorpade, A. Gennadios and M. A. Hanna, *Bioresour. Technol.*, 2001, **76**, 57.
14. K. L. G. Ho, A. L. Pometto, A. Gadea, J. A. Briceño and A. Rojas, *J. Environ. Polym. Degrad.*, 1999, **7**, 173.
15. L. Liu, S. Li, H. Garreau and M. Vert, *Biomacromolecules*, 2000, **1**, 350.
16. H. Tsuji, Y. Tezuka and K. Yamada, *J. Polym. Sci. Part B: Polym. Phys.*, 2005, **43**, 1064.
17. S. Li, M. Tenon, H. Garreau, C. Braud and M. Vert, *Polym. Degrad. Stab.*, 2000, **67**, 85.
18. M. Hakkarainen, S. Karlsson and A. C. Albertsson, *Polymers*, 2000, **41**, 2331.
19. M. Hakkarainen, *Adv. Polym. Sci.*, 2002, **157**, 113.
20. M. Shimao, *Biotechnology*, 2002, **12**, 242.
21. H. Nishida and Y. Tokiwa, *J. Environ. Polym. Degrad.*, 1993, **1**, 227.
22. J. Mergaert and J. Swings, *J. Ind. Microbiol.*, 1996, **17**, 463.
23. T. Suyama, Y. Tokiwa, P. Ouichanpagdee, T. Kanagawa and Y. Kamagata, *Appl. Environ. Microbiol.*, 1998, **64**, 5008.
24. C. A. Murphy, J. A. Cameron, S. J. Huang and R. T. Vinopal, *Appl. Environ. Microbiol.*, 1996, **62**, 456.
25. H. Nishida and Y. Tokiwa, *Chem. Lett.*, 1994, **8**, 1547.
26. R. P. Singh, J. K. Pandey, D. Rutot, Ph. Degée and Ph. Dubois, *Carbohydr. Res.*, 2003, **338**, 1759.
27. L. A. Utracki, *Clay-Containing Polymeric Nanocomposites*, Rapra Technology Limited, UK, 2004.
28. M. Pluta, A. Galeski, M. Alexandre, M. A. Paul and P. Dubois, *J. Appl. Polym. Sci.*, 2002, **86**, 1497.
29. J. H. Chang, Y. U. An and G. S. Sur, *J. Polym. Sci. Part B: Polym. Phys.*, 2003, **41**, 94.
30. N. Pantoustier, B. Lepoittevin, M. Alexandre, P. Dubois, D. Kubies, C. Calberg and R. Jérôme, *Polym. Eng. Sci.*, 2002, **42**, 1928.
31. Y. Di, S. Iannace, E. D. Maio and L. Nicolais, *J. Polym. Sci. Part B: Polym. Phys.*, 2003, **41**, 670.
32. S. Sinha Ray, K. Yamada, M. Okamoto, A. Ogami and K. Ueda, *Chem. Mater.*, 2003, **15**, 1456.
33. S. R. Lee, H. M. Park, H. Lim, T. Kang, X. Li, W. J. Cho and C. S. Ha, *Polymers*, 2002, **43**, 2495.
34. M. A. Paul, C. Delcourt, M. Alexandre, P. Degée, F. Monteverde and P. Dubois, *Polym. Degrad. Stab.*, 2005, **87**, 535.
35. J. A. Tetto, D. M. Steeves, E. A. Welsh and B. E. Powell, *Proceedings ANTEC'99*, 1999, 1628.

36. P. Maiti, C. A. Batt and E. P. Giannelis, *Polym. Mater. Sci. Eng.*, 2003, **88**, 58.
37. K. Fukushima, C. Abbate, D. Tabuani, M. Gennari and G. Camino, *Polym. Degrad. Stab.*, 2009, **94**, 1646.
38. K. Fukushima, D. Tabuani and G. Camino, *Mater. Sci. Eng. C*, 2009, **29**, 1433.
39. S. Sinha Ray and M. Okamoto, *Prog. Polym. Sci.*, 2003, **28**, 1539.
40. S. Li and S. McCarthy, *Biomaterials*, 1999, **20**, 35.
41. S. Sinha Ray, K. Yamada, M. Okamoto and K. Ueda, *Polymers*, 2003, **44**, 857.
42. K. Fukushima, C. Abbate, D. Tabuani, M. Gennari and G. Camino, *Mater. Sci. Eng. C*, 2010, **30**, 566.
43. H. Tsuji, K. Nakahara and K. Ikarashi, *Macromol. Mater. Eng.*, 2001, **286**, 398.

CHAPTER 6
Biodegradable Polyesters: Synthesis and Physical Properties

JASNA DJONLAGIC AND MARIJA S. NIKOLIC

Belgrade University, Faculty of Technology and Metallurgy, Karnegijeva 4, 11000 Belgrade, Serbia

6.1 Introduction

Biodegradable polymers have come into the focus of research and development as a proposed solution to the increasing environmental problems associated with the disposal of traditional commodity plastics.[1-10] The use of a polymeric material which will disintegrate in nature after it has served its function, in contrast to the currently used, long-lasting polymeric materials, would certainly be the best solution. From the view point of depletion of petrochemical resources, the production of such biodegradable polymeric material from renewable sources would positively contribute to the issue of sustainable material development.[11] Additionally, in the field of medicine, there is also a growing requirement for degradable polymeric materials which could temporarily serve their function, after which they would be eliminated from the body.[12-17] Thus, environmental protection and biomedicine are the two main fields of development and application of biodegradable polymers.

In order to be biodegradable, polymers have to contain groups in the polymer main chain which can undergo scission under biological conditions. Polyesters are a class of polymers which contain hydrolytically labile ester linkages. Thus, they are susceptible to degradation in biological environments, where the conditions for hydrolysis are readily met. Polyester hydrolysis can proceed with the help of enzymes produced by certain organisms; however, they can also degrade, at least partially, without the catalytic action of enzymes. Nevertheless, polyesters which predominately undergo abiotic hydrolysis are also termed biodegradable, since the degradation occurs in a biological environment. In the case of medical applications of biodegradable polymers, it is also important that the polymer is completely eliminated from the body, in which case it is termed bioresorbable.[18]

The classification of biodegradable polyesters can be made according to their origin: those of natural origin, such as poly(3-hydroxybutyrate), and those synthesized chemically. The subgroup of chemically synthesized polyesters contains the polyesters derived from renewable resources, such as poly(lactic acid) and those obtained from petrochemicals, such as poly(ε-caprolactone) and poly(butylene succinate) (Figure 6.1). Chemically synthesized polyesters can be obtained in two principal ways: by direct esterification reaction of hydroxy acids or diacids and diols, or by ring-opening polymerization, ROP, of lactones or cyclic diesters.

The polycondensation reaction is not a favorable way to obtain high molecular weight polyesters. The first investigated polyesters obtained in the direct polycondensation reaction from diols and diacids in pioneering work of Carothers had very low molecular weights of a few thousands, with a large proportion of oligomeric products.[19,20] The difficulty in obtaining polyesters in the polycondensation reaction lies in the strict requirement for stoichiometric amounts of diols and diacids and in equilibrium character of the reaction. Polycondensation reaction has to be driven to high yields in order to obtain high molecular weights by the continuous removal of the by-product from the highly viscous melt. Thus, beside high temperature which is favorable for the esterification reaction, long reaction times, due to progressive decrease in reactive chain ends, are necessary to drive the reaction to completion and so achieve a polyester with the desired properties. One more difficulty arising from

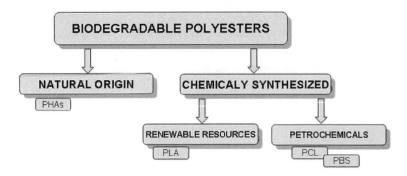

Figure 6.1 Classification of biodegradable polyesters according to their origin with selected examples: PHAs, poly(hydroxyalkanoate)s; PLA, poly(lactic acid); PCL, poly(ε-caprolactone); PBS, poly(butylene succinate).

the character of this reaction is the requirement for highly pure monomers, since monofunctional impurities can interfere with the course of reaction and greatly affect the molecular weight of the final product.

Compared to polycondensation reaction, ROP of cyclic esters proceeds under milder reaction conditions, does not require long reaction times and polymers of high molecular weights can be easily obtained. Additionally, polymer structure and molecular weights and weight distributions are more easily controlled in ROP reactions. Disadvantage of this process is the higher production costs compared to polycondensation reactions.

Aliphatic biodegradable polyesters are usually semicrystalline polymers with a range of melting temperatures. However, for the majority of polyesters, the melting and crystallization temperatures are low, which limits their processability and application. Among the different aliphatic polyesters, with a variety of physical properties, similarities with some commodity plastics, such as polyethylene (PE), poly(propylene) (PP), poly(styrene) (PS), or poly(ethylene terephthalate) (PET), are to be found. The biodegradability properties not only depend on chemical composition, but also on higher order structure, such as degree of crystallinity, crystalline form and crystalline structure. The biodegradability window of aliphatic polyesters is wide since the degradation properties can be tuned by varying the chemical structure, which also has an impact on the higher order structure.

In creating a biodegradable polyester with degradation behavior suitable for the intended temporary application, it is essential to understand the mode of degradation to which it is prone and how it can be controlled. Abiotic hydrolysis of the main chain ester bonds usually accompanies the biodegradation process of aliphatic polyesters and is of primary importance in medical applications. The rate of polyester chain cleavage can be controlled either by the material properties or by external factors influencing hydrolysis reaction such as pH or temperature. The erosion of polyester material, as a consequence of degradation, can proceed in two ways: surface or bulk erosion. Surface eroding polyesters lose their mass from the surface, which is accompanied by a thinning of the specimen. The molecular weight is not affected in this mode of erosion, since the oligomers formed at the surface can leave it by dissolution in the external medium. Bulk eroding polyesters degrade through the whole mass of the specimen, the shape and size of which is not affected, while the molecular weight decreases. For certain applications, it is important that polymer erosion proceeds in particular manner; a typical example being the controlled release of active substances from a polymer matrix, when surface erosion would be preferable. Surface or bulk erosion are only two ideal cases and the mode of erosion can be influenced by number of factors, such as the diffusivity of water and small oligomers through the polymer matrix, the morphology and size of the sample and the rate of degradation.[21,22] Autocatalysis of the hydrolytic degradation can also have an effect on the erosion of a polymer sample.

The degradation of a polyester in a biological environment is a complex process. Hydrolysis of the ester bond can be catalyzed by specific enzymes. Extracellular enzymes which catalyze hydrolytic degradation of polyesters are too large to penetrate into the polymer matrix and their action is restricted only to the surface. Microorganisms which promote polyester degradation can act in a mechanical, chemical or enzymatic way.[23] The polymeric material is first extracellularly fragmented into substances which can enter the cells of microorganisms, where they are finally mineralized. The actions of microorganisms depend on the polymer composition and structure and on external conditions, such as temperature and humidity. Thus, besides investigations of the degradation of polymers by pure enzymes or microorganisms under laboratory conditions, the ultimate fate of polymeric material under environmental conditions has also to be tested. A number of tests for polymers intended for use as environmentally friendly alternatives to currently used bioinert plastics have been developed to assess their biodegradability.[24]

In the following sections, the synthesis and properties, with particular reference to their biodegradability, of the main members of the biodegradable polyester family, which are the most addressed in the scientific community and are promising candidates for commercial application, will be presented. The names, acronyms and structural formulas of the most important aliphatic biodegradable polyesters are given in Table 6.1.

Table 6.1 Structural formulas and acronyms of the main members of the aliphatic biodegradable polyesters family.

Polymer	Repeating unit	Example (Acronym)
Poly(α-hydroxy acid)	$-[-O-CHR-C(=O)-]-$	$R=H$, poly(glycolic acid), PGA $R=CH_3$, poly(lactide), PLA
Poly(ω-hydroxyalkanoate)	$-[-O-(CH_2)_x-C(=O)-]-$	$x=3$, poly(γ-butyrolactone), PBL $x=4$, poly(δ-valerolactone), PVL $x=5$, poly(ε-caprolactone), PCL
Poly(hydroxyalkanoate) (microbial polyesters)	$-[-O-CHR-(CH_2)_x-C(=O)-]-$	$R=CH_3$, $x=1$, poly(3-hydroxybutyrate), P(3HB) $R=H$, $x=2$, poly(4-hydroxybutyrate), P(4HB)
Poly(alkylene dicarboxylate)	$-[-C(=O)-(CH_2)_x-C(=O)-O-(CH_2)_y-O-]-$	$x=2$, $y=2$, poly(ethylene succinate), PES $x=2$, $y=4$, poly(butylene succinate), PBS $x=4$, $y=4$, poly(butylene adipate), PBA

6.2 Poly(α-hydroxy acid)s

The main members of this class of polyesters are poly(glycolic acid) and poly(lactic acid) as well as their copolymers. The synthesis of these two polyesters is in many respects similar; however, their physical properties are very different, as will be presented in the following text.

6.2.1 Poly(glycolic acid)

6.2.1.1 Synthesis of Poly(glycolic acid)

Poly(glycolic acid), PGA, is the simplest poly(α-hydroxy acid). The monomer for PGA, glycolic acid, can be found in nature in trace amounts in sugarcane, beets and grapes and is industrially mainly produced form chloroacetic acid in the reaction with sodium hydroxide. It can be also produced in bio-based process from glucose and other sugars. PGA can be obtained by the polycondensation of glycolic acid but the resulting polyester is of low molecular weight as a consequence of the equilibrium nature of the step-growth polycondensation reaction. Thus, the preferred synthetic approach for the synthesis of PGA of high molecular weight is the ROP of the cyclic diester, glycolide (Scheme 6.1). The preferred catalysts are organo tin, antimony, zinc or lead.

The monomer glycolide is obtained from glycolic acid by heating under controlled conditions. The oligomeric poly(glycolic acid) formed at the beginning of the reaction at lower temperatures of around 180 °C undergoes an unzipping degradation at higher temperatures of 255–270 °C to form glycolide.[25] The further synthesis of PGA from glycolide proceeds through different reaction mechanisms depending on the initiator and reaction conditions. It is in many respects similar to the synthesis of poly(lactic acid) from lactide monomer.

6.2.1.2 Properties and Degradation of Poly(glycolic acid)

PGA is thermoplastic material with a high degree of crystallinity and, consequently, exhibits a high tensile modulus. The glass transition temperature of this polymer lies in the range 35–40 °C and the melting temperature is very high for an aliphatic polyester (200–225 °C). Due to its fiber-forming properties,

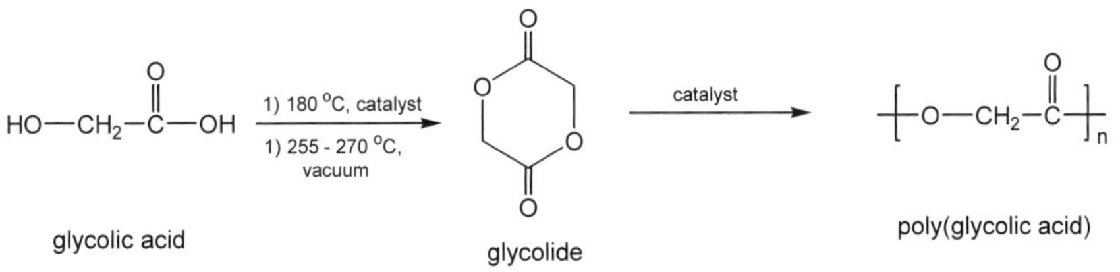

Scheme 6.1 Synthesis of poly(glycolic acid) from glycolic acid.

together with a high hydrolytic instability, it was the first synthetic polyester which appeared on the market as a biodegradable suture material. It was developed as the synthetic absorbable suture, Dexon®, in 1962, by the American Cyanamid Co.[25] Fibers made of PGA are very stiff, thus copolymerization is often performed to modify the mechanical properties of the resulting fibers.[16] The self-reinforced PGA, which refers to a composite material of oriented fibers within a matrix of the same chemical composition, exhibits extremely high modulus of around 12.5 GPa.[12,15] PGA is a hydrophobic polymer with a very low solubility in most common organic solvents but is soluble in hexafluoro-2-propanol.

The thermal degradation of PGA proceeds through random chain scission at lower and through specific end-chain scission at higher temperatures.[26] It shows better thermal stability than poly(lactic acid) and is, in a similar manner, sensitive to the presence of moisture.[27]

PGA readily hydrolyzes *via* random ester bond cleavage. Alkaline and strong acid media accelerate the degradation.[18] Hydrolytic degradation is also a main degradation pathway *in vivo*, which is accompanied with a rapid deterioration of the mechanical properties.[28,29] In some cases, the observed higher degradation rates *in vivo* compared to *in vitro* degradation were ascribed to the action of some enzymes such as some non-specific esterases and carboxyl peptidases, although some authors attribute such differences to physical and physiological factors.[18,30,31] In the course of degradation, PGA looses its strength in 1–2 months and is completely degraded in 6–12 months.[12] The glycolic acid produced by hydrolytic degradation of PGA is excreted in the urine. Glycolic acid can also be enzymaticaly converted to glycine, which either enters the citric acid cycle or is excreted in the urine.[15,27,30] PGA is considered to be completely biocompatible although higher concentrations of released glycolic acid in a rapid hydrolytic degradation of PGA can cause tissue damage.

6.2.2 Poly(lactic acid)

6.2.2.1 Synthesis of Poly(lactic acid)

Poly(lactic acid), PLA, is probably the most investigated and commercially used aliphatic biodegradable polyesters. It can be obtained from lactic acid in several different ways. Lactic acid is a chiral molecule which exists as two stereoisomers: naturally occurring L-lactic acid and D-lactic acid (Scheme 6.2).

L-lactic acid   D-lactic acid

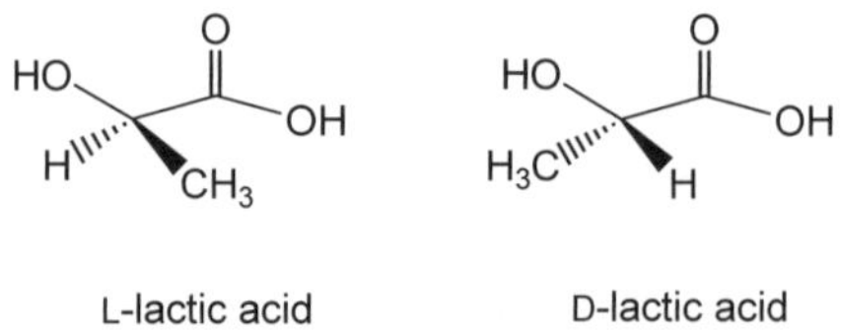

Scheme 6.2 Structural formulas of the two isomeric forms of lactic acid.

Biodegradable Polyesters: Synthesis and Physical Properties 155

This monomer can be obtained by a petrochemical route, through the hydrolysis of lactonitrile by a strong acid, in which case a racemic mixture of optically inactive lactic acid is obtained. In the preferred, biotechnological production of lactic acid, the monomer is obtained by fermentation of a carbohydrate carbon source from lactic bacteria or fungi and either enantiomer can be obtained in high purity. Thus, the monomer for the production of PLA is available from renewable and low-cost substances, such as wheat, rice bran, potato and corn starch and other biomasses.[32,33] One advantage of the chiral structure of lactic acid is the possibility of tailoring the final properties of the synthesized PLA through its stereostructure by simply changing the ratio of the D- and L-enantiomers in the polymer backbone. The polymerization of either enantiomer leads to semi-crystalline polymers (PLLA or PDLA), while amorphous PDLLA is obtained from a racemic mixture of the monomers and this difference in crystallinity has profound impact on all properties of PLA.

PLA can be obtained in two ways: through direct polycondensation of the hydroxy acid or by ROP of cyclic lactide monomer. The different reaction pathways to PLA are depicted in Scheme 6.3. Different nomenclatures of polymers obtained by the different routes are often observed in the literature: those obtained from lactic acid by direct polycondensation are referred to as poly(lactic acid), while those obtained from lactide monomer by ROP are referred to as poly(lactide). The general abbreviation used in both cases is PLA.

Direct polycondensation in the bulk through the reaction of hydroxy and carboxy functionalities of the lactic acid monomer with the elimination of water molecule has all the drawbacks associated with the character of an equilibrium step-growth polymerization. In addition, employing this polymerization method, the stereoregularity of the polymer cannot be controlled. This polymerization technique involves the use of a catalyst and reduced pressure.[34–37]

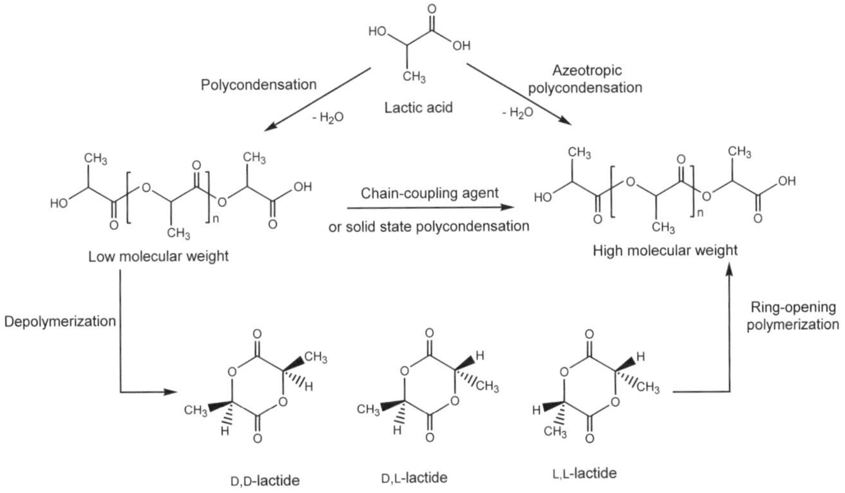

Scheme 6.3 Different routes to PLA of high molecular weight.

Oligomers obtained in the direct polycondensation reaction have a molecular weight of a few tens of thousands, although there are reports where higher molecular weight PLLA of up to 100 000 g mol^{-1} were obtained by the use of a binary catalyst based on Sn(II) activated with proton acids.[38] The oligomeric PLA obtained in the direct polycondensation process is of interest in drug release systems;[39,40] however, it is of little practical importance for the majority of other applications. Thus, further reaction steps are necessary to obtain a polymer of high molecular weight and useful physical and mechanical properties.

One way to increase the molecular weight of oligomeric PLA is to use chain-coupling agents. In order to use chain-coupling agents, which are bi/polyfunctional small compounds that react with one type of functional group, the oligomeric PLA with equal number of hydroxyl and carboxyl end functionalities has to be modified to contain only one type of these functional groups at the chain terminus.[41–43] Thus, polymerization of lactic acid is conducted in the presence of compounds which contain hydroxyl groups, such as 2-butene-1,4-diol, glycerol or 1,4-butanediol, which leads to a polymer with hydroxyl endgroups. In cases where an acid, such as maleic, succinic, adipic or itaconic acid, is added, carboxyl end functionalities will prevail in the final polymer. By the reaction of the so-obtained telechelic oligomers and suitable linking molecules, such as isocyanates or oxazolines, a polymer of high molecular weight with improved mechanical properties can be obtained.[44–46]

One more method to increase the molecular weight of oligomeric PLA is solid state polycondensation.[47–49] This method consists of heating a polymer obtained in an ordinary melt-polycondensation process in the form of pellet, chip or powder to a temperature below the melting temperature, in the presence of catalyst to promote further polycondensation. The by-product of the polycondensation is removed under reduced pressure or under an inert gas flow. The reaction is performed at temperatures above the glass transition temperature, to ensure sufficient chain mobility for the feasibility of the reaction which occurs in the amorphous regions of the material, but at a temperature which is low enough to ensure the absence of thermal or oxidative degradation and side reactions. Compared to melt polymerization, this process requires longer reaction times; however, the resulting polymer has a high molecular weight, reaching those obtained in the ring-opening polymerization, and reduced discoloration.

The problems associated with water removal in the direct polycondensation reaction of lactic acid can be overcome by the use of a solvent in an azeotropic condensation process.[50–52] In this process, after purification of the lactic acid under reduced pressure, a suitable low-boiling organic solvent, such as diphenyl ether, and a catalyst is added; the polymerization is promoted by the removal of the by-product in a refluxing system with molecular sieves. A polymer of molecular weight as high as 300 000 g mol^{-1} can be obtained in a single-step process. This type of process has been brought to an industrial stage by Mitsui Chemicals. The advantage of a solution-based process is the lower reaction temperature which prevents side reactions and degradation of the product, and leads to a polymer of high molecular weight. The main disadvantage of this process remains the use of organic solvents, which requires additional purification steps and recycling of the solvent.

ROP of lactide was already demonstrated by Carothers in 1932,[53] but the products had low molecular weight. New developments in the purification of lactide monomer and improvements in the synthesis techniques led recently to PLA with suitable properties being synthesized by this method. The lactide monomer was obtained from oligomeric PLA in an internal transesterification (back-biting) process performed at high temperatures and reduced pressure. This process, due to racemization, yields a mixture of the three possible dilactide forms: D,D-lactide, L,L-lactide and D,L-lactide (*meso*-lactide) (Scheme 6.3). The desired isomer can be obtained from this mixture by distillation and crystallization.

By the nature of the ROP of lactide, high molecular weight polymers with tailored stereostructures can be obtained in a short time. The polymerization of lactide proceeds by anionic, cationic or by a coordination–insertion mechanism (Scheme 6.4). The process of polymerization can be realized in bulk, solution, emulsion or dispersion.

In the cationic ROP, only triflic acid and trifluoromethanesulfonic acid have shown potential as initiators for the polymerization of lactide.[54] The polymerization starts by protonation or alkylation of carbonyl oxygen, which results in positively charged alkyl–oxygen bond and the propagation step involves cleavage of this bond. The PLA obtained by this route is optically active if the appropriate reaction temperature is chosen ($<50\ °C$); however, the product obtained at such temperatures is of low molecular weight.

In the anionic polymerization of lactide, the reaction is initiated by the nucleophilic attack of the negatively charged initiator on the carbonyl group followed by acyl–oxygen bond cleavage. Thus, the propagating species are alkoxide ions of very high basicity. The consequence of this could be deprotonation of the monomer with possible chain transfer leading to low molecular weight products. Racemization is also an undesirable effect associated with this mechanism of ROP. The best initiators for the anionic polymerization of lactides are alkali metal alkoxides.[55]

The ROP of lactide in the presence of metal compounds of tin, aluminum, zinc, titanium or zirconium as catalysts proceeds *via* the coordination–insertion mechanism. The initiators in this type of reaction are usually metal alkoxides. In the first step, temporary coordination of lactide through the carbonyl group with the metal in the initiator leads to increased nucleophilicity of the alkoxide and electrophilicity of the carbonyl group, thereby facilitating the insertion of the monomer into the metal–O bond.[56] This is the most investigated and applied method for the synthesis of PLA due to the mild reaction conditions, since the reaction proceeds *via* covalent species. High molecular weights of $200\,000\ \text{g mol}^{-1}$ are easily achievable with minimum side reactions and racemization.

The most frequently used initiator for the polymerization of lactide is stannous(II)-ethylhexanoate (tin(II)-octoate), $Sn(Oct)_2$. It is believed that the true

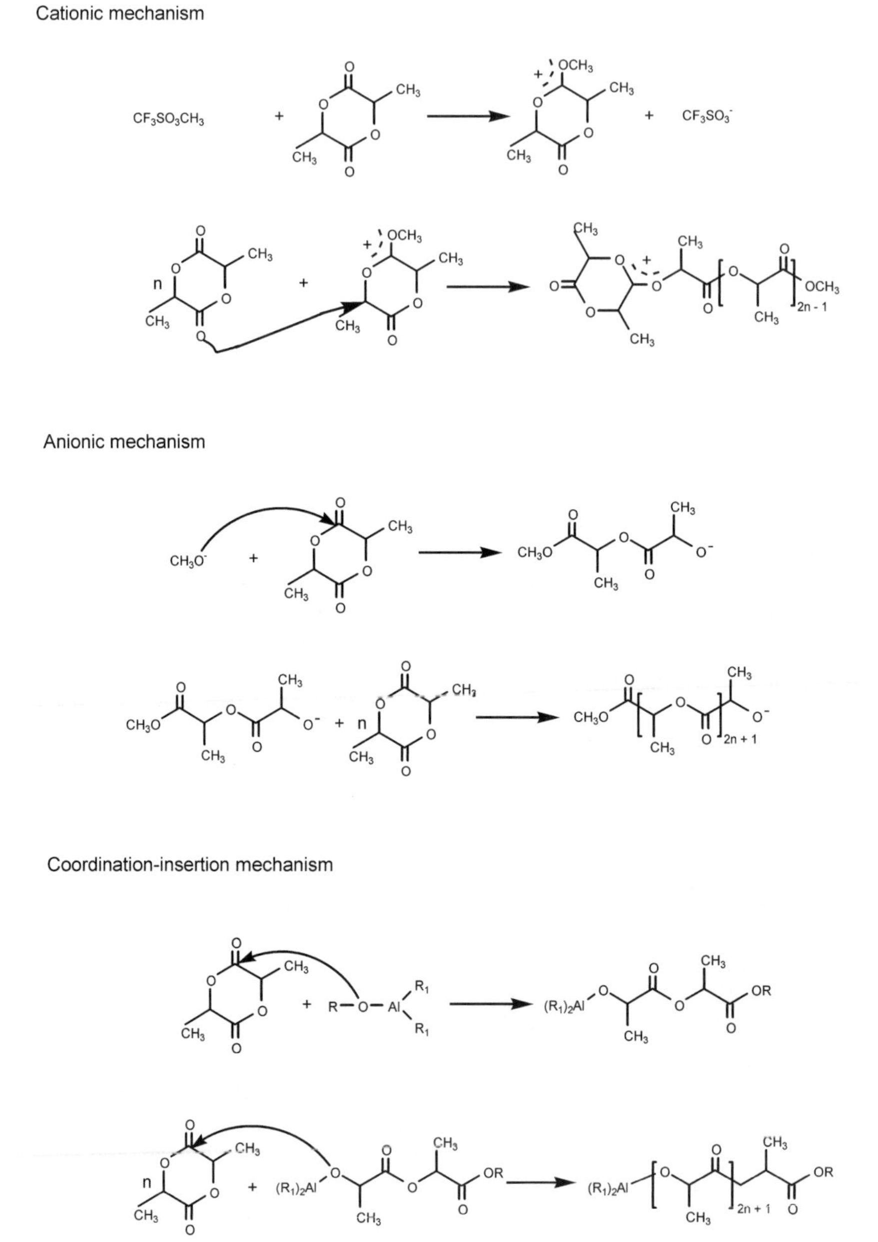

Scheme 6.4 Mechanisms of the cationic, anionic and coordination insertion ROP of lactide.

initiator in Sn(Oct)$_2$-initiated polymerizations is tin(II)-alkoxide formed in the first stages from Sn(Oct)$_2$ and small amounts of alcohols.[57,58] This initiator is used in the solvent-free process developed by Cargill for the industrial production of PLA.[59] There is concern about the use of tin compounds due to their toxicity; however, the organometallic tin(II)-octoate is of little toxicity and has been approved for use by the FDA.

6.2.2.2 Properties and Degradation of Poly(lactic acid)

The properties of PLA depend strongly on the ratio and distribution of the stereoisomers comprising the polyester chain. In addition to the dependence on stereoregularity, the properties of PLA depend on the molecular weight as well as on the annealing and processing conditions.[60–62] Homopolymers poly(L-lactide), PLLA and poly(D-lactide), PDLA, are semicrystalline. PLLA crystallizes in the α-, β- and γ-form, of which the α-form is the most common. PLLA in the α-form has a pseudo-orthorombic unit cell with lattice dimensions $a = 1.066$ nm, $b = 0.616$ nm and $c = 2.888$ nm.[63] By the introduction of other stereoisomer into the polymer chain, the crystallization is hindered and finally completely lost, resulting in an amorphous polymer.[64–66] Crystallization of stereocopolymers depends not only on the composition, but also on the distribution of stereoisomers within polymer chain.[67] The melting temperature of enantiomeric pure PLLA is 170–180 °C and the glass transition temperature is around 60 °C. The melting temperature decreases with the introduction of the second enantiomer and can be lowered by as much as 50 °C.[65,67] When PLLA is blended with PDLA, for certain blend compositions, a stereocomplex can be formed which is accompanied with an increase in the melting temperature of up to 50 °C compared to the homopolymers. The PLA stereocrystals have a triclinic unit cell with the lattice dimensions $a = b = 1.498$ nm, $c = 0.87$ nm, $\alpha = \beta = 90°$ and $\gamma = 120°$.[68] The glass transition temperature, besides molecular weight, is also dependent on the stereostructure of the PLA and is lower for stereocopolymers. The solubility of PLA also depends on composition. PLLA is soluble in chlorinated and fluorinated solvents, dioxane, furan and dioxolane. Stereocopolymer, PDLLA is also soluble in acetone, pyridine, ethyl lactate, tetrahydrofuran, xylene, ethyl acetate, dimethylsulfoxide, N,N-dimethylformamide and 2-butanone. Non-solvents for both PLAs are water, alcohols and hydrocarbons, such as hexane and heptane.

Semicrystalline PLLA in terms of mechanical properties, having a high tensile modulus of around 3–4 GPa, flexural modulus of 4–5 GPa, tensile strength of 50–70 MPa and low elongation at break of around 4%, is a stiff, brittle and hard polymer.[69–71] However, depending on the degree of crystallinity, molecular weight and stereoregularity, the mechanical properties can vary so that elastic and soft PLAs can be obtained.[60] The tensile strength of PLLA increased from 55 to 59 MPa with increasing molecular weight from 23 000 to 67 000 g mol^{-1}, whereas PDLLA stereocopolymers of molecular weight from 47 500 to 114 000 g mol^{-1} had lower tensile strengths in the range from 43–53 MPa. With the increase in crystallinity achieved by annealing, the mechanical properties of PLLA were further increased, especially the flexural and tensile modulus.[60] Due to its good mechanical properties, PLLA is a good candidate as a biomaterial for load bearing applications and for articles where rigidity and strength are required.[12] PLA, with its high modulus and low elongation at break has very similar mechanical properties to poly(styrene), which is a strong and relatively brittle material.[72] The mechanical properties of PLA recommend this polymer as a good replacement for non-degradable commodity plastics in applications such as packaging. However, in order to be used as a soft film, the addition of plasticizers to reduce T_g is necessary, and a number of plasticizers, preferably biodegradable, have been investigated in order to modify the properties of PLA.[69,73,74]

PLA is easily thermally degraded, which is a critical point in the processing of this polymer. According to Söldegård et al. pathways of PLA degradation can be differentiated into: thermohydrolysis, zip-like depolymerization, thermo-oxidative degradation and transesterification reactions.[70] Proposed pathways of thermal degradation of PLA are intra- and intermolecular ester exchange, cis-elimination, radical and concentrated non-radical reactions occurring above 200 °C and resulting in the formation of CO, CO_2, acetaldehyde and methylketene.[75] Accordingly, the thermal degradation of PLA is influenced by the presence of trace amounts of water, residual monomer and catalyst, molecular weight and by the type of end groups.[75–77] Thus, before thermal processing, well purified PLA has to be properly dried.[78]

PLA is highly prone to hydrolysis and this is the most important degradation pathway of this polymer. Hydrolytic degradation of PLA commences in the amorphous parts of a specimen, into which water can penetrate more easily, by the random scission of hydrolysable ester linkages. After the amorphous parts have been degraded, degradation proceeds in the crystalline domains from the edges toward the center. The increase in crystallinity, which is observed during the degradation, can be attributed to the reorganization of the remaining undegraded chains which can crystallize themselves.[79] Since the amorphous parts of the specimen are attacked first, the optical purity of the polymer, which influences the crystallinity, will determine the rate of hydrolysis of PLA and by changing the ratio of enantiomeric units, the desired rate of hydrolysis could be chosen.[80] In addition to chemical structure, other material properties, such as molecular weight, morphology, shape of the specimen, purity of the sample, pretreatment (e.g. annealing or drawing) influence the hydrolytic degradation. Additionally, external factors, such as temperature, pH, ionic strength and buffering capacity, can also influence the hydrolytic degradation of PLA.[18] The hydrolytic degradation is autocatalyzed by the carboxylic end groups formed by the scission of the ester linkages. During the course of hydrolysis, oligomers formed at the surface, being soluble in water, can easily leave the specimen by dissolution in the surrounding medium, which is not the case for those entrapped in the interior of the specimen. Thus, the concentration of carboxylic groups and, therefore, the rate of hydrolysis, are increased in the interior of the specimen. Consequentially, larger objects (like plates and beads) and objects with low porosity with a lower surface to volume ratio degrade faster than small ones (like films and microspheres) and more porous ones.[18,81–83]

Although PLA is not so prone to enzymatic degradation as other polyesters, several enzymes which catalyze the degradation of PLA have been isolated, e.g. pronase, proteinase K, bromelian and a number of esterase-type enzymes.[84,85] Tonomura tested 56 commercially available and industrially used proteases.

Among the tested enzymes, Savinase (Nord sibirsk) had the highest degrading activity toward PLA, with a specific activity corresponding to about one-half of that of proteinase K.[86] The most investigated enzyme, proteinase K, degrades preferentially L-lactyl units, being completely ineffective in the degradation of poly(D-lactide).[87,88] In addition, it was postulated that proteinase K preferentially degrades L–L, L–D and D–L bonds, as opposed to D–D ones.[89] It was shown that lipases degrade preferentially amorphous PLA.[90] Enzymatic degradation occurs at the surface of a specimen (homogenous erosion), due to the inability of the enzymes to penetrate the sample and perform their catalytic action in the interior. Swelling of the sample induced by water uptake is thus important in the enzymatic hydrolysis of PLA.[91]

PLA is the least susceptible to microbial attack of all aliphatic polyesters of natural or synthetic origin. The distribution and population of PLA-degrading microorganisms, as well as their percentage, in the soil environment is the smallest compared to those degrading other biodegradable polyesters.[92,93] Only oligomeric PLA is assimilated by some fungal strains.[94] It is thus believed that the degradation of PLA in the environment commences and is dominated by abiotic hydrolysis. However, the degradation is accelerated in the presence of microorganisms.[95–97] PLA is completely assimilated by microorganisms in the environment without any toxic effect; however, it was shown that the presence of other chain units, such as those from some isocyanate chain extenders, can have toxic effects.[98]

Hydrolysis is the main mode of degradation of PLA *in vivo*. Degradation under *in vivo* conditions is, as expected, influenced by the stereostructure of the polymer. While highly crystalline PLLA has been shown to remain for even 5–7 years in the body[16] and there are no evidence that it is completely degradable *in vivo*,[99] an amorphous PDLLA copolymer with 50% of the second enantiomer underwent a mass loss in 12–16 months.[15] Since the hydrolytic degradation in a case of these polymers leads to bulk erosion, the mechanical strength is lost more rapidly (1–2 months for PDLLA and 6 months for PLLA) without observable mass changes.[12] The hydrolytic degradation of PLA leads to lactic acid monomer, the normal metabolite in the human body, which is decomposed to water and CO_2 *via* the citric acid cycle.

6.3 Poly(ε-caprolactone)

6.3.1 Synthesis of Poly(ε-caprolactone)

Poly(ε-caprolactone), PCL, comprised of hexanoate repeating units, is the most important synthetic biodegradable aliphatic polyester among the poly(ω-hydroxyalkanoate)s (Table 6.1). It can be synthesized by polycondensation of the corresponding hydoxy acid: 6-hydoxycaproic (6-hydroxyhexanoic) acid. However, this method of synthesis has been less investigated for the production of PCL because of all the shortcomings inherent to the direct polycondensation reaction. Thus, poly(ε-caprolactone) is mainly produced by ROP of the cyclic monomer ε-caprolactone. This monomer is manufactured by the oxidation of cyclohexanone with peracetic acid (Scheme 6.5).

Depending on the employed initiator/catalyst system, the ROP of ε-caprolactone can proceed *via* anionic, cationic or coordination–insertion mechanisms (Scheme 6.6). The reaction can be performed in bulk, solution or as an emulsion or dispersion polymerization, and is in many respects similar to the ROP of the six-membered cyclic diester lactide.[100]

Initiators which have been used for the cationic polymerization of ε-caprolactone include protic acids, Lewis acids, stabilized carbocations and acylating agents. The mechanism involves the formation of charged species, followed by the ring opening of the activated lactone by the attack of another monomer and insertion into the chain. This type of polymerization is not very significant for the production of PCL. The reaction is difficult to control and transesterification reactions, which occur, limit the molecular weight of the obtained

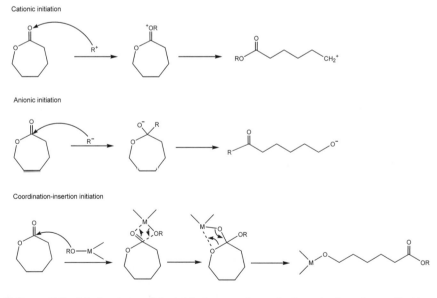

Scheme 6.5 Synthesis of ε-caprolactone from cyclohexanone.

Scheme 6.6 Mechanisms of the initiation step for cationic, anionic and coordination-insertion ROP of ε-caprolactone.

polymer. It can, however, be useful for the production of copolymers which can not be synthesized by other routes.[101]

The anionic polymerization of ε-caprolactone proceeds by the scission of the acyl–oxygen bond, leading to an alkoxide end group. Depending on the initiator and reaction conditions, the reaction may have living character. Initiators for this type of ε-caprolactone polymerization include alkali metals, alkali metal alkoxides, alkali metal naphthalenide complexes with crown ethers and alkaline metals in graphene. The disadvantage of this reaction is the occurrence of the intramolecular transesterification reaction. When potassium *tert*-butoxide was used in THF, the ester interchange reaction occurred leading to the formation of very large amounts of cyclic oligomers.[102] This can be suppressed by intercalation of alkaline metals in graphite and also by the use of lithium *tert*-butoxide in the apolar solvent, benzene.[103,104]

The coordination–insertion mechanism is the most frequently used to obtain high molecular weight PCL, with high conversions and under mild conditions. The reaction proceeds through coordination of the monomer to the active site of the initiators, acyl–oxygen bond scission and insertion into the metal–oxygen bond. Since the reaction involves covalent propagating species of low reactivity, side reactions are suppressed and the reaction can have the character of a living ROP. Initiators for this type of reaction include dibutyl zinc, alkoxides and halides of aluminum, magnesium, zinc and titanium, stannous chloride and octoate and some lanthanide compounds such as yttrium and lanthanum alkoxides.[101,104,105] Aluminum-based initiators provide good control over the reaction since transfer and termination are not significant. The reaction initiated by the most studied aluminum based initiator, aluminum isopropoxide, in toluene or THF at low temperatures between 0–25 °C has living character, which allows for the formation of block-copolymers with other cyclic esters.[106,107]

The most used initiator for the polymerization of ε-caprolactone is $Sn(Oct)_2$. It is effective, commercially available and soluble in most organic solvents and in lactones. It is believed that in the beginning stages of the reaction an equilibrimum is established between $Sn(Oct)_2$ and corresponding alkoxide, formed in a reaction of $Sn(Oct)_2$ and some hydrogen active compounds (*e.g.* alcohols) (Scheme 6.7). Alkoxide formed in this reaction initiates the polymerization. For reactions performed without the deliberate addition of hydrogen active compounds, the impurities present in $Sn(Oct)_2$ can be involved in reaction of alkoxide formation.[108,109] It was reported that molecular weights of 1×10^6 g mol^{-1} could be obtained using this initiator, which is also the limiting value of molecular weight which can be obtained by

Scheme 6.7 Formation of the active species from tin(II)-octoate in ROP of ε-caprolactone.

this procedure.[110] Sn(Oct)$_2$ is active at higher temperatures which promotes intermolecular and intramolecular transesterification reactions and broadens the molecular weight distribution.[105]

PCL can also be obtained by free radical polymerization of 2-methylene-1,3-dioxepane initiated with AIBN.[111,112] However, this reaction pathway is not of great importance except in the case of the synthesis of some copolymers.[113,114]

6.3.2 Properties and Degradation of Poly(ε-caprolactone)

PCL is a semicrystalline polymer with a melting temperature in the range 59–64 °C. The degree of crystallinity increases with decreasing molecular weight and can reach 80% for PCL of 5000 g mol^{-1} molecular weight in contrast to 40% for PCL with a molecular weight of 100 000 g mol^{-1}. PCL crystallizes with an orthorhombic unit cell with characteristic dimensions $a = 0.7496$ nm, $b = 0.4974$ nm and $c = 1.7197$ nm.[115] The glass transition temperature of PCL lies well below room temperature at around –60 °C. PCL has a low tensile strength, below 20 MPa, but an extremely high elongation at break, which can be up to 1000%.[12,105,116–118] Due to its high olefin content, PCL resembles linear low density polyethylene in some respects, *i.e.* waxy feel and similar tensile modulus.[6,116] The highly valuable property of PCL is its ability to form blends with a range of other polymers. Blends of PCL with other biodegradable polymers, such as PLA, PLGA and starch, have been investigated in order to control the degradation rate or for improved performance of the product.[10,101] In blends with non-degradable polymers, such as PVC, chlorinated PE and bisphenol A polycarbonate, PCL serves mainly as a plasticizer.[101,104] PCL is soluble in chloroform, dichloromethane, carbon tetrachloride, benzene, toluene, cyclohexanone and 2-nitropropane. It is slightly soluble in acetone, 2-butanone, ethyl acetate, dimethylformamide and acetonitrile and is insoluble in alcohols, petroleum ether and diethyl ether.

The thermal degradation of PCL proceeds *via* random chain cleavage through *cis* elimination and *via* specific chain end scission by unzipping from the ω-hydroxyl end of the polymer. These processes occur in bulk and also for the PCL in solution.[119–125] According to Persenaire *et al.*, an increase in the molecular weight improves the thermal stability of PCL due to the random nature of chain scission; the formation of the volatile products is more probable when the molecular weight is low.[120] In this proposed model of PCL thermal degradation, the unzipping process occurs in the second step during thermal treatment. According to Sivalingam *et al.*, random chain cleavage and unzipping occur as parallel rather than as consecutive steps, whereby random chain scission predominates at lower temperatures, while specific chain end scission predominates at higher temperatures.[121,122] There has also been a report on the single-step degradation of PCL involving only the unzipping process.[125] The thermal stability of PCL is reduced in the presence of traces of metals, such as Zn, which can promote unzipping depolymerization.[123]

PCL is prone to hydrolytic degradation, however, at a much slower rate of degradation compared to those of PLA and PGA. The mechanism of hydrolytic degradation is similar to the one observed for PLA and proceeds by random ester bond cleavage. It is a bulk process accompanied with a decrease in molecular weight during the hydrolysis and the rate by which the molecular weight decreases is independent of the shape of the specimen (surface/volume ratio).[101,126] The hydrolysis of the ester bonds is autocatalyzed by the carboxylic end groups.[127,128] Degradation commences in the amorphous parts and the crystallinity increases during degradation in the buffer solution in much the same manner as in the case of PLA. The rate of hydrolysis is influenced by pH and temperature, being more rapid in basic solutions at elevated temperatures.[129,130] Long-term investigations of the hydrolytic degradation of PCL films in fluids that resemble biological media showed that the degradation of PCL proceeds faster in such media compared to the hydrolysis in phosphate buffer solution.[131]

PCL is prone to degradation in the presence of enzymes, especially lipase-type enzymes.[18,132] Lipase-catalyzed hydrolytic degradation of PCL was investigated for a number of lipases, such as *Pseudomonas* lipase, *Rizophus arrhizus* and *Rizophus delemer*, which were shown to greatly increase the rate of hydrolytic degradation of this polyester.[133–140] Degradation of PCL by lipases is influenced by number of factors, such as molecular weight and crystallinity of the PCL, as well as the size of the spherulites and porosity of the sample.[134,135] Since erosion of the polymer in enzymatic degradation proceeds at the sample surface, the shape and size of the specimen influence the rate of degradation.[141] Cross-linking, which reduces the crystallinity, facilitates the hydrolytic degradation as a bulk process, due to the easier penetration of water into the amorphous phase.[142] On the other hand, enzyme-catalyzed degradation, as a surface process, is affected in the opposite direction, *i.e.* the rate of degradation is slower due to the formation of a network structure which reduces the access of the enzymes to the polymer.[143]

Microorganisms that degrade PCL are broadly distributed in nature.[92] PCL is readily degraded by bacterial and fungal strains,[144–147] and in different environments, such river water, compost, soil or sewage sludge.[130,148–152] Degradation by the microorganisms proceeds *via* surface erosion and is accompanied with rapid weight loss of the sample. Complete degradation of PCL within 14 days was shown to occur under composting conditions.[153] Low molecular weight degradation products, such as caprolactone, 6-hydroxyhexanoic acid, cyclic dimer and cyclic trimer, can be easily assimilated by composting microorganisms in the course of 2 weeks.[154]

PCL degradation *in vivo*, as in the case of abiotic hydrolytic degradation, is a slow process and it takes 2–3 years for complete degradation.[155] Degradation of PCL *in vivo* proceeds by chemical hydrolysis of the ester groups accompanied with a reduction in molecular weight. There is no weight loss in the early stages of degradation, which, together with molecular weight decrease, indicate bulk rather than surface erosion. This mode of degradation indicates that the degradation of PCL *in vivo* is not enzyme catalyzed in the early stages of degradation. In the second stage of PCL degradation, weight loss occurs, accompanied with a deceleration of the molecular weight decrease (rate of chain scission) due to the increased crystallinity of the sample.[101] When the molecular weight of PCL falls to 3000 g mol^{-1}, small pieces of PCL can be intracellulary degraded in the phagosomes of macrophages, giant cells or fibroblasts in less than 13 days at some implant sites.[156] The sole metabolite besides water was 6-hydroxycaproic acid derived from the complete hydrolysis of the polyester. The product of PCL hydrolysis, 6-hydroxycaproic acid and further coenzyme A, can enter the citric acid cycle and be eliminated from the body.[104] Some recent studies showed that PCL is excreted from the body immediately after it is metabolized and does not accumulate in any body organ.[157] Investigations on PCL-based scaffolds showed that the degradation pathway of these devices, under simulated physiological conditions and *in vivo*, follows the general trend of bulk chemical degradation.[158,159] Due to its slow degradation *in vivo*, compared to PLA and PGA, PCL is more suitable for long-term implants.[160]

6.4 Poly(hydroxyalkanoate)s

6.4.1 Synthesis of Poly(hydroxyalkanoate)s

Poly(hydroxyalkanoate), PHA, is a general name for a large family of polyesters which are produced directly by microorganisms from renewable resources. Being produced by microorganisms, they are inherently biodegradable. The first such polymer to be discovered by Lemoigne in the 1920s, produced by bacteria *Bacillus megaterium*, was poly(3-hydroxybutyrate), P(3HB), which is the PHA most often found in nature.[161] In 1974, with the discovery of the copolymer poly(3-hydroxybutyrate)-*co*-poly(3-hydroxyvalerate), P(3HB-*co*-3HV), it was recognized that other 3-hydroxyalkanoate units can be incorporated in the polyester chain.[162] By varying the type of microorganisms and carbon source, different monomers can participate in copolymer building. P(3HB) and other PHAs are stored in the microbe cells and serve as an intracellular energy and carbon storage material, similar to the starch and glycogen in other living systems. The formation of polymer is usually induced by the lack of essential nutrients such as sulfate, ammonium, phosphate, potassium, iron, magnesium or oxygen and in the presence of a source of excess carbon, such as sugars. Furthermore, some species which accumulate PHAs during growth, without nutrient limitations, have been found. The polymer accumulates in the cytoplasm of the cells in the form of discrete granules, the number and size of which can vary depending on the species. The size of the granules range from 0.2 to 0.5 μm and the polymer inside is in the amorphous state, which is easily transformed to crystalline form upon any physical treatment.[163,164]

The general formula of poly(hydroxyalkanoate)s is given in Scheme 6.8. The majority of PHAs have the same three-carbon backbone but differ in the length

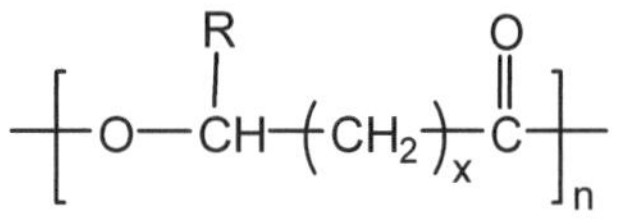

Scheme 6.8 Structural formula of poly(hydroxyalkanoate)s.

$$\left[O-\underset{\underset{R}{|}}{CH}-(CH_2)_x-\underset{\underset{}{\overset{O}{\|}}}{C} \right]_n$$

x = 1   R = H   Poly(3-hydroxypropionate)
        R = CH$_3$ Poly(3-hydroxybutyrate)
        R = CH$_2$CH$_3$ Poly(3-hydroxyvalerate)
        R = (CH$_2$)$_2$CH$_3$ Poly(3-hydroxyhexanoate)
        R = (CH$_2$)$_4$CH$_3$ Poly(3-hydroxyoctanoate)

x = 2   R = H   Poly(4-hydroxybutyrate)

x = 3   R = H   Poly(5-hydroxyvalerate)

of the alkyl groups attached at the β position. PHAs are divided into two classes: short chain length PHA (sclPHA, having from 3 to 5 carbon atoms in the monomeric unit) and medium chain length PHA (mclPHA, having from 6 to 14 carbon atoms in the monomer unit). Polymers with 4-hydroxybutyric acid moieties have also been prepared and are considered as good candidates for medical applications.[165,166] Some other PHAs, the structures of which are derived from 5-hydroxy and 6-hydroxy acid repeating units, have also been isolated. More than 100 monomers have been found to participate in the construction of different PHAs.[167,168]

Depending on the size and type of the pendant alkyl group and the composition of the polymer, the properties of PHAs can vary over a broad range, from rigid plastics to tough elastomers.[169] The versatility of PHAs can be increased by different types of chemical modifications due to possibility of introducing different functionalities, such as unsaturated groups as well as bromo-, hydroxyl-, methyl-branched and aromatic units, into the chemical structure of PHA by altering the type of carbon source and microorganisms used for their production.[170] For example, a biodegradable rubber of microbial origin was prepared by cross-linking of an unsaturated polyester.[171]

The molecular weights of PHAs range from 2×10^4 to 3×10^6 g mol^{-1} depending on the microorganisms type, growth conditions and type of carbon source.[163] By using a recombinant *Escherichia coli* under special conditions, ultra-high molecular weight P(3HB) with molecular weight $> 2 \times 10^7$ g mol^{-1} has been obtained.[172]

The biosynthetic production of PHAs is the most important and investigated, and is the only method that has been developed to commercialization. There are three possible metabolic pathways involved in the biosynthetic production of PHAs, which have been reviewed elsewhere.[173] In simplified form, the carbon source is converted to a hydroxyacyl-CoA thioester, the monomer which is polymerized to PHA, involving different metabolic pathways depending on the carbon source (Figure 6.2).

The polymerization reaction of hydroxyacyl-CoA thioester is catalyzed by the enzyme PHA synthase, the key enzyme in the production of PHA.[174] In early investigations of the mechanism of polymerization, which was broadly accepted in later mechanistic studies, it was proposed that two thiol groups in the active sites of PHA synthase were involved in the reaction.[174–177] The thiol groups form thioesters with the monomers and a thioester–oxyester interchange reaction occurs, leaving one thiol group free for bonding to another monomer. The propagation step involves bonding another monomer to the free thiol group and a subsequent thioester–oxyester interchange reaction, adding one more monomer unit to the growing chain (Scheme 6.9).

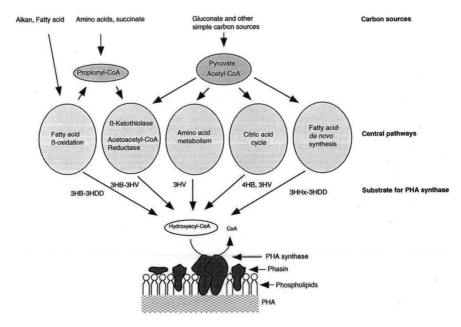

Figure 6.2 Metabolic routes towards substrates for PHA synthases (reprinted from Reference 174 with permission from Elsevier).

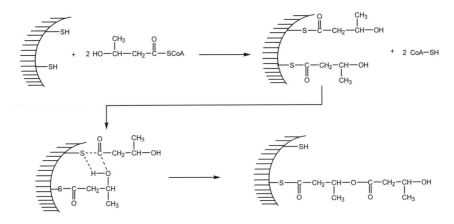

Scheme 6.9 Mechanism of PHA synthesis catalyzed by the two thiol groups in PHA synthase (adapted from References 161 and 175).

Some other reaction schemes based on covalent/non-covalent reaction pathways, which include only one active center on the PHA synthase have been proposed.[178] More than 50 different PHA synthases have been cloned and characterized.[179] These enzymes are located at the surface of the PHA granules. Due to enzyme stereospecificity, all the synthesized PHAs with a chiral carbon atom are isotactic, i.e. they are all comprised of monomers with the R-configuration. PHA synthases usually show broad substrate specificity and by modifying a substrate different copolymers can be obtained from a single strain. P(3HB-co-3HV) was prepared using glucose and propionic acid as the carbon source from *Ralstonia metallidurans* (formerly known as *Alcaligenes eutrophus*). By using a butyric acid and pentanoic acid as the carbon source, *Ralstonia metallidurans* can produce P(3HB-co-3HV) copolymers which contain up to 85 mol% of HV units.[180] After it was shown that *Pseudomonas oleovorans* grown on octane were able to produce poly(hydroxyoctanoate),[181] a number of mclPHA were produced from this microorganism with up to 12 carbon atoms by using different sodium salts of n-alkanoic acids as substrates.[182,183] Substrate specificity of PHA synthases is reflected in the fact that these enzymes can usually catalyze the polymerization of sclPHA or mclPHA.[174] However, some recent reports show that copolymers of sclPHA and mclPHA could also be produced.[184,185]

A large number of strains that accumulate PHAs have been isolated. However, only some of them are considered suitable for the production of PHA when all factors, such as the use of an inexpensive carbon source, growth rate, polymer synthesis rate and extent of polymer accumulation, are taken into account.[186,187] One of the most investigated is *Ralstonia metallidurans*, due to its ability to accumulate large amounts of P(3HB), up to 80% of the dry cell mass.[188,189] *Alcaligenes latus* and *Azotobacter vinelandii* were also considered interesting for further development for PHA production owing to their capability to produce PHA during the growth phase, which would shorten and simplify the production process.[190–192] *Methylotropus* were investigated in order to develop a methanol source-based process.[193,194] *Pseudomonas* species are of interest due to their ability to produce different types of mclPHAs, with a great variety of functional groups introduced into the polyester.[183,195,196] In order to achieve high production of P(3HB), in addition to wild type bacteria, recombinant types were also investigated, especially *Escherichia coli*.[197,198]

Investigations directed toward optimization of biotechnological processes to be performed using cheap and readily available carbon substrates are of great importance, since the final cost of the industrial production is mainly dictated by the cost of the carbon source. Those carbon sources which lead to structurally identical monomers are designated as related, while those which are structurally completely different from the generated monomers are classified as unrelated. Carbon sources can vary from simple carbohydrates, alkanes and fatty acids to plant oils and various other cheap carbon sources, such as wastewater from olive mills, molasses, whey, starchy wastewater, corn step liquor and palm oil mill effluent.[199]

Synthesis of PHA in a biotechnology-based method can be performed with batch, fed-batch and continuous cultures.[199,200] Extraction of PHA from the biomass can be performed in two ways and is an important step which determines the final economy of a process.[201] In the first approach, after sterilization, the cells are treated with a cocktail of enzymes (including proteases, nucleases and lysozymes) and an anionic detergent to remove proteins, nucleic acids and cell walls, leaving the PHA unchanged and ready to be separated. The second approach is a solvent-based extraction which relies on the PHA solubility in chloroform and other convenient organic solvents, such as dichloromethane, propylene carbonate and dichloroethane. PHA is recovered from chloroform solution by precipitation in methanol or ethanol. The so-obtained PHA is of high purity (more than 98%); however, large amounts of toxic solvents are consumed in this process (for the extraction of PHA a ratio of 20/1 for solvent/polymer is used), which greatly influences the cost of the production and is environmentally unfriendly.[186]

Efforts to synthesize P(3HB) without impurities of natural origin have led to the development of chemical routes to PHAs, which include the ROP of butyrolactone and other different lactones.[202,203] However, due to the inability to easily obtain optically active polyesters and the lower molecular weights of the obtained products compared to those achievable in the biosynthetic approaches, these routes to PHAs are not very significant, except for the preparation of otherwise unattainable copolymers.[204,205] On the other hand, racemic P(3HB) obtained by ROP is less crystalline, with elastomeric properties, and can be of interest in some medical applications, such as drug delivery.

Yet another synthetic approach to PHAs is *in vitro* synthesis by an isolated PHA synthases.[206–208] The advantages of this process include the ability to obtain polyesters in water at room temperature with a much easier purification procedure and free of biologically related impurities, which is of interest for biomedical applications. There is also a fundamental interest in this synthetic approach which allows mechanistic studies. Moreover, the possibility of synthesizing block copolymers is opened by this *in vitro* synthetic route. However, some recent studies have proven the possibility of also obtaining block copolymers in a living systems.[209] In this synthesis procedure, for one-enzyme systems, the substrates are coenzyme A thioesters of hydroxy fatty acids, the prices of which greatly limits further development of this process. The transfer to the more readily available substrates requires the addition of other enzymes, adding further demands and complexity to this approach to PHA production.[208]

Since the first successful accumulation of P(3HB) in the cytoplasm of the transgenic plant *Arabidopsis thaliana* in 1992, much effort has been spent on the further development of this intriguing pathway for the low cost and large scale production of PHAs from CO_2 and solar energy.[210,211] However, there are many requirements, ranging from production control and outcome to social impacts, which have to be addressed before the full commercialization of this process is enabled.

6.4.2 Properties and Degradation of Poly(hydroxyalkanoate)s

As already stated, the properties of PHAs can vary greatly depending on the type and amount of HA units of which the polyester chain is comprised. P(3HB) is a crystalline polyester which, due to its exceptional optical purity, exhibits high degrees of crystallinity up to 80%. Isotactic P(3HB) crystallizes in two modifications: an α- and β-form, of which the α-form is more common. P(3HB) in the α-form has an orthorhombic unit cell with lattice parameters $a = 0.576$ nm, $b = 1.320$ nm and $c = 0.596$ nm.[212] Its melting temperature is around 180 °C and its glass transition temperature is around 4 °C. P(3HB) crystallizes in the form of large spherulites which is a cause of brittleness in this polyester. The brittleness, which develops upon storage in a timescale of days, is the major drawback for the application of this polyester.[213,214] By the introduction of 3HV units, the brittleness is reduced. P(3HB-co-3HV) copolymers are highly crystalline polyesters with the crystallinity unaffected by copolymerization, a property which is ascribed to isodimorphism, i.e. the different polymeric units of this copolymer are found to co-crystallize.[215] These copolyesters have reduced melting temperatures compared to both homopolymers and also exhibit slow crystallization rates from the melt.[216,217] The minimum in the melting temperature of the copolymers is around 80 °C, when the amount of 3HV units is about 40 mol%.[215,218] Unlike the copolymers with 3HV units, copolymers with 4HB units exhibit reduced crystallinity compared to P(3HB), as low as 14 %.[218,219] A depression of the melting point is also observed for poly(3-hydroxybutyrate-co-4-hydroxybutyrate), P(3HB-co-4HB), copolymers. The glass transition temperature continuously decreases in both cases, reaching –8 °C for pure P(3HV) or –48 °C in the case of P(4HB).[163,220] Compared to the sclPHAs, the mclPHAs have lower melting temperatures of about 45–60 °C and glass transition temperatures in the range –25 °C to –50 °C.

In terms of mechanical properties, with a Young's modulus of about 3.5 GPa and a tensile strength of 40 MPa, P(3HB) is similar to PP, but is a stiff and brittle material (Table 6.2). It exhibits low elongations at break of about 5% (similar as PS). With the incorporation of 3HV units into the polyester chain, a

Table 6.2 Physical properties of selected poly(hydroxyalkanoate)s.[4,163,199,221]

Polymer	Melting temperature, °C	Glass transition temperature, °C	Young's modulus, GPa	Tensile strength, MPa	Elongation at break, %
P(3HB)	179	4	3.5	40	5
P(3HB-co-20%3HV)	145	–1	0.8	20	50
P(3HB-co-17%3HH)[a]	120	–2		20	850
P(4HB)	53	–48	149	104	1000
P(3HB-co-16%4HB)	150	–7		26	444
Poly(propylene)	176	–10	1.7	38	400
Poly(styrene)		100	3.2–5.0	30–60	3–5

[a]Poly(3-hyxroxybutyrate-co-3-hydroxyhexanoate)

tougher, less brittle and more ductile polymer with improved flexibility and impact strength is obtained. However, the susceptibility of the properties deterioration on aging can be even worse for P(3HB-co-3HV) copolymers than for the homopolymer. P(4HB) has a much higher tensile strength (104 MPa) compared to P(3HB) and is a strong thermoplastic material. It also has much higher elongations at break, as high as 1000%, which show the high ductility of this polymer.[219] For P(3HB-co-4HB) copolymers, the tensile strength decreases with the introduction of 4HB units up to 16 mol% to 26 MPa, and then increases reaching 104 MPa for the P(4HB). Copolymers, P(3HB-co-4HB), have higher elongations at break compared to P(3HB), over the whole composition range.

The mclPHAs, compared to the sclPHAs, are polymers with lower crystallinity, flexible and soft and exhibit behavior like those of thermoplastic elastomers, with low tensile strengths and high elongations at break.[222–224] Copolymers of sclPHA and mclPHA, with low contents of mclPHA units, have improved properties compared to brittle sclPHAs and exhibit properties similar to poly(propylene).

P(3HB) and copolymers P(3HB-co-3HV) possess piezoelectricity, a property which could be useful in medical applications for bone healing and repair through electrical stimulation of bone pins derived from these materials.[202]

P(3HB) is soluble in various solvents, such as chloroform, dichloromethane and alcohols with more than three carbon atoms. However, to completely solubilize crystalline P(3HB) even at a low concentration, refluxing in chloroform for up to 24 h is required. P(3HB) is sparingly soluble in toluene, pyridine, dioxane and octanol and is insoluble in water, lower alcohols, aromatic hydrocarbons, diethyl ether and hexane.

P(BHB) is unstable on heating and easily decompose at temperatures above its melting temperature with a rapid reduction in molecular weight.[225] Recent investigations on the thermal stability of P3HB and P(3HB-co-3HV) with 30 mol% of 3HV units revealed that upon continual heating in the temperature range from 250–400 °C, the copolymer showed better thermal stability.[226] Some P(3HB-co-3HV) copolyesters (3HV = 0–71 mol%) exhibited better thermal stability on prolonged heating above their respective melting temperatures and showed better promise for thermal processing than P(3HB), since their melting temperatures are much lower than that of P(3HB).[227] The main pyrolysis product of P(3HB) is crotonic acid produced by the *cis* elimination reaction, which also yields oligomeric products. The degradation products of P(3HB-co-3HV) were mainly propene, 2-butenoic acid, 2-pentenoic acid, propenyl-2-butenoate, butyl-2-butenoate and CO_2.[226,228]

As to PCL and PLLA, the hydrolytic degradation of P(3HB) is a slow process. A recent study showed that the weight loss of P(3HB) film in phosphate buffer solution at 37 °C was only 2.8% after 19 weeks of incubation.[229] The hydrolysis proceeds by random chain scission with hydroxy acid oligomers as the intermediate product and monomeric acid as the ultimate degradation product. The rate of P(3HB) hydrolysis is affected by a number of factors such as pH, temperature and crystallinity of the sample.[151,230] The degradation proceeds faster at higher temperatures and under alkaline conditions with 3-hydroxybutyric acid and also crotonic acid as major products of alkaline hydrolysis, owing to the different mechanistic pathways compared to neutral and acidic conditions.[230] Reports on the influence of the introduction of 3HV units into P(3HB) on the degradation rate of the copolyesters are often contradictory. While some studies showed that in the range of small amounts of 3HV (12 mol%), the hydrolytic degradation of the copolymers is favored, some studies on copolymers with higher amounts of 3HV units (45 and 71 mol%) found a decreasing hydrolysis rate.[229,231] As proposed by Albertsson et al., there might be some maximum in the 3HV content above which the hydrolysis is no longer favorable.[232]

PHAs are readily degraded in the presence of enzymes PHA depolymerase. Since PHAs are not water soluble and, as solids, can not be directly absorbed by microorganisms, many of them excrete specific enzymes PHA depolymerase. These enzymes degrade PHAs extracellulary to form oligomeric products which, being soluble and of low molecular weight, can easily enter a cell, where they are further metabolized. Enzymatic degradation of PHAs occurs at the surface of the polymer, where the enzymes are absorbed *via* a binding domain and are further involved in the PHA hydrolysis *via* a catalytic domain. The enzymatic degradation commences in the amorphous regions of the polymer specimen, while in the later stages both amorphous and crystalline phases are degraded without preference. Thus, the degree of crystallinity and crystallite size have a profound effect on the rate of degradation.[221,233,234] Some studies showed that an atactic, completely amorphous P(3HB) obtained by a chemical route is not prone to enzymatic attack and that the biodegradation can be induced by the presence of some crystalline phase. For example, natural P(3HB), which can provide appropriate enzyme binding sites, was shown to induce the degradation of amorphous P(3HB).[235] The substrate specificity of the catalytic domain of PHA depolymerases, investigated on the number of different aliphatic polyesters, was shown to be relatively narrow.[236] With the introduction of a second co-monomer unit, 3HA, the rate of degradation of the copolymers was increased by 5–10 times compared to P(3HB), however, only up to 10–20 mol% of the comonomer units.[224,237–239] A further increase in the co-monomer content decreased the degradation rate. It was also shown that an increase in the side-chain length at the β-carbon for copolymers with larger fractions of long HA units, further reduced the rate of ester bond hydrolysis by the enzyme.[240]

Microorganisms that can degrade PHAs are widely distributed in different environments. The percentage of P(3HB)-degrading microorganisms in the environment was estimated to be 0.5–9.6% of the total colonies.[241] Many of them, belonging to bacteria and fungal strains, have been isolated and investigated.[187] PHAs completely degrade to CO_2 and water in an aerobic environment and to methanol under anaerobic conditions. The degradation of PHAs was investigated in different environments, such as soil, sea water, compost and sewage sludge.[151,242–246] Complete degradation of P(3HB-*co*-3HV) copolymer was shown to occur in 6, 75 and 350 weeks in anaerobic sewage, soil and sea water, respectively.[186]

P(3HB) is also considered for medical applications. β-Hydroxybutyric acid, as a product of P(3HB) hydrolysis, can enter the metabolic pathways of organisms since it is a normal constituent of human blood. Degradation of PHA *in vivo* is slow due to the lack of suitable enzymes (PHA depolymerases). Some investigations on the fate of P(3HB) implants in mice have shown that the weight loss of the sample was 1.6% 6 months after implantation.[247] In terms of *in vivo* degradation, P(3HB-*co*-3HV) copolymers and their blends have better application promise compared to P(3HB).[202] Some recent reports on the degradation of microspheres of P(3HB-*co*-3HV) blended with PCL showed that the weight loss followed the order newborn calf serum > pancreatin > synthetic gastric juice > Hanks' buffer, which indicates some enzyme activity in addition to simple ester hydrolysis.[248] In another report, it was suggested that rat lipases may be involved in the hydrolysis of P(3HB) implants.[249] P(4HB) was degraded faster *in vivo* compared to P(3HB) and is considered to be a good candidate for biomedical applications.[166]

6.5 Poly(alkylene dicarboxylate)s

6.5.1 Synthesis of Poly(alkylene dicarboxylate)s

Poly(alkylene dicarboxylate) refers to a class of polyesters which are obtained from diols and diacids. A variety of homo and copolyesters can be obtained in a stepwise reaction starting from diacids and diols, allowing the properties to be tailored to fit the desired purpose. Aliphatic poly(alkylene dicarboxylate)s show good biodegradability properties; however, only a few combinations of diacids and diols give polyesters with melting and crystallization temperatures high enough for practical significance of these polyesters. One rare example of an aliphatic polyester having a melting temperature above 100 °C is poly(butylene succinate), PBS. For this reason, PBS, together with its copolyesters, is among the most investigated poly(alkylene dicarboxylate)s. In addition to 1,4-butanediol and succinic acid, a number of other diols, such as 1,2-ethylenediol, 1,3-propanediol and 1,6-hexanediol, and diacids, such as adipic and sebacic, have been used for the synthesis of biodegradable aliphatic homopolyesters and copolyesters.[250–257] Unlike aliphatic polyesters, aromatic polyesters are not prone to biodegradation. However, the introduction of some amount of aromatic residues into an aliphatic polyester chain can improve the unfavorable physical properties of aliphatic polyesters, while the biodegradability will be retained.[258,259] These aliphatic-aromatic polyesters comprise an important class of biodegradable polyesters for use as environmentally friendly alternatives to some commodity plastics.

The monomers used in the polycondensation reaction for the production of poly(alkylene dicarboxylate)s are basically from petrochemical sources. However, some of them can be obtained from renewable sources. For example, 1,3-propanediol can be produced by fermentation of glycerol, which is a by-product from biodiesel or plant oil production.[260] Succinic acid can be synthesized from glucose or whey by bacterial fermentation in very high yields.[261]

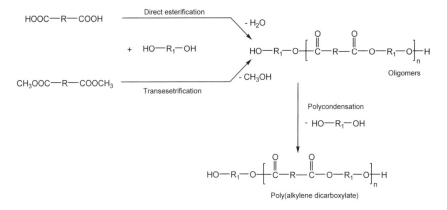

Scheme 6.10 Synthesis of poly(alkylene dicarboxylate)s by the polycondensation reaction.

Poly(alkylene dicarboxylate)s are obtained in polycondensation reaction which is usually performed as the direct polycondensation of diols with diacids or as a transesterification reaction, in which case the starting monomer is the dialkyl ester of the corresponding diacid (Scheme 6.10).

Polycondensation in bulk is performed as a two-stage reaction. In the first stage, the reaction temperature is raised controllably (up to 200–220 °C) with continual distillation of the formed by-product. At the end of the first stage, the reaction mixture contains small oligomers. In the second stage, when the by-product ceases to distil, the reaction temperature is increased (up to 230 °C) and vacuum is applied to promote diol removal and hence to increase the chain length. Even higher temperatures of the second stage could promote polycondensation; however, in the case of aliphatic polyesters, decomposition of the oligomers of small molecular weight could occur, which could reduce the molecular weight of the polyesters.[262] Nowadays, aliphatic polyesters with molecular weights which could provide good mechanical properties can be obtained by the use of highly effective catalysts, such as titanium tetrabutoxide (Ti(OBu)$_4$) or others based on germanium, tin and antimony.[256,263] In a kinetic investigation of polycondensation reaction for the synthesis of different poly(alkylene succinate)s, it was shown that the type of diol, *i.e.* ethylene glycol, butylene glycol or 1,3-propylene glycol, has an effect on the kinetics of reaction, but not on the molecular weight of the final product. Molecular weights of poly(alkylene succinate)s were positively affected by an increase in the polycondensation temperature and amount of catalyst (Ti(OBu)$_4$).[264,265] In another study of the titanium tetraisopropoxide-catalyzed synthesis of PBS, it was shown that for very high catalyst/monomer ratios, the molecular weight as well as the thermal properties of the obtained polyesters could deteriorate.[266]

Instead of vacuum for the removal of the by-product, azeotropic distillation in solution polymerization can be used.[253] Solid-state polymerization, which is often used to increase molecular weights of aromatic polyesters and PLA, is not applicable in case of aliphatic polyesters, due to their low melting temperatures.

Another approach to increase the molecular weight of polyesters obtained in direct polycondensation is to use chain-extenders, for example through the reaction of α,ω-functionalized polyesters with diisocyanates or oxazolines.[254,267,268] Using hexamethylene diisocyanate as a chain extender, an increase in the number average molecular weight from 33 000 to 72 000 g mol^{-1} for copolymers with butylene succinate and butylene adipate units was reported.[267]

As previously stressed, the advantage of the polycondensation reaction is the variety of monomers (aliphatic, aromatic, cyclic) which can be introduced into the polyester chain without great alteration of the reaction conditions. For example, segmented poly(ether ester)s, based on PBS hard segments and with polyethers in soft segments, used to increase hydrophilicity of copolyesters, were successfully obtained through transesterification polycondensation using Ti(OBu)$_4$ as a catalyst and the corresponding polyether macrodiols. Different polyethers such as poly(ethylene oxide) (PEO), poly(propylene oxide) (PPO) or poly(tetramethylene oxide) (PTMO), could be used as polyether component in the soft segments (Scheme 6.11).[269–272] Using the same reaction procedure, unsaturated groups through the reaction of dimethyl fumarate could be successfully incorporated into the PBS polyester chains (Scheme 6.12).[273]

The synthesis of polyesters by enzyme-catalyzed reactions are gaining considerable interest.[274–276] The milder reaction conditions, the absence of a toxic metal catalyst and the possibility of having good control over the polymer structure are obvious advantages of such a reaction pathway to polyesters. Simultaneously, the low molecular weights of the products and high cost of enzymes are disadvantage of such processes, indicating the need for further improvements in enzyme-catalyzed syntheses for the production of suitable products.

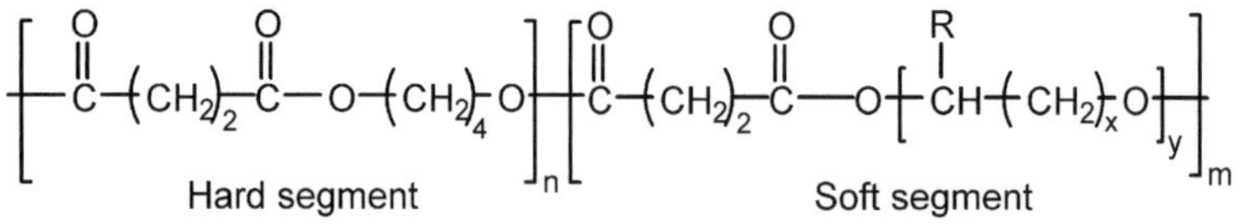

Scheme 6.11 Structural formula of poly(ester ether)s with PBS as the hard segments and different polyethers in soft segment: PEO (R = H, x = 1), PTMO (R = H, x = 3) or PPO (R = CH$_3$, x = 1).

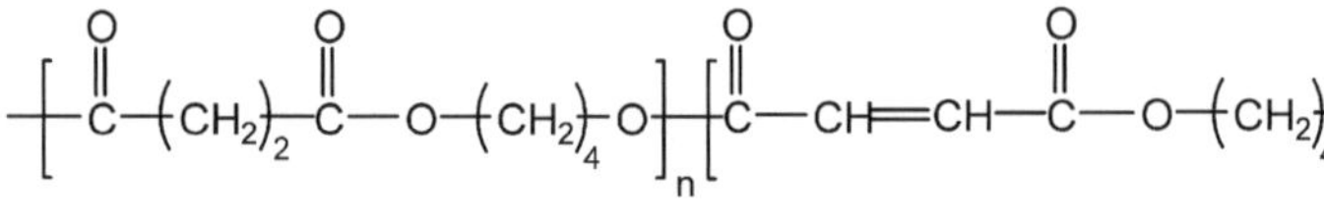

Scheme 6.12 Structural formula of unsaturated copolyesters of PBS and fumarate moieties.

6.5.2 Properties and Degradation of Poly(alkylene dicarboxylate)s

Aliphatic poly(alkylene dicarboxylate) homopolyesters are semicrystalline polymers with low melting temperatures. Only some polyesters based on succinic acid and oxalic acid have melting temperatures above 100 °C. The most investigated poly(alkylene dicarboxylate), PBS, has a melting temperature of 114 °C and a glass transition temperature of –43 °C. It is a highly crystalline polyester which can be found in two crystal modification, the α- and β-form. The more common, α-form, has a monoclinic unit cell with lattice dimensions $a = 0.523$ nm, $b = 0.912$, $c = 1.090$ nm and $\beta = 123.9°$.[277] Other very often investigated homopolyesters are poly(ethylene succinate) (PES), with $T_m = 104$ °C and $T_g = -6$ °C and poly(butylene adipate) (PBA), with $T_m = 60$ °C and $T_g = -60$ °C. Polyesters with 1,3-propanediol are the subject of intense investigation, since this monomer can be obtained from renewable sources. However, polyesters with this diol have low melting temperatures, below 60 °C.[255] In general, an odd-even effect is observed for the melting temperatures of homopolyesters, where odd number of methylene units in either the glycol or diacid component leads to a polymer with lower melting temperatures. Copolyesters obtained by variation of either the acid or diol component are usually crystalline and show a decrease in the melting temperature with the introduction of the second comonomer unit.[252,278] The dependencies of melting temperature versus composition very often exhibit a minimum for a certain copolyester composition.[250,251,256,279,280] This pseudoeutectic behavior is an indication of isodimorphism, *i.e.*, co-crystallization of different polymeric units. Isomorphism in a strict sense means that only one crystal structure is observed throughout the entire copolyester composition, while in isodimorphism, two or more crystalline phases are observed with a structural transition near the pseudoeutectic composition.[281] Copolyesters usually exhibit isodimorphism and in addition to the lowering of melting temperatures, the crystallinity is also decreased with increasing content of the minor copolyester component. This has an influence on the mechanical as well as the biodegradability properties of the copolyesters. There are also poly(alkylene dicarboxylate) copolymers based on sebacic and adipic acid and 1,6-hexanediol and 1,10-decanediol for which a high level of crystallinity was observed throughout the entire composition range.[282] Copolyesters with aromatic or some type of cyclic units exhibit higher melting temperatures, which depend on the type and content of the aromatic or cyclic comonomer.[258,283,284]

PBS is an opaque, soft and strong material which resembles LDPE. It has a tensile strength between those of PE and PP from 20 to 37 MPa, a Young's modulus between LDPE and HDPE of around 0.5 GPa and an elongation at break of 300% up to 560% for low molecular weight samples.[4,250,258,278,285] As expected, the mechanical properties of PBS improve with increasing molecular weight.[267] These mechanical properties together with excellent processability (commonly used conditions for the processing of polyolefins) recommend PBS and different types of its copolyesters as good replacements for non-biodegradable polymers such as LDPE or PP. Since the mechanical properties are very dependant on the crystallinity, these properties usually decrease with copolymerization. The decreasing trend in mechanical properties usually follows the decrease in the crystallinity with the incorporation of units of the second comonomer. This was observed for a number of copolyesters of PBS obtained by the introduction of different diols, such as 1,3-propanediol, or acids such as adipic acid.[250,278] The mechanical properties also decrease for copolyesters with aromatic acids, such as terephthalic acid, when they are present in small amounts; however, for higher amounts of aromatic residues, the mechanical properties progressively increase, but with the consequence of loss of biodegradability.[258,283] The elongation at break usually increases with the introduction of more flexible comonomer units, such as adipic acid moieties.[278,285]

PBS shows good thermal stability, much better than other biodegradable polyesters such as PLA, PHB and PCL, which is comparable to that of aromatic polyesters, despite its lower melting temperature. It was shown that the thermal degradation of high molecular weight PBS is not influenced by its molecular weight.[266] Thermal stability investigations of PBS during continuous heating showed that no appreciable degradation occurs up to 300 °C. It was proposed that the degradation follows two different mechanisms corresponding to weight loss in two stages. The activation energies which were determined for these two steps are similar to the ones calculated for the thermal degradation of the aromatic polyester poly(butylene terephthalate), PBT, and based on this, it was speculated that similar mechanisms were operative during the thermal degradation of these two polymers.[286] PES shows even better thermal stability compared to PBS, with the maximum decomposition rate occurring at 413 °C under applied conditions, which is higher than the corresponding temperature for the PBS of 399 °C. Overall, when these two polyesters were compared with poly(propylene succinate) (PPS), for which the corresponding temperature was determined to be 408 °C, it was concluded that the number of methylene groups in the diol component has a profound impact on the thermal stability of these polyesters. It was postulated that with increasing number of methylene groups, the thermal stability of these polyesters decreases.[286] However, this was not the case in a similar investigation performed on a series of poly(alkylene adipate)s, where the thermal stability of polyesters follows the trend poly(butylene adipate) (PBA) > poly(propylene adipate) (PPA) > poly(ethylene adipate) (PEA).[287] PBA is the most stable of these three polyesters and has thermal stability comparable to that of PBS. For all the mentioned polyesters, it was postulated that thermal degradation occurs in two consecutive steps, in which two different mechanisms are operative. Aliphatic copolyesters with two different acids or diols also exhibit good thermal stabilities.[288–292] No trend was found in the thermal stability of aliphatic copolyesters with changing composition and, due to the different starting premises (molecular weight, shape and crystallinity of the samples), it is difficult to draw a general conclusion. However, the thermal stability of some copolyesters was enhanced compared to that of the parent homopolyester.[280,290] Aromatic copolyesters are more stable than their aliphatic homologues. Hence, the thermal stability of aliphatic–aromatic copolyesters can be improved compared to the parent aliphatic homopolyester for a certain content of aromatic residues in the polymer chain.[293,294] It was shown that substitution of aromatic rings with the aliphatic cyclic ones can increase the thermal stability of the corresponding polyesters. Consequently, the introduction of cyclic units into an aliphatic polymer backbone can greatly increase thermal stability of random copolyesters.[284,295,296] As expected, the introduction of thermally labile ether bonds into polyesters deteriorates the thermal stability of the copolyesters.[292]

Hydrolytic degradation of PBS is slow, due to its hydrophobic nature and high crystallinity. It was reported that the residual mass of a PBS sample after incubation in a phosphate buffer solution for 15 weeks was 35%.[297] Hydrolytic degradation of PBS proceeds with an immediate and continuous decrease in the molecular weight, which is followed by weight loss in the later stages of degradation. This suggests a homogenous degradation mechanism similar to that of other aliphatic polyesters. The degradation rates are increased in more alkaline solutions.[252] The degree of crystallinity has a major influence on the degradation rate, since the degradation commences first in the amorphous parts of the polymer. In addition to crystallinity, the internal structure of the PBS crystallites also influence the progress of hydrolysis.[298] Other types of aliphatic polyesters exhibit the same tendencies. For copolyesters, the interplay between the usually reduced crystallinity of copolyesters and the chemical composition dependencies will result in an increased or decreased rate of hydrolysis for certain copolyester compositions.[252]

The degradation of poly(alkylene dicarboxylate)s is enhanced by lipase enzymes.[256,299,300] Lipases are water soluble and act on insoluble substrates in an effective manner only on an oil–water interface.[301] The enzymatic degradation rate depends on a number of factors such as chemical composition, hydrophylicity/hydrophobicity balance, degree of crystallinity, size of spherulites and lamellar thickness. The degradation rate also depends on external factors, such as the employed enzyme, temperature and pH at which the experiment is performed. It would be reasonable to expect that the polyesters with higher concentration of ester linkages within the polymer chain would exhibit higher rates of degradation, but this is only observed in some cases.[299] Moreover, by increasing the hydrophilicity of a polyester chain through the introduction of hydrophilic polyethers, *e.g.* PEO, the degradation of polyesters is enhanced.[269,270] However, the higher order structure directed by particular chemical structures seems to have greater influence on the degradation and often screens the chemical composition dependence. The degree of crystallinity is the factor which predominately determines the degradation rate in an unfavorable fashion. This is the main reason for usually observed higher rates of degradation for copolyesters compared to homopolyesters.[256,280,301] It was shown that chain mobility, which is correlated with the melting temperature of polyester, has a significant influence on the rate of enzymatic degradation. A dependence of degradation rate of polyesters on the temperature difference between the melting temperature of the polyester and temperature of experiment (ΔT_{mt}) was postulated by Marten et al.[302] These authors correlated ΔT_{mt} with chain mobility in the crystalline domains of the material, the degradation of which is the rate-determining step in an enzymatic degradation process of crystalline polyesters. This concept also applies to aliphatic–aromatic copolyesters, which showed reduced degradability as a consequence of more rigid chain structures.[303–305]

Different microorganisms can degrade poly(alkylene dicarboxylate)s. The percent of PBS-degrading microorganisms in soil was estimated to be 0.2–0.6% of the total colonies.[92] It was also reported that some fungal strains can degrade PBA, PEA and PPA.[93,306] Strain HT-6, isolated as an PBS-degrading microorganism, was able to assimilate 60% of PBS powder in 8 days and even 90% of PBS in emulsion.[307] For some isolated PBS-degrading strains, it was shown that degradation depends on the cell density and that degradation of PBS was suppressed if the strains were placed in non-sterile soil with high cell density level of the indigenous microorganisms.[308] Aliphatic–aromatic copolyesters are not as prone as aliphatic ones to degradation by microorganisms. However, Kleeberg et al. were able to isolate some *Thermomonosopra fusca* strains which were capable of degrading a PBA copolyester with terephthalic acid, PBAT, to the corresponding monomers.[309,310] It was shown that some microbial strains which were efficient in aliphatic polyester degradation can be efficient in the degradation of aliphatic–aromatic copolyesters in an emulsified form.[311]

The degradation of PBS and its various copolyesters was tested in different environments, such as soil, compost and sea water.[253,285,301,312] The degradation in soil was faster and less selective than degradation performed with the use of a single microbial strain or enzyme.[253,301] Since it is a surface phenomenon, the degradation greatly depends on the sample form.[312] For the degradation of PBS in a controlled compost, it was shown that the degradation proceeded slower with a longer incubation time compared to that of PCL.[313] The intermediate products in degradation of PBS are 1,4-butanediol and succinic acid, which are readily metabolized by microorganisms *via* the citric acid cycle. Degradation of copolyesters in different environments usually proceeds faster compared to that of homopolyesters.[253,285,301] For some PBS copolyesters with sebacic or adipic acid, it was shown that degradation in soil can be higher than that of the microbial polyesters P(3HB) and P(3HB-*co*-3HV).[301] Aliphatic–aromatic polyesters containing up to 60% of terephthalic acid in the total amount of acids can be degraded in soil and compost, whereby the degradation under composting conditions is much faster.[314,315] It was assumed that the long aromatic sequences in the oligomeric intermediates were hydrolyzed rather than metabolized, unlike monomeric products and aliphatic oligomers.[259]

Based on the low hydrolysis degradation rate of poly(alkylene dicarboxylate)s, it is to be expected that *in vivo* degradation would be a slow process. *In vivo* degradation investigations were mainly concentrated on aliphatic–aromatic polyesters modified by the introduction of PEO into the chain backbone.[316] For copolyesters containing PEO, in addition to hydrolytic degradation of the ester bonds, oxidative degradation of the aliphatic ether can also occur *in vivo*. As a response to foreign bodies, macrophages release oxidative agents which can initiate random degradation of the ether part of the polyester along the chain backbone.[317] For pure aliphatic polyesters based on PBS with PEO in soft segments, it was shown that *in vivo* degradation proceeds faster with decreasing length (molecular weight) of the PEO, due to the higher concentration of hydrolysable ester bonds in copolyesters with a shorter polyether block.[318] By replacing the aliphatic succinate residues with aromatic terephthalate, the degradation is retarded.[318] Degradation of copolyesters based on PBT and PEO in soft segments is incomplete and fragments rich in PBT may remain in the body.[319] However, it was shown in a number of studies that all these segmented aromatic–aliphatic polyesters and also aliphatic polyesters such as PBS and PPS have good biocompatibility.[251,297,316,318–320]

6.6 Application of Biodegradable Polyesters

Biodegradable polymers have been developed for two main fields of application: ecological and medical. The requirements for such applications are different in some respects and stricter in the case of medical applications. While some biodegradable polyesters can be successfully applied either in solving ecological problems or medical ones, some of them can be used for both applications. Some examples of the previously discussed polyesters are given in Figure 6.3 with reference to their possible applications.

6.6.1 Ecological Applications

From an environmental point of view, solving solid waste problems associated with the use of biostable polymers by their replacement with biodegradable polyesters is obviously a good choice. Recycling of polymers is also of

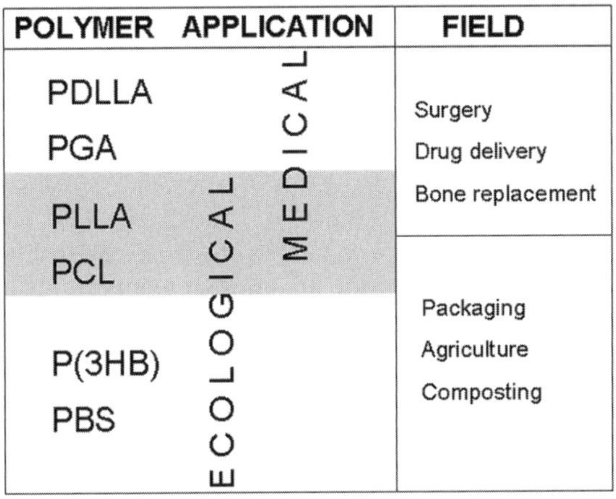

POLYMER	APPLICATION	FIELD
PDLLA	MEDICAL	Surgery
PGA		Drug delivery
PLLA	MEDICAL / ECOLOGICAL	Bone replacement
PCL		
P(3HB)	ECOLOGICAL	Packaging
PBS		Agriculture
		Composting

Figure 6.3 Possible applications of biodegradable polyesters.

importance as a solution to solid waste management and is often the first option to be considered. However, many polymers can not be easily separated for recycling or can not be recycled at all. In addition, the recycling of polymeric materials can not be repeated indefinitely. In such cases, biodegradable polymers are good alternative. In nature, the biodegradation of polyesters proceeds through the action of microorganisms which excrete enzymes capable of hydrolyzing ester bonds. Some of the discussed polyesters, such as PLA, are, on the contrary, degraded mainly through abiotic routes. When soluble fractions are formed, they can be further metabolized in the microbial cells. The ultimate products of such processes are CO_2, CH_4, water and biomass. The polymer itself along with its additives and also any intermediate product must be completely harmless for the environment. The time of degradation should fit the desired application and environment in which the polymer is to be degraded, *i.e.*, soil, compost, sea water, *etc*. Polyesters must possess satisfactory mechanical properties and processability suitable for the intended application. Biodegradable polyesters for ecological applications are produced by a number of companies worldwide and are already on the market. Different grades of PLA are produced by NatureWorks LLC-Cargill under the trade name Ingeo™ Biopolymer. Mitsui Chemicals Inc. in Japan produces PLA under the trade name LACEA®. PCL is produced by Perstorp UK under the trade name CAPA®. It can also be found under the trade name LACTEL®, produced by the Durest Corporation, for use in pharmacy and for different medical devices. PHAs are produced under a trade name Biomer® by Biomer, Germany. These polymers are also produced by Biocycle in Brazil under the trade name Biocycle®. Different grades of PBS and its copolymer with adipic acid, PBSA, are produced by ShowaDenko, Japan under the trade name Bionolle®. Similar copolymers are produced by NaturePlast, France under the trade name NATUREPLAST® PBE. This company also produces different grades of PLA and PHA. Aliphatic–aromatic polyesters are produced by a number of companies, such as BASF under the trade name Ecoflex®. DuPont produces aliphatic–aromatic copolyesters under the trade name Biomax® PTT using 1,3-propanediol from renewable sources. Some selected properties of commercially available polyesters aimed at replacing biostable commodity plastics are presented in Table 6.3.

The presented polyesters have a range of properties which classifies them as plastic materials similar to LDPE or PET. They can be used for packaging products such as flexible packaging, food trays, cosmetic bottles and composting bags. In agriculture, they can be used for delivery systems for fertilizers and pesticides or, often blended with some other materials such as starch, for mulch films. The broader use of biodegradable polyesters in ecological applications is hampered by their high price compared to commodity plastics. It is to be expected that the prices will be reduced with the further development of the production processes and increase in the production volume. In addition, governmental involvement and the role of the public in the promotion of environmentally friendly materials can be expected to enhance the use of biodegradable polyesters.

Table 6.3 Properties of commercial biodegradable polyesters.

Property	PLA Ingeo3051D (NatureWorks)	PCL CAPA6500 (Perstorp)	PHB Biocycle1000 (Biocycle)	PBS Bionolle#1020 (ShowaDenko)	PBSA Bionolle#3020 (ShowaDenko)	PBAT Ecoflex FBX 7011 (BASF)	LDPE	PET
Melting temparature, °C	150–165	60–62	170–175	114–115	93–95	110–115	108	255
Glass transition temperature, °C	55–65	−60		−32	−45		−120	78
Density, g cm^{-3}	1.25		1.20	1.26	1.23	1.25–1.27	0.92	1.4
Tensile yield strength, MPa	48	17.5	32	34	19	35/44	12	58
Flexular modulus, MPa	3828	411	2200	580	340	95/80[a]	176	2900[a]
Elongation, %	2.5	>700	4.0	320	400	560/710	400	300

[a]Tensile modulus of elasticity.

6.6.2 Medical Applications

Biodegradable polymeric materials comprise a large group of biomaterials, *i.e.*, materials intended to interface with biological systems. Whenever a biomaterial is required for a limited period, the use of polymers which can biodegrade is beneficial, since there is no need for surgical removal of the implant after it has fulfilled its function. In addition to biodegradability within the timeframe which suites the desired healing process, biodegradable polymers for use as biomaterials have to meet other requirements. The polymer and its degradation products should not be toxic and should not evoke any immunogenic response. The mechanical properties should match the desired application requirements, with the dynamics of the loss in mechanical properties through the course of degradation in harmony with the healing process. The material should also be sterilizable. Biocompatibility is of primary importance in such applications. Since biodegradable polymers do not stay in the body for long period, the biocompatibility limitation, which should be proven for an extended period, is more easily meet, than in the case of permanent implants.

Biodegradable polyesters, which meet all above-mentioned criteria, can find application as temporary scaffolds, in drug delivery or in tissue engineering. The first application of biodegradable polyesters was as surgical sutures developed in the 1960s. The polyesters usually employed for this application are PGA (Dexon®) and copolymer PGLLA (Vicryl®). The disadvantage of the metal implants used for fixation during the healing of damaged bone is that they have to be removed in a second surgery. There is also a danger that the bone can be refractured, since after removal of a metal implant there can be a period of bone weakness. Biodegradable polymers which gradually lose their mechanical properties with the simultaneous transfer of stress to the damaged area and complete absorption after completion of healing are seen as good candidates to be used in this area of medicine. A number of commercial products, made mainly of PLA and PGA, are now on the market.[16] In drug delivery systems, the biodegradable polymer serves as a matrix into which an active substance is dispersed. Besides the protective function of such a matrix, in the case of biodegradable polymers, the degradation of the matrix can control the release profile of a drug and be finally decomposed, eliminating the need for surgical removal. The polymer carrier can be in different forms, such as microspheres, nanospheres, beads, cylinders or discs. Among the biodegradable polyesters, PLA, PGA and their copolymers have been the subject of the most intensive research in this field.[321] For long-term delivery devices, PCL has been extensively investigated due to its slower degradation profile.[322] In tissue engineering, where a scaffold has the role of a support for a growing tissue during healing, the use biodegradable polyesters have also been considered. Processability of polyesters used for these purposes has to allow for the production of devices of complex shapes and porosity which will support cell growth and proliferation. Emerging trends in the design of biodegradable polyester devices for tissue engineering is to incorporate bioactive substances which will promote cell growth and tissue repair. Polyesters which are used for this biomedical application include PLA, PGA, PCL and their various copolymers.[17,30]

6.7 Future Trends in Biodegradable Polyesters

Biodegradable aliphatic polyesters are a large family of polymers with a range of properties which recommend them for various applications. They are already materials of relevance in different areas of everyday life and also as high-performance materials for medical applications. Investigations on improved synthetic procedures and performance of these polymers will continue to be the focus of scientific and industrial attention. Progress in synthetic procedures for condensation polymers toward chemoselective, regioselective and stereoselective polycondensation reactions is expected. Enzyme-catalyzed polycondensation and ROP reactions are also fields of intense research. For microbial polyesters, bioengineering is expected to lead to more efficient production and design of novel polyester structures with tailored properties. Advances in the area of the formation of new chemical structures through different functionalization or through the building of complex molecular architectures will certainly open new possibilities for the application of these polymers. New emerging fields in the design of biodegradable polyester materials is the formulation of nano-biocomposites. The main objective is to achieve major improvements in the properties of nano-biocomposites with the addition of small amounts of nano-filler. It is to be expected that the biodegradable polyester market will see a steady growth in the future with all these advances, together with growing ecological awareness.

References

1. R. Chandra and R. Rustgi, *Prog. Polym. Sci.*, 1998, **23**, 1273.
2. G. Scott, Why Biodegradable Polymers? in *Degradable Polymers: Principles and Application*, ed. G. Scott, Kluwer Academic Publishers, 2002, p. 1.
3. *Handbook of Biodegradable Polymers*, ed. C. Bastioli, Rapra Technology Limitid, Shawbury, Shrewsbury, Shropshire, SY4 4NR, UK, 2005.
4. M. Bahattacharya, R. L. Reis, L. Correlo and L. Boesel, Material Properties of Biodegradable Polymers, in *Biodegradable Polymers for Industrial Applications*, ed. R. Smith, Woodhead Publishing Limited, Cambrige, England, 2005, p. 336.
5. A. Wendy, A. Allan and T. Brian, *Polym. Int.*, 1998, **47**, 89.
6. C. Emo and S. Roberto, *Adv. Mater.*, 1996, **8**, 305.
7. A. U. B. Queiroz and F. P. Collares-Queiroz, *Polym. Rev.*, 2009, **49**, 65.
8. S. J. Huang, Biodegradable Polymers in *Encyclopedia of Polymer Science and Engineering*, ed. H. F. Mark, N. M. Bikales, C. G. Overberger, G. Menges and J. I. Kroschwitz, John Wiley & Sons, New York, 1985, Vol. 2, p. 220.

9. D. Satyanarayana and P. R. Chatterji, *J. Macromol. Sci. Rev. Macromol. Chem. Phys.*, 1993, **C33**(3), 349.
10. L. Avérous, *J. Macromol. Sci.-Pol. R*, 2004, **44**, 231.
11. M. Stefan, *Angew. Chem.*, 2004, **43**, 1078.
12. L. S. Nair and C. T. Laurencin, *Prog. Polym. Sci.*, 2007, **32**, 762.
13. M. Vert, *Biomacromolecules*, 2004, **6**, 538.
14. M. Vert, *Prog. Polym. Sci.*, 2007, **32**, 755.
15. P. B. Maurus and C. C. Kaeding, *Oper. Tech. Sports Med.*, 2004, **12**, 158.
16. J. C. Middleton and A. J. Tipton, *Biomaterials*, 2000, **21**, 2335.
17. M. Monique and W. H. Dietmar, *Polym. Int.*, 2007, **56**, 145.
18. Suming Li and M. Vert, Biodegradation of Aliphatic Polyesters in *Degradable Polymers: Principles and Application*, ed. G. Scott, Kluwer Academic Publishers, 2002, p. 71.
19. W. H. Carothers and J. A. Arvin, *J. Am. Chem. Soc.*, 1929, **51**, 2560.
20. W. H. Carothers and J. W. Hill, *J. Am. Chem. Soc.*, 1932, **54**, 1559.
21. A. Göpferich, *Biomaterials*, 1996, **17**, 103.
22. F. von Burkersroda, L. Schedl and A. Göpferich, *Biomaterials*, 2002, **23**, 4221.
23. N. Lucas, C. Bienaime, C. Belloy, M. Queneudec, F. Silvestre and J.-E. Nava-Saucedo, *Chemosphere*, 2008, **73**, 429.
24. A. A. Shah, F. Hasan, A. Hameed and S. Ahmed, *Biotechnol. Adv.*, 2008, **26**, 246.
25. D. K. Gilding and A. M. Reed, *Polymer*, 1979, **20**, 1459.
26. G. Sivalingam and G. Madras, *Polym. Degrad. Stab.*, 2004, **84**, 393.
27. S. Nagarajan and B. S. R. Reddy, *J. Sci. Ind. Res.*, 2009, **68**, 993.
28. I. Lautiainen, H. Miettinen, A. Mäkelä, P. Rokkanen and P. Törmälä, *Clin. Mater.*, 1994, **17**, 197.
29. J. Vasenius, P. Helevirta, H. Kuisma, P. Rokkanen and P. Törmälä, *Clin. Mater.*, 1994, **17**, 119.
30. P. A. Gunatillake, R. Adhikari and N. Gadegaard, *Eur. Cells Mater.*, 2003, **5**, 1.
31. N. Ashammakhi and P. Rokkanen, *Biomaterials*, 1997, **18**, 3.
32. D. Garlotta, *J. Polym. Environ.*, 2001, **9**, 63.
33. K. Madhavan Nampoothiri, N. R. Nair and R. P. John, *Bioresour. Technol.*, 2010, **101**, 8493.
34. J. L. Espartero, I. Rashkov, S. M. Li, N. Manolova and M. Vert, *Macromolecules*, 1996, **29**, 3535.
35. K. Hiltunen, J. V. Seppala and M. Harkonen, *Macromolecules*, 1997, **30**, 373.
36. S.-H. Hyon, K. Jamshidi and Y. Ikada, *Biomaterials*, 1997, **18**, 1503.
37. S. I. Moon and Y. Kimura, *Polym. Int.*, 2003, **52**, 299.
38. S. I. Moon, C. W. Lee, M. Miyamoto and Y. Kimura, *J. Polym. Sci. Pol. Chem.*, 2000, **38**, 1673.
39. J. Mauduit, N. Bukh and M. Vert, *J. Control. Release*, 1993, **23**, 209.
40. Y. Zhao, Z. Wang and F. Yang, *J. Appl. Polym. Sci.*, 2005, **97**, 195.
41. K. Hiltunen, M. Harkonen, J. V. Seppala and T. Vaananen, *Macromolecules*, 1996, **29**, 8677.

42. K. Hiltunen and J. V. Seppälä, *J. Appl. Polym. Sci.*, 1998, **67**, 1011.
43. K. Hiltunen and J. V. Seppälä, *J. Appl. Polym. Sci.*, 1998, **67**, 1017.
44. J. Ren, Q. F. Wang, S. Y. Gu, N. W. Zhang and T. B. Ren, *J. Appl. Polym. Sci.*, 2006, **99**, 1045.
45. J. Tuominen and J. V. Seppala, *Macromolecules*, 2000, **33**, 3530.
46. W. Zhong, J. Ge, Z. Gu, W. Li, X. Chen, Y. Zang and Y. Yang, *J. Appl. Polym. Sci.*, 1999, **74**, 2546.
47. T. Maharana, B. Mohanty and Y. S. Negi, *Prog. Polym. Sci.*, 2009, **34**, 99.
48. S. I. Moon, C. W. Lee, I. Taniguchi, M. Miyamoto and Y. Kimura, *Polymer*, 2001, **42**, 5059.
49. S.-I. Moon, I. Taniguchi, M. Miyamoto, Y. Kimura and C.-W. Lee, *High Perform. Polym.*, 2001, **13**, S189.
50. M. Ajioka, H. Suizu, C. Higuchi and T. Kashima, *Polym. Degrad. Stab.*, 1998, **59**, 137.
51. K. W. Kim and S. I. Woo, *Macromol. Chem. Phys.*, 2002, **203**, 2245.
52. M. Ajioka, K. Enomoto, K. Suzuki and A. Yamaguchi, *J. Polym. Environ.*, 1995, **3**, 225.
53. W. H. Carothers, G. L. Dorough and F. J. v. Natta, *J. Am. Chem. Soc.*, 1932, **54**, 761.
54. H. R. Kricheldorf and R. Dunsing, *Macromol. Chem. Phys.*, 1986, **187**, 1611.
55. A. Bhaw-Luximon, D. Jhurry, N. Spassky, S. Pensec and J. Belleney, *Polymer*, 2001, **42**, 9651.
56. H. R. Kricheldorf, *Chemosphere*, 2001, **43**, 49.
57. H. R. Kricheldorf, I. Kreiser-Saunders and A. Stricker, *Macromolecules*, 2000, **33**, 702.
58. A. Kowalski, A. Duda and S. Penczek, *Macromolecules*, 2000, **33**, 7359.
59. J. Lunt, *Polym. Degrad. Stab.*, 1998, **59**, 145.
60. G. Perego, G. D. Cella and C. Bastioli, *J. Appl. Polym. Sci.*, 1996, **59**, 37.
61. T. Miyata and T. Masuko, *Polymer*, 1998, **39**, 5515.
62. H. Tsuji and Y. Ikada, *Polymer*, 1995, **36**, 2709.
63. S. Sasaki and T. Asakura, *Macromolecules*, 2003, **36**, 8385.
64. S. Baratian, E. S. Hall, J. S. Lin, R. Xu and J. Runt, *Macromolecules*, 2001, **34**, 4857.
65. H. Tsuji and Y. Ikada, *Macromolecules*, 1992, **25**, 5719.
66. S. Brochu, R. E. Prud'homme, I. Barakat and R. Jerome, *Macromolecules*, 1995, **28**, 5230.
67. J.-R. Sarasua, R. E. Prud'homme, M. Wisniewski, A. Le Borgne and N. Spassky, *Macromolecules*, 1998, **31**, 3895.
68. L. Cartier, T. Okihara and B. Lotz, *Macromolecules*, 1997, **30**, 6313.
69. S. J. Huang, Poly(lactic acid) and Copolyesters in *Handbook of Biodegradable Polymers*, ed. C. Bastioli, Rapra Technology Limitid, Shawbury, Shrewsbury, Shropshire, SY4 4NR, UK, 2005, p. 287.
70. A. Södergård and M. Stolt, *Prog. Polym. Sci.*, 2002, **27**, 1123.
71. A. P. Gupta and V. Kumar, *Eur. Polym. J.*, 2007, **43**, 4053.
72. J. R. Dorgan, H. Lehermeier and M. Mang, *J. Polym. Environ.*, 2000, **8**, 1.

73. O. Martin and L. Avérous, *Polymer*, 2001, **42**, 6209.
74. S. Jacobsen and H. G. Fritz, *Polym. Eng. Sci.*, 1999, **39**, 1303.
75. F. D. Kopinke, M. Remmler, K. Mackenzie, M. Möder and O. Wachsen, *Polym. Degrad. Stab.*, 1996, **53**, 329.
76. D. Cam and M. Marucci, *Polymer*, 1997, **38**, 1879.
77. I. C. McNeill and H. A. Leiper, *Polym. Degrad. Stab.*, 1985, **11**, 309.
78. L. T. Lim, R. Auras and M. Rubino, *Prog. Polym. Sci.*, 2008, **33**, 820.
79. D. Cam, S.-H. Hyon and Y. Ikada, *Biomaterials*, 1995, **16**, 833.
80. M. Vert, S. Li and H. Garreau, *J. Control. Release*, 1991, **16**, 15.
81. S. Li, H. Garreau and M. Vert, *J. Mater. Sci.- Mater. M.*, 1990, **1**, 198.
82. I. Grizzi, H. Garreau, S. Li and M. Vert, *Biomaterials*, 1995, **16**, 305.
83. K. A. Athanasiou, J. P. Schmitz and C. M. Agrawal, *Tissue Eng.*, 1988, **4**, 53.
84. D. F. Williams, *Eng. Med.*, 1981, **10**, 5.
85. H. Fukuzaki, M. Yoshida, M. Asano and M. Kumakura, *Eur. Polym. J.*, 1989, **25**, 1019.
86. Y. Oda, A. Yonetsu, T. Urakami and K. Tonomura, *J. Polym. Environ.*, 2000, **8**, 29.
87. M. S. Reeve, S. P. McCarthy, M. J. Downey and R. A. Gross, *Macromolecules*, 1994, **27**, 825.
88. R. T. MacDonald, S. P. McCarthy and R. A. Gross, *Macromolecules*, 1996, **29**, 7356.
89. S. Li, M. Tenon, H. Garreau, C. Braud and M. Vert, *Polym. Degrad. Stab.*, 2000, **67**, 85.
90. Y. Tokiwa and A. Jarerat, *Biotechnol. Lett.*, 2004, **26**, 771.
91. S. Li, A. Girard, H. Garreau and M. Vert, *Polym. Degrad. Stab.*, 2000, **71**, 61.
92. Y. Tokiwa and B. Calabia, *J. Polym. Environ.*, 2007, **15**, 259.
93. D. Y. Kim and Y. H. Rhee, *Appl. Microbiol. Biotechnol.*, 2003, **61**, 300.
94. A. Torres, S. M. Li, S. Roussos and M. Vert, *Appl. Environ. Microbiol.*, 1996, **62**, 2393.
95. M. Hakkarainen, S. Karlsson and A. C. Albertsson, *Polymer*, 2000, **41**, 2331.
96. M. Hakkarainen, S. Karlsson and A. C. Albertsson, *J. Appl. Polym. Sci.*, 2000, **76**, 228.
97. G. Kale, R. Auras, S. P. Singh and R. Narayan, *Polym. Test.*, 2007, **26**, 1049.
98. J. Tuominen, J. Kylmä, A. Kapanen, O. Venelampi, M. Itävaara and J. Seppälä, *Biomacromolecules*, 2002, **3**, 445.
99. H. Pistner, H. Stallforth, R. Gutwald, J. Mühling, J. Reuther and C. Michel, *Biomaterials*, 1994, **15**, 439.
100. K. Stridsberg, M. Ryner and A.-C. Albertsson, *Adv. Polym. Sci.*, 2002, **157**, 41.
101. C. G. Pitt, Poly-ε-Caprolactone and Its Copolymers in *Biodegradable Polymers as Drug Delivery Systems*, ed. M. Casin and R. Langer, Marcel Dekker, New York, 1990, p. 71.
102. K. Ito, Y. Hashizuka and Y. Yamashita, *Macromolecules*, 1977, **10**, 821.

103. I. B. Rashkov, I. Gitsov, I. M. Panayotov and J. Pascault, *J. Polym. Sci. Pol. Chem.*, 1983, **21**, 923.
104. A.-C. Albertsson and I. Varma, *Adv. Polym. Sci.*, 2002, **157**, 1.
105. M. Labet and W. Thielemans, *Chem. Soc. Rev.*, 2009, **38**, 3484.
106. C. Jacobs, P. Dubois, R. Jerome and P. Teyssie, *Macromolecules*, 1991, **24**, 3027.
107. A. Loefgren, A. C. Albertsson, P. Dubois, R. Jerome and P. Teyssie, *Macromolecules*, 1994, **27**, 5556.
108. A. Kowalski, A. Duda and S. Penczek, *Macromol. Rapid Commun.*, 1998, **19**, 567.
109. S. Penczek, A. Duda, A. Kowalski, J. Libiszowski, K. Majerska and T. Biela, *Macromol. Symp.*, 2000, **157**, 61.
110. A. Duda, S. Penczek, A. Kowalski and J. Libiszowski, *Macromol. Symp.*, 2000, **153**, 41.
111. W. J. Bailey, Z. Ni and S. R. Wu, *J. Polym. Sci. Pol. Chem.*, 1982, **20**, 3021.
112. Y. Hiraguri, K. Katase and Y. Tokiwa, *J. Macromol. Sci. A: Pure Appl. Chem*, 2005, **42**, 901.
113. J. Xu, Z. L. Liu and R. X. Zhuo, *J. Appl. Polym. Sci.*, 2007, **103**, 1146.
114. F. Sanda and T. Endo, *J. Polym. Sci. Pol. Chem.*, 2001, **39**, 265.
115. H. Bittiger, R. H. Marchessault and W. D. Niegisch, *Acta Cryst.*, 1970, **B26**, 1923.
116. I. Yoshito and T. Hideto, *Macromol. Rapid Comm.*, 2000, **21**, 117.
117. P. Bordes, E. Pollet and L. Avérous, *Prog. Polym. Sci.*, 2009, **34**, 125.
118. R. A. Gross and B. Kalra, *Science*, 2002, **297**, 803.
119. G. Sivalingam and G. Madras, *Polym. Degrad. Stab.*, 2003, **80**, 11.
120. O. Persenaire, M. Alexandre, P. Degée and P. Dubois, *Biomacromolecules*, 2001, **2**, 288.
121. G. Sivalingam, S. P. Vijayalakshmi and G. Madras, *Ind. Eng. Chem. Res.*, 2004, **43**, 7702.
122. G. Sivalingam, R. Karthik and G. Madras, *J. Anal. Appl. Pyrol.*, 2003, **70**, 631.
123. H. Abe, N. Takahashi, K. J. Kim, M. Mochizuki and Y. Doi, *Biomacromolecules*, 2004, **5**, 1480.
124. S. Ting-Ting, J. Heng and G. Hong, *Polym.-Plast. Technol.*, 2008, **47**, 398.
125. Y. Aoyagi, K. Yamashita and Y. Doi, *Polym. Degrad. Stab.*, 2002, **76**, 53.
126. D. R. Chen, J. Z. Bei and S. G. Wang, *Polym. Degrad. Stab.*, 2000, **67**, 455.
127. H. Antheunis, J.-C. van der Meer, M. de Geus, A. Heise and C. E. Koning, *Biomacromolecules*, 2010, **11**, 1118.
128. H. Antheunis, J.-C. van der Meer, M. de Geus, W. Kingma and C. E. Koning, *Macromolecules*, 2009, **42**, 2462.
129. W. P. Ye, F. S. Du, W. H. Jin, J. Y. Yang and Y. Xu, *React. Funct. Polym.*, 1997, **32**, 161.
130. A. C. Albertsson, R. Renstad, B. Erlandsson, C. Eldsäter and S. Karlsson, *J. Appl. Polym. Sci.*, 1998, **70**, 61.

131. J. Peña, T. Corrales, I. Izquierdo-Barba, A. L. Doadrio and M. Vallet-Regí, *Polym. Degrad. Stab.*, 2006, **91**, 1424.
132. Y. Tokiwa and T. Suzuki, *Nature*, 1977, **270**, 76.
133. A. Iwamoto and Y. Tokiwa, *Polym. Degrad. Stab.*, 1994, **45**, 205.
134. H. Tsuji and T. Ishizaka, *J. Appl. Polym. Sci.*, 2001, **80**, 2281.
135. M. Mochizuki, M. Hirano, Y. Kanmuri, K. Kudo and Y. Tokiwa, *J. Appl. Polym. Sci.*, 1995, **55**, 289.
136. Z. Gan, Q. Liang, J. Zhang and X. Jing, *Polym. Degrad. Stab.*, 1997, **56**, 209.
137. L. Liu, S. Li, H. Garreau and M. Vert, *Biomacromolecules*, 2000, **1**, 350.
138. H. Fukuzaki, M. Yoshida, M. Asano, M. Kumakura, T. Mashimo, H. Yuasa, K. Imai and Y. Hidetoshi, *Polymer*, 1990, **31**, 2006.
139. Y. Hou, J. Chen, P. Sun, Z. Gan and G. Zhang, *Polymer*, 2007, **48**, 6348.
140. S. Li, L. Liu, H. Garreau and M. Vert, *Biomacromolecules*, 2003, **4**, 372.
141. M. Funabashi, F. Ninomiya and M. Kunioka, *J. Polym. Environ.*, 2007, **15**, 7.
142. A. Höglund, M. Hakkarainen and A.-C. Albertsson, *J. Macromol. Sci. A: Pure Appl. Chem.*, 2007, **44**, 1041.
143. D. Darwis, H. Mitomo, T. Enjoji, F. Yoshii and K. Makuuchi, *Polym. Degrad. Stab.*, 1998, **62**, 259.
144. T. Suyama, Y. Tokiwa, P. Ouichanpagdee, T. Kanagawa and Y. Kamagata, *Appl. Environ. Microbiol.*, 1998, **64**, 5008.
145. J. G. Sanchez, A. Tsuchii and Y. Tokiwa, *Biotechnol. Lett.*, 2000, **22**, 849.
146. C. A. Murphy, J. A. Cameron, S. J. Huang and R. T. Vinopal, *Appl. Environ. Microbiol.*, 1996, **62**, 456.
147. Y. Oda, H. Asari, T. Urakami and K. Tonomura, *J. Ferment. Bioeng.*, 1995, **80**, 265.
148. Y. Doi, K.-i. Kasuya, H. Abe, N. Koyama, I. Shin-ichi, T. Koichi and Y. Yoshida, *Polym. Degrad. Stab.*, 1996, **51**, 281.
149. A. Ohtaki, N. Akakura and K. Nakasaki, *Polym. Degrad. Stab.*, 1998, **62**, 279.
150. A. Ohtaki, N. Sato and K. Nakasaki, *Polym. Degrad. Stab.*, 1998, **61**, 499.
151. M. Hakkarainen, *Adv. Polym. Sci.*, 2002, **157**, 113.
152. C. Eldsäter, B. Erlandsson, R. Renstad, A. C. Albertsson and S. Karlsson, *Polymer*, 2000, **41**, 1297.
153. C. M. Buchanan, D. D. Dorschel, R. M. Gardner, R. J. Komarek and A. W. White, *J. Macromol. Sci. A: Pure Appl. Chem.*, 1995, **32**, 683.
154. M. Hakkarainen and A. C. Albertsson, *Macromol. Chem. Phys.*, 2002, **203**, 1357.
155. R. A. P. A. Gunatillake, *Eur. Cells Mater.*, 2003, **5**, 1.
156. S. C. Woodward, P. S. Brewer, F. Moatamed, A. Schindler and C. G. Pitt, *J. Biomed. Mater. Res.*, 1985, **19**, 437.
157. H. Sun, L. Mei, C. Song, X. Cui and P. Wang, *Biomaterials*, 2006, **27**, 1735.
158. C. X. F. Lam, D. W. Hutmacher, J. T. Schantz, M. A. Woodruff and S. H. Teoh, *J. Biomed. Mater. Res. A*, 2009, **90A**, 906.

159. X. F. L. Christopher, *et al.*, *Biomed. Mater.*, 2008, **3**, 034108.
160. M. A. Woodruff and D. W. Hutmacher, *Prog. Polym. Sci.*, 2010, **35**, 1217.
161. R. W. Lenz and R. H. Marchessault, *Biomacromolecules*, 2005, **6**, 1.
162. L. L. Wallen and W. K. Rohwedder, *Environ. Sci. Technol.*, 1974, **8**, 576.
163. K. Sudesh, H. Abe and Y. Doi, *Prog. Polym. Sci.*, 2000, **25**, 1503.
164. G. N. Barnard and J. K. Sanders, *J. Biol. Chem.*, 1989, **264**, 3286.
165. Y. Doi, M. Kunioka, Y. Nakamura and K. Soga, *Macromolecules*, 1988, **21**, 2722.
166. D. P. Martin and S. F. Williams, *Biochem. Eng. J.*, 2003, **16**, 97.
167. B. Witholt and B. Kessler, *Curr. Opin. Biotech.*, 1999, **10**, 279.
168. Y. B. Kim and R. W. Lenz, *Adv. Biochem. Eng. Biot.*, 2001, **71**, 51.
169. I. Noda, P. R. Green, M. M. Satkowski and L. A. Schechtman, *Biomacromolecules*, 2005, **6**, 580.
170. B. Hazer and A. Steinbüchel, *Appl. Microbiol. Biotechnol.*, 2007, **74**, 1.
171. G. J. M. de Koning, H. M. M. van Bilsen, P. J. Lemstra, W. Hazenberg, B. Witholt, H. Preusting, J. G. van der Galiën, A. Schirmer and D. Jendrossek, *Polymer*, 1994, **35**, 2090.
172. S. Kusaka, H. Abe, S. Y. Lee and Y. Doi, *Appl. Microbiol. Biotechnol.*, 1997, **47**, 140.
173. L. Jingnan, R. C. Tappel and C. T. Nomura, *Polym. Rev.*, 2009, **49**, 226.
174. B. H. A. Rehm and A. Steinbüchel, *Int. J. Biol. Macromol.*, 1999, **25**, 3.
175. Y. Kawaguchi and Y. Doi, *Macromolecules*, 1992, **25**, 2324.
176. Y. Jia, T. J. Kappock, T. Frick, A. J. Sinskey and J. Stubbe, *Biochemistry*, 2000, **39**, 3927.
177. S. Zhang, S. Kolvek, R. W. Lenz and S. Goodwin, *Biomacromolecules*, 2003, **4**, 504.
178. J. Stubbe and J. Tian, *Nat. Prod. Rep.*, 2003, **20**, 445.
179. A. Steinbüchel and S. Hein, *Adv. Biochem. Eng. Biotechnol.*, 2001, **71**, 81.
180. Y. Doi, A. Tamaki, M. Kunioka and K. Soga, *J. Chem. Soc., Chem. Commun.*, 1987, **21**, 1635.
181. M. J. de Smet, G. Eggink, B. Witholt, J. Kingma and H. Wynberg, *J. Bacteriol.*, 1983, **154**, 870.
182. R. A. Gross, C. DeMello, R. W. Lenz, H. Brandl and R. C. Fuller, *Macromolecules*, 1989, **22**, 1106.
183. H. Brandl, R. A. Gross, R. W. Lenz and R. C. Fuller, *Appl. Environ. Microbiol.*, 1988, **54**, 1977.
184. M. Liebergesell, F. Mayer and A. Steinbüchel, *Appl. Microbiol. Biotechnol.*, 1993, **40**, 292.
185. H. Matsusaki, H. Abe and Y. Doi, *Biomacromolecules*, 2000, **1**, 17.
186. S. Y. Lee, *Biotechnol. Bioeng.*, 1996, **49**, 1.
187. S. Khanna and A. K. Srivastava, *Process Biochem.*, 2005, **40**, 607.
188. T. Fukui and Y. Doi, *Appl. Microbiol. Biotechnol.*, 1998, **49**, 333.
189. H. W. Ryu, S. K. Hahn, Y. K. Chang and H. N. Chang, *Biotechnol. Bioeng.*, 1997, **55**, 28.
190. F. Wang and S. Y. Lee, *Appl. Environ. Microbiol.*, 1997, **63**, 3703.
191. W. J. Page and O. Knosp, *Appl. Environ. Microbiol.*, 1989, **55**, 1334.

192. E. Grothe and Y. Chisti, *Bioproc. Biosyst. Eng.*, 2000, **22**, 441.
193. A. Yezza, D. Fournier, A. Halasz and J. Hawari, *Appl. Microbiol. Biotechnol.*, 2006, **73**, 211.
194. J. Schrader, M. Schilling, D. Holtmann, D. Sell, M. V. Filho, A. Marx and J. A. Vorholt, *Trends Biotechnol.*, 2009, **27**, 107.
195. B. Füchtenbusch and A. Steinbüchel, *Appl. Microbiol. Biotechnol.*, 1999, **52**, 91.
196. C. Guo-Qiang, X. Jun, W. Qiong, Z. Zengming and H. Kwok-Ping, *React. Funct. Polym.*, 2001, **48**, 107.
197. F. Liu, W. Li, D. Ridgway, T. Gu and Z. Shen, *Biotechnol. Lett.*, 1998, **20**, 345.
198. S. J. Park, W. S. Ahn, P. R. Green and S. Y. Lee, *Biotechnol. Bioeng.*, 2001, **74**, 82.
199. E. Akaraonye, T. Keshavarz and I. Roy, *J. Chem. Technol. Biotechnol.*, 2010, **85**, 732.
200. M. Zinn, B. Witholt and T. Egli, *Adv. Drug Deliver. Rev.*, 2001, **53**, 5.
201. P. Suriyamongkol, R. Weselake, S. Narine, M. Moloney and S. Shah, *Biotechnol. Adv.*, **25**, 148.
202. C. W. Pouton and S. Akhtar, *Adv. Drug Deliver. Rev.*, 1996, **18**, 133.
203. M. Juzwa and Z. Jedliński, *Macromolecules*, 2006, **39**, 4627.
204. M. Okada, *Prog. Polym. Sci.*, 2002, **27**, 87.
205. S. Bloembergen, D. A. Holden, T. L. Bluhm, G. K. Hamer and R. H. Marchessault, *Macromolecules*, 1987, **20**, 3086.
206. T. U. Gerngross, K. D. Snell, O. P. Peoples, A. J. Sinskey, E. Csuhai, S. Masamune and J. Stubbe, *Biochemistry*, 1994, **33**, 9311.
207. A. Steinbüchel, B. Füchtenbusch, V. Gorenflo, S. Hein, R. Jossek, S. Langenbach and B. H. A. Rehm, *Polym. Degrad. Stab.*, 1998, **59**, 177.
208. A. Steinbüchel, *Macromol. Biosci.*, 2001, **1**, 1.
209. E. N. Pederson, C. W. J. McChalicher and F. Srienc, *Biomacromolecules*, 2006, **7**, 1904.
210. Y. Poirier, D. E. Dennis, K. Klomparens and C. Somerville, *Science*, 1992, **256**, 520.
211. Y. Poirier, *Adv. Biochem. Eng. Biotechnol.*, 2001, **71**, 209.
212. J. Cobntbekt and R. H. Mabchessault, *J. Mol. Biol.*, 1972, **71**, 735.
213. B. L. Hurrell and R. E. Cameron, *J. Mater. Sci.*, 1998, **33**, 1709.
214. F. Biddlestone, A. Harris, J. N. Hay and T. Hammond, *Polym. Int*, 1996, **39**, 221.
215. M. Di Lorenzo, M. Raimo, E. Cascone and E. Martuscelli, *J. Macromol. Sci. Phys.*, 2001, **40**, 639.
216. W. J. Orts, R. H. Marchessault and T. L. Bluhm, *Macromolecules*, 1991, **24**, 6435.
217. S. Bloembergen, D. A. Holden, G. K. Hamer, T. L. Bluhm and R. H. Marchessault, *Macromolecules*, 1986, **19**, 2865.
218. M. Kunioka, A. Tamaki and Y. Doi, *Macromolecules*, 1989, **22**, 694.
219. Y. Saito and Y. Doi, *Int. J. Biol. Macromol.*, 1994, **16**, 99.

220. S. Philip, T. Keshavarz and I. Roy, *J. Chem. Technol. Biotechnol.*, 2007, **82**, 233.
221. Y. Doi, S. Kitamura and H. Abe, *Macromolecules*, 1995, **28**, 4822.
222. K. Sudesh and Y. Doi, Polyhydroxyalkanoates in *Handbook of Biodegradable Polymers*, ed. C. Bastioli, Rapra Technology Limitid, Shawbury, Shrewsbury, Shropshire, SY4 4NR, UK, 2005, p. 219.
223. K. D. Gagnon, R. W. Lenz, R. J. Farris and R. C. Fuller, *Macromolecules*, 1992, **25**, 3723.
224. H. Preusting, A. Nijenhuis and B. Witholt, *Macromolecules*, 1990, **23**, 4220.
225. K. J. Kim, Y. Doi and H. Abe, *Polym. Degrad. Stab.*, 2006, **91**, 769.
226. S. D. Li, J. D. He, P. H. Yu and M. K. Cheung, *J. Appl. Polym. Sci.*, 2003, **89**, 1530.
227. M. Kunioka and Y. Doi, *Macromolecules*, 1990, **23**, 1933.
228. F. D. Kopinke, M. Remmler and K. Mackenzie, *Polym. Degrad. Stab.*, 1996, **52**, 25.
229. H. Liu, M. Pancholi, J. Stubbs and D. Raghavan, *J. Appl. Polym. Sci.*, 2010, **116**, 3225.
230. J. Yu, D. Plackett and L. X. L. Chen, *Polym. Degrad. Stab.*, 2005, **89**, 289.
231. Y. Doi, Y. Kanesawa, M. Kunioka and T. Saito, *Macromolecules*, 1990, **23**, 26.
232. C. Eldsäter, S. Karlsson and A.-C. Albertsson, *Polym. Degrad. Stab.*, 1999, **64**, 177.
233. G. Tomasi, M. Scandola, B. H. Briese and D. Jendrossek, *Macromolecules*, 1996, **29**, 5056.
234. H. Abe, Y. Doi, H. Aoki and T. Akehata, *Macromolecules*, 1998, **31**, 1791.
235. M. Scandola, M. L. Focarete, G. y. Adamus, W. Sikorska, I. Baranowska, S. Swierczek, M. Gnatowski, M. Kowalczuk and Z. Jedlinski, *Macromolecules*, 1997, **30**, 2568.
236. K.-i. Kasuya, T. Ohura, K. Masuda and Y. Doi, *Int. J. Biol. Macromol.*, 1999, **24**, 329.
237. H. Abe and Y. Doi, *Biomacromolecules*, 2001, **3**, 133.
238. H. Abe and Y. Doi, *Int. J. Biol. Macromol.*, 1999, **25**, 185.
239. H. Abe, Y. Doi, H. Aoki, T. Akehata, Y. Hori and A. Yamaguchi, *Macromolecules*, 1995, **28**, 7630.
240. Y. Kanesawa, N. Tanahashi, Y. Doi and T. Saito, *Polym. Degrad. Stab.*, 1994, **45**, 179.
241. Y. Tokiwa and B. P. Calabia, *Biotechnol. Lett.*, 2004, **26**, 1181.
242. J. Mergaert, A. Webb, C. Anderson, A. Wouters and J. Swings, *Appl. Environ. Microbiol.*, 1993, **59**, 3233.
243. H. Tsuji and K. Suzuyoshi, *J. Appl. Polym. Sci.*, 2003, **90**, 587.
244. H. Tsuji and K. Suzuyoshi, *Polym. Degrad. Stab.*, 2002, **75**, 347.
245. C. L. Yue, R. A. Gross and S. P. McCarthy, *Polym. Degrad. Stab.*, 1996, **51**, 205.

246. A. Ohtaki and K. Nakasaki, *Waste Manage. Res.*, 2000, **18**, 184.
247. S. Gogolewski, M. Jovanovic, S. M. Perren, J. G. Dillon and M. K. Hughes, *J. Biomed. Mater. Res.*, 1993, **27**, 1135.
248. T. W. Atkins and S. J. Peacock, *J. Biomat. Sci.- Polym. E.*, 1996, **7**, 1075.
249. M. Löbler, M. Saß, P. Michel, U. T. Hopt, C. Kunze and K. P. Schmitz, *J. Mater. Sci.- Mater. M.*, 1999, **10**, 797.
250. G. Z. Papageorgiou and D. N. Bikiaris, *Biomacromolecules*, 2007, **8**, 2437.
251. G. Z. Papageorgiou and D. N. Bikiaris, *Macromol. Chem. Physc.*, 2009, **210**, 1408.
252. Y. Yoo, M.-S. Ko, S.-I. Han, T.-Y. Kim, S. Im and D.-K. Kim, *Polym. J.*, 1998, **30**, 538.
253. C. Zhu, Z. Zhang, Q. Liu, Z. Wang and J. Jin, *J. Appl. Polym. Sci.*, 2003, **90**, 982.
254. H. Shirahama, Y. Kawaguchi, M. S. Aludin and H. Yasuda, *J. Appl. Polym. Sci.*, 2001, **80**, 340.
255. S. S. Umare, A. S. Chandure and R. A. Pandey, *Polym. Degrad. Stab.*, 2007, **92**, 464.
256. M. Mochizuki, K. Mukai, K. Yamada, N. Ichise, S. Murase and Y. Iwaya, *Macromolecules*, 1997, **30**, 7403.
257. G. Z. Papageorgiou, D. N. Bikiaris, D. S. Achilias, S. Nanaki and N. Karagiannidis, *J. Polym. Sci. Pol. Phys.*, 2010, **48**, 672.
258. M. Nagata, H. Goto, W. Sakai and N. Tsutsumi, *Polymer*, 2000, **41**, 4373.
259. R.-J. Müller, I. Kleeberg and W.-D. Deckwer, *J. Biotechnol.*, 2001, **86**, 87.
260. T. Willke and K. Vorlop, *Eur. J. Lipid Sci. Technol.*, 2008, **110**, 831.
261. H. Song and S. Y. Lee, *Enzyme Microb. Technol.*, 2006, **39**, 352.
262. K. Chrissafis, K. M. Paraskevopoulos and D. N. Bikiaris, *Thermochim. Acta*, 2006, **440**, 166.
263. T. Yokozawa, N. Ajioka and A. Yokoyama, *Adv. Polym. Sci.*, 2008, **217**, 1.
264. D. N. Bikiaris and D. S. Achilias, *Polymer*, 2006, **47**, 4851.
265. D. N. Bikiaris and D. S. Achilias, *Polymer*, 2008, **49**, 3677.
266. J. Yang, S. Zhang, X. Liu and A. Cao, *Polym. Degrad. Stab.*, 2003, **81**, 1.
267. V. Tserki, P. Matzinos, E. Pavlidou and C. Panayiotou, *Polym. Degrad. Stab.*, 2006, **91**, 377.
268. C. Q. Huang, S. Y. Luo, S. Y. Xu, J. B. Zhao, S. L. Jiang and W. T. Yang, *J. Appl. Polym. Sci.*, 2010, **115**, 1555.
269. D. Pepic, E. Zagar, M. Zigon, A. Krzan, M. Kunaver and J. Djonlagic, *Eur. Polym. J.*, 2008, **44**, 904.
270. D. Jovanovic, M. S. Nikolic and J. Djonlagic, *J. Serb. Chem. Soc.*, 2004, **69**, 1013.
271. D. Pepic, M. Radoicic, M. S. Nikolic and J. Djonlagic, *J. Serb. Chem. Soc.*, 2007, **72**, 1515.
272. D. Pepic, M. S. Nikolic and J. Djonlagic, *J. Appl. Polym. Sci.*, 2007, **106**, 1777.

273. M. S. Nikolic, D. Poleti and J. Djonlagic, *Eur. Polym. J.*, 2003, **39**, 2183.
274. S. Kobayashi, H. Ritter, D. Kaplan, H. Uyama and S. Kobayashi, Enzymatic Synthesis of Polyesters *via* Polycondensation in *Enzyme-Catalyzed Synthesis of Polymers*, Springer Berlin/Heidelberg, 2006, p. 133.
275. S. Kobayashi, *Macromol. Rapid Commun.*, 2009, **30**, 237.
276. I. K. Varma, A.-C. Albertsson, R. Rajkhowa and R. K. Srivastava, *Prog. Polym. Sci.*, 2005, **30**, 949.
277. Y. Ichikawa, H. Kondo, Y. Igarashi, K. Noguchi, K. Okuyama and J. Washiyama, *Polymer*, 2000, **41**, 4719.
278. V. Tserki, P. Matzinos, E. Pavlidou, D. Vachliotis and C. Panayiotou, *Polym. Degrad. Stab.*, 2006, **91**, 367.
279. G. Montaudo and P. Rizzarelli, *Polym. Degrad. Stab.*, 2000, **70**, 305.
280. M. S. Nikolic and J. Djonlagic, *Polym. Degrad. Stab.*, 2001, **74**, 263.
281. P. Pan and Y. Inoue, *Prog. Polym. Sci.*, 2009, **34**, 605.
282. G. J. Howard and S. Knutton, *Polymer*, 1968, **9**, 527.
283. U. Witt, R.-J. Müller, J. Augusta and H. Widdecke, *Macromol. Chem. Phys.*, 1994, **105**, 793.
284. C. Berti, A. Celli, P. Marchese, E. Marianucci, S. Sullalti and G. Barbiroli, *Macromol. Chem. Phys.*, 2010, **211**, 1559.
285. T. Fujimaki, *Polym. Degrad. Stab.*, 1998, **59**, 209.
286. K. Chrissafis, K. M. Paraskevopoulos and D. N. Bikiaris, *Thermochim. Acta*, 2005, **435**, 142.
287. T. Zorba, K. Chrissafis, K. M. Paraskevopoulos and D. N. Bikiaris, *Polym. Degrad. Stab.*, 2007, **92**, 222.
288. C.-J. Tsai, W.-C. Chang, C.-H. Chen, H.-Y. Lu and M. Chen, *Eur. Polym. J.*, 2008, **44**, 2339.
289. H. Y. Lu, M. Chen, C. H. Chen, J. S. Lu, K. C. Hoang and M. Tseng, *J. Appl. Polym. Sci.*, 2010, **116**, 3693.
290. C. H. Chen, H. Y. Lu, M. Chen, J. S. Peng, C. J. Tsai and C. S. Yang, *J. Appl. Polym. Sci.*, 2009, **111**, 1433.
291. C.-H. Chen, J.-S. Peng, M. Chen, H.-Y. Lu, C.-J. Tsai and C.-S. Yang, *Colloid Polym. Sci.*, 2010, **288**, 731.
292. A. Cao, T. Okamura, K. Nakayama, Y. Inoue and T. Masuda, *Polym. Degrad. Stab.*, 2002, **78**, 107.
293. F. Li, X. Xu, Q. Li, Y. Li, H. Zhang, J. Yu and A. Cao, *Polym. Degrad. Stab.*, 2006, **91**, 1685.
294. M. Soccio, L. Finelli, N. Lotti, M. Gazzano and A. Munari, *J. Polym. Sci. Pol. Phys.*, 2007, **45**, 310.
295. H. C. Ki and O. Ok Park, *Polymer*, 2001, **42**, 1849.
296. C. Berti, A. Celli, P. Marchese, G. Barbiroli, F. Di Credico, V. Verney and S. Commereuc, *Eur. Polym. J.*, 2009, **45**, 2402.
297. H. Li, J. Chang, A. Cao and J. Wang, *Macromol. Biosci.*, 2005, **5**, 433.
298. K. Cho, J. Lee and K. Kwon, *J. Appl. Polym. Sci.*, 2001, **79**, 1025.
299. M. Mochizuki and M. Hirami, *Polym. Adv. Technol.*, 1997, **8**, 203.
300. Y. Tokiwa, T. Suzuki and K. Takeda, *Agric. Biol. Chem.*, 1986, **50**, 1323.

301. P. Rizzarelli, C. Puglisi and G. Montaudo, *Polym. Degrad. Stab.*, 2004, **85**, 855.
302. E. Marten, R.-J. Müller and W.-D. Deckwer, *Polym. Degrad. Stab.*, 2003, **80**, 485.
303. E. Marten, R.-J. Müller and W.-D. Deckwer, *Polym. Degrad. Stab.*, 2005, **88**, 371.
304. Y. Tokiwa and T. Suzuki, *J. Appl. Polym. Sci.*, 1981, **26**, 441.
305. L. Zhao and Z. Gan, *Polym. Degrad. Stab.*, 2006, **91**, 2429.
306. N. Ishii, Y. Inoue, K.-i. Shimada, Y. Tezuka, H. Mitomo and K.-i. Kasuya, *Polym. Degrad. Stab.*, 2007, **92**, 44.
307. H. Pranamuda, Y. Tokiwa and H. Tanaka, *Appl. Environ. Microbiol.*, 1995, **61**, 1828.
308. M. Abe, K. Kobayashi, N. Honma and K. Nakasaki, *Polym. Degrad. Stab.*, 2010, **95**, 138.
309. I. Kleeberg, C. Hetz, R. M. Kroppenstedt, R.-J. Muller and W.-D. Deckwer, *Appl. Environ. Microbiol.*, 1998, **64**, 1731.
310. U. Witt, T. Einig, M. Yamamoto, I. Kleeberg, W. D. Deckwer and R. J. Müller, *Chemosphere*, 2001, **44**, 289.
311. T. Nakajima-Kambe, F. Ichihashi, R. Matsuzoe, S. Kato and N. Shintani, *Polym. Degrad. Stab.*, 2009, **94**, 1901.
312. J. H. Zhao, X. Q. Wang, J. Zeng, G. Yang, F. H. Shi and Q. Yan, *J. Appl. Polym. Sci.*, 2005, **97**, 2273.
313. M. Kunioka, F. Ninomiya and M. Funabashi, *Int. J. Mol. Sci.*, 2009, **10**, 4267.
314. R.-J. M. U. Witt and W.-D. Deckwer, *J. Macromol. Sci. A: Pure Appl. Chem.*, 1995, **A32**(4), 851.
315. R.-J. Müller, Aliphatic-Aromatic Polyesters in *Handbook of Biodegradable Polymers*, ed. C. Bastioli, Rapra Technology Limited, Shawbury, Shrewsbury, Shropshire, SY4 4NR, UK, 2005, p. 303.
316. R. Shi, D. Chen, Q. Liu, Y. Wu, X. Xu, L. Zhang and W. Tian, *Int. J. Mol. Sci.*, 2009, **10**, 4223.
317. K. Sutherland, J. R. Mahoney, A. J. Coury and J. W. Eaton, *J. Clin. Invest.*, 1993, **92**, 2360.
318. R. van Dijkhuizen-Radersma, J. R. Roosma, J. Sohier, F. L. A. M. A. Péters, M. van den Doel, C. A. van Blitterswijk, K. de Groot and J. M. Bezemer, *J. Biomed. Mater. Res. A*, 2004, **71A**, 118.
319. A. A. Deschamps, A. A. van Apeldoorn, H. Hayen, J. D. de Bruijn, U. Karst, D. W. Grijpma and J. Feijen, *Biomaterials*, 2004, **25**, 247.
320. L.-C. Wang, J.-W. Chen, H.-L. Liu, Z.-Q. Chen, Y. Zhang, C.-Y. Wang and Z.-G. Feng, *Polym. Int.*, 2004, **53**, 2145.
321. F. Mohamed and C. F. van der Walle, *J. Pharm. Sci.*, 2008, **97**, 71.
322. V. R. Sinha, K. Bansal, R. Kaushik, R. Kumria and A. Trehan, *Int. J. Pharm.*, 2004, **278**, 1.

CHAPTER 7
Synthesis and Characterization of Thermoplastic Agro-polymers

C. J. R. VERBEEK AND J. M. BIER

School of Engineering, University of Waikato, Private Bag 3105, Hamilton 3240, New Zealand

7.1 Introduction

Petrochemical polymers have become ubiquitous for their excellent properties and durability. Unfortunately, they also create an enormous environmental burden. Motivations behind sustained research in reducing dependence on polymers from petrochemical sources are similar to those in energy research; a decreasing fossil fuel supply with a corresponding price increase and a widespread awareness of sustainability.[1] As a result, finding new uses for agricultural commodities has become an important area of research.

Petroleum-based materials could potentially be replaced with renewable and biodegradable materials such as polysaccharides or proteins.[2] Biodegradable materials are capable of undergoing microbial or enzymatic degradation,[3] although this does not does not necessarily imply renewability, and several petroleum-based biodegradable polymers are commercially available. Biodegradable polymers are generally classified as outlined in Figure 7.1.[3-5] In some cases blends between natural and synthetic polymer are considered a separate class.[6]

The potential use of agro-polymers in the plastics industry has long been recognized. Agro-polymers are extracted from either plants or animals. They could contribute to a reduction of dependence on fossil resources as agricultural resources are generally sustainable. Some of the polymers in this family

---

RSC Green Chemistry No. 12
A Handbook of Applied Biopolymer Technology: Synthesis, Degradation and Applications
Edited by Sanjay K. Sharma and Ackmez Mudhoo
© Royal Society of Chemistry 2011
Published by the Royal Society of Chemistry, www.rsc.org

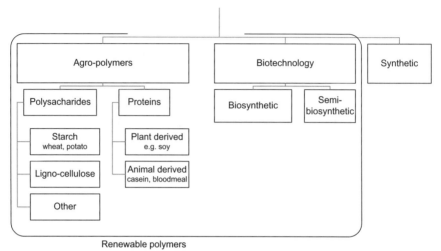

Figure 7.1 Classification of biodegradable polymers.

can be processed directly into thermoplastic materials; however, most require chemical modification. A further benefit is that these polymers are often by-products of other agricultural activities. Common characteristics of agro-polymers are their hydrophilicity, fast degradation rate and sometimes unsatisfactory mechanical properties, particularly in wet environments.[4,5] These polymers can be considered an innovative and sustainable approach to reduce reliance on petrochemical polymers.[3] The main technological challenge is to successfully modify the properties of these materials to account for deficiencies such as brittleness, water sensitivity, and low strength.

7.1.1 Polysaccharides

Polysaccharides are carbohydrate polymers, formed from either mono- or disaccharides through glycosidic bonds. The most significant polysaccharides of relevance here are cellulose, starch, lignin, pectin and chitin. The structures of some of these polymers are shown in Figure 7.2.

Cellulose is one of the most abundant biopolymers and forms the major constituent of plant cell walls. Currently cellulose is only processed thermoplastically after chemical modification;[2] cellulose acetate would be the most significant product of this nature.

Starch is produced by many plants as a source of stored energy and is found in plant roots, stalks, crop seeds and crops such as rice, corn, wheat, tapioca and potato. It consists of linear amylose and branched amylopectin.[2] Starch is recovered by wet grinding, sieving and drying as native starch.[7]

Lignins are aromatic amorphous polymers and are obtained from almost all types of natural plants, in association with cellulose.[8] Lignins are recovered

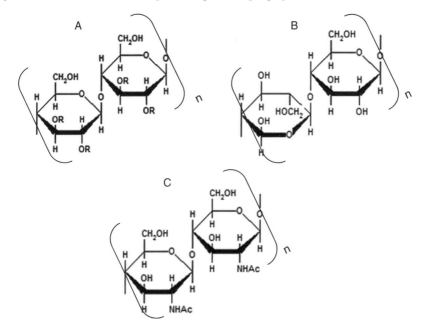

Figure 7.2 Monomers of polysaccharides. A: cellulose (R = H) or cellulose acetate (R = Ac); B: starch; C: chitin.

from cellulose by strong acid/base treatment or by high pressure steam followed by solvent extraction.[8] Modified lignins have been compounded or grafted with thermoplastics. The benefit of the ligneous residues is that they are very polar whilst having a hydrophobic skeleton. As a result, thermoplastic properties are dependent on the presence of a polar plasticizer such as water.[9] Interest in lignin as a polymeric material is mainly because it is readily available, although molecular mass and structure may vary considerably depending on the chemical processes followed during extraction.[1]

Pectins are polysaccharides that contain 1,4-linked α-D-galactosyluronic acid residues. Pectin substances are present in various proportions in plants and act as an intercellular cement. Their gelling properties allow the formation of films which are often used for as fruit or vegetable coating.[9]

Chitin (poly-N-acetyl-D-glucosamine, 2-acetamido-2-deoxy-1,4-β-D-glucan) is one of the three most abundant polysaccharides (along with starch and cellulose) and is extracted from shells of crustaceans and cell walls of fungi.[2] Chitosan is produced by deacetylation of chitin and can be spun into fibres, cast into films, or precipitated in a variety of micro-morphologies. Major applications are in biomaterials, pharmaceuticals, cosmetics, metal ion sequestration, agriculture and the food industry.[1,2]

Of the polysaccharide family, starch would be the only substance that is widely used as a thermoplastic without chemical modification and will be discussed in further sections.

7.1.2 Proteins

Proteins are natural polymers which contribute to biological functions within a cell, along with other biological macromolecules such as polysaccharides and nucleic acids. Proteins are classified as condensation polymers because their synthesis involves elimination of water to produce peptide bonds between amino acids. As well as being important for biological systems, proteins are also used in the medicine, food and materials industries. Due to the diverse building blocks of proteins and their unique structure, a large variety of biodegradable materials can be produced offering a wide range of functional properties.[10]

Proteins are readily available as by-products or wastes of the agricultural and horticultural industries. As a result, proteins from plants (wheat gluten, soy, sunflower and corn) and animals (gelatin, keratin, casein, whey and bloodmeal) have been manufactured into plastics. Many studies have been carried out using casting and compression moulding techniques. However, commercial viability of protein-based plastics hinges on utilization of common synthetic processing techniques such as extrusion and injection moulding.[11]

In the next section, synthesis of thermoplastics from starch and proteins is discussed. Only these two polymers are included in the discussion as these form the majority of potential thermoplastics produced from renewable, agricultural sources. Polymers requiring chemical modification of the monomer are specifically excluded as this chapter focuses on direct utilization of natural resources.

7.2 Synthesis

It is well known that polymers can generally be classified as either thermoset, thermoplastic, or rubbery. Most agro-polymers do not behave thermoplastically without some additives and would typically degrade before a flowable melt can be formed. It is therefore necessary to consider the synthesis route to render these materials ready for thermoplastic processing.

Polymer processing is concerned with the mixing and shaping of polymeric materials to form them into useful products. Processing usually involves the application of heat and pressure, with the specific method depending on whether the material is thermoplastic or thermoset.[10] Thermoplastic processing involves melting a polymer, followed by shaping and finally cooling the material in its new form. The softening temperature and atmospheric stability of the polymer, as well as the geometry and size of the finished product, are important. The heat required for melting can be supplied by radiation, conduction, or mechanical work. The most important thermoplastic processing techniques are extrusion, post-die processing, thermo-forming and injection moulding. The largest volume of thermoplastics is probably processed by means of extrusion.[10]

An extruder consists of a heated, fixed metal barrel that contains either one or two screws. The screws act by conveying the material through the heated barrel, inducing shear forces and increasing pressure along the barrel before leaving the extruder at the die. Process variables include the feed material's composition, screw speed, barrel temperature profile, feed rates and die size and shape. The degree of screw fill, specific mechanical energy input (SME), torque, pressure at the die, residence time and product temperature are influenced by these process variables.[11]

7.2.1 General Considerations

In order to reach a flowable state, amorphous polymer chains need to be above their glass transition and crystalline regions need to be at a temperature above the melting point. Polymer chains are typically linked by a multitude of interactions such as hydrogen bonding, hydrophobic interactions and other weak van der Waal's forces. When sufficient thermal energy is supplied and these interactions are overcome, chains will have the mobility necessary to allow them to move in relation to each other, or flow.[9] Chain entanglements and secondary interactions are what differentiate synthetic polymers from other low molecular mass organic substances. Inter- and intra-molecular bonds, as well as chain entanglements, prevent chain slippage, thus leading to the superior properties of polymers.[10]

It has been found that by heating hydrophilic polymers (such as starch and protein) in closed volumes in the presence of water, homogeneous melts may be formed which can be processed like conventional petrochemical-based thermoplastics.[12] The transformation of agro-polymers into a thermoplastic is often more complex than conventional thermoplastic processing. The diversity of polymer structures and interactions, as well as the dependence of their structure on the extraction technique followed complicates processing. Furthermore, hydrogen bonding is strongly water-sensitive which makes processing this class of polymers even more complex.[9]

7.2.2 The Role of Additives

Thermoplastic properties of agro-polymers depend on the presence of water or other plasticizers. Agro-polymers have relatively low degradation temperatures and the energy required to disrupt intermolecular bonding is close to the energy leading to degradation.[9] Plasticizers improve processability by interposing themselves between the polymer chains and alter the forces holding the chains together. This occurs through two mechanisms, lubrication and increasing free volume. For example, with proteins small molecules are easily incorporated into the protein matrix, shown by the high plasticizing effect of water and glycerol. Water is considered a natural plasticizer of proteins and is used extensively in protein extrusion. When plasticizers are compared based on the mass fraction in a bioplastic, low molecular mass compounds such as water will be present at larger numbers compared to high molecular mass compounds. Every plasticizer molecule can interact with a protein chain, which implies that at equal mass fractions, water is normally more efficient than other plasticizers.[11]

Most agro-polymers contain polar functionalities forming hydrogen bonding with water molecules through hydroxyl groups or amines. Many interactions with water are possible, with variable intensity, and typically define the polymer's water-holding capacity.[9] Water is often considered an essential constituent of agro-polymers and makes processing delicate. It is the interaction with water and between chains that influence structure–property relationships which makes thermal analysis a useful method to assess the role of water in agro-polymers.[5]

The low boiling temperature of water is a serious complication during processing as evaporation typically occurs before adequate processing temperatures for low enough viscosity are reached.[9] Alternatively, hydrophilic compounds such as polyols, carbohydrates and amines may also interact with polar groups in agro-polymers, thereby plasticizing the material. Some examples are glycerol, sorbitol, saccharose, urea, triethylene glycol and polyethylene glycol.[11] These molecules must be polar to ensure compatibility with polymers and small enough to penetrate the macromolecular network. These plasticizers also allow preparation of compounds which can be preserved before further processing because equilibrium moisture content in agro-polymers (with ambient air) is not sufficient to allow a flowable melt.

7.2.3 Starch

One specific thermoplastic agro-polymer of interest here is thermoplastic starch (TPS). Starch can only be converted into a thermoplastic material in the presence of plasticizers using heat and shear.[1,6] The benefit of TPS is its renewability, compostability and relatively low price compared with synthetic thermoplastics.[8] One problem with the use of starch is its high glass transition temperature (T_g). Brittleness also increases with time due to free volume relaxation and retrogradation. In order to increase flexibility and processability, plasticizers such as water, glycol, sorbitol, urea, amide, sugars and quaternary amines have been used in TPS.[6] Dimensional stability and mechanical properties of thermoplastic starch are highly dependent on moisture content.[1] Unfortunately, the excessive hydrophilicity of starch is not significantly reduced by polyol plasticizers.

In its native state, starch is semi-crystalline (about 20–45%) and water insoluble. Native starch granules typically have dimensions ranging from 0.5 to 175 μm and appear in a variety of shapes. It is composed of linear (amylose) and branched (amylopectin) polymers of α-D-glucose. Amylose has a molecular mass of about 10^5–10^6 g mol^{-1}, while amylopectin has a molecular mass in the range 10^7–10^9 g mol^{-1}. Starch rich in amylose is usually preferred for conversion to TPS as the linearity of amylose improves the processability of starch even though it is present as a minor component (between 20 and 30wt%). The ratio of amylose to amylopectin depends on the source and age of the starch, and can also be influenced by the extraction process.[4,9,13,14]

TPS is obtained after disruption and plasticization of native starch's granular structure into a homogeneous amorphous phase by applying thermo-mechanical energy in a continuous extrusion process. Water and glycols are commonly used plasticizers, although urea and formamide have also been explored.[15,16] A chemical agent, such as urea, favours granule destruction by disrupting hydrogen bonding in the crystallites.[9] Thermal processing of TPS involves multiple chemical and physical reactions:[13]

- water diffusion
- granule expansion
- gelatinization
- decomposition/destruction
- melting and crystallization.

Gelatinization is particularly important and it is the basis of thermoplastic synthesis. The destruction of crystalline regions in starch granules is an irreversible process that includes granular swelling, crystalline melting and molecular solubilization.[13] As a consequence of gelatinization, the melting temperature (T_m) and the glass transition (T_g) are reduced. Excess water is generally required to ensure dispersion and a high degree of gelatinization. If using less water in the formulation (<20 wt%), the melting temperature (T_m = 220–240 °C) tends to be close to the degradation temperature. To overcome this, a non-volatile plasticizer (glycerol or other polyols) is added to decrease the glass transition temperature.[8] During extrusion, shear forces physically tear apart starch granules, allowing faster transfer of water into the interior molecules. At low moisture content, small amounts of gelatinized, melted, as well as fragmented starch (due to degradation or decomposition) exist simultaneously.[13]

Starch processing is much more complicated and difficult to control than conventional polymer processing due to its unique phase transitions, high viscosity, water evaporation and fast retrogradation. However, with proper formulation development (plasticizer selection, *etc.*) and suitable processing conditions, many of these challenges can be overcome.[6,13]

One of the challenges of TPS is the result of the structures of amylose and amylopectin. Amylose is considered a linear structure and amylopectin branched. Superior mechanical properties can be obtained if the structure is formed at low water contents and if it is based on the branched component of amylopectin.[12] These macromolecules are able to create short range interactions (van der Waal's and hydrogen bonding) between macromolecules or functional groups (*e.g.* OH). However, chain entanglements develop if chains are long enough to develop long range interactions leading to higher elongation at break values. Neither amylose nor amylopectin develop significant long range interactions; these materials are not cohesive enough. In addition, hydrophilicity also enables the system to partially crystallize (or retrograde) as a function of the water content.[6,15]

Melting and mixing are the two main objectives of compounding or extrusion processing, and process parameters are typically adjusted to minimize chain degradation. However, fragmentation is unavoidable, with the extent depending on operating conditions, such as screw speed, temperature, moisture content and starch type. Controlling degradation is important since it influences processing properties such as viscosity, but it can also affect the performance of final products.[13]

After processing, TPS shows ageing as evident from an increase in mechanical properties such as tensile modulus. Below the T_g it shows physical ageing accompanied by material densification, while above the T_g, recrystallization is even more pronounced. The re-ordering of gelatinized or thermoplastic starch into crystalline regions is known as retrogradation. The retrogradation kinetics depend on chain mobility, plasticizer type and content.[8]

7.2.4 Proteins

It has been said that in order for proteinaceous bioplastics to be commercially feasible, they need to be processable using equipment currently used for synthetic thermoplastics. Proteinaceous bioplastics are often brittle and water sensitive, and overcoming this is one of the driving forces behind research in this field. Physiochemical properties and processing conditions are often governed by the protein's structural properties, and therefore also final material properties.[17]

A synthetic polymer consists of identical monomers, covalently bonded in a long chain. Unlike synthetic polymers, proteins are complex hetero-polymers, consisting of up to 20 different amino acid repeat units. The amino acid repeat unit contains two carbon atoms as well as nitrogen, differing only in their functional side groups. In its natural environment, a protein will be folded into secondary, tertiary and quaternary structures stabilized through hydrophobic interactions, hydrogen bonding and electrostatic interactions between amino acid functional groups. The folded conformation is a delicate balance of these interactions.[18] Once folded, the structure may be stabilized further with strong covalent cross-links. Due to the diverse building blocks of proteins and their unique structure, a large variety of biodegradable materials can be produced offering a wide range of functional properties.[11]

It is well known that the visco-elastic behaviour of amorphous or semi-crystalline polymers can be divided into five regions; these are the glassy, leathery, rubbery rubbery-flow and viscous states. The transformation from one region to another is dependent on temperature, while the temperature at which each transition occurs is dependent on the polymer structure. Processing can only be done above temperatures corresponding to the rubbery-flow region. Most literature on proteinaceous bioplastics suggests that processing be done above the protein's softening point, which would imply a temperature well above the T_g.

Extrusion and injection moulding of polymers require that a viscous melt be formed by the polymer upon the addition of heat. That implies that interactions between chains are sufficiently low to allow relative movement of chains, but some interaction is required to impart some degree of melt strength in the material. Synthetic polymers generally satisfy these requirements. In the absence of strong intermolecular forces (hydrogen bonds), chain entanglements and van der Waal's forces are the most important mechanisms that impart a good melt strength.

Small changes in environmental conditions, such as increasing temperature, pressure, change of pH or addition of chemicals, can disrupt a protein's folded conformation and this is called denaturing. Denaturation is a unique property of proteins and can be defined as the modification of secondary, tertiary or quaternary structures of a protein molecule. This is not to be confused with degradation, which is the loss of primary structure, or breaking of covalent peptide bonds. For a protein to behave like a synthetic polymer, the protein chain is required in an extended conformation enabling the formation of sufficient chain entanglement. In order to do this, multiple non-covalent and covalent interactions need to be reduced, allowing chains to unfold and form new interactions and entanglements.[10,11]

Unlike starch, where destructuring leads to an amorphous fluid, denaturation of proteins exposes core structural groups which can be more hydrophobic than surface groups (depending on the environment). In a polar environment the structure forms with polar residues in the surface and hydrophobic ones inside, and vice versa. Furthermore, with an increase in temperature, chains become more mobile but their movement is restricted because of newly formed intermolecular forces. Hydrophobic interaction intensifies due to temperature and denaturation which is followed by coagulation.[9]

Hydrogen bonds, ionic interactions, hydrophobic interactions and covalent disulfide bonds are affected during denaturation, causing an unravelling of protein structure into a random coil conformation. Denaturation can have several important consequences, such as increase in viscosity of protein solutions, decrease of solubility due to exposure of hydrophobic groups, increase of reactivity of side groups, altered sensitivity to enzymatic proteolysis and altered surfactant properties. Denaturing exposes functional groups of the amino acid side chains, thereby introducing new interactions by means of hydrophobic, hydrogen or ionic bonding. Although some of these effects can be seen as negative, denaturation and the consequences thereof are important for protein processing.

Understanding how to process agro-polymers with similar properties to synthetic polymers requires an understanding of how these forces are manifested in proteins. The processability of proteins depends on their transition from the glassy to rubbery and viscous-flow states. These transitions are achieved with judicial application of heat, pressure, shear, chemical additives and plasticizers. Specific amino acid residues (primary protein structure) and initial structure (natural protein state) of the protein will influence each of these factors. Based on the structure of proteins and the requirement for thermoplastic processing, three broadly categorized processing requirements have been identified:[11]

- breaking of intermolecular bonds (non-covalent and covalent) that stabilize proteins in their native form by using chemical or physical means
- arranging and orientating mobile chains in the desired shape
- enabling formation of new intermolecular bonds and interactions to stabilize the three-dimensional structure.

In order to fulfil these requirements, a typical formulation of a proteinaceous bioplastic would include some or all of the following components:

- plasticizers or a combination of plasticizers to promote flowability
- reactive additives to promote cross-link reduction, such as sodium sulfite or sodium bisulfite[19,20]
- additives such as pigments, preservatives (sodium benzoate), foaming agents, bleaching agents (hydrogen peroxide, calcium carbonate, barium peroxide),[21] titanium oxide, sodium bisulfate[22] or anti-foaming agents
- lubricants or extrusion aids such as soybean oil, fatty acids and vegetable oils[23] as well as fused silica and ammonium hydroxide[24]
- extenders such as fibres and clay, to improve mechanical properties
- modifiers to impart water resistance.

The processes occurring during extrusion are considered an equilibrium reaction between temperature-induced polymerization and shear-induced de-polymerization. In this context, polymerization means the formation of intermolecular forces or covalent cross-links while de-polymerization may also imply protein degradation. It was found that high temperatures and high specific mechanical energy (SME) input could induce excessive cross-linking and/or degradation of protein chains. At high temperatures and low moisture or plasticizer content, viscous heat dissipation could increase the likelihood of protein degradation.[11]

Denaturing, cross-linking and plasticization are probably the most important aspects of protein processing. The melt temperature and the temperature dependence of viscosity can be affected by the type and the amount of plasticizer. Generally, increasing the amount of plasticizer will lower the melt temperature and viscosity of the blend.[25] The extruder is particularly suitable for processing proteins which do not aggregate excessively and therefore maintain a lower constant processing torque.

7.3 Characterization

7.3.1 Overview of Characterization

Selecting appropriate biopolymers for commercial use is a complicated task and involves careful consideration of the properties of the polymeric material. In the context of this chapter, the meaning of the term 'Characterization of Agro-polymers' is twofold. Not only are the techniques used for characterization important, but also how they define the characteristics of agro-polymers. Typically, chemical structure and morphology as well as mechanical and thermal properties are important. In addition to these, aspects that need consideration are the shape and form of raw materials as well as their quality, supply, cost and physical properties.[26] Obtaining optimal material properties in a final product requires appropriate processing which is dependent on structure. Structure, property and processing cannot be considered independently, as illustrated in Figure 7.3.

The structure of polymers is described at multiple levels. Biochemists traditionally refer to the primary, secondary and tertiary structure of native proteins. Similarly for thermoplastics there are multiple levels of structural information, such as chain architecture and crystallinity. All of these contribute to observed material properties and can be manipulated by processing. In Figure 7.4, the general structure of proteins, starch and synthetic polymers are compared. At the monomeric level starch is made up of repeating units of glucose and exists in two forms – branched (amylopectin) or unbranched (amylose). In native starch these are arranged in granules consisting of hard and soft regions.[27] The blocklets in the hard shell are made up of crystalline amylopectin surrounded by amorphous amylose. Processing disrupts these ordered regions.[28] Proteins consist of a chain of peptide linkages with various amino acid side groups present. These side groups contribute to a range of interactions which, combined with hydrogen bonding around the polypeptide backbone, lead to different secondary structures.[11] Synthetic plastics include a range of materials with different monomers and linkages. Olefins, or vinyl-like polymers, such as polypropylene are hydrophobic, whereas nylons contain amide linkages similar to the peptide backbone in proteins which are hydrophilic and form hydrogen bonds with water.[29] Vinyl-type polymers form helical conformations similar to α-helices in proteins and many petroleum-derived

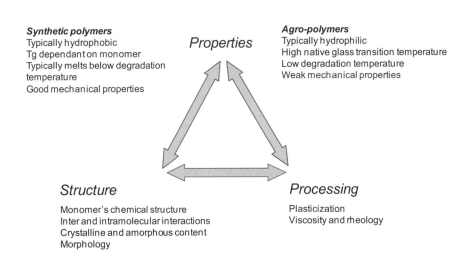

Figure 7.3 Structure–property–processing relationship.

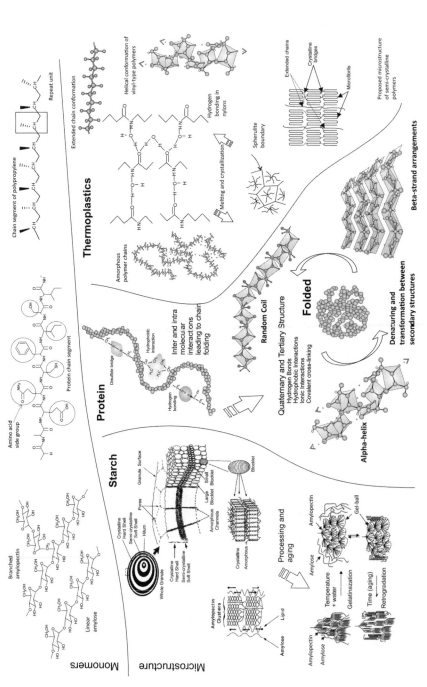

Figure 7.4 Monomers and microstructure of starch, protein and examples of synthetic thermoplastics. Microstructure of starch reprinted from *Carbohydrate Polymers*, **32**(3–4), D. J. Gallant, B. Bouchet, and P. M. Baldwin, Microscopy of starch: Evidence of a new level of granule organization, page No. 188, Copyright (1997), with permission from Elsevier.[27] Gelatinization and retrogradation of starch reprinted with kind permission from Springer Science + Business Media: *Journal of Materials Science Microstructure and mechanical properties of*

polymers are semi-crystalline and contain spherulites.[30,31] Within these there are crystalline regions surrounded by an amorphous region with similar morphology to the amylopectin clusters in starch.[29] The techniques discussed in this section are used to assess these structural aspects and how they define the characteristics of thermoplastic starch or proteins.

Techniques such as Fourier transform infrared spectroscopy (FT-IR) and nuclear magnetic resonance (NMR) are used in determining information about the chemical structure of the monomers and the nature of covalent bonds between them. Molecular mass and molecular mass distribution, as well as chemical nature of side groups, determine the interaction between polymer chains. Where interactions between chains lead to ordered regions, crystalline phases are observed, whilst other less ordered regions are said to be amorphous. X-ray diffraction is often used to assess structural information, such as the degree of crystallinity and specific crystal structures while microscopy techniques, such as SEM or TEM, are used to determine morphology.

All these different levels of structure contribute to the quality of the final product, as defined by it mechanical and physical properties. Important mechanical and physical properties are summarized in Table 7.1.[26]

As well as being rate dependent, the mechanical properties of polymers show significant temperature-related effects. Of particular importance for

Table 7.1 Mechanical and physical properties of polymers.[26,32]

Mechanical properties		Physical properties	
Tension, compression, torsion and flexural	Yield, modulus, stress at break, Poisson's ratio, elongation at break and toughness	Density	kg/m^3
Hardness	Resistance to permanent deformation by indentation	Melting temperature	Crystalline melting
Fatigue	Susceptibility to failure due to cyclic loading	Glass transition	Amorphous relaxation
Creep	Permanent elongation due to a static load over a long period	Heat deflection temperature	Similar to softening point
Impact	Dynamic loading	Thermal decomposition temperature	Thermally induced degradation
Viscoelasticity	Time-temperature effects under dynamic/cyclic loading	Specific heat, thermal conductivity and thermal expansion	Important process design parameters
		Optical properties	Colour, reflectance and opacity
		Chemical resistance	Resistance to chemical reactions during and after processing
		Water absorption	Measures material-solvent interaction

thermoplastic processing is the glass transition of amorphous regions. At low temperatures polymers exhibit glassy properties, as chains are not particularly mobile, effectively frozen into place amongst the other chains. As temperature increases, the chains become more mobile, and are able to slide past each other more quickly in response to applied forces. The transition from the glassy state to the rubbery state is influenced by chain architecture and micro-structure. Agro-polymers, especially proteins, have severely more complicated structures than synthetic polymers, and relating this relaxation behaviour to structure can be challenging. In addition, starch and proteins are highly hydrophilic and their mechanical and physical properties are highly dependent on moisture content. The rate at which water is absorbed or lost as well as equilibrium moisture content are important characteristics of agro-polymers. Other properties of particular interests for natural polymer thermoplastics include thermal decomposition, thermally induced cross-linking (thermosetting behaviour) and crystallization.

Effective characterization is more than just assessing these properties, but also understanding the effect of processing parameters such as the method used, temperature, conditioning time and additives.

7.3.2 Mechanical Behaviour

Mechanical properties can probably be regarded as the most significant aspect of materials, as adequate mechanical properties are required for any application. This is not to say that other physical properties, such as density and thermal conductivity, can be neglected as these are also vital in materials selection. In understanding the mechanical properties of agro-polymers it is useful to compare their behaviour to that of synthetic polymers.

Mechanical properties are essentially a description of a material's response to an external load and its ability to reversibly or irreversibly deform as a result of it. These properties are generally evaluated by standardized methods. Mechanical properties of natural polymers (or agro-polymers) are not characterized any differently than conventional polymers and important properties include strength, Young's modulus, stress at yield, elongation and toughness (energy to break), *etc.*

Polymers typically behave visco-elastically, that is their mechanical properties are time and temperature dependent. However, the properties mentioned above are measured almost instantaneously and it is assumed that the material behaves elastically, or more importantly linear elastic if a Young's modulus is considered. In Figure 7.5, typical stress–strain diagrams are shown for brittle, plastic and highly elastomeric behaviour, as observed in many synthetic polymers. This behaviour is dependent on temperature as well as the amount of plasticizer (or other additives).[32,33]

Synthetic polymers have an exceptionally broad range of mechanical properties. Highly stiff and brittle behaviour is observed, but also more ductile materials that can show extensions of up to 1000%. Figure 7.6 presents a summary of some commodity polymers' relationship between Young's

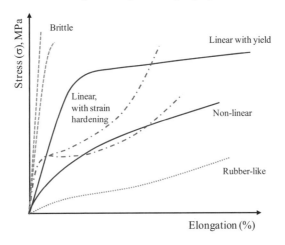

Figure 7.5 Idealized stress strain behaviour of typical agro-polymers.

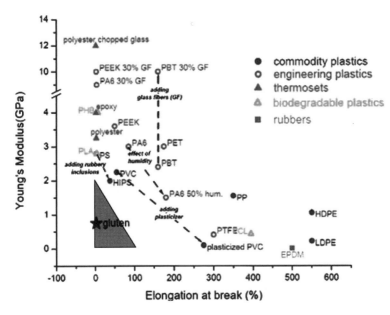

Figure 7.6 Mechanical properties of biomaterials compared to currently available synthetic polymers.[34] Adapted with permission from B. Lagrain, B. Goderis, K. Brijs and J. A. Delcour, *Biomacromolecules*, 2010, **11**, 533–541. Copyright 2010 American Chemical Society. The shadowed triangle shows systems of soy,[35] sunflower isolate,[36,37] starch/zein,[38] soy/corn starch,[39] whey,[40] zein,[41] and wheat protein.[42]

modulus and elongation at break.[34] Also shown in Figure 7.6 is thermoplastic gluten, relative to synthetic polymers.[34] Although gluten is not representative of all agro-polymers, it does fall in the broad region of where agro-polymers could be expected (indicated by shaded triangle).

The mechanical properties of polymers are largely associated with distribution and concentration of inter- and intra-molecular forces. Extrusion and other thermal processing techniques lead to structural rearrangements and new interactions which can be adjusted with the use of plasticizers and chemical additives. In general, true plasticizers will increase the flexibility of a moulded product, imparting greater extensibility. On the other hand, increasing molecular interactions will result in a material with higher tensile strength and stiffness. Furthermore, harsh processing conditions can lead to degradation, adversely effecting mechanical properties.[11]

The mechanical properties of agro-polymers are generally a lot weaker than those of commodity engineering plastics. Their properties are also strongly dependent on moisture content as well as other plasticizers because of their characteristic hydrophilicity. Apart from these, the most prevalent factors affecting the mechanical properties include processing conditions and composition.

Starch and protein agro-polymers show a variety of stress–strain behaviour; from rubber-like to highly brittle. The specific behaviour is determined by the factors mentioned earlier. Idealized stress–strain diagrams for a few selected systems are shown in Figure 7.5. As an example, wheat gluten plasticized with glycerol showed linear behaviour with yield, while cross-linking eliminated the yield point.[43] Zein plastics have been shown to exhibit behaviour that is similar to thermoplastic-rubber blends; that is effectively linear behaviour followed by strain hardening.[44] It has been shown for feathermeal-based plastics that the yield point is associated with breaking secondary interactions, such as hydrophobic interactions. If the load is removed before break, the deformation across the yield point was reversible due to chain refolding.[45] Non-linear behaviour, shown in Figure 7.6, has also been observed for whey protein isolate films[42] and wheat flower polymers.[46] Transition between different behaviour is often observed and depends on moisture content, plasticizer content and additives.

Microscopic techniques including scanning electron microscopy (SEM), transition electron microscopy (TEM) and light microscopy are used to study structure and morphology at higher levels. SEM is used to study fracture morphology after tensile testing.[47–49] SEM is also used to investigate phase morphology and generate emulsification curves in the study of blends, such as thermoplastic starch blended with polyethylene,[50] soy protein blended with polylactide[51] or soy protein and polyester.[52]

When electron microscopy is used to study the morphology of agro-materials, the surface roughness is often taken as an indication of the failure mode. In the example shown (Figure 7.7), soy protein isolate (SPI) was plasticized with stearic acid and glycerol.[53] SPI resin with 30% glycerol (Figure 7.7A) showed lower roughness at the fracture surface than the other two specimens containing 25% stearic acid (Figure 7.7B). The height and contour of the asperities at the fractured surface of the resin containing stearic acid also indicated that the presence of stearic acid resulted in a more ductile failure.

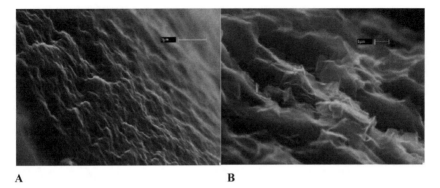

Figure 7.7 Fracture surface of SPI plasticized with glycerol, with (B) and without (A) stearic acid.[53] Reprinted from *Industrial Crops and Products*, **21**(1), P. Lodha and A. N. Netravali, Thermal and mechanical properties of environment-friendly 'green' plastics from stearic acid modified-soy protein isolate, 49–64, Copyright (2005), with permission from Elsevier.

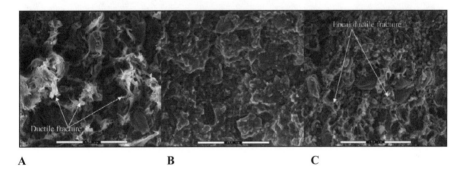

Figure 7.8 Tensile fracture surfaces of (A) glycerol, (B) sorbitol and (C) polyester amide plasticized soy-based agro-polymers.[48] Reprinted with permission from P. Tummala, W. J. Liu, L. T. Drzal, A. K. Mohanty and M. Misra, *Ind. Eng. Chem. Res.*, 2006, **45**, 7491–7496. Copyright 2006 American Chemical Society.

In a further example, the tensile fracture surfaces for soy protein plasticized by glycerol, sorbitol and polyester amide are shown in Figure 7.8. It can be seen that glycerol plasticized samples showed ductile fracture features with a coarse surface. Sorbitol plasticized samples showed brittle fracture features with relatively smooth surfaces. These samples also had higher strength and stiffness, but lower elongation at break. Using polyester amide as a plasticizer resulted in local ductile fracture features and the samples had moderate mechanical properties.[48]

If soy protein is plasticized with acetamide, it was shown that brittle fracture dominated up to 20% acetamide, as evident from the sharp ridges on the SEM images (Figure 7.9). As plasticizer content increased the surface become

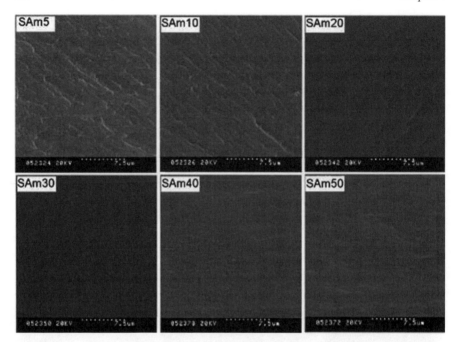

Figure 7.9 SEM images of cross sections of soy plastic sheets plasticized with different amounts of acetamide.[54] Liu, D. and L. Zhang: Structure and properties of soy protein plastics plasticised with acetamide. *Macromolecular Materials and Engineering* 2006, **291**, 820–828. Copyright Wiley-VCH Verlag GmbH & Co. KGaA. Reproduced with permission.

smoother, but above 30% the surface again appeared fluctuant, most likely because of excess plasticizer.[54]

Similar observations have been made for starch plasticized with glycerol, xylitol, sorbitol or maltitol. Fractured surfaces are shown in Figure 7.10 and it was observed that the surface of glycerol plasticized material (A) was rough. This was likely due to glycerol- and amyl pectin-rich domains forming a heterogeneous material. The fractured surface of xylitol plasticized system (B) appeared different with a rough structured surface, while sorbitol (C) and maltitol (D) plasticized surfaces were perfectly smooth.[55]

Unfortunately, overall mechanical property improvements in agro-polymers have been slow. A large range of starch-based plastics have been surveyed in terms of strength at break versus their elongation at break.[15] It has been shown that the mechanical properties follow a master curve that is almost exclusively dependent on the amount of plasticization (Figure 7.11).[15] The situation for protein-based plastics is not much different, and tensile strengths seldom exceed 25 MPa, with elongation at break being less than 140%.[35–38,41,42,56] As with starch-based materials, this behaviour is mostly determined by the amount of plasticizer. Very little research has been successful in developing high strength agro-polymer, with corresponding high elongation.

Synthesis and Characterization of Thermoplastic Agro-polymers 215

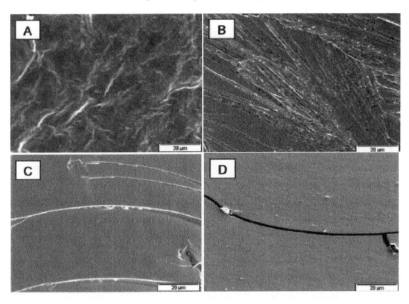

Figure 7.10 Scanning electron micrographs of the fractured surface of (A) glycerol, (B) xylitol, (C) sorbitol and (D) maltitol plasticized starch. Reprinted with permission from A. P. Mathew and A. Dufresne, *Biomacromolecules*, 2002, **3**, 1101–1108. Copyright 2002 American Chemical Society.

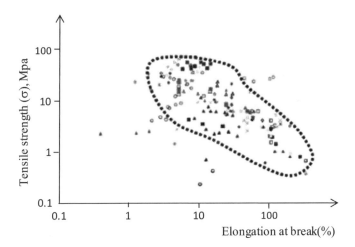

Figure 7.11 Tensile strength versus elongation at break for a variety of starch based agro-polymers.[15] N. Follain, C. Joly, P. Dole and C. Bliard: Mechanical properties of starch-based materials. I. Short review and complementary experimental analysis. *Journal of Applied Polymer Science*, 2005, **97**(5), 1783–1794. Copyright John Wiley and Sons. Reproduced with permission.

7.3.2.1 Factors Affecting Mechanical Properties

Moisture Content. The moisture content or water uptake of agro-polymers is an important topic and just about every research paper dealing with agro-polymers also considers moisture content.[44,55,57–63] Moisture content is important from at least two perspectives. Most importantly, the moisture content directly influences the mechanical properties of the plastic, typically increasing elongation and reducing strength by effectively plasticizing the material.[4,10,11,44] Secondly, in protein-based systems, the ability to absorb water could also be an indication of cross-link density.[10,59]

Water is required in the manufacture of both thermoplastic starch and thermoplastic protein. For starch, it ensures gelatinization and the formation of a continuous thermoplastic phase. In proteins, it lowers the glass transition and denaturing temperature,[10,11] allowing processing. However, dry resin ensures dimensional stability directly after injection moulding and over time.

The equilibrium moisture content of hydrophilic agro-polymers depends on the relative humidity of the environment and is typically described by the Guggenheim-Anderson-de Boer (GAB) and Brunauer-Emmett-Teller (BET) equations.[63,64] Equilibrium moisture content is also influenced by the type and amount of plasticizer in the system. Careful selection of additives and plasticizers could control the equilibrium moisture content of these materials, but only to a limited extent. The equilibrium moisture content of starch as a function of relative humidity and plasticizer type and content has been determined. It was found that the equilibrium moisture content for starch can be as high as 30% at 70% relative humidity, but, by using different plasticizers, this number could be significantly reduced.[55,62] In Figure 7.12, a phase diagram for a water/glycerol/starch system (as a function of plasticizer content) is shown which highlights the importance of the interactions between water, plasticizer and polymer.[65]

Plasticization. Plasticizers are typically high boiling substances and are usually good solvents for the polymer. The mechanism of how plasticizers work is typically described using the free volume theory.[66] Free volume can be increased by increasing the number of chain ends, or by decreasing the molecular mass. Alternatively, flexible side chains also increases free volume. This is called internal plasticization, and the free volume is spatially fixed with regard to the polymer molecule. Changing functional groups on the protein chain is not a common method of altering T_g, although it has been shown that after reacting proteins with aldehydes the processability was improved, but additional plasticizers were still required.[67]

External plasticization is the addition of a small molecule that can increase free volume at any location along the polymer chain and is proportional to the amount of this molecule added. The increased free volume enables more chain movement, by the mechanisms mentioned earlier, thereby decreasing the T_g.

The chemical nature of the plasticizer will strongly influence its efficiency. Aspects such as polarity, hydrogen bonding capability and density will

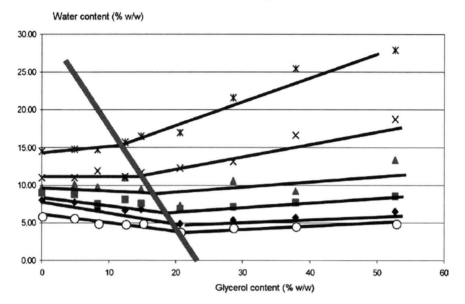

Figure 7.12 Equilibrium moisture content in plasticized starch as a function of plasticizer content at different relative humidity[65] (○ - 11%, ♦ - 33%, ■ - 44%, ▲ - 58%, X - 68%, *- 80%). Reprinted from *Food Chemistry*, **96**(3), L. Godbillot, P. Dole, C. Joly, B. Rogé and M. Mathlouthi, Analysis of water binding in starch plasticized films, 380–386. Copyright (2006), with permission from Elsevier.

determine how it functions as a plasticizer. However, chain flexibility may also permit protein chains to associate tightly with each other, leading to more intermolecular interactions; efficient plasticizers for proteins are therefore those that also disrupt these interactions.[33,66]

The basic rationale of plasticization is that plasticizers in agro-polymers can attract the water molecules around them, reduce the intermolecular interactions between the agro-polymer chains, and then increase the flexibility of the product. The effective working parts or active sites of plasticizers in agro-polymers are believed to be their hydrophilic parts such as hydroxyl groups. Hydroxyl groups are considered to develop hydrogen bonds between polymer–water–plasticizer or polymer–plasticizer, replacing the polymer–polymer interactions in the unplasticized agro-polymer materials.[64]

Ultimately, the amount and type of plasticizer is important in determining the mechanical properties of agro-polymers.[14,40,56,59,61,68–72] Although plasticization is required for processing the net result is a reduction in mechanical properties, as pointed out in Figure 7.11.

Chemical Composition. Along with additives such as plasticizers and moisture content, the chemical structure of the macromolecules used as agro-polymers also affects their mechanical properties. For starch, the amylopectin

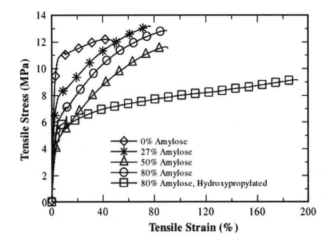

Figure 7.13 TS of thermoplastic starch containing different amounts of amylose.[74] Reprinted from *Carbohydrate Polymers*, **78**(4), A. L. Chaudhary, P. J. Torley, P. J. Halley, N. McCaffery and D. S. Chaudhary, Amylose content and chemical modification effects on thermoplastic starch from maize – Processing and characterisation using conventional polymer equipment, 917–925. Copyright (2009), with permission from Elsevier.

to amylose ratio is important,[15,68,69,73,74] whereas the primary structure or amino acid sequence of proteins are important.[10,11] As an example, the tensile strength of thermoplastic starch with varying amounts of amylose is shown in Figure 7.13. It can be seen that a more ductile material was obtained when the amylose content was higher.

Short range interactions, such as van der Waal's and hydrogen bonding, develop at low distance between functional groups (*e.g.* OH) and if these are strong enough the strength of the material at a low elongation will be high.[15] On the other hand, long-distance interactions are required for a large elongation at break; this is mostly chain entanglement and is dependent on molecular mass. Starch[15] and protein-based thermoplastics often suffer from not being able to develop sufficient chain entanglements and therefore have a low elongation at break.

In protein and starch based polymers, new interactions often form after processing. For starch, ageing or retrogradation causes chain rearrangement, often resulting in embrittlement.[35,68,69,72,74] For proteins the level (intra- and intermolecular) at which cross-links are formed determine the tensile properties of protein based plastics.[11,43,57,75–77] Cross-linking can be a result of the chemical nature of the protein or can be induced to improve mechanical properties.[43]

7.3.3 Thermal Properties

The thermal behaviour of agro-polymers is complicated relative to conventional polymers. As with mechanical properties, the thermal properties of

Synthesis and Characterization of Thermoplastic Agro-polymers 219

proteins and starches depend on moisture content, which may change during heating.[73] In addition, the composition of agro-polymers also influences their thermal behaviour. For example, the aforementioned variable ratio of amylose to amylopectin in starch also gives rise to different thermal properties.[14] Proteins are heteropolymers with a combination of hydrophobic, hydrophilic, acidic and basic side chains and have a wide range of different intermolecular interactions compared to synthetic homopolymers.

7.3.3.1 Thermal Analysis Techniques

Thermal analysis refers to a range of techniques were one or more properties of a material is measured as a function of temperature. Common techniques, their abbreviations and their application are listed in Table 7.2. Of these, DSC, DMA and TGA are widely used for characterizing thermal transitions and thermal stability.

DSC. DSC measures the difference in heat flow between a specimen and a reference as a function of temperature. Specimens are placed in a special pan, which can either be sealed or not. The reference is usually an identical, but

Table 7.2 Summary of common thermal analysis techniques.[78,79]

Technique	Abbreviations	Measures	Reveals
Thermogravimetry or thermal gravimetric analysis	TG or TGA	Mass change	Decomposition temperature Oxidation temperature Volatilisation of moisture and plasticiser Moisture content
Differential thermal analysis	DTA	Temperature difference	Exothermic and endothermic thermal events
Differential Scanning Calorimetry	DSC	Heat flow difference	Heat capacity Phase changes (melting etc) Glass transition
Dilatometry	TD	Length or volume change	Creep, thermal expansion
Thermomechanical analysis	TMA	Deformation under constant load	Creep, thermal expansion, heat deflection temperarure
Dynamic mechanical analysis or Dynamic mechanical thermal analysis	DMA or DMTA	Deformation under oscillating load	Viscoelastic properties Glass transition
Dielectric Analysis	DES	Dielectric properties	Glass transition
Simultaneous DSC/TGA or DTA/TGA	SDT	Mass change and heat flow or temperature difference	Causes of mass loss events.

empty pan. In heat flux DSC both pans are heated in the same oven and the temperature difference between them is used to determine heat flow to the sample. Exothermic events cause the sample pan temperature to be higher than the reference, whilst endothermic events cause the sample temperature to be lower at any given point in time. In power compensation DSC the pans are held in separate ovens and the difference in electrical work needed to keep the temperature increasing at the same rate is used to determine the heat flow. Exothermic events cause the sample pan to require less energy input and endothermic events cause it to require more. The main output of a DSC experiment is a curve representing heat flow versus temperature. Depending on the level of analysis required, the first or second derivative may also be useful.

As a polymer is heated through its glass transition, a sudden change in heat capacity occurs. The heat flow measured by the DSC is proportional to the specific heat capacity and hence reveals this transition.[79] Figure 7.14 shows the identification of this region in scans of thermoplastic starch plasticized with glycerol and containing different amounts of citric acid.

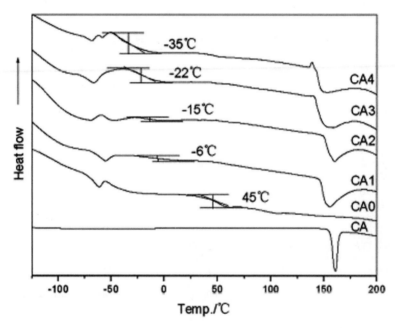

Figure 7.14 Identification of the glass transition as a step change in heat flow, CA = citric acid, CA0 = TPS with 30 pph glycerol, CA1 = TPS with 30 pph glycerol and 10 pph citric acid, CA2 = TPS with 30 pph glycerol and 20 pph citric acid, CA3 = TPS with 30 pph glycerol and 30 pph citric acid, CA4 = TPS with 30 pph glycerol and 40 pph citric acid.[70] Reprinted from *Carbohydrate Polymers*, **69**(4), R. Shi, Z. Z. Zhang, Q. Y. Liu, Y. M. Han, L. Q. Zhang, D. F. Chen and W. Tian, Characterization of citric acid/glycerol co-plasticized thermoplastic starch prepared by melt blending, 248–755. Copyright (2007), with permission from Elsevier.

Table 7.3 Typical DSC responses to thermal events in polymeric samples.[78,79]

Thermal Event	Effect on plot of heat flow vs temperature
Glass transition	Step change in heat flow
Crystallisation	Exothermic peak
Melting	Endothermic peak
Denaturing (in proteins)	Endothermic peak
Degradation (in inert atmosphere)	Endothermic trend
Oxidation	Exothermic trend

It would be better to measure the glass transition by cooling, but for historical reasons, it is usually measured in heating.[79] The typical DSC responses to this and other thermal events are listed in Table 7.3.

DSC is sensitive to the thermal history of the sample. A common approach is to heat and cool a specimen prior to scanning for thermal events during a second heating cycle. The first heating scan provides information about the thermal history (processing or ageing) of the sample. The cooling and second heating cycle are then performed at known thermal history. Transitions, such as the glass transition, are then determined from the results of the second scan.[80] This may pose a concern for proteins as heating during the first scan may affect protein conformation and hydration. An alternative approach is to use modulated DSC, in which the heating rate oscillates.[81] This allows separation of reversible events (glass transition and melting or fusion) and non-reversible events (oxidation, curing, relaxation and cold crystallization) which is complicated as these may overlap.[78,81] For starch this can be particularly useful for studying the multiple transitions that occur during gelatinization, although the onset and peak temperatures appear lower than those observed in conventional DSC.[82]

One other complication is that results from a DSC experiment depend on the rate of temperature change. For this reason the temperature ramp rate should always be reported with the results. It is also advisable to complement the analysis with other techniques, such as DMA.

DMA. In DMA a specimen is subjected to an oscillating force and the material's response is recorded as a function of temperature or frequency. Different instrument geometries allow testing of fibres, films, bars or even powders. Optimal choice of testing mode and sample size depends on sample stiffness and geometry.[83]

The main outputs of a DMA scan (as a function of temperature) are:

- the elastic (or storage) modulus (E'), representing the elastic component of the material's response, or energy stored by reversible deformation of the material
- the loss modulus (E''), representing the viscous component of the material's response, or energy dissipated as heat by molecular rearrangement

- the damping coefficient (tan δ); this is the tangent of the phase lag between applied force and the sample responding and is equal to E''/E'.

The glass transition region is identified as the region in which the elastic modulus drops rapidly by around two orders of magnitude and there is a local maximum in the loss modulus. These contribute to a clear peak in the plot of tan δ versus temperature.[14,48,80,84–86] The glass transition temperature could also be reported as the peak in the loss modulus or the onset of drop in storage modulus.[47,87]

DMA results for the glass transition are frequency dependent. For this reason the frequency of testing should always be reported. It is also good practice to test at more than one frequency and compare results with those from other techniques, such as DSC. For agro-polymers, the loss of moisture during scanning may also contribute to higher observed glass transition temperatures than those determined using DSC in sealed pans.[58]

Below the T_g, other thermal transitions relating to short range movements within polymer chains are identified as regions of rapid drop in storage modulus, but these are less dramatic than the glass transition. Above the glass transition, protein denaturing has been associated to a minimum in tan δ.[88] Figure 7.15 shows examples of plots of tan δ versus temperature for a thermoplastic protein and a thermoplastic starch. For the soy protein, β transitions can clearly be seen, along with a peak for ice melting at high water contents. The low temperature peak observed for the amylose films is likely due to the glycerol-rich phase, whereas the shifting upper peak will be the glass transition of the plasticized amylose.

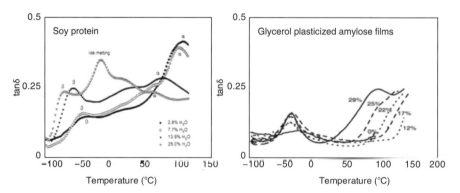

Figure 7.15 Plots of tan δ versus temperature for DMA scans of extruded soy protein sheets at different moisture contents[58] (reprinted from *Polymer*, **42**(6), J. Zhang, P. Mungara and J. Jane, Mechanical and thermal properties of extruded soy protein sheets, 2569–2578. Copyright (2001), with permission from Elsevier) and amylose films plasticized with different glycerol contents[89] (reprinted from *Carbohydrate Polymers*, **22**(3), G. K. Moates, T. R. Noel, R. Parker and S. G. Ring, Dynamic mechanical and dielectric characterisation of amylose-glycerol films, 247–253. Copyright (2001), with permission from Elsevier.)

TGA. In TGA mass loss is measured as a function of temperature. The sample is subjected to both a controlled temperature profile and a controlled atmosphere.[79] This reveals information such as the temperatures where moisture, solvents and plasticizers are volatilized from the material, as well as mass loss due to combustion or pyrolysis. Many instruments offer combined SDT analysis which allows for easier identification of the cause of mass loss events.[78] Additionally, FT-IR analysis can confirm volatilized species and decomposition products associated with mass loss events.[86] One of the most common applications of TGA is determination of thermal stability by heating to high temperatures in an inert atmosphere, usually nitrogen.[79] If a polymer is to be used in air, it is also important to determine its thermo-oxidative stability.[79] An example of the output of a TGA experiment is shown in Figure 7.16 for soy protein isolate with plasticizer made up of different ratios of ε-caprolactone to glycerol in both air and an inert atmosphere. The mass loss is usually presented as a function of temperature, and the first derivative is also plotted against temperature to more clearly determine the onset of thermal events.

For characterization of new materials it is advisable to utilize TGA prior to techniques such as DSC or DMA where thermal decomposition should be avoided.

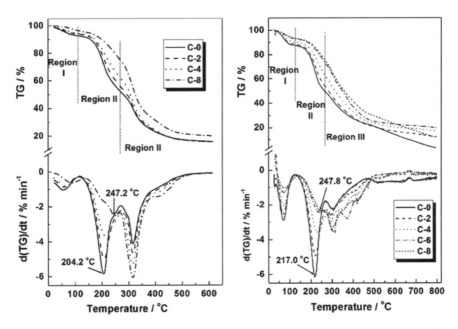

Figure 7.16 Percentage mass loss and first derivative TGA thermograms of SPI with 40 wt% plasticizer containing ε-caprolactone/glycerol ratios of 0:8 (C0), 2:6 (C2), 4:4 (C4), 6:2 (C6) and 8:0 (C8) in N_2 (left) and air (right).[86] Reprinted with permission from P. Chen, H. Tian, L. Zhang and P. R. Chang, *Ind. Eng. Chem. Res.*, 2008, **47**, 9389–9395. Copyright 2008 American Chemical Society.

Other Techniques. Differential temperature analysis (DTA) is similar to heat flux DSC. The difference in temperature is reported, but not converted into heat flow. This gives a similar shaped curve, but only provides qualitative information rather than the quantitative information available from DSC. Its main advantage is when run simultaneously to TGA, it can provide insight into the cause of mass loss events; *e.g.* exothermic (*i.e.* burning) or endothermic (*i.e.* evaporation). Additionally, DTA instruments can operate over a wider temperature range than DSC.[78] Although this may be an advantage for thermal analysis of some materials (*e.g.* metals) that melt at high temperatures, the relatively low degradation temperature of agro-polymers falls within the range of a typical DSC instrument.

In thermo-mechanical analysis (TMA) the change in dimensions under a minimal load is measured as a function of temperature.[83] This reveals the coefficient of thermal expansion (CTE) and can also be used to detect other transitions, such as the glass transition.

The heat deflection temperature (or heat distortion temperature) is an important material property mostly used to determine a material's useful temperature operating range. It refers to the temperature below which a moulded object can hold its own shape. It can be determined using a dynamic mechanical analyser set to apply a constant force. The HDT was determined for blends of plasticized soy flour (52% protein) and polyamide (nylon) as 45 °C when plasticized with 20 wt% sorbital, 35 °C when plasticized with 20 wt% glycerol, and 39 °C when plasticized with 10 wt% of each.[48] When such a blend is used to make composites with natural fibres, increasing content of natural fibres also increased the HDT.

Phase transition analysis uses an apparatus similar to a closed-chamber capillary rheometer. It measures displacement at a constant pressure. It yields glass transition results consistent with DMA, DSC and capillary rheometry.[80]

7.3.3.2 The Glass Transition

In both thermoplastic starch and thermoplastic proteins, the glass transition depends on numerous factors including moisture content, presence of plasticizers or other additives and processing history.

Reported glass transition temperatures for gelatinized starches are variable and depend on the methods and conditions used.[82] An extensive list of glass transition temperatures for thermoplastic starches and polymer blends containing thermoplastic starches of different compositions was compiled by Arvanitoyannis and Kassaveti.[90] These ranged from below −60 °C up to over 200 °C. The large range of contributing factors complicates a clear comparison between the T_g of materials produced in different studies, but do not preclude assessing the effect of specific conditions or additives.

The effect of sugar addition on the T_g of starch/glycerol/water systems was investigated using DMA. Adding sugar lowered the T_g and the elastic modulus suggesting that the presence of natural sugars in starch sources should be factored into polymer design.[72] The T_g was also observed to decrease with an increase in amylose to amylopectin ratio.[14]

Glass transitions of dry, unplasticized proteins are typically in the range of 120–250 °C depending on protein structure.[91] For highly plasticized proteins, T_g can be well below ambient temperatures. For example, DSC of compression moulded soy protein sheets plasticized with 50–70 g ethylene glycol per 100 g soy protein revealed the T_g to be in the range of –90 to –70 °C.[92]

Miscibility of blends can be characterized by examining the glass transition with either DMA or DSC. Fully miscible blends show a single T_g, intermediate of those of the individual components, or phases. Two glass transitions corresponding to the individual components are seen in immiscible blends. Partial mixing is indicated by shifting of the distinct T_g values closer together.[79] The effect of different plasticizers or miscibility agents on the blend can be assessed in this way. Soy protein plastics plasticized with sorbitol were found to be more miscible with a polyester amide than those plasticized with glycerol.[48]

Bulk proteinaceous material used for agro-polymer production may contain more than one kind of protein subunit and, like blends, this may be reflected in multiple glass transitions related to the different subunits. Modulated DSC of plasticized, mechanically processed gluten revealed three glass transitions: one near –30 °C associated with molecular motion of free glycerol, along with one at 40 °C and one near 75 °C associated with the plasticized glutenin and gliadin gluten subunits respectively.[88] Likewise, DMA of thermoplastic soy protein isolates prepared with urea showed two glass transitions; a low T_g relating to 7s globulin, and a higher T_g relating to 11s globulin.[87] Depending on the instrument used, the resolution of the glass transition may or may not distinguish such phases. For a thermoplastic produced from soy protein isolate and 25 wt% glycerol, two glass transitions were observed in DMA, but only one was seen in DSC.[47]

7.3.3.3 Denaturing, Destructuring, Gelatinization and Physical Ageing

In starch, disruption of the ordered granular structure when heated in a solvent (usually water) is called destructuration or gelatinization. The temperature at which this occurs increases as the amount of solvent is decreased and is seen as an endothermic peak during the first scan in DSC.[93] Destructuration is an irreversible transition and, after cooling, an amorphous entanglement of amylose and amylopectin is formed.

An irreversible exothermic event has also been seen with some plasticizers. Mixtures of amylopectin and glycerol exhibited such a peak at 95–110 °C, mixtures of amylose and glycerol at 95–105 °C, mixtures of amylopectin and ethylene glycol at 70–85 °C and mixtures of amylose and ethylene glycol at 65–75 °C.[94] These peaks appear to be due to the formation of hydrogen bonding between the plasticizer and starch molecules and are not dependent on the crystallinity present in the polysaccharides.[94]

DSC is used by biochemists to quantitatively analyse the thermodynamics of protein folding and unfolding.[95] Any conformational change leading to disruption of protein functionality is called denaturation. Depending on conditions and the extent of conformational change, denaturing may be a reversible or irreversible process.[96] Historically, much work on protein denaturing has focussed on proteins in solution, rather than solid materials.[97] Like the glass transition, the denaturation temperature of a protein is dependent on water content, shifting to higher temperatures with decreasing water content.[98] In solution, protein denaturing typically occurs below 100 °C.[98] In thermoplastic proteins, denaturing occurs above the glass transition and for dry proteins this may be above 200 °C at temperatures close to degradation.[11]

Protein denaturing is seen as an endothermic peak in DSC. As with the glass transition, the phenomenon may be observed at multiple temperatures for blends containing multiple phases. Soy proteins exhibit two such peaks related to the denaturing of the 7s globulin and 11s globulin subunits.[87,98] Increased concentration of urea decreases the size of these peaks, confirming urea has partially or, at higher concentrations, fully denatured the subunits.[87]

Although, strictly speaking, denaturing refers to disruption of the native secondary, tertiary and quaternary structures, similar interactions are present in processed thermoplastic proteins. Depending on the protein used and the processing conditions with which it was plasticized, thermoplastic proteins may or may not exhibit a 'denaturing' peak in DSC. Cast films prepared from wheat gluten in the presence of sodium sulfite and ethanol exhibited an endothermic peak in modulated DSC, whereas thermo-mechanically prepared films prepared with glycerol in a torque rheometer did not.[88]

DSC was used to characterize soy protein plasticized with glycerol and modified with functional monomers that had been extruded and compression moulded. An endothermic peak corresponding to protein denaturing was seen at 152 °C in the absence of functional monomers. This temperature decreased after reactive extrusion with maleic anhydride or glycidyl methacrylate, but increased with styrene addition.[99] This provided information regarding the interaction of these monomers with the protein. For example, styrene is hydrophobic and does not interact with hydroxyl groups on the protein.

Ordered secondary structures like α-helixes and β-sheets occurring in thermoplastically processed proteins are effectively forms of crystallinity. In analysis of extruded feather keratin, which used sodium sulfite to break disulfide bridges, a shift from low crystallinity to higher crystallinity was seen between DSC scans of samples with 4% sodium sulfite and samples with 5% sodium sulfite.[100] Proteins may also contribute to the formation of ordered regions in polymers/protein blends. DSC analysis of crystal formation in soy protein/poly(butylene succinate) (PBS) blends showed soy protein both induced and accelerated PBS crystallization.[101]

An endothermic peak, just before the T_g, is seen for the first DSC scan in many proteins containing gluten subunits and has been suggested to be due to physical ageing.[80] Physical ageing is a characteristic of polymers in their glassy state and is a slow transformation of amorphous regions to a more ordered structure.[79] It is accompanied by a reduction in internal energy and free volume and when heated through the glass transition causes an endothermic peak in DSC analysis.[47] Physical ageing is also readily apparent in thermoplastic starch as a form of retrogradation below the T_g.[102] This increase in crystallinity causes brittleness and limits the industrial applications of thermoplastic starches.[74] Retrogradation is studied with X-ray scattering techniques and is discussed further in Section 7.3.3.7.

Physical ageing has been observed for soybean protein plasticized with 25 wt% glycerol through enlargement of an endothermic peak at around 50–60 °C in the DSC scan of samples stored for increasing lengths of time at 50% RH.[47]

7.3.3.4 Mass Loss and Degradation

Thermal degradation of plasticized proteins, as determined by TGA, generally consists of four events that often overlap:[11]

(1) water elimination
(2) plasticizer elimination or decomposition
(3) weak bond cleavage contributing to peptide bond cleavage
(4) stronger bond cleavage leading to total degradation.

Soy proteins plasticized with ε-caprolactone/glycerol exhibit three mass loss regions under nitrogen atmosphere. Mass loss up to 120 °C was due to loss of moisture; between 120 °C and 260 °C due to volatilization of plasticizers and beyond 250 °C due to rupture of peptide bonds. Higher ε-caprolactone content imparted greater thermal stability, reducing the magnitude of the mass loss in the first two regions and shifted the volatilization of glycerol to a higher temperature.[86]

Soy proteins plasticized with acetimide exhibited similar regions, but had greater stability than those plasticized with glycerol.[54] Thermoplastic zein plasticized with polyethylene glycol (PEG) 400 only exhibited two significant mass loss regions, as the volatilization of the PEG occurred at similar temperatures to protein degradation.[103]

Acetylated gluten, zein, pea and soy proteins were found to be less stable than their unmodified counterparts. Nevertheless, the decomposition temperatures of the modified proteins are still above the softening temperatures required to process them.[104] Sorbitol plasticized soy protein/polyester blends exhibited greater thermal stability than glycerol plasticized soy protein/polyester blends.[48]

Thermal degradation of thermoplastic starch occurred after moisture loss and the degradation temperature was independent of the original moisture content in TGA.[105] Sealed high pressure DSC pans were used to study degradation of starch at high pressure and constant moisture content, simulating extrusion conditions. The onset of degradation and its broadness was dependent on moisture content which indicated that the decomposition mechanism was different from that occurring in unsealed TGA.[105]

7.3.3.5 The Importance of Moisture

The strong dependence of agro-polymer properties on moisture poses a particular challenge for thermal analysis of these materials. Some argue that the loss of plasticizer or moisture while testing at elevated temperatures implies that material properties are changing during testing and a different material is effectively tested.[106] These losses are an important consideration when designing applications for a material and are relevant. Nevertheless, techniques for containing water in the sample as it is heated are used. Coating samples with silicone oil was used to prevent moisture loss during DMA scans of gluten, corn zein and whey plasticized with water.[80]

Sealed pans can be used in DSC to restrict water loss.[73] However, data is more reliable when there is a larger contact area between the sample and the pan, giving preference to standard pans when moisture (or other volatile) loss is not significant.[79] This is rarely the case for agro-polymers and sealed pans are preferred.

Hermetically sealed aluminium pans are common and usually able to withstand internal pressures of up to 0.2 MPa.[79] In DSC analysis of starch, these pans ruptured at 110 °C, limiting the range of the experiment.[73] The use of high pressure stainless steel pans allowed scanning up to 300 °C; however, the extra mass of these pans reduced the accuracy to which changes in heat capacity of the sample could be detected.[73] It is recommended that moisture sensitive agro-polymers are tested in more than one kind of pan. When investigating the effect of water content on thermal properties of sunflower proteins both sealed aluminium pans and stainless steel pans with O-rings were used. The use of stainless steel pans allowed detection of a denaturing peak between 120 °C and 189 °C with moisture contents between 0% and 30 wt%.[107]

Alternatively, higher heating rates could reduce the affect of moisture loss on DSC results.[107] An advanced application of this is hyper DSC, which scans at very high heating rates and allows greater clarity in determination of the glass transition of thermoplastic starch than standard or modulated DSC.[105]

In agro-polymers, especially proteins, it is often necessary to distinguish between bound water (non-freezable) and free (freezable) water. Bound and freezable water content can be determined from water melting peaks detected in DSC.[108] The area of the endothermic peak represents the heat of fusion of the freezable fraction; division by the heat of fusion of pure water yields the mass of free water. Subtracting that value from the total water content of the sample gives the mass of bound water.[109]

The glass transition of sunflower proteins (by DSC) reduced from 180.8 °C with no water present to 5.3 °C with 26.12 wt% water. Up to 26.12 wt%, no endothermic peak was observed, indicating all moisture was bound. At 50 wt% water, a broad endothermic peak associated with the melting of freezable water obscured the glass transition.[107]

Similarly, in extruded soy proteins a strong endothermic peak for freezable water was observed in polymers containing 26 wt% water, but not at lower water contents.[58] It appeared that above a critical moisture content, the glass transition does not continue to decrease but merges with the peak for water crystallizing.[58] The glass transition for these extruded soy protein sheets detected by DMA was lower than that detected in sealed DSC pans which may be due to water loss during the DMA scan.[58] On the other hand, some researchers have suggested moisture loss is more significant beyond the glass transition.[106]

DSC was used to determine the T_g of starch plasticized with either glycerol or xylital at different free water contents. Glycerol was shown to be the better plasticizer at lower water content, but xylital was found to be better at higher water content.[71] It has been postulated that in corn starch systems water exists in three phases: bound, loosely bound and free water.[110]

The hydrophilic nature of starch and proteins can be explained in terms of the chemical nature of their monomers. These contain many hydrophilic groups capable of hydrogen bonding. It is hydrogen bonding that dominates interactions with water, most plasticizers and most inter- and intra-molecular interactions in agro-polymers. Proteins also have a variety of side groups with different functionalities which may give rise to hydrophobic interactions and covalent cross-linking. These interactions give rise to ordered regions within polymers that may be characterized with techniques such as Fourier transform infrared spectroscopy (FT-IR) and X-ray scattering (XRS).

7.3.3.6 FT-IR

FT-IR is a technique used to investigate changes in the nature of chemical bonds. Different covalent interactions stretch, vibrate or bend at specific frequencies and allow the identification of specific interactions. With the exception of exact optical isomers, the same IR spectrum will not be observed for compounds with different structures.[111] Complete determination of the cause of every peak and elucidation of the exact chemical structure is not practical for mixtures of complex macromolecules. Rather, changes in the location or magnitude of characteristic peaks provide information about structural and chemical changes occurring during processing. Important changes after thermoplastic processing investigated by FT-IR include changes in secondary structures (α-helices and β-sheets or turns) in proteins and the interaction between chains and plasticizers in both starch and proteins.

Secondary Structure in Proteins. Characteristic peaks related to vibrations in peptides (forming the protein backbone) are shown in Table 7.4. Different molecular geometries and hydrogen bonding present in a protein's secondary structure contribute to different C=O stretching frequencies in the amide I region.[112] Deconvolution of this region allows estimation of protein secondary structure with good approximation.[112]

In solutions, the solvent dominates the FT-IR spectrum and must be subtracted, often causing distortions. In particular, the signal of water overlaps with the amide I region. For this reason, FT-IR of proteins is often performed

Table 7.4 Molecular motions responsible for characteristic FTIR absorbance peaks in peptide links.[112]

Molecular motion	Characteristic frequency or frequency ranges (cm^{-1})	Designation
N-H stretching	3300	Amide A
	3100	Amide B
C=O stretching	1600 – 1690	Amide I
C–N stretching and N–H bending	1480 – 1575	Amide II
	1229 – 1301	Amide III
O–C–N bending	625 –767	Amide IV
Out of plane N–H bending	640 – 800	Amide V
Out of plane C=O bending	537 – 606	Amide VI
Skeletal torsion	200	Amide VII

using dried protein powders in potassium bromide (KBr) pellets.[113] Plastic films may be mounted directly in FT-IR spectrophotometers.

Although the native secondary structure is denatured during thermoplastic processing, secondary structure elements may contribute to physical fixation as the processed material cools.[97] Changes to secondary structure induced by thermoplastic processing depend on the protein's primary structure, processing conditions and additives such as plasticizers or denaturing agents. In general, extrusion appears to favour an increase in ordered β-sheet regions at the expense of α-helices, but this is not always the case.[11] Time resolved FT-IR of soybean protein films held at 100 °C showed changes in the amide I region over time that may be attributed to an increase in β-turn or weak β-sheet structures.[114] The peaks in the amide I region overlap and some form of deconvolution is needed to distinguish the combined peak into separate peaks relating to different secondary type structures, as shown in Figure 7.17.

The amide I region for hot pressed films of egg albumin, lactalbumin, feather keratin and wheat gluten were analysed with deconvolution software. Increased order was observed in the form of β-sheets as glycerol content increased up to a critical value for each protein. Beyond this, critical value order decreased.[115] The result of the deconvolution for wheat gluten is summarized in Figure 7.18.

FT-IR analysis of film blown thermoplastic zein plasticized with polyethylene glycol highlighted the interdependence of structure and processing. Different batches of zein powder contained different ratios of β-sheets to α-helices and the best blown films were prepared from those with largest relative α-helical content. In turn, processing increased the α-helical content and decreased the presence of ordered β-sheet regions.[103] Cross-links induced by γ-radiation in whey, casein and soya films appeared to have reduced β-sheet regions.[116]

FT-IR analysis of thermoplastic sheets containing soy protein isolate and glycerol showed no change in the characteristic peaks of the individual components indicating that no covalent interactions formed between them.[117] In contrast, soy protein plastics plasticized with both ε-caprolactone and glycerol did not show peaks characteristic of caprolactone, indicating it had been consumed.[86]

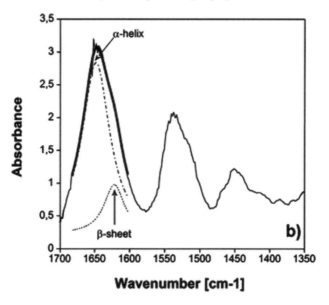

Figure 7.17 FTIR spectrogram of thermoplastic zein protein. The combined peak for the amide I region is separated as shown into component peaks relating to different secondary structures.[103] M. Oliviero, E. D. Maio and S. Iannace, Effect of molecular structure on film blowing ability of thermoplastic zein. *Journal of Applied Polymer Science*, 2010, **115**(1), 277–287. Copyright John Wiley and Sons. Reproduced with permission.

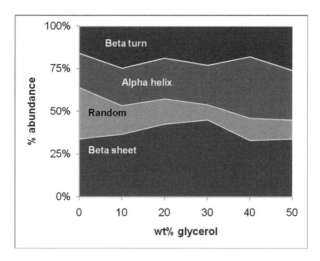

Figure 7.18 Secondary structure content of hot pressed wheat gluten films with glycerol as a plasticizer estimated by deconvolution of amide I region in FTIR spectrum.[115]

Starch/plasticizer Interactions. Peaks observed in thermoplastic starch between 992 and 1200 cm^{-1} have been associated with interaction between starch molecules and plasticizer and can be used to evaluate the efficiency of different plasticizers.[68]

Within this region, a peak at 1020 cm^{-1} can be attributed to C–O stretching in C–O–C bonds for native starch, while peaks at 1081 and 1156 cm^{-1} are indicative of C–O stretching in C–O–H bonds.[118] In thermoplastic starch, hydrogen bonding between starch and plasticizer causes absorbance to shift to lower wavenumbers.[118,119] Greater reduction in wavenumber is indicative of stronger plasticizer/starch interactions.[68] This shifting is apparent in the FT-IR spectra for native starch and TPS plasticized with ethanolamine, as shown in Figure 7.19. The effects of some plasticizers and ageing on these peaks are listed in Table 7.5.

As seen in Table 7.5, the peak for C–O–C stretching separates into two peaks as thermoplastic starch (10 wt% 30 pph glycerol) ages. This split peak is also seen in TPS with 30 pph glycerol and 1 pph citric acid prior to ageing and is unchanged after 70 days.[102]

In thermoplastic starch processed with 100 parts hydrous starch (20 % water content) and 30 pph glycerol and higher concentrations citric acid (10–40 pph),

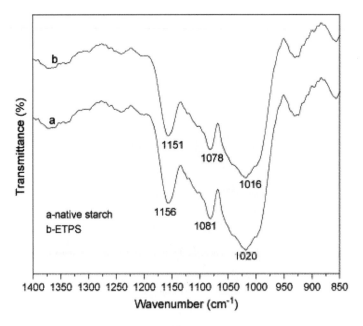

Figure 7.19 FTIR spectra of starch and ethanolamine plasticized thermoplastic starch (ETPS). Peaks which shift indicating H-bonding with plasticizer are labelled.[119] Reprinted from *Polymer Degradation and Stability*, **90**(3), M. F. Huang, H. G. Yu and X. F. Ma, Ethanolamine as a novel plasticiser for thermoplastic starch, 501–507, Copyright (2005), with permission from Elsevier.

Table 7.5 Shifts in some FITR peaks indicative of plasticiser H-bonding interactions in TPS.[70,102,118–120]

	Peak Postitions (cm^{-1})		
	C–O in C–O–C	C–O in C–O–H	
Native starch[102,118–120]	1020	1081	1156
TPS plasticised with Formamide[118]	1012	1077	1155
Ethanolamine[119]	1016	1078	1151
Ethylenebisformamide[120]	1016	1078	1150
Glycerol (30 pph) (after extrusion)[102]	1014	1078	1150
Glycerol (70 days aging)[102]	1016 and 996	1078	1150
Glycerol (30 pph) with citric acid (1 pph) (After extrusion)[102]	1012 and 990	1078	1150
Glycerol with citric acid (70 days aging)[102]	1012 and 990	1078	1150
Glycerol with high citric acid[70]	1024		1149

the peak for C–O stretching in C–O–C bonds was seen at 1024 cm^{-1}. Residual citric acid was removed by washing in deionized water after processing and before FT-IR. This peak decreased in size relative to the peak at 1149 cm^{-1} (C–O stretching in C–O–H bonds) as citric acid content increased, indicative of a reduction in molecular mass.[70]

Other peaks observed in thermoplastic starch provide further information about interactions between polysaccharide chains, water and plasticizer. A peak at 1644 cm^{-1} corresponds to water strictly bonded to starch.[68] A peak at 3389 cm^{-1} in native cornstarch is ascribed to free, intermolecular and intramolecular bound hydroxyl groups and decreased to 3325 cm^{-1} in TPS plasticized with ethanolamine.[119] Again, as with the peaks in Table 7.5, this indicates H-bonding interactions forming with the plasticizer at the expense of interactions between chains. Other researchers found a similar peak at 3413 cm^{-1} for native starch.[69] After processing with high glycerol contents, this peak shifted to lower wavenumbers and decreased in intensity. As thermoplastic starch aged (retrograded), this peak shifted back towards that seen in native starch. After 70 days, the peak had returned to 3413 cm^{-1} for TPS with 30% glycerol.[69] For higher glycerol contents this shifting slowed, suggesting the glycerol restricted retrogradation.[69]

A peak at 2931 cm^{-1} representing C–H stretching in CH_2 groups did not shift in TPS plasticized with ethanolamine, indicating the plasticizer did not interact with these groups.[119]

7.3.3.7 XRD/XRS

X-ray diffraction (XRD) and X-ray scattering (XRS) are techniques in which a sample is exposed to X-rays and the resultant scatter pattern is interpreted to provide information about the spatial arrangement of atoms. Strictly speaking, X-ray diffraction refers to the patterns of constructive and destructive interference due to the scattering of rays by crystal planes. Scattering is a more general term also applicable to amorphous materials which show broad 'halo' peaks, rather than sharp peaks caused by diffraction. In practice, similar instrumentation is used for both techniques and the terms are often used interchangeably.[78] Wide angle X-ray diffraction (WAXD) and wide angle X-ray scattering (WAXS) (scattering angle $2\theta > 5°$) are most commonly used for characterization of crystallinity in polymers.[78]

For thermoplastic starch and proteins, X-ray scattering techniques provide information about conformational changes induced by processing or ageing as well as information about the compatibility of plasticizers and cross-linking agents. Figure 7.20 shows examples of the plots of intensity versus scattering angle that are obtained as thermoplastic starches age and as plasticizer content changes for wheat gluten.

The original crystalline order of native starch is destroyed in the production of thermoplastic starch, but even below the T_g, chains exhibit enough mobility and re-order over time.[93] This re-forming of ordered crystalline regions is called retrogradation and can be observed using WAXS.[68,69,72,74,119] Different plasticizers and plasticizer content affect the rate of retrogradation and the type of crystallinity that forms. The types of crystallinity that can be identified using XRD are listed in Table 7.6.

Maize starch plasticized with 30 wt% glycerol displayed a shift from type-A crystallinity to type-V_H and type-V_A. Plasticization with 30 wt% of a mixture

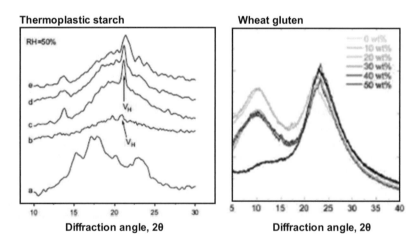

Figure 7.20 WAXS of (a) native starch and glycerol plasticized thermoplastic starch stored for 0, 30, 60 and 90 days respectively (b–e)[119] (Reprinted from *Polymer Degradation and Stability*, **90**(3), M. F. Huang, H. G. Yu and X. F. Ma, Ethanolamine as a novel plasticiser for thermoplastic starch, 501–507, Copyright (2005), with permission from Elsevier) and wheat gluten plasticized with different glycerol contents.[115] (Reprinted with permission from A. I. Athamneh, M. Griffin, M. Whaley and J. R. Barone, *Biomacromolecules*, 2008, **9**, 3181–3187. Copyright 2008 American Chemical Society.)

Table 7.6 Types of crystallinity in Starch.[68–69,93]

Crystallinity type	Description
A type (characteristic of maize)	Double helical structure characteristic of maize starch.[68] Denser and less hydrated than B type.[93]
B type (characteristic of potato)	Double helical structure.[69]
C type (characteristic of cassava)	Intermediate form between A and B.[93]
V type (complex with other molecules)	Amorphous or crystalline complex with other molecules (eg lipids in native starch, plasticiser or other polymer in TPS or TPS blends). Insoluble in water.[93] Single helical structure.[69] Can be either V_H (hydrated) or V_A (anhydrous) depending on moisture content.
E type (present in extruded starch)	Unstable form in TPS at low moisture. Rearranges to V type during conditioning.[93]

of urea and formamide produced an amorphous material, which was resistant to retrogradation.[68] Corn starch plasticized with ethanolamine has some type-V_H crystallinity induced during processing, but demonstrated no observable crystallinity after 30, 60 or 90 days at 50% relative humidity.[119] This is because ethanolamine forms stronger hydrogen bonds with starch than glycerol, restricting molecular rearrangements and the formation of new hydrogen bonding between starch molecules.[119] The same effect was seen using formamide as a plasticizer.[118] When glycerol is used as plasticizer, addition of a small amount of citric acid can strengthen the hydrogen bonding and restrict retrogradation.[102] This may be due to esterified citric acid groups bound to starch, inhibiting inter- and intra-molecular hydrogen bonding, while their free carboxyl groups increase solubility.[70]

XRD analysis during ageing of thermoplastic starch with high glycerol content (up to 60 wt%) showed that increased glycerol was helpful for forming single helix crystals (V-type complexes), but limited the formation of double helix B-type crystallinity.[69] V-type complexes may be useful when water resistance is a desired property as they are insoluble, even at high pressure and temperature.[93]

Storage conditions also affect the rate and type of retrogradation. Extruded corn meal (31 wt% moisture stored at 6 °C for 204 days) exhibited 31 % B and V_H-type crystallinity, whereas reducing the water content to 22 wt% lowered crystallinity to 3 %, mainly V_H-type. However, the T_g of the formulations was different; the former was stored above its T_g, while the latter was stored below its T_g.[106] As well as detecting crystallinity during ageing, WAXS is also used to investigate interactions between polymers and additives. Small molecules used as plasticizers may form crystals in the solid phase, giving rise to crystalline peaks of the pure plasticizer. If these peaks are absent in the thermoplastic it shows that the additive has dispersed throughout the polymer. For example, citric acid and glycerol were used in thermoplastic starch, where it was shown that the characteristic peaks of the plasticizers disappeared only if homogeneously distributed.[70]

Likewise, good compatibility was observed between soy protein isolate and acetamide used as plasticizer. The clear peaks observed for crystalline acetamide are not seen when mixed with soy protein isolate.[54] Starch blended with 30 wt% polystyrene and extruded to make loose fill packing foams formed new crystalline peaks not characteristic of either component.[121]

Alternatively, X ray scattering can demonstrate that cross-linking or reactive agents do not induce conformational changes. Comparison of WAXS plots of native starch and starch modified with sodium trimetaphosphate and sodium tripolyphosphate under alkaline conditions showed that no structural changes occurred other than cross-linking.[122]

X-ray scattering has also been used to investigate conformational changes in proteins. This is a different application than protein crystallography of highly purified protein crystals. Comparison between zein and thermoplastic zein suggested that alpha helical structures survived thermoplastic processing, but that inter-helical packing is disrupted.[103] Comparison between glycerol/SPI mixtures before and after compression moulding showed a smaller degree of crystallinity before processing, implying that heat and pressure induced rearrangement characteristic of strong intermolecular forces.[123] The same effect is seen compression moulded sheets of soy protein isolate and ethylene glycol.[92]

7.3.3.8 Other Spectroscopic Techniques

Other spectroscopic techniques are sometimes used to observe structural changes and polymer/plasticizer interactions. Vibrational circular dichroism spectroscopy has been used to investigate denaturing of protein films formed from bovine serum albumin.[124] Raman spectroscopy was used to investigate secondary structure changes in feather keratin extruded with sodium sulfite and revealed a transition from α-helices to β-sheets at sodium sulfite concentrations less than 4 wt%.[100] At higher sodium sulfite concentrations increased crystallinity was seen. Solid state NMR was used to investigate the interaction between dry starch and plasticizers, including glycerol and ethylene glycol, revealing a decrease in plasticizer mobility after heat treatment.[94] Although these are powerful techniques they are not as commonly used with thermoplastic starches and thermoplastic proteins and are beyond the scope of this chapter.

7.4 Conclusions

Thermoplastic agro-polymers are synthesized from starch and proteins. Both starch and protein experience a barrier to thermoplastic processing due to only a small difference between glass transition and degradation temperatures. For dry, unplasticized starch or protein, degradation occurs prior to the formation of a flowable melt. To overcome this, synthesis requires addition of plasticizers to reduce the T_g to temperatures which enable processing. For starch, the crystalline structure of starch granules also needs to be disrupted to allow processing and this is called gelatinization. Similarly, in proteins, formation of thermoplastic melt disrupts higher order structures and this is called denaturing.

After processing, both starch and protein agro-polymers suffer the same drawback. Due to their hydophilicity, water is usually an effective plasticizer and is required for processing. Unfortunately, this hydrophilicity also causes the mechanical properties of products made from agro-polymers to depend on moisture content. Exposure to atmospheres of varying relative humidity leads to undesirable changes in properties. Furthermore, agro-polymers are often characterized by inferior mechanical properties compared to synthetic thermoplastics. These properties also depend on plasticizer content, although ductility is obtained at the expense of strength.

The most common techniques used for characterization of thermal properties and structure are TGA, DSC, DMA, FT-IR and XRD. Each of these techniques is used to elucidate specific aspects of agro-polymers.

TGA is used to assess thermal stability of agro-polymers. Starch and proteins degrade at temperatures around 200–250 °C which is not much higher than their glass transition temperatures. The T_g of agro-polymers in their native state and after thermoplastic processing is often determined by DSC and DMA. These techniques can also be used to determine the temperatures at which gelatinization, denaturing and crystallization occur. As with mechanical properties, thermal properties are dependent on moisture content and care needs to be taken to restrict or account for moisture loss by evaporation during thermal analysis.

FT-IR and XRD are used to assess the structural changes induced during thermoplastic processing and the interaction between plasticsers and polymer chains. For proteins the native secondary, tertiary and quaternary structures are destroyed by processing, but new secondary structure-like elements may be induced. Thermoplastic starch is semi-crystalline and tends to retrograde to higher crystalline content. XRD is extremely useful to determine the degree of crystallinity in starch.

The structure–property–processing relationship of agro-polymers is highly interdependent. A considerable research effort has been devoted to firstly obtain processible thermoplastics. Now, sustained efforts are required to overcome the water sensitivity and time-dependent properties of these materials.

References

1. A. Gandini, *Macromolecules*, 2008, **41**, 9491.
2. M. Flieger, M. Kantorova, A. Prell, T. Rezanka and J. Votruba, *Folia Microbiol.*, 2003, **48**, 27.
3. F. Chivrac, E. Pollet and L. Averous, *Mater. Sci. Eng., R*, 2009, **67**, 1.
4. L. Averous and P. J. Halley, *Biofuels Bioprod. Biorefin.*, 2009, **3**, 329.
5. L. Yu, K. Dean and L. Li, *Prog. Polym. Sci.*, 2006, **31**, 576.
6. X. L. Wang, K. K. Yang and Y. Z. Wang, *J. Macromol. Sci., Polym. Rev.*, 2003, **C43**, 385.
7. D. Le Corre, J. Bras and A. Dufresne, *Biomacromolecules*, **11**, 1139.
8. L. Averous, *J. Macromol. Sci., Polym. Rev.*, 2004, **C44**, 231.

9. A. Rouilly and L. Rigal, *J. Macromol. Sci., Polym. Rev.*, 2002, **C42**, 441.
10. C. J. R. Verbeek and L. E. van den Berg, *Recent Pat. Mater. Sci.*, 2009, **2**, 171.
11. C. J. R. Verbeek and L. E. van den Berg, *Macromol. Mater. Eng.*, 2010, **295**, 10.
12. R. F. T. Stepto, *Macromol. Symp.*, 2006, **245**, 571.
13. H. S. Liu, F. W. Xie, L. Yu, L. Chen and L. Li, *Prog. Polym. Sci.*, 2009, **34**, 1348.
14. R. A. de Graaf, A. P. Karman and L. Janssen, *Starch-Starke*, 2003, **55**, 80.
15. N. Follain, C. Joly, P. Dole and C. Bliard, *J. Appl. Polym. Sci.*, 2005, **97**, 1783.
16. I. Siro and D. Plackett, *Cellulose*, 2010, **17**, 459.
17. B. Cuq, N. Gontard and S. Guilbert, *Cereal Chem.*, 1998, **75**, 1.
18. R. H. Garrett and C. M. Grisham, *Biochemistry*, 2nd edn., Brooks/Cole-Thomson Learning, Pacific Grove, 1999.
19. *NZ Pat.*, NZ551531, 2009.
20. *US Pat.*, US5665152, 1997.
21. *WIPO Pat.*, WO2006017481A2, 2006.
22. *US Pat.*, US5710190, 1998.
23. *US Pat.*, US5523293, 1996.
24. *US Pat.*, US3615715, 1971.
25. *US Pat.*, US5882702, 1999.
26. K. A. Rosentrater and A. W. Otieno, *J. Polym. Environ.*, 2006, **14**, 335.
27. D. J. Gallant, B. Bouchet and P. M. Baldwin, *Carbohydr. Polym.*, 1997, **32**, 177.
28. L. Yu and G. Christie, *J. Mater. Sci.*, 2005, **40**, 111.
29. B. L. Deopura and I. Textile *Polyesters and Polyamides*, Woodhead Publishing in association with the Textile Institute; CRC Press, Cambridge, England and Boca Raton, FL, 2008.
30. J. Candlin, in *The Chemical Industry*, ed. A. Heaton, Chapman & Hall, London, 1994.
31. A. Heaton, *The Chemical Industry*, 2nd edn., Chapman & Hall, London, 1994.
32. W. D. Callister, *Materials Science and Engineering, An Introduction*, 6 edn., John Wiley & Sons, Inc., New York, 2003.
33. I. M. Ward and D. W. Hadley, *An Introduction to the Mechanical Properties of Solid Polymers*, 1st edn., Wiley, 1993.
34. B. Lagrain, B. Goderis, K. Brijs and J. A. Delcour, *Biomacromolecules*, 2010, **11**, 533.
35. H. C. Huang, T. C. Chang and J. Jane, *J. Am. Oil Chem. Soc.*, 1999, **76**, 1101.
36. O. Orliac, F. Silvestre, A. Rouilly and L. Rigal, *Ind. Eng. Chem. Res.*, 2003, **42**, 1674.
37. A. Rouilly, A. Mériaux, C. Geneau, F. Silvestre and L. Rigal, *Polym. Eng. Sci.*, 2006, **46**, 1635.

38. S. Lim and J. Jane, *J. Environ. Polym. Degrad.*, 1994, **2**, 111.
39. J. U. Otaigbe, H. Goel, T. Babcock and J. Jane, *J. Elastomers Plast.*, 1999, **31**, 56.
40. V. M. Hernandez-Izquierdo, D. S. Reid, T. H. McHugh, J. D. J. Berrios and J. M. Krochta, *J. Food Sci.*, 2008, **73**, E169.
41. D. J. Sessa, G. W. Selling, J. L. Willett and D. E. Palmquist, *Ind. Crops Prod.*, 2006, **23**, 15.
42. N. Leblanc, R. Saiah, E. Beucher, R. Gattin, M. Castandet and J.-M. Saiter, *Carbohydr. Polym.*, 2008, **73**, 548.
43. S. Sun, Y. Song and Q. Zheng, *Food Hydrocolloids*, 2007, **21**, 1005.
44. Q. Wu, H. Sakabe and S. Isobe, *Polymer*, 2003, **44**, 3901.
45. S. Sharma, J. N. Hodges and I. Luzinov, *J. Appl. Polym. Sci.*, 2008, **110**, 459.
46. M. Mastromatteo, S. Chillo, G. G. Buonocore, A. Massaro, A. Conte, A. Bevilacqua and M. A. D. Nobile, *J. Food Eng.*, 2009, **92**, 467.
47. X. Mo and X. Sun, *J. Polym. Environ.*, 2003, **11**, 15.
48. P. Tummala, W. J. Liu, L. T. Drzal, A. K. Mohanty and M. Misra, *Ind. Eng. Chem. Res.*, 2006, **45**, 7491.
49. Z. K. Zhong and S. X. Sun, *J. Appl. Polym. Sci.*, 2003, **88**, 407.
50. A. Taguet, M. A. Huneault and B. D. Favis, *Polymer*, 2009, **50**, 5733.
51. J. W. Zhang, L. Jiang and L. Y. Zhu, *Biomacromolecules*, 2006, **7**, 1551.
52. P. Mungara, T. Chang, J. Zhu and J. Jane, *J. Polym. Environ.*, 2002, **10**, 31.
53. P. Lodha and A. N. Netravali, *Ind. Crops Prod.*, 2005, **21**, 49.
54. D. Liu and L. Zhang, *Macromol. Mater. Eng.*, 2006, **291**, 820.
55. A. P. Mathew and A. Dufresne, *Biomacromolecules*, 2002, **3**, 1101.
56. V. M. Hernandez-Izquierdo and J. M. Krochta, *J. Food Sci.*, 2008, **73**, 30.
57. S. N. Swain, K. K. Rao and P. L. Nayak, *J. Appl. Polym. Sci.*, 2004, **93**, 2590.
58. J. Zhang, P. Mungara and J. Jane, *Polymer*, 2001, **42**, 2569.
59. M. Pommet, A. Redl, S. Guilbert and M.-H. Morel, *J. Cereal Sci.*, 2005, **42**, 81.
60. R. Mani and M. Bhattacharya, *Eur. Polym. J.*, 1998, **34**, 1467.
61. M. E. R. Ortiz, E. San Martin-Martinez and L. P. M. Padilla, *Starch-Starke*, 2008, **60**, 577.
62. D. Lourdin, L. Coignard, H. Bizot and P. Colonna, *Polymer*, 1997, **38**, 5401.
63. A. Hochstetter, R. A. Talja, H. J. Helén, L. Hyvönen and K. Jouppila, *LWT – Food Sci. Technol.*, 2006, **39**, 893.
64. Y. Zhang and J. Han, *J. Food Sci.*, 2008, **73**, E313.
65. L. Godbillot, P. Dole, C. Joly, B. Rogé and M. Mathlouthi, *Food Chem.*, 2006, **96**, 380.
66. A. Kumar and R. K. Gupta, *Fundamentals of Polymers*, 1st edn., McGraw-Hill International Editions, 1998.
67. *US Pat.*, US2238307, 1941.
68. R. Zullo and S. Iannace, *Carbohydr. Polym.*, 2009, **77**, 376.

69. R. Shi, Q. Y. Liu, T. Ding, Y. M. Han, L. Q. Zhang, D. F. Chen and W. Tian, *J. Appl. Polym. Sci.*, 2007, **103**, 574.
70. R. Shi, Z. Z. Zhang, Q. Y. Liu, Y. M. Han, L. Q. Zhang, D. F. Chen and W. Tian, *Carbohydr. Polym.*, 2007, **69**, 748.
71. D. S. Chaudhary, *J. Appl. Polym. Sci.*, 2010, **118**, 486.
72. E. M. Teixeira, A. L. Da Róz, A. J. F. Carvalho and A. A. S. Curvelo, *Carbohydr. Polym.*, 2007, **69**, 619.
73. L. Yu and G. Christie, *Carbohydr. Polym.*, 2001, **46**, 179.
74. A. L. Chaudhary, P. J. Torley, P. J. Halley, N. McCaffery and D. S. Chaudhary, *Carbohydr. Polym.*, 2009, **78**, 917.
75. C. M. Vaz, P. F. N. M. van Doeveren, G. Yilmaz, L. A. de Graaf, R. L. Reis and A. M. Cunha, *J. Appl. Polym. Sci.*, 2005, **97**, 604.
76. J. A. G. Areas, *Crit. Rev. Food Sci. Nutr.*, 1992, **32**, 365.
77. H. Madeka and J. L. Kokini, *J. Am. Oil Chem. Soc.*, 1996, **73**, 433.
78. Y. Leng, *Materials Characterization: Introduction to Microscopic and Spectroscopic Methods*, J. Wiley, Singapore and Hoboken, NJ, 2008.
79. J. D. Menczel and R. B. Prime, *Thermal Analysis of Polymers: Fundamentals and Applications*, John Wiley, Hoboken, NJ, 2009.
80. C. Bengoechea, A. Arrachid, A. Guerrero, S. E. Hill and J. R. Mitchell, *J. Cereal Sci.*, 2007, **45**, 275.
81. C. H. Tang, S. M. Choi and C. Y. Ma, *Int. J. Biol. Macromol.*, 2007, **40**, 96.
82. F. W. Xie, W. C. Liu, P. Liu, J. Wang, P. J. Halley and L. Yu, *Starch-Starke*, 2010, **62**, 350.
83. K. P. Menard, *Dynamic Mechanical Analysis: A Practical Introduction*, 2nd edn., CRC Press, Boca Raton, FL, 2008.
84. M. I. Beck, I. Tomka and E. Waysek, *Int. J. Pharm.*, 1996, **141**, 137.
85. W. Thakhiew, S. Devahastin and S. Soponronnarit, *J. Food Eng.*, 2010, **99**, 2164.
86. P. Chen, H. Tian, L. Zhang and P. R. Chang, *Ind. Eng. Chem. Res.*, 2008, **47**, 9389.
87. X. Q. Mo and X. Z. Sun, *J. Am. Oil Chem. Soc.*, 2001, **78**, 867.
88. A. Jerez, P. Partal, I. Martinez, C. Gallegos and A. Guerrero, *Biochem. Eng. J.*, 2005, **26**, 131.
89. G. K. Moates, T. R. Noel, R. Parker and S. G. Ring, *Carbohydr. Polym.*, 2001, **44**, 247.
90. I. S. Arvanitoyannis and I. Kassaveti, in *Biodegradable Polymer Blends and Composites from Renewable Resources*, ed. L. Yu, Wiley, Hoboken, NJ, 2009, pp. xi, 487.
91. S. Guilbert and B. Cuq, in *Handbook of Biodegradable Polymers*, eds. C. Bastioli and L. Rapra Technology, Rapra Technology, Shrewsbury, 2005, pp. xviii, 534.
92. Q. Wu and L. Zhang, *Ind. Eng. Chem. Res.*, 2001, **40**, 1879.
93. A. J. F. Carvalho, in *Monomers, Polymers and Composites from Renewable Resources*, eds. B. Mohamed Naceur, G. Alessandro, Elsevier, Amsterdam, 2008, pp. 321.

94. A. L. M. Smits, P. H. Kruiskamp, J. J. G. van Soest and J. F. G. Vliegenthart, *Carbohydr. Polym.*, 2003, **53**, 409.
95. C. K. Larive, S. M. Lunte, M. Zhong, M. D. Perkins, G. S. Wilson, G. Gokulrangan, T. Williams, F. Afroz, C. Schoneich, T. S. Derrick, C. R. Middaugh and S. Bogdanowich-Knipp, *Anal. Chem.*, 1999, **71**, 389R.
96. L. Zhang and M. Zeng, in *Monomers, Polymers and Composites from Renewable Resources*, eds. B. Mohamed Naceur and G. Alessandro, Elsevier, Amsterdam, 2008, pp. 479.
97. L. A. De Graaf, *J. Biotechnol.*, 2000, **79**, 299.
98. N. Kitabatake, M. Tahara and E. Doi, *Agric. Biol. Chem.*, 1989, **53**, 12012.
99. W. J. Liu, A. K. Mohanty, P. Askeland, L. T. Drzal and M. Misra, *J. Polym. Environ.*, 2008, **16**, 177.
100. J. R. Barone, W. F. Schmidt and N. T. Gregoire, *J. Appl. Polym. Sci.*, 2006, **100**, 1432.
101. Y. D. Li, J. B. Zeng, W. D. Li, K. K. Yang, X. L. Wang and Y. Z. Wang, *Ind. Eng. Chem. Res.*, 2009, **48**, 4817.
102. J. G. Yu, N. Wang and X. F. Ma, *Starch-Starke*, 2005, **57**, 494.
103. M. Oliviero, E. D. Maio and S. Iannace, *J. Appl. Polym. Sci.*, 2010, **115**, 277.
104. S. Brauer, F. Meister, R. P. Gottlober and A. Nechwatal, *Macromol. Mater. Eng.*, 2007, **292**, 176.
105. X. X. Liu, L. Yu, H. S. Liu, L. Chen and L. Li, *Polym. Degrad. Stabil.*, 2008, **93**, 260.
106. J. L. Brent, S. J. Mulvaney, C. Cohen and J. A. Bartsch, *J. Cereal Sci.*, 1997, **26**, 313.
107. A. Rouilly, O. Orliac, F. Silvestre and L. Rigal, *Polymer*, 2001, **42**, 10111.
108. Z. K. Zhong and X. S. Sun, *J. Appl. Polym. Sci.*, 2001, **81**, 166.
109. D. J. Muffett and H. E. Snyder, *J. Agric. Food Chem.*, 1980, **28**, 13035.
110. Z. Zhong and X. S. Sun, *J. Food Eng.*, 2005, **69**, 453.
111. J. L. Koenig, *Spectroscopy of Polymers*, 2nd edn. Elsevier, New York, 1999.
112. J. Kong and S. Yu, *Acta Biochim. Biophys. Sin.*, 2007, **39**, 549.
113. L. A. Forato, R. Bernardes and L. A. Colnago, *Anal. Biochem.*, 1998, **259**, 136.
114. K. Tian, D. Porter, J. Yao, Z. Shao and X. Chen, *Polymer*, 2010, **51**, 2410.
115. A. I. Athamneh, M. Griffin, M. Whaley and J. R. Barone, *Biomacromolecules*, 2008, **9**, 3181.
116. M. Lacroix, T. C. Le, B. Ouattara, H. Yu, M. Letendre, S. F. Sabato, M. A. Mateescu and G. Patterson, *Radiat. Phys. Chem*, 2002, **63**, 827.
117. P. Guerrero, A. Retegi, N. Gabilondo and K. de la Caba, *J. Food Eng.*, 2010, **100**, 145.
118. X. F. Ma and J. G. Yu, *J. Appl. Polym. Sci.*, 2004, **93**, 1769.

119. M. F. Huang, H. G. Yu and X. F. Ma, *Polym. Degrad. Stabil.*, 2005, **90**, 501.
120. J. H. Yang, J. G. Yu and X. F. Ma, *Carbohydr. Polym.*, 2006, **63**, 218.
121. H. A. Pushpadass, G. S. Babu, R. W. Weber and M. A. Hanna, *Packag. Technol. Sci.*, 2008, **21**, 171.
122. W.-J. Lee, Y.-N. Youn, Y.-H. Yun and S.-D. Yoon, *J. Polym. Environ.*, 2007, **15**, 35.
123. Q. Wu and L. Zhang, *J. Appl. Polym. Sci.*, 2001, **82**, 3373.
124. G. Shanmugam and P. L. Polavarapu, *Biophys. Chem.*, 2004, **111**, 73.

CHAPTER 8
Degradable Bioelastomers: Synthesis and Biodegradation

Q. Y. LIU,[1] L. Q. ZHANG[2] AND R. SHI[3]

[1] Beijing University of Aeronautics and Astronautics, School of Chemistry and Environment, Xueyuan Road of Haidian District, 100191, Beijing, People's Republic of China; [2] Beijing University of Chemical Technology, College of Materials Science and Engineering, Beisanhuan East Road of Chaoyang District, 100029, Beijing, People's Republic of China; [3] Laboratory of Bone Tissue Engineering of Beijing Research Institute of Traumatology and Orthopaedics, Beijing 100035, People's Republic of China

8.1 Character, Definition and Category of Degradable Bioelastomers

Biodegradable polymers such as polyesters have been paid much attention, and have been used to reduce environmental pollution from material wastes.[1] They can be applied in diverse fields such as packaging, paper coating, fibres and biomedicine.[2,3] Accompanying the biological degradation, which is mainly hydrolysis, microbial degradation or oxidation degradation, biodegradable polymers will become water, carbon dioxide and other eco-friendly and non-toxic compounds. Most biodegradable polymers[4,5] have broad applications in biomedical fields, and are used as surgical sutures,[6] matrices for drug delivery[7] and tissue engineering scaffolds,[8] being called degradable polymeric biomaterials.

Degradable bioelastomers are a significant branch of degradable polymeric biomaterials which possess four features different from other biomaterials: (1) stable cross-linked structures of three-dimensional networks (chemical cross-linking or physical cross-linking); (2) certain flexibility and elasticity providing mechanical stimulation for tissue engineering constructs; (3) appropriate mechanical properties especially matching with soft tissues of bodies; (4) easily adjustable and designable biodegradability by tuning the cross-linking density.

Degradable bioelastomers have presented potential applications particularly in the fields of tissue engineering and controlled drug delivery. As the scaffolds in tissue engineering, degradable bioelastomers are required to support cell-oriented growth aimed at the generation of replacement tissues, and to degrade at a suitable degradation rate matching with the generation of new tissues. Their role of mechanical stimulation in the engineering of elastic soft tissues has been demonstrated recently to be important.[9–11] As the carriers in drug delivery, degradable bioelastomers can be driven to release drugs by osmotic pressure at a nearly constant rate,[12] which is the same as the traditional silicon rubbers.[13] They do not need subsequent surgical removal after being implanted in bodies, in common with other degradable biomaterials.

Firstly, based on the typical descriptions of biomaterials,[14,15] a degradable bioelastomer is defined as a biodegradable elastomeric biomaterial, either alone or as a part of a complex system, being used to direct the course of any therapeutic or diagnostic procedure by control of interactions with components of living systems in human or veterinary medicine. Secondly, in view of the ASTM definition of 'elastomer', a degradable bioelastomer takes on the characteristics of elastomer materials, whose glass transition temperature (T_g) is lower than room and body temperature, and which has the ability of resilience to at least its 1.25-fold original length within 1 minute when it is stretched to its 1.5-fold original length and then released. Thirdly, as a very successful biomaterial, the meaning of a degradable bioelastomer also includes bionics, biomimetics and being bio-inspired. That is to say, a degradable bioelastomer will show similar molecular structures and physicochemical properties to the replacement tissue or its surrounding tissue, which will endow it with the biocompatibility, mainly referring to blood compatibility, tissue compatibility and biomechanical compatibility. For example, as well as collagen and elastin, which are cross-linked polymers that provide elasticity to the natural extracellular matrix (ECM), degradable bioelastomers should present matched mechanical properties with the tissues or tissue components as shown in Table 8.1.[16–22]

Degradable bioelastomers are generally classified as thermoplastic and thermoset degradable bioelastomers. Thermoplastic degradable bioelastomers such as segment polyurethanes including biodegradable structures are usually phase-separated block copolymers cross-linked by physical interaction. The soft and rubber-like segments (with a low T_g) give them flexibility, while glassy or crystallizable segments (with a high T_g or melting temperature) provide strength and stiffness.[23,24] Thermoset degradable bioelastomers are often cross-linked by covalent bonding, which present very stable network structures.[25,26] Thermoplastic degradable bioelastomers have the advantage of being easily fabricated by melt and solvent processing methods; however, their hard regions with high T_g or having been crystallized often make them biodegrade in a slow rate and heterogeneous fashion.[27,28] Thermoset degradable bioelastomers are

Table 8.1 Mechanical properties of selected human soft tissues and tissue components.

Tissues and tissue components	Young's modulus (MPa)	Tensile strength (MPa)	Elongation at break (%)
Vascular wall elastin	0.3–0.6	0.3–0.6	100–220
Vascular wall collagen	$1 \times 10^2 - 2.9 \times 10^3$	5–500	5–50
Relaxed smooth muscle	0.006		300
Contracted smooth muscle	0.01–1.27		300
Knee articular cartilage	2.1–11.8		
Carotid artery	0.084 ± 0.022		
Aortic valve leaflet	15 ± 6		21 ± 12
Pericardium	20.4 ± 1.9		34.9 ± 1.1
Cerebral artery	15.7		50
Cerebral vein	6.85		83
Ureter		$0.47^a/1.43^b$	$98^a/36^b$
Trachea		$0.35^a/2.16^b$	$81^a/61^b$
Inferior vena cava		$3.03^a/1.17^b$	$51^a/84^b$
Ascending aorta		$1.07^a/0.069^b$	$77^a/81^b$

Note: ^{a} = transverse, ^{b}b = longitudinal.

very hard to be processed to the desired shapes under the help of heat and solvent again after solidification, but their biodegradation rate is more uniform, and the remaining dimension is more stable.

8.2 Requirements of Degradable Bioelastomers

8.2.1 Safety

Both degradable bioelastomers and nondegradable bioelastomers should be of high purity, and possess optimal physicochemical properties when they are applied in biomedical fields in order to guarantee their safety. The monomers used to prepare bioelastomers need to be inexpensive, non-toxic and easy to use. When these bioelastomers are implanted in bodies, adverse events such as thrombosis, cell injuries, plasma and protein degeneration, enzyme inactivation, electrolyte disturbances/imbalances, inflammation, carcinogenesis, toxication and allergic reactions should be avoided.[29]

For degradable bioelastomers, more and higher requirements are concerned. Varied additives used during preparation and processing, such as initiators, catalysts and cross-linking agents, need to be non-toxic, or they need to be removed completely from the final products. When they are used as tissue engineering scaffolds and drug delivery carriers, their degradation rate should match with the rate of tissue regeneration and drug release. They need to be completely degraded or absorbed in bodies after a successful therapy, so that no substances causing irritation or inflammation remain. Furthermore, the intermediate products from degradable bioelastomers during biodegradation should be non-toxic.

8.2.2 Biodegradation

In order to endow bioelastomers with biodegradability in a particular biological situation, the bioelastomers should be first designed to contain easily bond-cleavage segments in their molecular structure. The segments, including ester, anhydride, amide and orthoester units *etc.* shown in Scheme 8.1, have been proved to be biodegradable under the biological conditions. Then, the biodegradation mechanism of degradable bioelastomers such as hydrolysis, enzyme-catalysed and oxidation degradation should be considered. Commonly, hydrolysis is of more concern. Anhydride and orthoester units are the most easily hydrolysable, being followed by ester and amide units. Thirdly, the factors influencing the biodegradation, such as molecular structures, compositions, cross-linking density, crystallinity, molecular weight, hydrophilicity/hydrophobicity, sample shape and morphology, ambient temperature and pH value, need to be clearly understood. The bioelastomers with lower cross-linking density, lower crystallinity, lower molecular weight, strong hydrophilicity, thinner section and porous structure usually biodegrade faster.[30]

The biodegradation of bioelastomers mainly involves surface degradation (or surface erosion) and bulk degradation, and the most direct method to describe this is to record their mass losses and observe their shape and morphology after a period of *in vitro* or *in vivo* degradation. The surface degradation usually occurs when the degradation rate is faster than the rate of water diffusion into bioelastomers. The surface degradation rate is easily predictable because the biodegradation first happens on the surface of the materials and the shape is usually kept well all along. The surface degradation of bioelastomers is particularly desirable for applications in controlled drug delivery systems because the rate of drug release from them can be directly determined by the degradation rate of the materials. Bulk degradation occurs if the degradation and weight loss are correlated with the rate of water penetration into the bulk of bioelastomers. When bulk degradation occurs, the mass losses are fast throughout the whole samples, and the original sizes of the samples are usually kept for a long time and then suddenly cracks completely. The bulk degradation of bioelastomers is more suited to applications in drug burst release, and sometimes is also suitable for tissue engineering scaffolds.

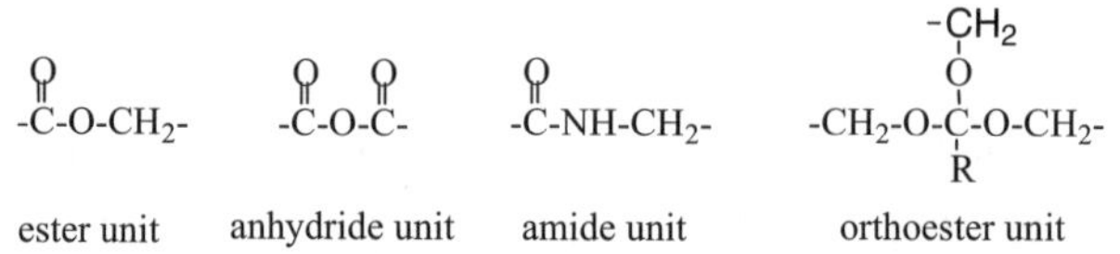

$$-\overset{O}{\underset{\|}{C}}-O-CH_2- \qquad -\overset{O}{\underset{\|}{C}}-O-\overset{O}{\underset{\|}{C}}- \qquad -\overset{O}{\underset{\|}{C}}-NH-CH_2- \qquad -CH_2-O-\overset{\overset{-CH_2}{\underset{|}{O}}}{\underset{\underset{R}{|}}{C}}-O-CH_2-$$

ester unit   anhydride unit   amide unit   orthoester unit

Scheme 8.1 Functional units in molecular segments being capable of biodegradation.

8.2.3 Cross-linking

Biodegradable polymers can be prepared both by the refining and modifying of natural polymers such as polysaccharides, cellulose and proteins, and by chemically synthesizing routes such as polycondensation, ring-opening polymerization and enzyme-catalysed polymerization.[1] All these preparation methods of biodegradable polymers, especially the chemically synthesizing methods, can be also used to prepare degradable bioelastomers. In addition, adopting thermally and photo-initiated free-radical polymerization is a good choice, too.

No matter what the preparation method is, it is important that cross-linking structures with three-dimensional networks be formed in bioelastomers, as well as the biodegradable segments that should be achieved in the molecular chains and which will endow the materials with excellent flexibility, high resilience and good biodegradation. By introducing cross-links into a polymer chain, physical properties such as the crystallinity, the melting point, the glass transition temperature and solubility will be affected. The biodegradability of the polymer will also be affected. The mechanical properties such as tensile strength, impact strength and modulus of the polymer will be enhanced after the polymer is cross-linked.

The cross-linking forms of degradable bioelastomers are mainly classified as physical and chemical cross-linking. The physical cross-linking usually originates from the crystalline regions or the regions with high glass transition temperatures, which will endow degradable bioelastomers with good processing properties (thermoplasticity) under the help of heat and solvents. The physically cross-linked bioelastomers are often segmented copolymers chiefly synthesized by polycondensation, ring-opening and enzyme catalysis. The chemical cross-linking usually derives from the monomers or segments with multifunctional groups such as hydroxyl, carboxyl and double bonds, which will endow degradable bioelastomers with good structural stability under conditions of heating and stretching. The chemically cross-linked bioelastomers chiefly are the thermoset or double-bond cured polymers synthesized by polycondensation, thermally or photo-initiated free-radical polymerization.

The formation of chemical cross-linking networks by photo-cross-linking reactions is very important. The advantages of photo-cross-linking over other cross-linking techniques include high curing rates at room temperature, spatial control of polymerization and solidification of the networks *in vivo* with minimal heat production. Especially seen from the clinical angle, the photo-initiated cross-linking is very attractive because it minimizes patient's discomfort, risk of infection, scar formation and cost of therapy.

8.3 Synthesis and Biodegradation of Degradable Bioelastomers

Many degradable bioelastomers have been reported to be synthesized by varied synthesis methods, and biodegradation is a very significant property. In this section, degradable bioelastomers are chiefly classified as follows: segmented polyurethane bioelastomers; poly(ε-caprolactone) related bioelastomers; polylactide related bioelastomers; polycarbonate related bioelastomers; poly-(glycerol sebacate) bioelastomer and its derivatives; citric acid related polyester bioelastomers; poly(ether ester) bioelastomers; and poly(ester amide) bioelastomers. Their synthesis and biodegradation are specially introduced, accompanied by many typical examples. Finally, other several novel degradable bioelastomers are also mentioned.

8.3.1 Degradable Segmented Polyurethane Bioelastomers

8.3.1.1 Summary

Polyurethane (PU) bioelastomers are one of the most traditional polymeric biomaterials, being either biostable or biodegradable, and have been applied in many biomedical fields including ligament and meniscus reconstruction, blood-contacting materials, infusion pumps, heart valves, insulators for pacemaker leads and nerve guidance channels.[31] Here, degradable PU bioelastomers are especially discussed, most of which belong to segmented polyurethanes (SPU) with good processing ability.

Degradable SPU bioelastomers are usually synthesized by two steps, firstly preparing prepolymers terminated by isocyanate groups through reacting polyester or polyether diols with diisocyanate monomers, and then extending the prepolymers using chain extenders such as diol or diamine monomers. They may be named as poly(ester-urethanes) or poly(ether-urethanes), and consist of soft segments (polyester or polyether segments) and hard segments (the segments formed by the diisocyanates and chain extenders), which often results in micro-phase separation structures. The micro-phase separation structures usually endow SPU bioelastomers with good haemocompatibility. They can be easily processed using heat and solvents, and their structures and properties can be adjusted in a wide range based on the different soft segments, diisocyanates and chain extenders.

When synthesizing degradable SPU bioelastomers, poly(ε-caprolactone) (PCL), poly(ethylene glycol) (PEO), polyhydroxyalkanote (PHA), poly(lactic acid) (PLA) and their copolymers all which are terminated by hydroxyl groups, are usually first prepared, and then used as the polyester or polyether soft segments because they are biocompatible and easily hydrolytically and enzymatically degraded.[32] 1,4-butane diisocyanate (1,4-BDI) can be a good non-toxic alternative to obtain hard segments because the degradation products are expected to be 1,4-butanediamine (putrescine), a non-toxic diamine that is essential for cell growth and differentiation.[33–35] L-lysine ethyl ester diisocyanate (LDI) is also a good choice because the degradation products are non-cytotoxic components.[36] Furthermore, 1,6-hexamethylene diisocyanate (1,6-HDI) and isophorone diisocyanate (IPDI) are sometimes adopted. The molecular structures of the above mentioned four aliphatic diisocyanate monomers are shown in Scheme 8.2. Aromatic diisocyanate monomers may be cautiously considered for use for the preparation of hard segments because the toxicity of the monomers and degradation intermediates is still a very controversial problem. The biodegradation of SPU bioelastomers is controlled mainly by polyester or polyether soft segments, and the ones with amorphous and hydrophilic soft segments degrade more quickly than those with semicrystalline and hydrophobic soft segments.

8.3.1.2 Synthesis and Biodegradation

Uncatalysed synthesis of the degradable SPU bioelastomers using PCL as soft segments, 1,4-BDI and 1,4-butanediol (BDO) is described in Scheme 8.3.[37] The PCL polyester diols with molecular weights of 750–2800 g mol^{-1} were first synthesized by the ring-opening polymerization of ε-caprolactone (ε-CL)

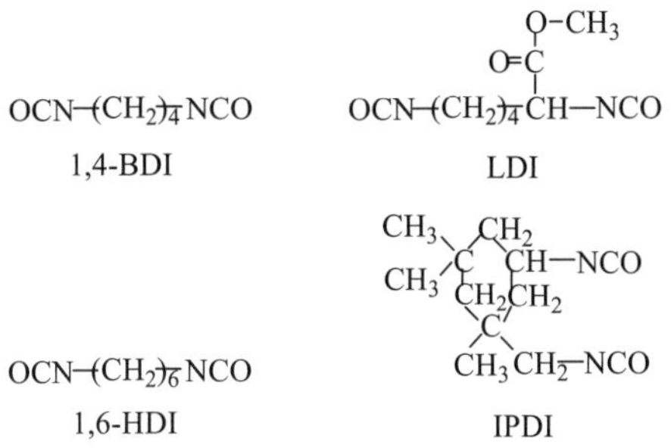

Scheme 8.2 Molecular structures of the referred four diisocyanate monomers.

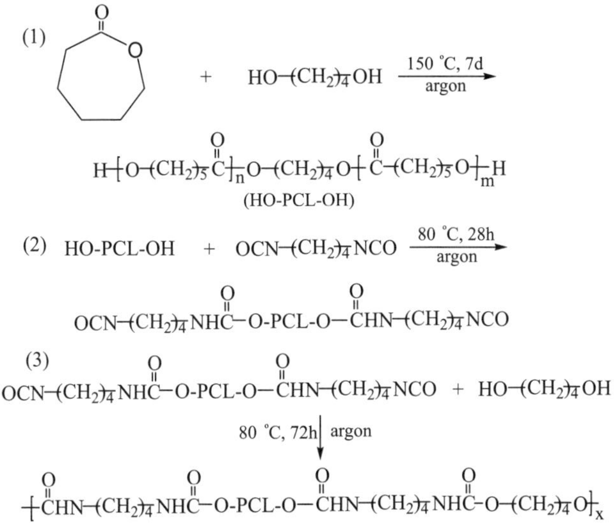

Scheme 8.3 Uncatalysed synthesis of the degradable SPU bioelastomers from PCL polyester diol, 1,4-BDI and BDO.

initiated with BDO. Catalyst-free synthesis of the SPU bioelastomers not only weakened the several side reactions which often existed during the catalysed synthesis, but also improved the materials' biocompatibility. T_g of the bioelastomers ranged from −45.9 to −58.5 °C. Experiments on the bioelastomers as meniscal replacements in dogs' knees were conducted,[38] which demonstrated that they were able to prevent rupture of the sutures out of the meniscus scaffolds during the implantation, and they degraded relatively slowly without swelling during application.

The degradable SPU bioelastomers from PCL polyester diol, 1,4-BDI and putrescine or lysine ethyl ester chain extenders at the molar ratio of 1:2:1 were synthesized with the catalysts of stannous octoate ($Sn(Oct)_2$) as shown in Scheme 8.4.[39] Incubation in phosphate buffer solutions for 8 weeks resulted in mass losses above 50% when using lysine ethyl ester as chain extender, while the mass losses were above 10% when using putrescine as chain extender. The degradation products were shown to have no toxicity to human endothelial cells. When they were endowed with an ability to induce local angiogenesis by controlled release of a basic fibroblast growth factor (bFGF), the SPU scaffolds degraded slightly slower in comparison with the ones loaded with bFGF probably because of the lower water absorption in phosphate buffer solution at 37 °C, as shown in Figure 8.1.[40]

The degradable SPU bioelastomers from PCL-b-PEO-b-PCL triblock copolymers as soft segments, 1,4-BDI and peptide Ala-Ala-Lys (AAK) chain extenders were prepared.[41] PCL-b-PEO-b-PCL polyester diols were first synthesized by ring-opening polymerization of ε-CL initiated by poly(ethylene glycol) with $Sn(Oct)_2$ as catalyst at 120 °C for 24 h under nitrogen atmosphere, then the SPU bioelastomers were prepared according to the reaction

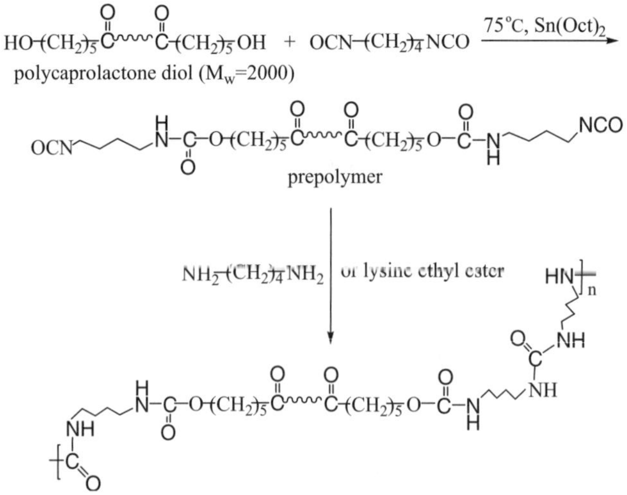

Scheme 8.4 Synthesis of the degradable SPU bioelastomers from PCL polyester diol, 1,4-BDI and putrescine or lysine ethyl ester.

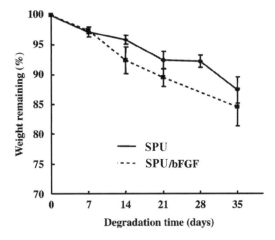

Figure 8.1 Effect of degradation time on the weight remaining for the SPU and SPU/bFGF bioelastomer scaffolds in phosphate buffer solutions (pH = 7.4, 37 °C). Adapted from Guan et al.[40]

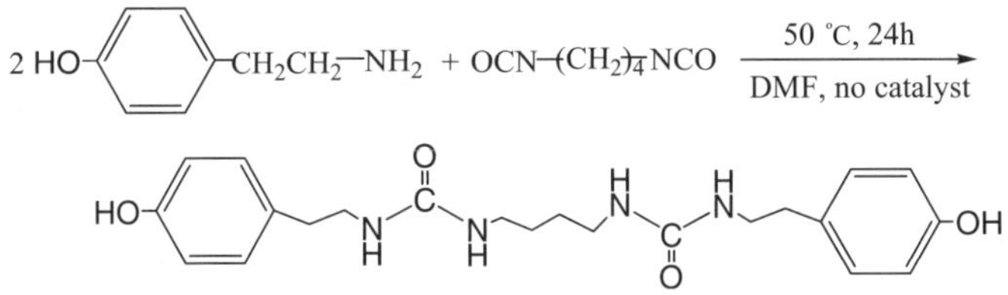

Scheme 8.5 Synthesis of the tyramine-1,4-BDI-tyramine adducts from tyramine and 1,4-BDI.

stoichiometry of 2:1:1 (BDI/PCL-b-PEO-b-PCL/AAK) in addition of Sn(Oct)$_2$ in dimethyl sulfoxide (DMSO) at 75 °C. The bioelastomers presented a T_g lower than −54 °C. Their mass losses after 56 day degradation in phosphate buffer solution without elastase at 37 °C were within 10–18%, while the corresponding mass losses with elastase increased and were within 19–35%, which demonstrated that the introduction of AAK sequences made the bioelastomers show elastase sensitivity. In another study done by the same research team, putrescine was used to replace AAK as chain extenders for preparing degradable SPU bioelastomers.[42]

Tyramine-1,4-BDI-tyramine as chain extender was used to prepare a degradable SPU bioelastomer.[43] The tyramine-1,4-BDI-tyramine adducts were first synthesized as described in Scheme 8.5, based on the fact that the reactivity of the amine was 1000–2000 times higher than that of phenol at 25 °C. Then, PCL polyester diols were fabricated from BDO initiators and ε-caprolactone monomers with Sn(Oct)$_2$ as catalyst at 135 °C for 24 h under argon atmosphere. Thirdly, the prepolymers were produced from PCL polyester diol and 1,4-BDI at the molar ratio 2:1 under 70 °C in DMF. Finally, the bioelastomers were

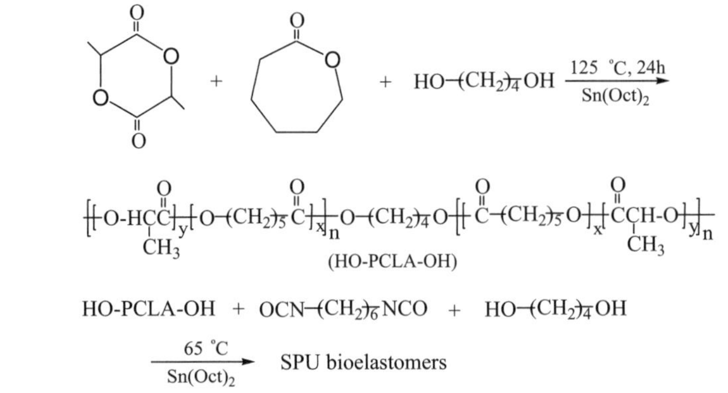

Scheme 8.6 Synthesis of the degradable SPU bioelastomers from PCLA polyester diol, 1,6-HDI and BDO.

obtained through extending the prepolymers in use of tyramine-1,4-BDI-tyramine adducts at 70 °C in DMF. T_g of the bioelastomers was about −52 to −55 °C. The degradable SPU bioelastomers from poly(ε-caprolactone-co-lactide) (PCLA) diols, 1,6-HDI and BDO were prepared as shown in Scheme 8.6.[44] ε-CL was randomly copolymerized with L-lactide (L-LA) lactide initiated by BDO and catalysed by Sn(Oct)$_2$ to eliminate the crystallinity of PCL. T_g of the bioelastomers ranged from −7.71 to −46.3 °C. The degradable SPU bioelastomers from BDO and LDI were synthesized as follows:[45] the prepolymers terminated by isocyanate groups were first synthesized by the reaction between BDO and LDI with dibutyltin dilaurate (DBTDL) as catalyst at 40 °C for 30 min in toluene, and then PCL polyester diols with molecular weight of 2000 g mol^{-1} were added to continuously react for 22 h. T_g of the bioelastomers was about −33 °C.

The PHA-based degradable SPU bioelastomers from oligo[(R,S)-3-hydroxybutyrate] diol (PHB polyester diol), 1,6-HDI and four different aliphatic diol monomers (1,4-butanediol, 1,6-hexanediol, 1,8-octanediol and 1,10-decanediol) were prepared.[46] PHB polyester diols were first synthesized by the transesterification and condensation reaction between ethyl(R,S)-3-hydroxybutyrate and the four different aliphatic diols with dibutyltin oxide as the catalyst in bulk at 100 °C for 3 h under a gentle stream of argon and further at 110 °C for 5 h with gradual reduction of the pressure to 0.5 mmHg, as shown in Scheme 8.7. The bioelastomers were prepared from 1,6-HDI and PHB polyester diols alone, or PHB polyester diols mixed with PCL, poly(butylene adipate) (PBA) or poly(diethylene glycol adipate) (PDEGA) polyester diols using dibutyltin dilaurate as catalyst at 80 °C in 1,2-dichloroethane. T_g of the bioelastomers varied from −54 to −23 °C. The preliminary study on their biodegradation was carried out in microbially active leaf compost stored in an environmental chamber maintained at around 20 °C and 95% relative humidity, and the weight loss was found to increase with the increasing of PHB contents in the bioelastomers.

Through inducing gemini quaternary ammonium side groups (GA8) on hard segments, a cationic degradable SPU bioelastomer was prepared.[47] L-lysine-derivatized diamines containing GA8 were first synthesized, then a series of SPU bioelastomers were prepared from PCL polyester diols, LDI, BDO and GA8 with N,N-dimethyl acetamide (DMAc) as solvent and Sn(Oct)$_2$ as catalyst, which were terminated by methoxyl-poly(ethylene glycol). T_g of the bioelastomers ranged from −34.9 to −37.8 °C. *In vitro* degradation in phosphate buffer solutions showed that there was not much difference between enzymatic and hydrolytic degradation, especially for the bioelastomers with higher gemini content. The degradation rate of the bioelastomers increased with the gemini contents, and their weight losses reached nearly 50% and 100% when the gemini molar contents were correspondingly 70% and 100% after the hydrolytic and enzymatic degradation with Lipase AK for 12 h. The bioelastomer sample displayed a rough and porous surface after 12 h of enzymatic degradation or 24 h of hydrolytic degradation, where the pores and cracks were then observed growing into the film to present a channel-like structure after 24 h of enzymatic degradation as shown in Figure 8.2. Another waterborne degradable PU bioelastomer was prepared by the same research team using IPDI, PCL, PEO, BDO and L-lysine as raw materials,[48] but the chemical crosslinking networks formed in it.

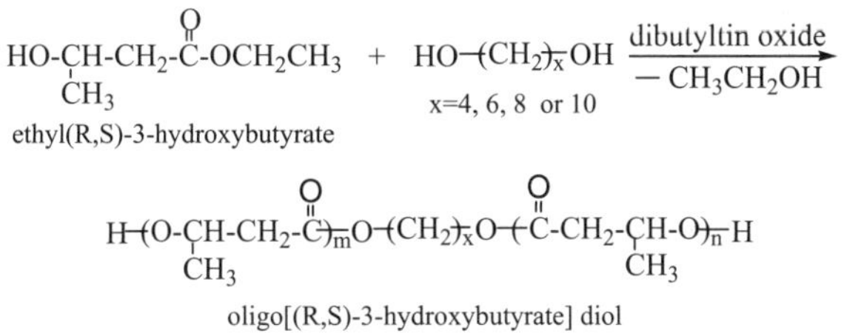

Scheme 8.7 Synthesis of the PHB polyester diols from ethyl(*R*,*S*)-3-hydroxybutyrate and aliphatic diols.

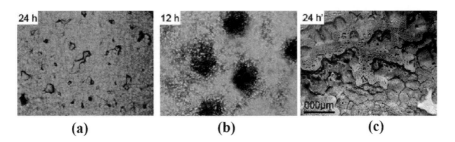

(a)   (b)   (c)

Figure 8.2 Polarizing light microscopy (PLM) images of the bioelastomer sample during degradation in phosphate buffer solutions (pH = 7.4, 37 °C) at different degradation time: (a) hydrolytic degradation for 24 h, (b) enzymatic degradation for 12 h and (c) enzymatic degradation for 24 h. Adapted from Ding *et al.*[47]

8.3.2 Poly(ε-caprolactone) Related Bioelastomers

8.3.2.1 Summary

Poly(ε-caprolactone) (PCL) is a rubbery semicrystalline polymer at physiological temperature, whose T_g is usually lower than −60 °C, being different from other biodegradable aliphatic polyesters such as polyglycolide (PGA) and polylactide (PLA) whose T_g are separately about 38 °C and 55 °C. These polyesters have been applied in biomedical fields such as tissue engineering and drug delivery. PCL polymers as well as PGA and PLA degrade mainly by hydrolysis yielding hydroxyl carboxylic acids, which in most cases are non-toxic and completely metabolized by the body. But PCL degrades very slowly and its complete biodegradation *in vivo* may take 2–3 years. In order to adjust the degradability of the PCL polymers without sacrificing their biocompatibility, PCL segments are often hybridized with other segments such as PLA to achieve linear PCL copolymers. The linear PCL copolymers with high amounts of ε-CL are generally crystalline, while the amorphous copolymers display high mobility because of low T_g and hardly retain their shape at body temperature.

The PCL segments can also react with other segments such as polybasic carboxylic acid, polymers including double bonds, PGA and PLA, remaining the elasticity of PCL segments, to prepare PCL related bioelastomers mainly by polycondensation, copolymerization, thermally and photo initiated polymerization. For example, by the way of introducing a cross-linkable moiety such as acryloyl or methacryloyl group into PCL based prepolymers and then polymerizing them with peroxide or photo initiation such as ultraviolet (UV) irradiation, the double bond cured PCL related bioelastomers can be achieved. PCL related bioelastomers have been studied to use as drug carriers and soft tissue engineering scaffolds such as articular cartilage[49] and blood vessel[50] because of their good elasticity, biocompatibility and biodegradability.

8.3.2.2 Synthesis and Biodegradation

The photo-cured PCL-related bioelastomers were prepared by Nagata *et al.*, as shown in Scheme 8.8.[51] Firstly, non-cured PCL-based copolymers including double bonds were synthesized from PCL diols (M_w: 1250, 2000, 3000 g mol^{-1}), adipic acid and 4-hydroxycinnamic acid; secondly, the copolymers were photo-cured by ultraviolet light (UV) irradiation using a wavelength above 280 nm for obtaining the bioelastomers. T_g of the bioelastomers ranged from −52 to 64 °C. In phosphate buffer solutions with *Ps. cepacia* lipase, the weight loss of the bioelastomers decreased significantly with increasing the photo-curing time, as shown in Figure 8.3, while their weight loss increased remarkably with increasing M_w of PCL-diols. When M_w of PCL-diols was 1250 g mol^{-1}, their mass losses reached 30–93% after 10 day incubation.

The thermoset PCL-related bioelastomers from PCL diol (M_w: 530, 1250 and 2000 g mol^{-1}), tricarballylic acid and meso-1,2,3,4-butanetetracarboxylic acid were prepared by polycondensation as shown in Scheme 8.9.[52] The prepolymers

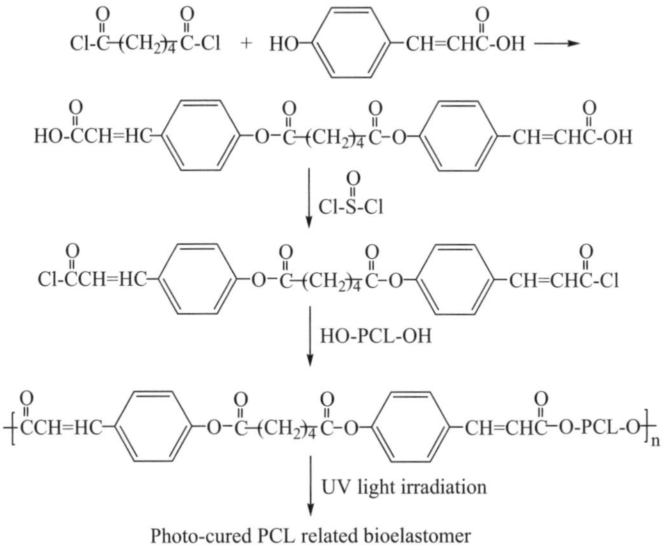

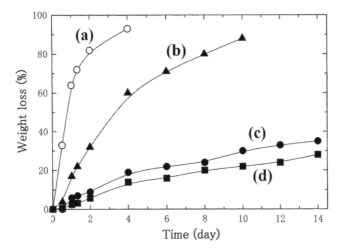

Scheme 8.8 Synthesis of the photo-cured PCL-related bioelastomers from PCL diols, adipic acid and 4-hydroxycinnamic acid.

Figure 8.3 Effect of photo-curing time on the weight losses of the PCL-related bioelastomers from PCL diols (M_w 1250 g mol^{-1}) in phosphate buffer solutions (pH = 7.4, 37 °C) with *Ps. cepacia* lipase: (a) 10 min, (b) 30 min, (c) 60 min and (d) 90 min. Adapted from Nagata and Sato[51]

were first synthesized at 260 °C under nitrogen atmosphere, then they were cast on an aluminium plate with a 17wt% DMF solution at 80 °C and further post-polymerized at 280 °C. T_g of the bioelastomers was within −49 to 67 °C. Their complete degradation in phosphate buffer solutions with *Rh. delemar* lipase was

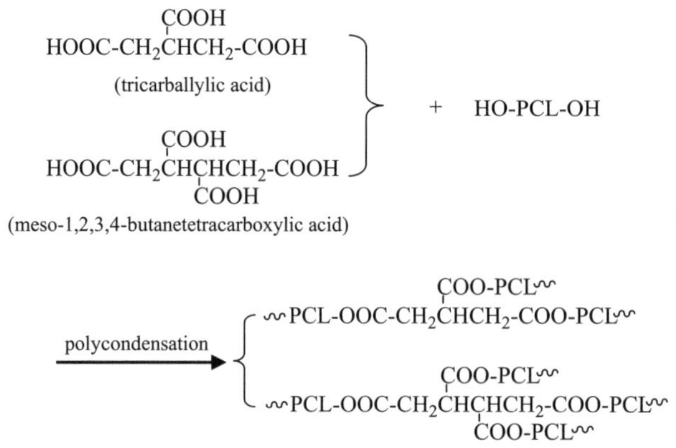

COOH
HOOC-CH$_2$CHCH$_2$-COOH
(tricarballylic acid)

\} + HO-PCL-OH

COOH
HOOC-CH$_2$CHCHCH$_2$-COOH
COOH
(meso-1,2,3,4-butanetetracarboxylic acid)

polycondensation →

~PCL-OOC-CH$_2$CHCH$_2$-COO-PCL~
            COO-PCL~

~PCL-OOC-CH$_2$CHCHCH$_2$-COO-PCL~
            COO-PCL~
            COO-PCL~

Scheme 8.9 Synthesis of the thermoset PCL-related bioelastomers from PCL diols and tricarballylic acid or meso-1,2,3,4-butanetetracarboxylic acid.

observed within 12 h, while no significant weight loss was observed in the absence of lipase. The enzymatic degradation rate of the bioelastomers became faster when the molecular weight of PCL diols increased.

Another photo-cured PCL-related bioelastomer was prepared by inducing acrylate groups at the end of the linear poly(ε-caprolactone-co-lactide-co-glycolide) diols and then curing of them through UV irradiation with 2,2-dimethoxy-2-phenylacetophenone (Irgacure 651) as photoinitiator.[53] Poly(ε-caprolactone-co-lactide-co-glycolide) diols were prepared by the ring-opening polymerization of ε-CL, L-LA and glycolide (GA), which was initiated by tetra(ethylene glycol) in the presence of Sn(Oct)$_2$ at 145 °C for 24 h under argon atmosphere. The poly(ε-caprolactone-co-lactide-co-glycolide) diacrylates (M_n = 1800, 4800 and 9300 g mol^{-1}) were synthesized by reacting poly(ε-caprolactone-co-lactide-co-glycolide) diols with acryloyl chloride in addition of triethylamine using dichloromethane as solvents at 0 °C for 6 h and further at room temperature for 18 h. T_g of the bioelastomers ranged from –0.4 to –47.2 °C. They presented a two-stage *in vitro* degradation. In the first stage, the bioelastomers' weight and strain remained almost constant, but a linear decrease in the Young's modulus and ultimate stress were observed; in the second stage, which began when the Young's modulus dropped below 1 MPa, there was a rapid weight loss and strain increase. The mass losses of the bioelastomers after 8 week degradation in phosphate buffer solution were lower than 9%, as shown in Figure 8.4.

The thermoplastic PCL-related bioelastomers were prepared as shown in Scheme 8.10, which were poly(L-lactide)-PCL (PLLA-PCL) multiblock copolymers.[54] The bischloroformates of carboxylated PLLA were first synthesized mainly by two steps, preparing PLLA diols initiated by 1,6-hexanediol with stannous octoate as catalyst at 130 °C for 5 h, and synthesizing acylhalide (COCl) group terminated PLLA at 60 °C for about 3 h from dicarboxylated PLLA which was prepared from PLLA diol and succinic anhydride. Then,

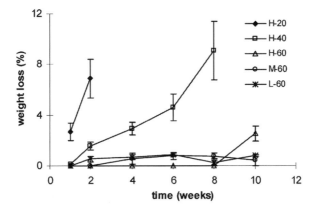

Figure 8.4 Weight losses of the photo-cured PCL-related bioelastomers during *in vitro* degradation in phosphate buffer solutions (pH = 7.4, 37 °C). The high molecular weight (9300 g mol^{-1}), medium molecular weight (4800 g mol^{-1}), and low (1800 g mol^{-1}) molecular weight oligomers are denoted by H–x, M–x and L–x, respectively, where x represents the percent content of ε-CL. Source: Shen et al.[53]

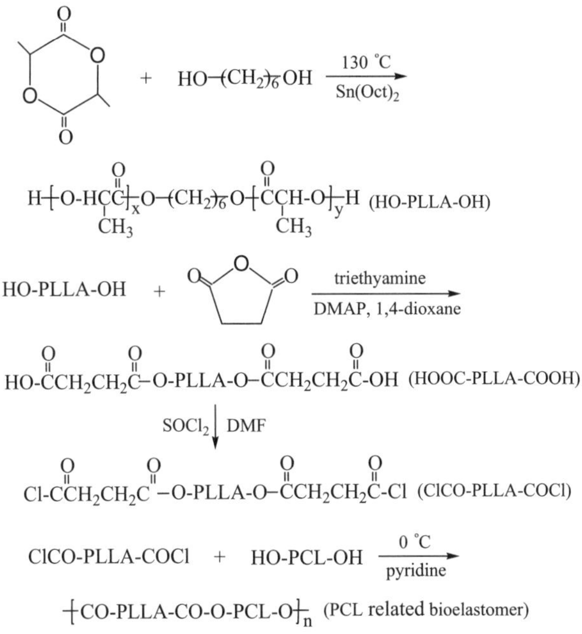

Scheme 8.10 Synthesis of the thermoplastic PCL-related bioelastomers from PLLA-PCL multiblock copolymers.

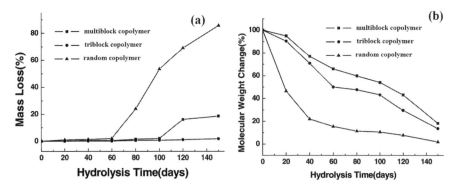

Figure 8.5 Mass loss (a) and molecular weight change (b) of the thermoplastic PCL related bioelastomers during *in vitro* degradation in phosphate buffer solutions (pH = 7.4, 37 °C). Adapted from Jeon et al.[54]

PLLA-PCL copolymers were achieved by using the coupling reaction between COCl-terminated PLLA with PCL-diols in the presence of pyridine and ice bath. T_g of the bioelastomers ranged between −8.5 and 63.5 °C. The mass losses of the bioelastomers after 150 day degradation in phosphate buffer solution were within 1–90%, and the random copolymers degraded more rapidly than the triblock and multiblock copolymers in view of mass loss and molecular weight change as shown in Figure 8.5.

A series of thermoplastic PCL-related bioelastomers from poly(L-lactic acid)-poly(ε-caprolactone)-poly(L-lactic acid) (PLA-PCL-PLA) triblock copolymers and 1,6-HDI were designed,[55] which were similar to SPU bioelastomers. The triblock copolymers composed of PCL flexible segments and PLA blocks with molecular weight of 550–6000 g mol^{-1}, were first synthesized by the ring opening polymerization of L-lactide initiated by PCL diols with Sn(Oct)$_2$ as catalyst at 145 °C for 1 h, and then they reacted with 1,6-HDI at 82 °C for 3 h under a dry nitrogen atmosphere to achieve the final bioelastomers. T_g of the bioelastomers was around −30 °C. When the molecular weight of PLA segments was 2000 g mol^{-1}, the initial molecular weight of the bioelastomers achieved was 190 000 g mol^{-1}, which separately decreased to 83 000 g mol^{-1}, 48 000 g mol^{-1} and 9500 g mol^{-1} after *in vitro* degradation of 1, 3 and 6 month as shown in Figure 8.6.

The thermoset PCL-related bioelastomers from star-poly(ε-caprolactone-co-D,L-lactide) and 2,2-bis(ε-CL-4-yl)-propane were synthesized by Amsden et al.[56,57] As the first step, the star-poly(ε-caprolactone-co-D,L-lactide) copolymers were prepared by ring opening polymerization of ε-CL with D,L-lactide (D,L-LA) using glycerol as initiator and stannous 2-ethylhexanoate as catalyst; as the second step, the bioelastomers were achieved by reacting the copolymers with different ratios of 2,2-bis(ε-CL-4-yl)-propane in the presence of ε-CL as a solvent and co-monomer at 140 °C. T_g of the bioelastomers was about −32 °C. After a 12 week *in vitro* degradation in phosphate buffer solutions, none of the bioelastomers were completely degraded, and they demonstrated a bulk

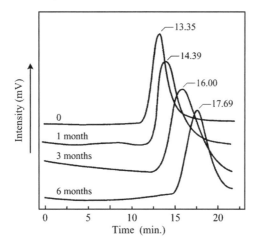

Figure 8.6 Gel permeation chromatography (GPC) curves as a function of their *in vitro* degradation in phosphate buffer solutions (pH = 7.4, 37 °C) for the PCL-related bioelastomer in which the molecular weight of PLA segments is 2000 g mol^{-1}. Source: Cohn and Salomon[55]

hydrolysis mechanism. A photo-cured PCL-related bioelastomer from an acrylated star-poly(ε-caprolactone-co-D,L-lactide) prepolymers was synthesized by UV initiation by the same research team.[58] T_g of the bioelastomers was about −8 °C. Over a degradation period of 12 weeks in phosphate buffer solutions, the bioelastomers exhibited little mass losses, no appreciable dimension change and little decrease in mechanical strength.

8.3.3 Polylactide-related Bioelastomers

8.3.3.1 Summary

Polylactide is a kind of biodegradable polyester that possesses many desirable properties such as non-toxicity, hydrolyzability and biocompatibility for use for varied biomedical purposes such as sutures, fracture fixation, oral implant and drug delivery microspheres.[59] It is popularly synthesized by the ring-opening polymerization of lactide monomers which are the cyclic dimers of lactic acid. Polymerization of racemic D,L-lactide typically results in atactic, amorphous polymers named poly(D,L-lactide) ($T_g \approx 60$ °C), whereas polymerization of L-lactide or D-lactide results in isotactic, semicrystalline polymers called poly(L-lactide) or poly(D-lactide) ($T_m \approx 180$ °C).[60,61] PLA fractures at very low strains (about 3%) after stretch, so it is unsuitable for use in numerous applications where elasticity and ductility are required. With the goal of extending the utility of PLA materials, the modification by plasticization, blending and incorporation into block copolymers can be used to enhance their properties, especially the flexibility.

ABA triblock copolymers that contain immiscible segments where A is a 'hard', high T_g or semicrystalline polymer and B is a 'soft' amorphous, low T_g polymer can behave as thermoplastic elastomers. Polylactide-containing ABA triblock copolymers are usually synthesized by the ring-opening polymerization of lactide initiated by hydroxyl-capped macromolecular monomers such as PEO, in which A is the PLA segment and B is the hydroxyl-capped macromolecular segment. When the polylactide-containing ABA triblock copolymers present good elasticity, biodegradability and biocompatibility which can be applied in biomedical fields, they are called PLA related bioelastomers.

8.3.3.2 Synthesis and Biodegradation

The thermoplastic PLA-related bioelastomers from PEO were prepared by a two-stage reaction.[62] PLA-PEO-PLA triblock copolymers were first synthesized by the ring-opening polymerization of LA initiated by PEO diols (M_w = 1000, 3200, 6000 and 10 000 g mol^{-1}) with Sn(Oct)$_2$ as catalyst at 145 °C, and then the copolymers achieved were conducted by the chain extension using 1,6-HDI as extender at 82 °C. When the molecular weight of PEO diols was 6000 g mol^{-1}, the bioelastomers presented no T_g in the measurement range of –120 to 200 °C, while they possessed low melting temperature around 28 °C and low fusion heat of 16 J g^{-1} which contributed to the soft segment. When the molecular weight of PEO diols was 1000 g mol^{-1}, the bioelastomers presented a PEO-related T_g around –20 °C and a second one around 31 °C coming from the PLA amorphous phase.

The renewable-resource thermoplastic PLA-related bioelastomers from polymenthide (PM), which is a noncrystalline, amorphous and degradable polyester with T_g of about –25 °C, were prepared by three steps as shown in Scheme 8.11.[63] Polymenthide diols (M_n = 20 000–55 000 g mol^{-1}) were first synthesized by the ring-opening polymerization of menthide in the presence of diethylene glycol (DEG) with diethyl zinc (ZnEt$_2$) as catalyst at 100 °C in toluene under an inert atmosphere, and then the polymenthide diols reacted with triethylaluminum for 30 min to form the corresponding aluminium alkoxide (AlEt$_3$) macroinitiators, and finally D,L-lactide was added into the macroinitiators to react for 50 min at 90 °C for achieving polylactide-b-polymenthide-b-polylactide (PLA-b-PM-b-PLA) bioelastomers. The bioelastomers behaved micro-phase separation structures with two T_g values. One T_g corresponding to the PM blocks was about 22 °C, and another T_g corresponding to the PLA blocks ranged from 21 to 53 °C depending on the molecular weights of the PLA blocks. The hydrolytic degradation experiments of the bioelastomers in phosphate buffer solutions demonstrated that they lost mass very little during the first 11 weeks of degradation, and a slight increase of mass loss was observed at week 21, being able to reach 25% at week 37, as shown in Figure 8.7.[64] While the mass loss of PLA was over 80% at week 21, a small mass loss of 8% at week 45 was observed for PM.

The thermoplastic PLA-related bioelastomers from bishydroxy-terminated [R,S]-PHB which presented elastomeric properties because of its low

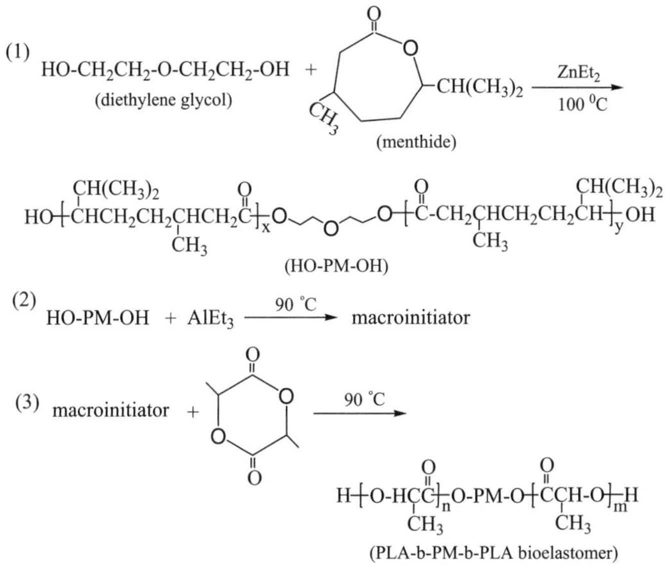

Scheme 8.11

(1) HO-CH$_2$CH$_2$-O-CH$_2$CH$_2$-OH + [menthide structure] -CH(CH$_3$)$_2$ $\xrightarrow{\text{ZnEt}_2}{100\,°C}$
(diethylene glycol)

HO-[CHCH$_2$CH$_2$CHCH$_2$C(=O)]$_x$-O~O~O-[C(=O)-CH$_2$CHCH$_2$CH$_2$CH]$_y$-OH
with CH(CH$_3$)$_2$ and CH$_3$ substituents
(HO-PM-OH)

(2) HO-PM-OH + AlEt$_3$ $\xrightarrow{90\,°C}$ macroinitiator

(3) macroinitiator + [lactide structure] $\xrightarrow{90\,°C}$

H-[O-HCC(=O)]$_n$-O-PM-O-[C(=O)CH-O]$_m$-H
CH$_3$ CH$_3$
(PLA-b-PM-b-PLA bioelastomer)

Scheme 8.11 Synthesis of the thermoplastic PLA-related bioelastomers from polymenthide and D,L-lactide.

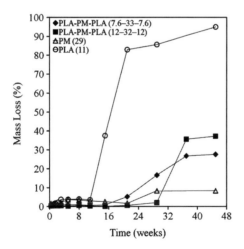

Figure 8.7 Mass loss–degradation time curves of the thermoplastic PLA-related bioelastomers with different compositions of PLA-PM-PLA in phosphate buffer solutions (pH = 7.4, 37 °C) in comparison with PLA and PM. Source: Wanamaker et al.[64]

crystallinity, were prepared as shown in Scheme 8.12, being actually PLA-PHB-PLA copolymers.[65] [R,S]-PHB diols were first synthesized by the ring-opening polymerization of [R,S]-b-butyrolactone in the presence of BDO with 1,3-dichlorotetrabutyldistannoxane as catalyst at 100 °C, and then the [R,S]-PHB

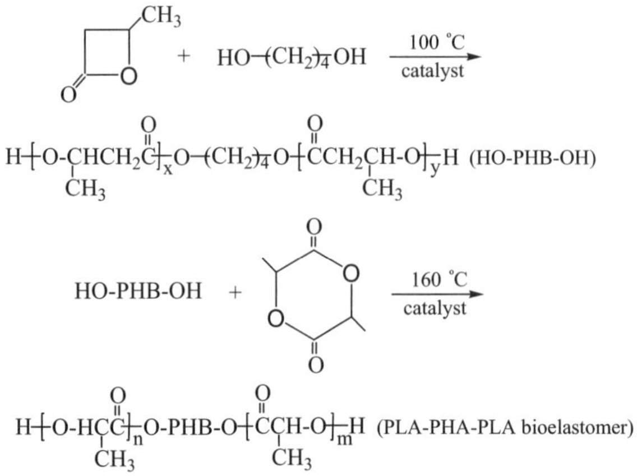

Scheme 8.12 Synthesis of the thermoplastic PLA-related bioelastomers from bishydroxy-terminated [RS]-PHB and L-lactide.

diols reacted with L-lactide using $Sn(Oct)_2$ as catalyst at 160 °C for obtaining the final bioelastomers. T_g of the bioelastomers ranged from −0.6 to 4.8 °C, corresponding to the PHB segments. The biodegradation of the bioelastomers were not reported.

The thermoplastic PLA-related bioelastomers from poly(1,3-trimethylene carbonate) (poly(TMC)), which was a rubbery and amorphous polymer with low T_g of about −25 °C, were prepared as shown in Scheme 8.13, being poly(LA-TMC-LA) copolymers.[66] The α,ω-hydroxy terminated poly(TMC) diols were first synthesized by the ring-opening polymerization of 1,3-trimethylene carbonate with 1,6-hexanediol as initiator and $Sn(Oct)_2$ as catalyst for 3 days at 130 °C, and then the LA monomers (D-lactide, L-lactide or D,L-lactide) were added into the poly(TMC) diols to polymerize for 3 days at 130 °C with $Sn(Oct)_2$ for obtaining the final bioelastomers. T_g of the bioelastomers ranged from −19.5 to 8.2 °C. Diethylene glycol instead of 1,6-hexanediol was first used to synthesize poly(TMC) diols at 140 °C with $Sn(Oct)_2$ as catalyst, and then the diols were further polymerized with LA monomers in the same conditions, following chain extension conducted with 1,6-HDI, other poly-(LA-TMC-LA) bioelastomers were prepared,[67] which possessed T_g of −11.1 to −1.8 °C. In vitro degradation tests demonstrated that the weight loss of the bioelastomers was 5–7% after 8 week degradation, and their molecular weight dropped rapidly within a week to about half of original molecular weight, and then seemed to be balanced as shown in Figure 8.8.

Polyisoprene (PI) polymers with low T_g were used as soft segments for preparing the thermoplastic PLA related bioelastomers which were synthesized by reacting α,ω-hydroxyl polyisoprene (HO-PI-OH) with lactide using $AlEt_3$ as catalyst in toluene at 90 °C as shown in Scheme 8.14.[68] The bioelastomers actually were a kind of PLA-PI-PLA copolymer, which could be achieved in larger quantities and at higher reaction concentrations after using aluminium

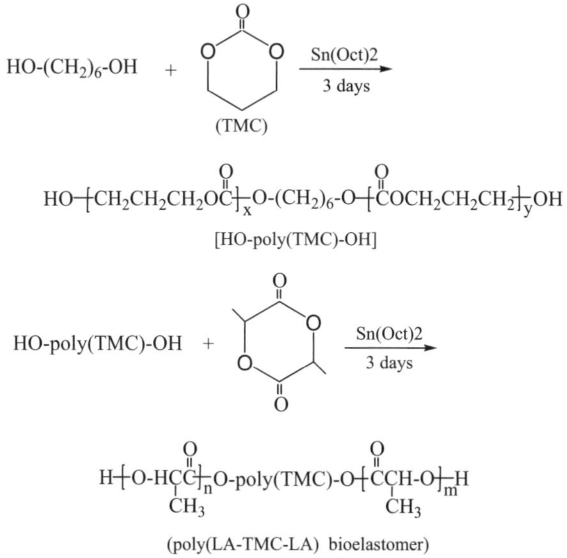

Scheme 8.13 Synthesis of the thermoplastic PLA-related bioelastomers from poly(TMC) and lactide.

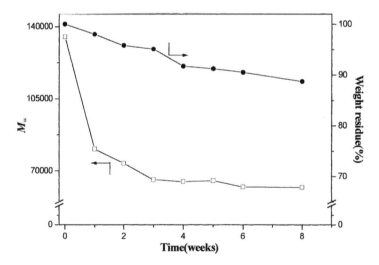

Figure 8.8 Weight remaining and molecular weight change of the poly(LA-TMC-LA) bioelastomers during *in vitro* degradation in phosphate buffer solutions (pH = 7.4, 40 °C). Source: Kim and Lee.[67]

triisopropoxide (Al(iOPr)$_3$) instead of AlEt$_3$.[69] T_g of the bioelastomers was about −62 °C, corresponding to PI segments.

The thermoplastic PLA-related bioelastomers from poly(butylene succinate) (PBS) which usually presented a T_g of near −40 °C, were prepared, being

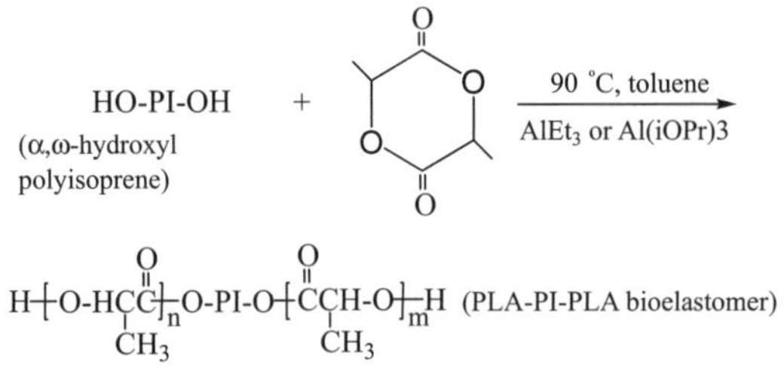

Scheme 8.14 Synthesis of the thermoplastic PLA-related bioelastomers from α,ω-hydroxyl polyisoprene and lactides.

poly(L-lactide-b-butylene succinate-b-L-lactide) (PLLA-b-PBS-b-PLLA) copolymers.[70] The PBS prepolymers terminated by hydroxyl groups were first synthesized through the esterification of succinic acid and 1,4-butanediol for 4 h at 210 °C under nitrogen atmosphere at the molar ratio of 5:6 (succinic acid/1,4-butanediol), and then a further polycondensation for 4 h at 210 °C under the vacuum pressure lower than 0.5 mmHg again. Secondly, a series of new PLLA-b-PBS-b-PLLA bioelastomers with various PLLA block lengths were prepared by ring-opening polymerization of L-lactide using the synthesized hydroxyl-capped PBS prepolymers ($M_n = 4900$ g mol^{-1}) as the macroinitiator with Sn(Oct)$_2$ as catalyst. T_g of the bioelastomers ranged from −29.7 to 13.4 °C. The bioelastomers were self-assembled *in situ* into biodegradable microparticles with diameters of 480–660 nm in acetonitrile solvent.[71]

8.3.4 Polycarbonate-related Bioelastomers

8.3.4.1 Summary

Polycarbonate (PC) is a flexible biodegradable polymer with low glass transition temperature and low Young's modulus, which degrade by an enzymatic surface erosion process both *in vitro* and *in vivo*. Aliphatic PC derived from ring-opening polymerization of the corresponding six-membered cyclic monomer has been found to possess favourable biocompatibility, low cytotoxicity and well tuneable biodegradability, which has been extensively studied as a biomedical and environment friendly material.[72,73] Most of aliphatic PC materials, such as poly(ethylene carbonate) (PEC), poly(1,2 propylene carbonate) (PPC), poly(trimethylene carbonate) (PTMC) and poly[2,2-(2-pentene-1,5-diyl) trimethylene carbonate] (PHTC), have appeared as synthetic rubbers or semicrystalline polymers in a relaxed solid state.[74,75] But they usually present increased crystallinity when they are in the stretched state, so the modifications of these PC polymers are required to decrease their crystallization and improve their flexibility.

The modifications of aliphatic PC polymers are often carried out by the ring-opening copolymerization of cyclic carbonate monomers and other kinds of cyclic monomers such as glycolide, lactide and ε-caprolactone. In addition, considering the low creep compliance of linear aliphatic PC polymers limits

Degradable Bioelastomers: Synthesis and Biodegradation 265

their application in long-term dynamic cell culture, especially poly(trimethylene carbonate), the way of preparing covalent networks from PC macromers to improve their creep-resistance by thermally or photo-initiated cross-linking reactions can also be chosen to modify PC polymers. After modifying the PC polymers, if the materials achieved possess good elasticity, biocompatibility and biodegradation, and can be applied in biomedical materials, they are called PC-related bioelastomers which are actually the PC copolymers and sometimes chemically cross-linked PC polymers.

8.3.4.2 Synthesis and Biodegradation

The thermoplastic PC-related bioelastomers from cyclic carbonates and hydroxyl capped PBS were prepared,[76] being poly(butylene succinate-co-cyclic carbonate) copolymers. During preparation, five kinds of six-membered cyclic carbonate monomers of trimethylene carbonate (TMC), 1-methyl-1,3-trimethylene carbonate (MTMC), 2,2-dimethyl-1,3-trimethylene carbonate (DMTMC), 5-benzyloxytrimethylene carbonate (BTMC) and 5-ethyl-5-benzyloxymethyl trimethylene carbonate (EBTMC) were first prepared from ethyl chloroformate and the corresponding diols in tetrahydrofuran (THF) as shown in Scheme 8.15. Then, PBS diols were synthesized by polycondensation of succinic acid and excess BDO at 210 °C for 2 h under nitrogen atmosphere. Finally, the poly-(butylene succinate-co-cyclic carbonate) bioelastomers were achieved by the ring-opening polymerization of carbonates in the presence of PBS diols for about 2 h at 210 °C using titanium tetraisopropoxide (Ti(i-OPr)$_4$) as catalyst under vacuum. T_g of the bioelastomers was between −35.8 and −28.3 °C.

The thermoset PC-related bioelastomers from multi-carboxylic acids were synthesized by a melt polycondensation.[77] The multifunctional carboxylic acids referred to tricarballylic acid (Y_t) and trimesic acid (Y), and polycarbonate diols presented the molecular weights of 1000 g mol^{-1} and 2000 g mol^{-1} (named as PCD$_{1000}$ and PCD$_{2000}$). The molecular structures of Y_t, Y and the polycarbonate

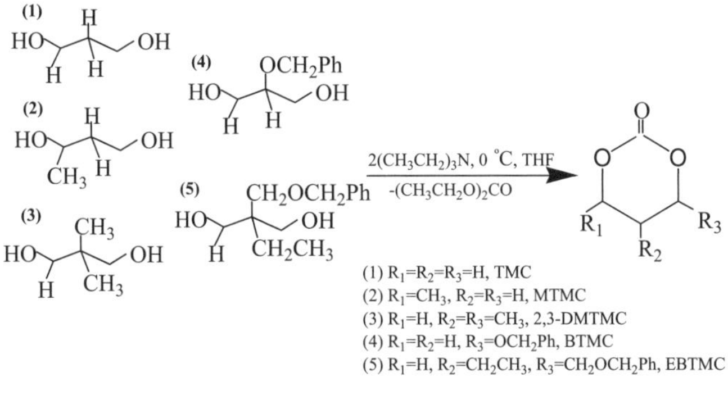

Scheme 8.15 Synthesis of five kinds of six-membered cyclic carbonates from the corresponding diols.

diol prepared are shown in Scheme 8.16. The prepolymers were first prepared in a stream of nitrogen at 260 °C for 20–40 min in PCD_{1000} system and for 90–180 min in PCD_{2000} system. Then, the prepolymers were dissolved in DMF and post-polymerized at 270 °C for 40–80 min to obtain the final products. T_g of the bioelastomers was in the range −51 to −41 °C. The degradability of the bioelastomers were evaluated in phosphate buffer solutions with *Rh. Delemar* lipase, and the results demonstrated that the degradation mechanism was enzyme-catalyzed hydrolysis from the ester linkages between Y_t or Y and PCD as well as the carbonate linkages in the polycarbonate segments. The weight losses of the bioelastomers increased almost linearly with degradation time, being lower than 10% after 5 day degradation, as shown in Figure 8.9.

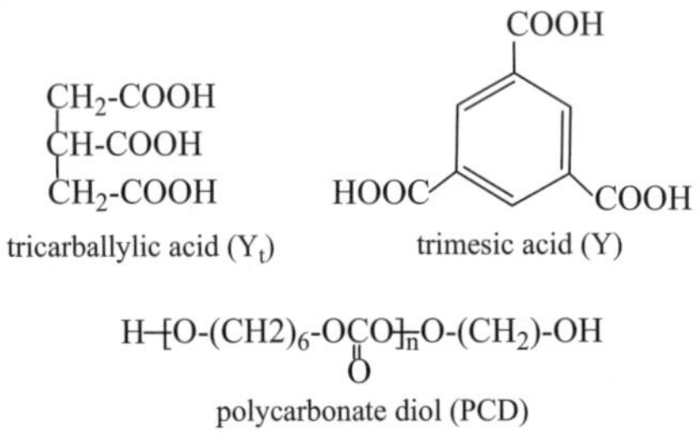

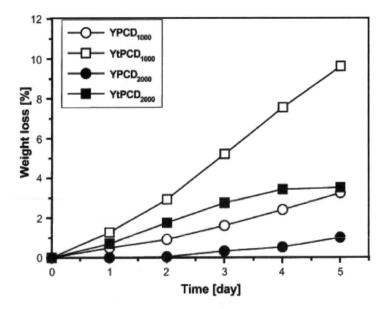

Scheme 8.16 Molecular structures of tricarballylic acid, trimesic acid and polycarbonate diol used in the preparation of thermoset polycarbonate related bioelastomers.

Figure 8.9 Weight loss of the thermoset PC-related bioelastomers against degradation time in phosphate buffer solutions (pH = 7.2, 37 °C) with *Rh. delemar* lipase. Source: Nagata et al.[77]

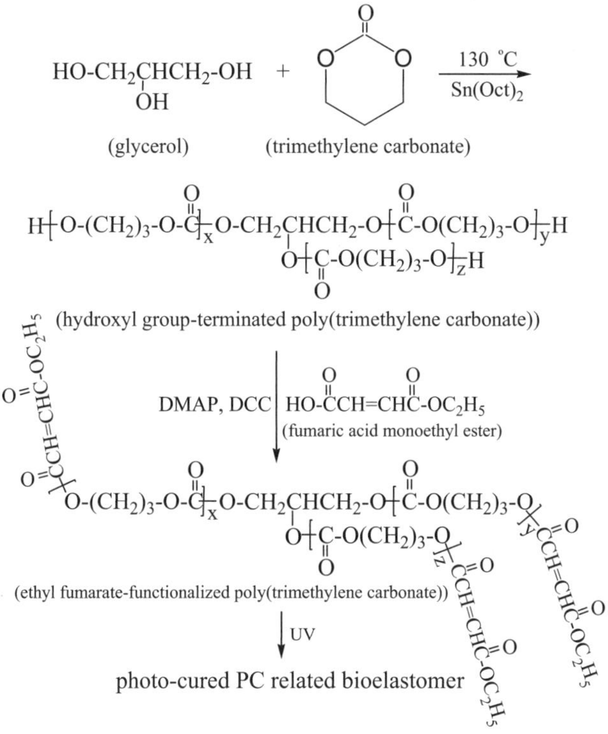

Scheme 8.17

Synthesis of the photo-cured PC-related bioelastomers from ethyl fumarate-functionalized poly(trimethylene carbonate) prepolymers.

The photo-cured PC-related bioelastomers from ethyl fumarate-functionalized poly(trimethylene carbonate) oligomers were prepared as shown in Scheme 8.17.[78] Photo-cross-linkable ethyl fumarate-functionalized poly(trimethylene carbonate) prepolymers with the molecular weights ranging from 4500 g mol^{-1} to 13 900 g mol^{-1} were first synthesized by reacting three-armed hydroxyl group-terminated poly(trimethylene carbonate) oligomers with fumaric acid monoethyl ester at room temperature using N,N-dicyclohexylcarbodiimide as coupling agent and 4-dimethylamino pyridine as catalyst. Then, the prepolymers were photo-cured by UV-initiated radical polymerization. The bioelastomers presented rubber-like behaviour with T_g varying between −18 °C and −13 °C. Their degradation mechanism belonged to enzyme-catalyzed degradation, but no more detailed degradation studies were reported.

8.3.5 Poly(glycerol sebacate) Bioelastomer and its Derivatives

8.3.5.1 Summary

Glycerol is a key component in the synthesis of phospholipids, and sebacic acid is the natural metabolic intermediate in the oxidation of fatty acid;[79,80] in

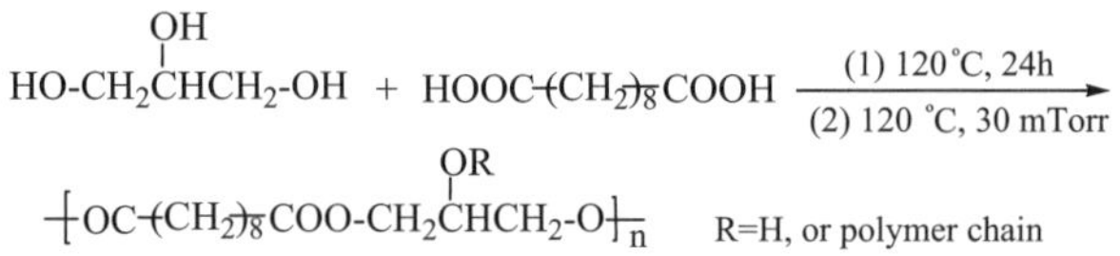

$$\text{HO-CH}_2\overset{\overset{\text{OH}}{|}}{\text{CH}}\text{CH}_2\text{-OH} + \text{HOOC-(CH}_2)_8\text{COOH} \xrightarrow[\text{(2) 120 °C, 30 mTorr}]{\text{(1) 120 °C, 24h}}$$

$$-\!\!\left[\text{OC-(CH}_2)_8\text{COO-CH}_2\overset{\overset{\text{OR}}{|}}{\text{CH}}\text{CH}_2\text{-O}\right]_n \quad \text{R=H, or polymer chain}$$

Scheme 8.18 Synthesis of the thermoset PGS bioelastomers from glycerol and sebacic acid.

addition, glycerol and copolymers containing sebacic unit have been approved for use in medical fields.[81] Based on the facts, using glycerol and sebacic acid as the monomers to prepare degradable bioelastomers is a very good strategy. The thermoset degradable poly(glycerol sebacate) (PGS) bioelastomers are the most excellent representative, and have been synthesized by directly melt polycondensing glycerol with sebacic acid at the molar ratio of 1/1, as shown in Scheme 8.18.[82] They are being studied for varied potential applications in soft tissue engineering such as vascular regeneration, myocardial tissue repairing and retinal tissue engineering.[83–89] They have also been researched as drug carriers.[90] T_g of the thermoset PGS bioelastomers were below –80 °C. Their degradation *in vitro* was difficult to correlate with the *in vivo* degradation, and the mass loss reached 15% after 10 weeks degradation in phosphate buffer solutions, whereas the complete degradation was observed after 6 weeks *in vivo*. Unlike poly(DL-lactide-co-glycolide) which degraded mostly by bulk degradation, *in vivo* degradation of the PGS bioelastomers was dominated by surface erosion as indicated by linear mass loss with time, preservation of implant geometry, better retention of mechanical strength, absence of surface cracks, and minimal water uptake.[91]

When designing the thermoset PGS bioelastomers, five hypotheses have been put forward: (1) endowing them with good mechanical properties by covalent cross-linking and hydrogen bonding; (2) making them form three-dimensional networks through the reaction between the trifunctional and difunctional monomers; (3) controlling their cross-linking density at a low value to make them very elastic; (4) intensifying their hydrolysis while weakening the enzyme-catalyzed degradation by inducing ester bonds in molecular chains; (5) lowering the probability of their heterogeneous degradation by creating cross-linking bonds which are the same as the ester bonds of the main chains. The derivatives stand for the bioelastomers which are designed and prepared because of the inspiration from the above hypotheses and the PGS bioelastomers.

8.3.5.2 Synthesis and Biodegradation

The thermoplastic PGS (TM-PGS) bioelastomers consisting of sols and gels were prepared by directly condensing glycerol with sebacic acid according to different molar ratios of 2:2, 2:2.5, 2:3, 2:3.5 and 2:4 (glycerol/sebacic acid) at 130 °C under a pressure of 1 kPa,[92] and which could be shaped by hot pressing under 90 °C and 15 MPa. The sol contents of the bioelastomers were above 60%, and their T_g ranged from –25.1 to –32.2 °C. Their mass losses reached

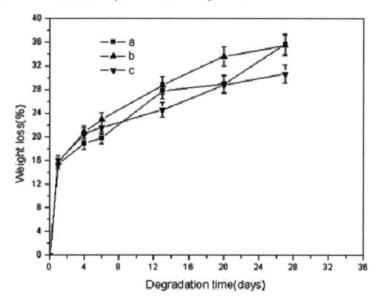

Figure 8.10 Weight loss–degradation time curves of the TM-PGS bioelastomers prepared by two steps using the PGS prepolymers with different molecular weights (phosphate buffer solutions, pH = 7.4, 37 °C): (a) M_n = 1681 g mol^{-1}, (b) M_n = 2426 g mol^{-1}, (c) 4429 g mol^{-1}. Source: Liu et al.[94]

16–37% after 7.5 day in vitro degradation. To improve the physicochemical properties of the TM-PGS bioelastomers, they were again prepared by the two-step method.[93,94] Firstly, non-cross-linked PGS prepolymers at the molar ratio of 1:1 (glycerol/sebacic acid) were synthesized through polycondensation at 130 °C under 1 kPa; secondly, the TM-PGS bioelastomers were achieved by adding sebacic acid into the prepolymers at the total molar ratio of 2:2.5 to continuously react in the same conditions, which could be processed to a desired shape by hot pressing under 130 °C and 15 MPa. T_g of the bioelastomers was within the range –22.5 and –22.6 °C, and they still consisted of gels and sols. The increased cross-linking density and decreased sol contents resulted in the decreased degradation rate with the mass loss lower than 36% after 28 day in vitro degradation, as shown in Figure 8.10.

The photo-cured PGS bioelastomers from PGS prepolymers and acryloyl chloride were prepared.[95] Firstly, the PGS prepolymers were synthesized by the polycondensation of equimolar glycerol and sebacic acid at 120 °C under vacuum; then, the PGS prepolymers were acrylated at 0 °C by the addition of acryloyl chloride with equimolar amount of triethylamine to achieve poly-(glycerol sebacate) acrylate (PGSA); finally, PGSA was cured through the initiation of UV light with 0.1% (w/w) photo-initiator 2,2-dimethoxy-2-phenylacetophenone. T_g of the bioelastomers ranged from –32.2 to –31.1 °C. Their in vitro enzymatic and hydrolytic degradation was dependent on the acrylation degree of PGSA, and they degraded only 10% after 10 weeks in phosphate

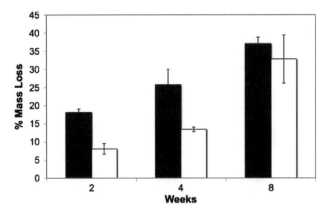

Figure 8.11 Mass loss of the Acr-PGS bioelastomers during the *in vivo* (black, dorsal pocket of male Sprague-Dawley rats) and *in vitro* (white, phosphate buffer solutions, pH = 7.2, 37 °C) degradation at 2, 4 and 8 weeks. Source: Ifkovits *et al.*[96]

buffer solutions. Another cured PGS bioelastomer from acrylated-poly(glycerol sebacate) (Acr-PGS) was prepared by adopting both redox and photo-initiated free radical polymerization.[96] The Acr-PGS prepolymers were first synthesized, and then were cured by redox initiation from benzoyl peroxide or *N,N*-dimethyl-*p*-toluidine, or by UV light with 2,2-dimethoxy-2-phenylacetophenone as photo-initiator for achieving the Acr-PGS bioelastomers. The *in vivo* and *in vitro* mass losses of the Acr-PGS bioelastomers separately reached the maximum of about 37% and 33% after 8 weeks of degradation, as shown in Figure 8.11.

By using polyol monomers, such as xylitol, sorbitol, mannitol and maltitol, instead of glycerol, a series of thermoset degradable poly(polyol sebacate) (PPS) bioelastomers were synthesized under the conditions of 120–150 °C and vacuum.[97] The molecular structures of the polyol monomers and corresponding PPS bioelastomers are shown in Scheme 8.19. T_g of the bioelastomers ranged from 7.3 to 45.6 °C. Mass losses were detected for all PPS bioelastomers by *in vitro* degradation tests, which were lower than 22% after 105 days. Their degradation rates under physiological conditions were slower than *in vivo* degradation rates. By introducing lactic acid into the system of glycerol and sebacic acid, the degradable poly(glycerol-sebacate-lactic acid) (PGSL) bioelastomers[98,99] were synthesized at 140–150 °C under nitrogen atmosphere and vacuum according to the molar ratios of 1:1:0, 1:1:0.25, 1:1:0.5 and 1:1:1 (glycerol/sebacic acid/lactic acid). The crystallization temperature of the bioelastomers ranged from −22.72 to −33.97 °C, which made the crystal phases of the bioelastomers melt to soft segments at room temperature. When the molar ratio was 1:1:0.25, *in vitro* degradation rate of the PGSL bioelastomers was the highest. The surface erosion first appeared during a short-term degradation, and then a combination of bulk degradation and surface erosion occurred after a long-term degradation.

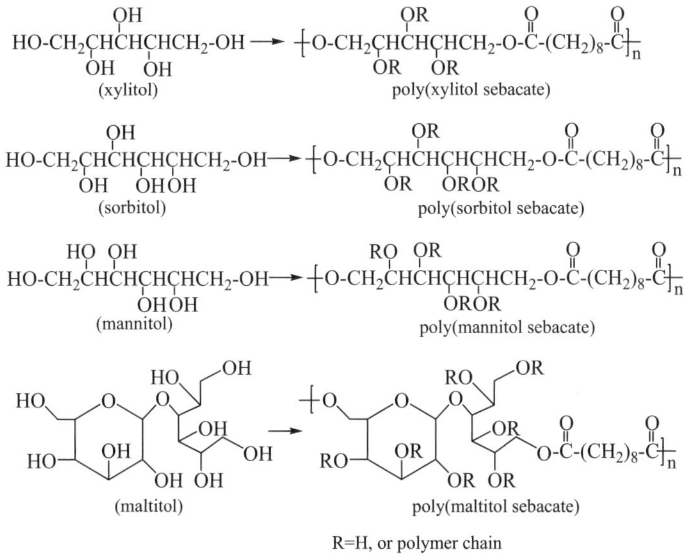

Scheme 8.19 Molecular structures of the polyol monomers and corresponding PPS bioelastomers.

8.3.6 Citric Acid-related Polyester Bioelastomers

8.3.6.1 Summary

Citric acid is a multifunctional monomer with three carboxyl groups and one hydroxyl group, being a non-toxic metabolic product of the bodies, readily available and inexpensive, which has been applied in biomedical fields. Citric acid with three carboxyl groups can react with the monomers with hydroxyl groups to form biodegradable polyesters. When the polyesters from citric acid present the characteristic of bioelastomers such as good elasticity, biodegradation and biocompatibility, they are called citric acid-related polyester bioelastomers. The remained functional groups such as hydroxyl and carboxyl in citric units may greatly contribute to the formation of the polyester three-dimensional networks by hydrogen bonding interaction. Poly(1,8-octanediol-co-citrate) (POC) is the pioneer of citric acid-related polyester bioelastomers, which has been synthesized by directly condensing 1,8-octanediol with citric acid under mild conditions without addition of any catalysts and cross-linking reagents.[100] Note that 1,8-octanediol is the largest aliphatic diol that is water soluble with no toxicity.

When designing the poly(diol citrate) bioelastomers including POC, the four rationales have been put forward as follow: (1) the use of non-toxic, readily available and inexpensive monomers; (2) incorporation of homogeneous biodegradable cross-links to confer elasticity to the resulting materials and leave behind some unreacted functional groups, which could be used for surface modifications; (3) the availability of various diols which provided flexibility to

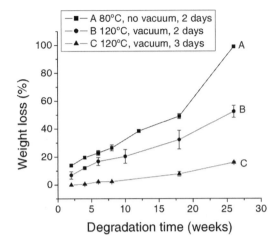

Figure 8.12 Weight loss of the POC bioelastomers prepared under different conditions during degradation in phosphate buffer solutions (pH = 7.4, 37 °C). Source: Yang et al.[100]

tune the mechanical and degradation properties of the resulting materials; (4) the establishment of intermolecular hydrogen bonding interactions, which would contribute to the mechanical properties of elastomers.

To achieve the POC bioelastomers, POC prepolymers with M_n of 1085 g mol^{-1} (M_w = 1088 g mol^{-1}) were first synthesized by the condensation of 1,8-octanediol and citric acid at 140 °C, and then the prepolymers were further post-polymerized at 120 °C, 80 °C, 60 °C or even 37 °C under vacuum or no vacuum for times ranging from 1 day to 2 weeks. T_g was observed to be between −10 °C and 0 °C. The biodegradation of the POC bioelastomers could be adjusted by controlling the reaction conditions, and the bioelastomers prepared under mild conditions (low temperature, e.g. 60 °C or 80 °C, no vacuum) had a significantly faster degradation rate than those which were prepared under relatively tougher conditions (high temperature, e.g. 120 °C, 2 Pa vacuum), as shown in Figure 8.12. The POC bioelastomers have been studied for use for coating expanded polytetrafluoroethylene (ePTFE) vascular grafts,[101] in vivo vascular tissue engineering,[102] cartilage tissue engineering,[103] cardiac tissue engineering[104] and substrate-mediated gene delivery.[105] Furthermore, the POC bioelastomer composites reinforced by hydroxyapatites (HA) have been prepared for the potential use in orthopaedic surgery such as graft interference screws.[106]

8.3.6.2 Synthesis and Biodegradation

Using other aliphatic diols instead of 1,8-octanediol, such as 1,6-hexanediol, 1,10-decanediol and 1,12-dodecanediol, to react with citric acid by

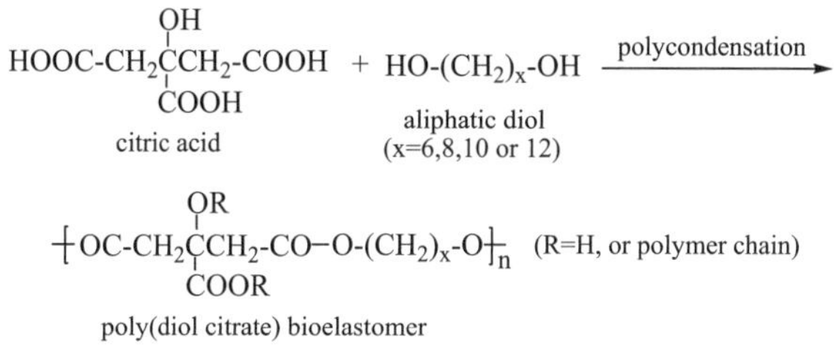

Scheme 8.20

$$\text{HOOC-CH}_2\underset{\underset{\text{COOH}}{|}}{\overset{\overset{\text{OH}}{|}}{\text{C}}}\text{CH}_2\text{-COOH} + \text{HO-(CH}_2)_x\text{-OH} \xrightarrow{\text{polycondensation}}$$

citric acid aliphatic diol (x=6,8,10 or 12)

$$\{\text{OC-CH}_2\underset{\underset{\text{COOR}}{|}}{\overset{\overset{\text{OR}}{|}}{\text{C}}}\text{CH}_2\text{-CO-O-(CH}_2)_x\text{-O}\}_n \quad (\text{R=H, or polymer chain})$$

poly(diol citrate) bioelastomer

Scheme 8.20 Synthesis of poly(diol citrate) bioelastomers from citric acid and varied aliphatic diols.

polycondensation, a series of degradable poly(diol citrate) bioelastomers such as poly(1,6-hexanediol-co-citrate) (PHC), poly(1,10-decanediol-co-citrate) (PDC) and poly(1,12-dodecanediol-co-citrate) (PDDC) were prepared as shown in Scheme 8.20.[107] T_g of the poly(diol citrate) bioelastomers was between −5 °C and 10 °C. The mass losses of the bioelastomers could reach 68% after 4 week degradation in phosphate buffer solutions at 37 °C. The *in vitro* degradation rate could be adjusted by varying the number of methylene units in the diol monomers, and the diols with decreasing number of methylene units resulted in increasing or faster degradation rates.

The double bond-cured POC bioelastomers were prepared by the method of incorporating acrylate or fumarate moieties into the molecular chains to provide secondary cross-linkable networks derived from free-radical polymerization.[108] Firstly, citric acid-based unsaturated prepolymers from citric acid, 1,8-octanediol and glycerol 1,3-diglycerolate diacrylate or bis(hydroxypropylfumarate) were synthesized by polycondensation for 40 min at 130 °C in nitrogen atmosphere as shown in Scheme 8.21. Secondly, the unsaturated prepolymers were cured for 12 h using benzoyl peroxide as catalyst at 80 °C, and then further cross-linked for 24 h at 120 °C, finally followed by the reaction for another 24 h at 120 °C under the pressure of 2 Pa to achieve the final bioelastomers. T_g of the bioelastomers was between −12.7 °C and −1.6 °C. The acrylated POC bioelastomers lost 11–14% of their masses after 1 month and about 20% after 2 months during the *in vitro* degradation, while the fumarate-based POC bioelastomers degraded faster, by 18–20% after 1 month and about 30% after 2 months.

The citric acid-related polyester bioelastomers from PEO were prepared, being named poly(ethylene glycol-co-citrate) (PEC) bioelastomers.[109] The PEC prepolymers were first synthesized by the direct condensation of PEO (molecular weight of 200 g mol^{-1}) and citric acid at 140 °C, and then they were post-polymerized and cross-linked in the mould at 120 °C to achieve the PEC bioelastomers. T_g of the PEC bioelastomers ranged from −17.6 to 2.9 °C. The PEC bioelastomers presented very good degradability, with weight losses over 60% after 96 h *in vitro* degradation, which could be tuned by controlling the post-polymerizing time as shown in Figure 8.13.

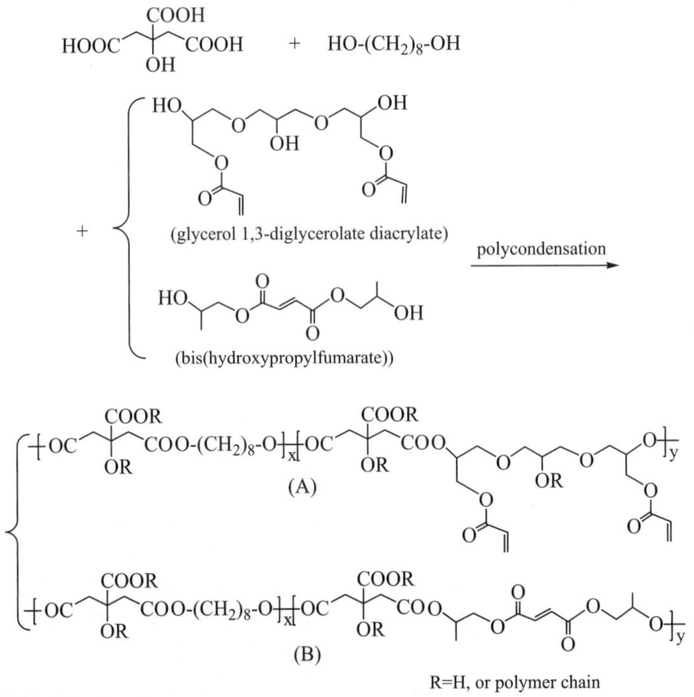

HOOC–C(COOH)(OH)–COOH + HO-(CH$_2$)$_8$-OH

+ {
(glycerol 1,3-diglycerolate diacrylate)

(bis(hydroxypropylfumarate))
}

→ polycondensation →

(A) acrylated prepolymers

(B) fumarate-containing prepolymers

R=H, or polymer chain

Scheme 8.21 Synthesis of the citric acid based unsaturated prepolymers from citric acid 1,8-octanediol and glycerol 1,3-diglycerolate diacrylate or bis(hydroxypropylfumarate): (A) acrylated prepolymers; (B) fumarate-containing prepolymers.

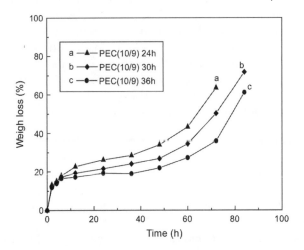

Figure 8.13 Weight loss of the PEC bioelastomers during degradation in phosphate buffer solutions (pH = 7.4, 37 °C) at the molar ratio of 10/9 (PEO/citric acid) with different post-polymerizing time. Source: Ding et al.[109]

By the polycondensation of 1,2-propanediol, sebacic acid and citric acid with no catalyst, the poly((1,2-propanediol-sebacate)-citrate) (PPSC) bioelastomers were prepared.[110] Poly(1,2-propanediol-sebacate) diols with M_w of 658 g mol^{-1} were first synthesized by polycondensation for 12 h at 140 °C under atmospheric pressure and further for 4 h under the pressure of 2 kPa. Then, the PPSC prepolymers were prepared by condensing poly(1,2-propanediol-sebacate) diols and citric acid for 12 h at 140 °C. Finally, the PPSC bioelastomers were achieved through the post-polymerization of the prepolymers for 12–36 h at 120 °C. T_g of the PPSC bioelastomers ranged from −13.8 to 9.8 °C. The mass losses of the PPSC bioelastomers were more than 60% after 20 day *in vitro* degradation. In order to improve the mechanical properties and biocompatibility of the PPSC bioelastomers, the nano-hydroxyapatite/PPSC bioelastomer composites were prepared.[111]

The poly(glycerol-sebacate-citrate) (PGSC) bioelastomers were prepared by the thermal curing of the mouldable mixtures consisting of citric acid and poly(glycerol-sebacate) prepolymers at molar ratios of 4/4/1 and 4/4/0.6 (glycerol/sebacic acid/citric acid).[112,113] The synthesis of the PGSC bioelastomers is shown in Scheme 8.22. T_g of the PGSC bioelastomers ranged from −27.7 to −12 °C. The PGSC bioelastomers presented a rapid mass loss during 2 day *in vitro* degradation, but the mass losses were lower than 20% after 4 weeks; furthermore, the degradation could be adjusted by controlling the thermal-curing time and molar ratio, as shown in Figure 8.14. Multi-walled carbon nanotubes (MWCNTs) were introduced to prepare the MWCNTs/PGSC bioelastomer composites.[114]

The poly(1,8-octanediol-citrate-sebacate) (POCS) bioelastomers were prepared by condensation of 1,8-octanediol, citric acid and sebacic acid according to the molar ratio of 1:1 [1,8-octanediol/(citric acid + sebacic acid)].[115] The POCS prepolymers were first synthesized at 140–145 °C, then the prepolymers were dissolved in dioxane for obtaining prepolymer solutions; finally the prepolymer solutions were cast into the mould and left in an oven at 80 °C for solvent evaporation and further polyesterification to achieve the POCS bioelastomers. T_g of the POCS bioelastomers ranged from −37 to −7 °C.

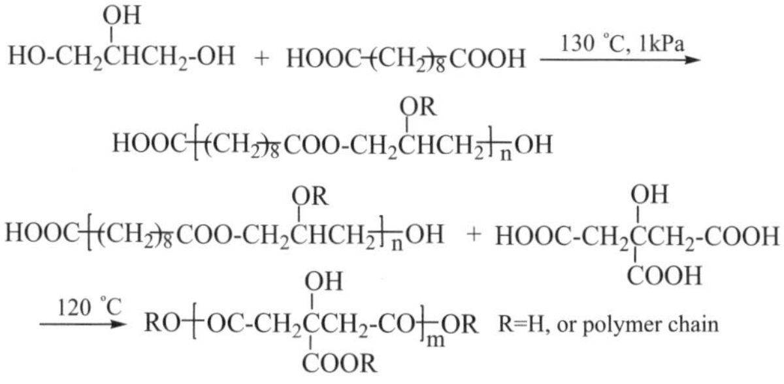

Scheme 8.22 Synthesis of the PGSC bioelastomers from glycerol, sebacic acid and citric acid.

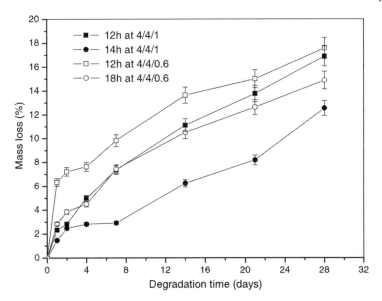

Figure 8.14 Mass loss–degradation time curves of the PGSC bioelastomers with different thermal-curing times at molar ratios of 4/4/1 and 4/4/0.6 in phosphate buffer solutions (pH = 7.4, 37 °C). Source: Liu et al.[112]

The POCS bioelastomers including more sebacic units in the molecular chains degraded more slowly in phosphate buffer solutions, and the mass losses could reach 64.1% after 6 h.

The urethane-doped POC bioelastomers were prepared as follows:[116] in the first step, citric acid and 1,8-octanediol were bulk polymerized at 140 °C at the molar ratio of 1:1.1 for achieving POC prepolymers; in the second step, the POC prepolymers were dissolved in 1,4-dioxane to form a 3% (wt/wt) solution, and reacted with HDI with stannous octoate as catalyst (0.1%wt) at 55 °C for preparing urethane-doped POC prepolymers at different feeding ratios of 1:0.6, 1:0.9 and 1:1.2 (POC prepolymer/HDI); in the third step, the urethane-doped POC prepolymers were cast into a Teflon mould and allowed to dry with a laminar airflow until all the solvents were evaporated, and then were further maintained in an oven at 80 °C to obtain the cured urethane-doped POC bioelastomers. T_g of the bioelastomers ranged from 0.64 to 5.20 °C. The urethane-doped POC bioelastomers with higher isocyanate content exhibited faster degradation rates, and their mass losses were lower than 16% after 2 months in phosphate buffer solutions.

8.3.7 Poly(ether ester) Bioelastomers

Poly(ether ester) bioelastomers are composed of polyether soft segments and polyester hard segments, and exhibit a thermoplastic property. They present a

Degradable Bioelastomers: Synthesis and Biodegradation 277

wide range of mechanical properties and adjustable biodegradability as good as those of the degradable SPU bioelastomers. Poly(ether ester) bioelastomers are commonly synthesized by polycondensation with titanate as catalysts, and the micro-phase separated structures are often formed in the materials especially when they possess high contents of hard segments and high molecular weights of soft segments.

The thermoplastic poly(ethylene glycol)/poly(butylene terephthalate) (PEG/PBT) is the most representative and mature poly(ether ester) bioelastomer, whose molecular structure is shown in Scheme 8.23, being usually synthesized by ester exchange methods.[117–119] T_g of the PEG/PBT bioelastomers is lower than 30 °C. Their in vitro and in vivo degradation happens both by hydrolysis and oxidation, in which hydrolysis is the main degradation mechanism.[120,121] The bioelastomers do not induce any adverse affects on the surrounding tissues and show a satisfactory biocompatibility. The PEG/PBT bioelastomers have been studied for use as skin substitutes,[122] elastomeric bioactive coatings on load-bearing dental and hip implants,[123] adhesion barriers,[124] bone replacements[125] and lysozyme delivery carriers.[126] When the phosphate groups are introduced into their molecular chains, another poly(ether ester) bioelastomer similar to the PEG/PBT bioelastomers are achieved, whose molecular structure is shown in Scheme 8.24, having been studied as nerve guide conduit materials[127] and drug delivery carriers.[128]

By bulk polymerization with a two-step melt polycondensation, a type of aliphatic poly(ether ester) bioclastomer called poly(butylene succinate)-co-poly(propylene glycol) (PBS-co-PPG) were prepared.[129] In the first step, succinic acid reacted with 1,4-butanediol at 180 °C until the theoretical amount of water was removed for achieving the desired PBS; in the second step, poly(propylene glycol) (PPG) was added into the reaction system, and the polycondensation was carried out at 220–230 °C with titanium butoxide as catalyst under the pressure of 10–15 Pa for obtaining the final PBS-co-PPG bioelastomers. The molecular structure of the poly(ether ester) bioelastomers is shown in Scheme 8.25. T_g of the bioelastomers ranged from −60.7 to −46.6 °C. As shown in Figure 8.15, the weight losses of the bioelastomers could reach 33%

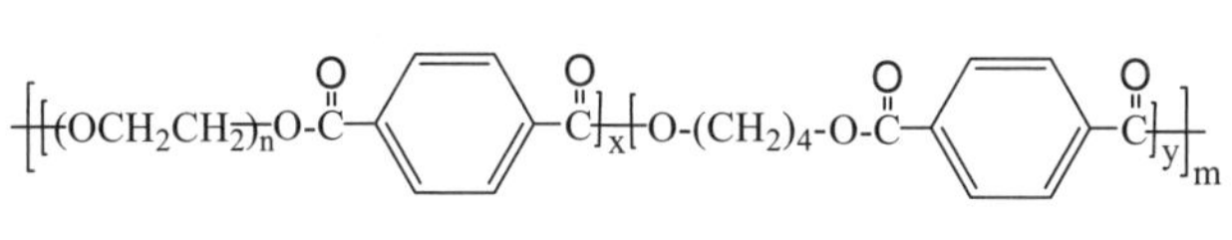

Scheme 8.23 Molecular structure of the poly(ethylene glycol)/poly(butylene terephthalate) (PEG/PBT) bioelastomers.

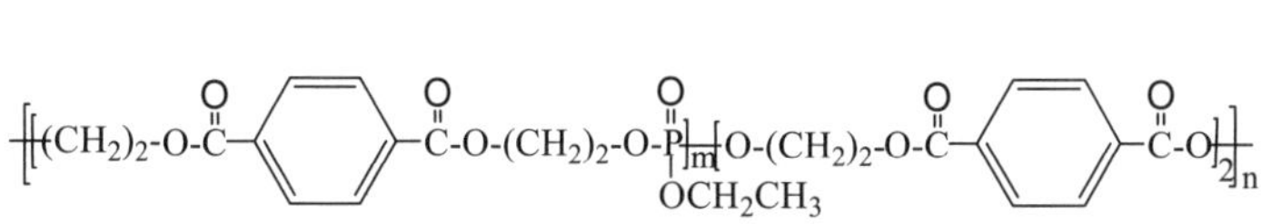

Scheme 8.24 Molecular structure of the phosphate-introduced poly(ether ester) bioelastomers.

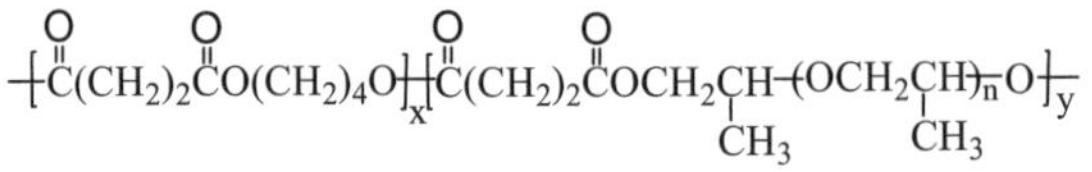

$$-\left[\overset{O}{\overset{\|}{C}}(CH_2)_2\overset{O}{\overset{\|}{C}}O(CH_2)_4O\right]_x\left[\overset{O}{\overset{\|}{C}}(CH_2)_2\overset{O}{\overset{\|}{C}}OCH_2\underset{CH_3}{\overset{}{C}}H(OCH_2\underset{CH_3}{\overset{}{C}}H)_{\overline{n}}O\right]_y$$

Scheme 8.25 Molecular structure of the poly(butylene succinate)-co-poly(propylene glycol) (PBS-co-PPG) bioelastomers.

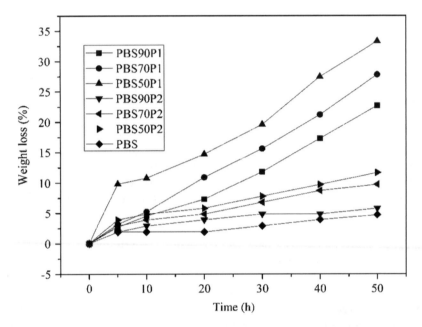

Figure 8.15 Weight loss of the PBS-co-PPG bioelastomers and PBS during degradation in phosphate buffer solutions (pH = 6.86, 45 °C): in the abbreviation of PBSxPy, "x", "P" and "y" separately represent the mass fraction of PBS such as 90%, PPG soft segment and molecular weight of PPG (1 stands for 1000 g mol^{-1}, and 2 stands for 2000 g mol^{-1}). Source: Huang et al.[129]

after 50 hour degradation in phosphate buffer solutions containing the lipase from *Pseudomonas*, and so degraded faster than PBS homopolyesters. Their weight losses increased as the mass fractions of PPG increased and the molecular weights of PPG decreased.

8.3.8 Poly(ester amide) Bioelastomers

Polyamides generally possess good mechanical properties and thermal stability due to the formation of strong hydrogen bonds between amide groups, while polyesters usually present good flexibility, solubility and hydrolysis ability. Biodegradable poly(ester amide) (PEA) materials are copolymers consisting of polyester and polyamide segments, simultaneously holding the characteristic of polyesters and polyamides, which are usually synthesized by the polycondensation of dicarboxylic acids, diamines, amino acids and diols of polyether or polyester.[130–135] Biodegradable PEA materials have been applied as tissue engineering scaffolds,[136] non-viral gene delivery reagents[137] and delivery carriers of proteins,[138] and their properties can be adjusted over a wide range by controlling the category and composition of polyester and polyamide segments. When biodegradable PEA materials show good elasticity, biocompatibility and biodegradation as biomaterials, they are called PEA bioelastomers.

Depending on the monomers that provide amide linkages, the PEA bioelastomers may be classified as either amino acid-related (such as glycine and lysine) and non-amino acid-related (such as aliphatic diamine and amino alcohol) PEA bioelastomers. The non-amino acid-related PEA bioelastomers usually degrade mainly by the hydrolysis of polyester segments, while the polyamide segments are comparatively stable. Amino acid units can be incorporated into the PEA molecular chains to create segments that are easily attacked by enzymes for improving the biodegradability of the PEA bioelastomers.

PEA bioelastomers made using the amino alcohol 1,3-diamino-2-hydroxypropane (DAHP) were synthesized as shown in Scheme 8.26.[139] DAHP, glycerol (G) or D,L-threitol (T), and sebacic acid (SA) at the molar ratio of 2:1:3 (DAHP/G/SA or DAHP/T/SA) were first added into a flask to react for 3 h at 120 °C under a nitrogen blanket; then they were allowed to react for another 9 h at 120 °C under a pressure of approximately 50 mTorr; finally, they were spread onto glass slides and cured for either 24 h or 48 h at 170 °C under approximately 50 mTorr for achieving poly(DAHP-G-SA) and poly(DAHP-T-SA) PEA bioelastomers. T_g of the two bioelastomers ranged from 33.7 °C to 48.0 °C. The poly(DAHP-G-SA) bioelastomers cured for 24 h and 48 h

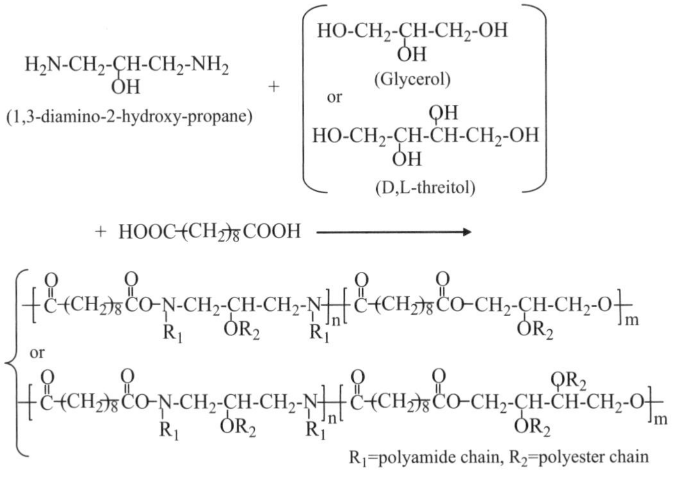

Scheme 8.26 Synthesis of the poly(ester amide) bioelastomers from 1,3-diamino-2-hydroxy-propane, glycerol or D,L-threitol and sebacic acid.

separately exhibited 97.0% and 44.3% mass losses after 6 week *in vitro* degradation (sodium acetate buffer solution, pH = 5.2, 37 °C), while the poly(DAHP-T-SA) bioelastomers cured for 24 h and 48 h lost 70.4% and 42.8% of mass, respectively. The *in vivo* degradation half-lives of the bioelastomers were up to 20 months.

8.3.9 Other Novel Degradable Bioelastomers

Bile acids have shown great promise for preparing biodegradable polymers that have been applied in controlled drug delivery.[140,141] The degradable main-chain bile acid-based bioelastomers were synthesized by entropy-driven ring-opening metathesis polymerization.[142] Two kinds of cyclic bile acids (38-membered ring and 35-membered ring) as shown in Scheme 8.27 were first synthesized in relatively high yields (73% and 59%, respectively) from their corresponding dienes at high dilution. Then, the bile acid-based bioelastomers with M_n of 58 700–151 500 g mol^{-1} as shown in Scheme 8.28 were prepared by entropy-driven ring-opening polymerization of the two cyclic bile acids at high concentrations using the highly efficient and stable second generation Grubbs catalyst. T_g values of the two corresponding bioelastomers achieved were separately 2.3 °C and 14.6 °C. Preliminary experiments demonstrated that the bioelastomers degraded slowly over a period of several months in phosphate buffer solutions.

The degradable polyester bioelastomers were prepared by the polycondensation of malic acid and 1,12-dodecandiol at the molar ratio of 1:1 and 1:2 (malic acid/1,12-dodecandiol).[143] The prepolymers from the two monomers were first synthesized at 140 °C, and then were poured into a mould for post curing at 160 °C to achieve the final bioelastomers. The mass losses of the bioelastomers with different curing times were lower than 9% after 30 day degradation in phosphate buffer solutions (pH = 7.4, 37 °C), and the long curing time caused their degradation rate to decrease.

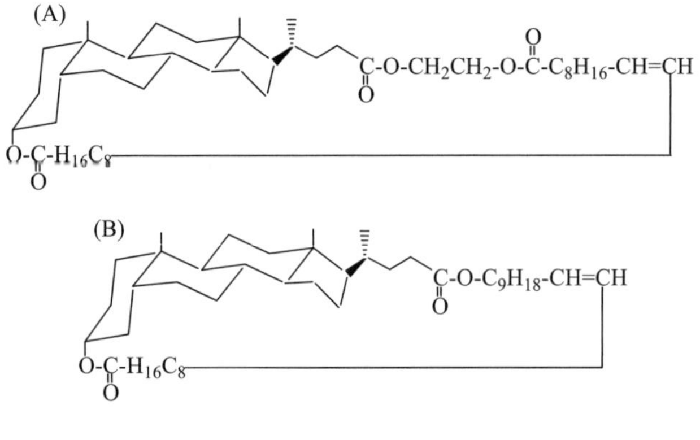

Scheme 8.27 Molecular structures of two kinds of cyclic bile acids: (A) 38-membered ring, (B) 35-membered ring.

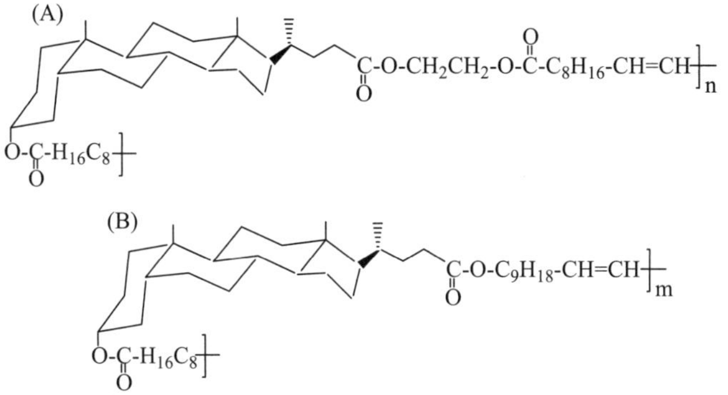

Scheme 8.28 Molecular structures of two kinds of bile acid-based bioelastomers: (A) from 38-membered cyclic bile acid, (B) from 35-membered cyclic bile acid.

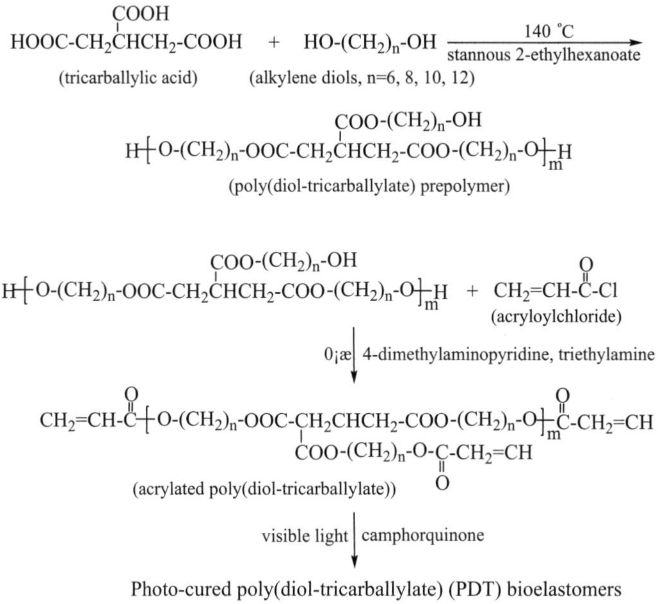

Scheme 8.29 Synthesis of the photo-cured poly(diol-tricarballylate) (PDT) bioelastomers from tricarballylic acid, alkylene diols and acryloylchloride.

The photo-cured poly(diol-tricarballylate) (PDT) degradable bioelastomers were prepared based on the polycondensation reaction between tricarballylic acid and alkylene diols such as 1,6-hexanediol, 1,8-octanediol, 1,10-decanediol and 1,12-dodecanediol, followed by acrylation and photo-cross-linking, as shown in Scheme 8.29.[144] The PDT prepolymers, such as poly(1,6-hexane diol-co-tricarballylate) (PHT), poly(1,8-octane diol-co-tricarballylate) (POT), poly(1,10-decane diol-co-tricarballylate) (PDET) and poly(1,12-dodecane diol-co-tricarballylate) (PDDT), were first synthesized at 140 °C with stannous 2-ethylhexanoate as catalyst under vacuum (50.8 mmHg). Then, acrylation of the PDT prepolymers was carried out by reacting acryloyl chloride with the terminal hydroxyl groups of the prepolymers in the presence of triethylamine and 4-dimethylaminopyridine in an ice bath. Finally, the acrylated poly(diol-tricarballylate) was cured by exposing it to visible light using camphorquinone as photo-initiator for achieving the final bioelastomers. T_g of the PHT, POT, PDET and PDDT bioelastomers were respectively –32 °C, –25 °C, –24 °C and –19 °C. The degradation tests in phosphate buffer solutions demonstrated that their weight losses were lower than 35% after 12 weeks, and the PDT bioelastomers from the alkylene diols with shorter chains demonstrated the lower weight losses.

The dual cured poly(octamethylene maleate (anhydride) citrate) (POMaC) degradable bioelastomers from maleic anhydride, citric acid and 1,8-octanediol were prepared as shown in Scheme 8.30; they could be cross-linked by the carbon–carbon linkage using UV irradiation and/or by ester bonding using polycondensation.[145] The POMaC prepolymers were first synthesized by carrying out a controlled condensation of the three monomers for 3 h at 140 °C under a nitrogen atmosphere according to the total molar ratio of 1:1 (acid/diol). Then, the POMaC prepolymers were cured with 2-hydroxy-1-[4(hydroxyethoxy)phenyl]-2-methyl-1 propanone (Irgacure 2959) (1 wt%) as photo-initiator in dimethyl sulfoxide by UV light, or they were cross-linked by post-polymerization at 80 °C for achieving the POMaC bioelastomers. The photo-cured POMaC bioelastomers with higher maleic anhydride ratios resulted in slower degradation rates, and the additional cross-linking through ester bond formation also resulted in slower degradation rates. For example, the photo-cured POMaC bioelastomer at the molar ratio of 6:4 (citric acid/maleic anhydride) was shown to degrade 77.50% after 10 weeks, while the dual cured POMaC bioelastomer at the molar ratio of 2:8 degraded only 18.45%.

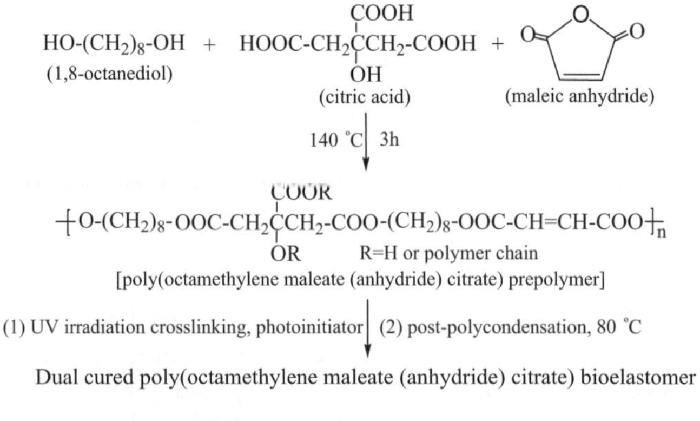

COOH
|
HO-(CH$_2$)$_8$-OH + HOOC-CH$_2$CCH$_2$-COOH + (maleic anhydride)
(1,8-octanediol) |
                        OH
                   (citric acid)

140 °C | 3h

COOR
|
$+$O-(CH$_2$)$_8$-OOC-CH$_2$CCH$_2$-COO-(CH$_2$)$_8$-OOC-CH=CH-COO$+_n$
    OR   R=H or polymer chain
[poly(octamethylene maleate (anhydride) citrate) prepolymer]

(1) UV irradiation crosslinking, photoinitiator | (2) post-polycondensation, 80 °C

Dual cured poly(octamethylene maleate (anhydride) citrate) bioelastomer

Scheme 8.30 Synthesis of the dual cured poly(octamethylene maleate (anhydride) citrate) bioelastomers from maleic anhydride, citric acid and 1,8-octanediol.

The poly(triol α-ketoglutarate) degradable bioelastomers such as poly(glycerol α-ketoglutarate) (PGa), poly(1,2,4-butanetriol α-ketoglutarate) (PBa) and poly(1,2,6-hexanetriol α-ketoglutarate) (PHa) were prepared by the polycondensation of α-ketoglutaric acid and the corresponding triols of glycerol, 1,2,4-butanetriol and 1,2,6-hexanetriol as shown in Scheme 8.31.[146] The poly(triol α-ketoglutarate) prepolymers were first synthesized by condensation of α-ketoglutaric acid and one triol according to the molar ratio of 1:1 (α-ketoglutaric acid/diol) for 1 h at 125 °C, and then the poly(triol α-ketoglutarate) bioelastomers were achieved by curing the prepolymers at either 60 °C, 90 °C or 120 °C for times ranging from 6 h to 7 days. The poly(triol α-ketoglutarate) bioelastomers could be easily modified through the reaction between ketone groups and oxyamines, hydrazines or hydrazides at physiological conditions without catalysts and co-reagents. The poly(triol α-ketoglutarate) bioelastomers all exhibited 100% mass losses in relatively short periods lower than 28 days in phosphate buffer solutions, and a wide range of degradation rates could be obtained by controlling the curing conditions, as seen from Figure 8.16.

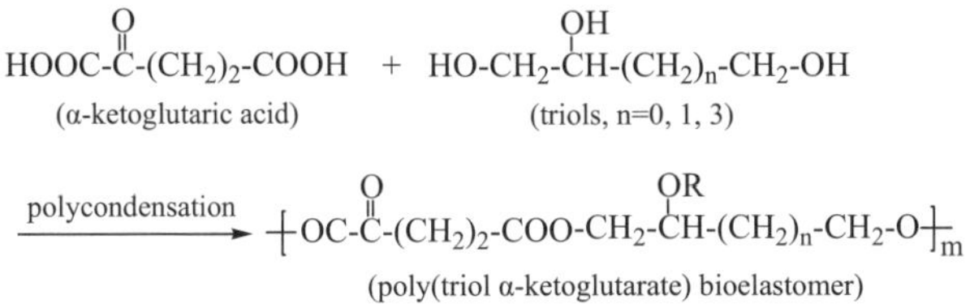

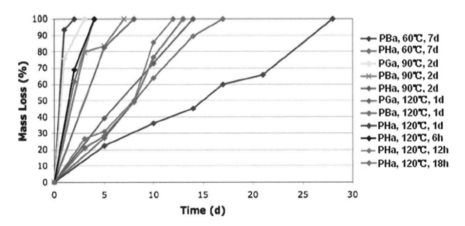

Scheme 8.31 Synthesis of the poly(triol α-ketoglutarate) bioelastomers from α-ketoglutaric acid and triols of glycerol, 1,2,4-butanetriol and 1,2,6-hexanetriol.

Figure 8.16 Mass loss of the poly(triol α-ketoglutarate) bioelastomers such as PGa, PBa and PHa during degradation in phosphate buffer solutions (pH = 7.4, 37 °C) under the conditions of different curing temperature and time. Adapted from Barrett and Yousaf.[146]

8.4 Conclusions

It can be seen that degradable bioelastomers such as degradable SPU bioelastomers, PCL-related bioelastomers, PLA-related bioelastomers, PC-related bioelastomers, PGS bioelastomer and its derivatives, citric acid-related polyester bioelastomers, poly(ether ester) bioelastomers, PEA bioelastomers and so on, have been developed rapidly as an important branch of degradable polymeric biomaterials, which are mainly applied in soft tissue engineering and controlled drug delivery. They can be prepared by varied synthesis methods such as polycondensation, ring-opening polymerization, thermally and photo-initiated radical polymerization. It will be popular to simultaneously adopt several synthesis methods to prepare a kind of bioelastomer, which may endow the bioelastomer with very good designability. Biodegradation is a very important property of degradable bioelastomers, which is often discussed in the studies. The degradation rates of degradable bioelastomers are usually able to be adjusted in a wide range by controlling their molecular structure, segment length and composition, molar ratio of monomers, reaction temperature, polymerization time, curing method and degradation condition, *etc*.

The SPU, PCL-related, PLA-related and PC-related bioelastomers, possessing the greatest design potential and most variable biodegradability, are the most feasible biomaterials to be largely produced by varied mature methods and technologies, and to be broadly applied in biomedical fields. Traditional poly(ether ester) and PEA bioelastomers have been used in biomedical fields, but the novel poly(ether ester) and PEA bioelastomers are less studied. The PGS bioelastomer and its derivatives, and citric acid-related polyester bioelastomers are being developed at a very fast rate, which seem to stand for the most active research aspect of degradable bioelastomers because the preparation methods usually are relatively more direct and simple. Other novel bioelastomers such as bile acid-based, PDT, POMaC and poly(triol α-ketoglutarate) bioelastomers have the most potential to be deeply researched in biomedical fields, and can be synthesized by some unique methods. In the future, degradable bioelastomers will be sure to play a more important role in degradable polymeric biomaterials.

Acknowledgements

This work was supported by Project of National Science Foundation for Young Scientists of China (Grant No. 51003003) and Key Project of Natural Science Foundation of China (Grant No. 50933001).

References

1. M. Okada, *Prog. Polym. Sci.*, 2002, **27**, 87.
2. W. Amass, A. Amass and B. Tighe, *Polym. Int.*, 1998, **47**, 89.
3. J. C. Middleton and A. J. Tipton, *Biomaterials*, 2000, **21**, 2335.

4. J. P. Fisher, T. A. Holland, D. Dean and A. G. Mikos, *Biomacromolecules*, 2003, **4**, 1335.
5. Y. Iwasaki, C. Nakagawa, M. Ohtomi, K. Ishihara and K. Akiyoshi, *Biomacromolecules*, 2004, **5**, 1110.
6. Y. Baimark, R. Molloy, N. Molloy, J. Siripitayananon, W. Punyodom and M. Sriyai, *J. Mater. Sci. Mater. Med.*, 2005, **16**, 699.
7. B. Naeye, K. Raemdonck, K. Remaut, B. Sproat, J. Demeester and S. C. De Smedt, *Eur. J. Pharm. Sci.*, 2010, **40**, 342.
8. J. H. Liao, X. A. Guo, D. Nelson, F. K. Kasper and A. G. Mikos, *Acta Biomaterialia*, 2010, **6**, 2386.
9. L. E. Niklason, J. Gao, W. M. Abbott, K. K. Hirschi, S. Houser, R. Marini and R. Langer, *Science*, 1999, **284**(5413), 489.
10. B. S. Kim and D. J. Mooney, *J. Biomech. Eng.*, 2000, **122**, 210.
11. S. D. Waldman, C. G. Spiteri, M. D. Grynpas, R. M. Pilliar and R. A. Kandel, *Tissue Eng.*, 2004, **10**(9–10), 1323.
12. B. Amsden and Y. L. Cheng, *J. Control. Release*, 1995, **33**, 99.
13. R. Schirrer, P. Thepin and G. Torres, *J. Mater. Sci.*, 1992, **27**, 3424.
14. D. F. Williams, *Biomaterials*, 2009, **30**, 5897.
15. A. de Mel, G. Jell, M. M. Stevens and A. M. Seifalian, *Biomacromolecules*, 2008, **9**, 2969.
16. J. E. Puskas and Y. H. Chen, *Biomacromolecules*, 2004, **5**, 1141.
17. D. E. T. Shepherd and B. B. Seedhom, *Rheumatology*, 1999, **38**, 124.
18. A. Thambyah, A. Nather and J. Goh, *Osteoarthritis Cartilage*, 2006, **14**, 580.
19. B. S. Gupta and V. A. Kasyanov, *J. Biomed. Mater. Res.*, 1997, **34**, 341.
20. A. Balguid, M. P. Rubbens, A. Mol, R. A. Bank, A. J. J. C. Bogers, J. P. Van Kats, B. A. J. M. De Mol, F. P. T. Baaijens and C. V. C. Bouten, *Tissue Eng.*, 2007, **13**, 1501.
21. J. M. Lee and D. R. Boughner, *Circ. Res.*, 1985, **57**, 475.
22. K. L. Monson, W. Goldsmith, N. M. Barbaro and G. T. Manley, *J. Biomech. Eng.*, 2003, **125**, 288.
23. A. A. Deschamps, D. W. Grijpma and J. Feijen, *J. Biomater. Sci. Polym. Ed.*, 2002, **13**, 1337.
24. A. A. Deschamps, D. W. Grijpma and J. Feijen, *Polymer*, 2001, **42**, 9335.
25. M. D. Lang, R. P. Wong and C. C. Chu, *J. Polym. Sci. Part A: Polym. Chem.*, 2002, **40**, 1127.
26. M. A. Carnahan and M. W. Grinstaff, *J. Am. Chem. Soc.*, 2001, **123**, 2905.
27. G. G. Pitt, M. M. Gratzl, G. L. Kimmel, J. Surles and A. Sohindler, *Biomaterials*, 1981, **2**, 215.
28. R. F. Storey and T. P. Hickey, *Polymer*, 1994, **35**, 830.
29. R. Yoda, *J. Biomater. Sci. Polym. Ed.*, 1998, **9**, 561.
30. A. Gopferich, *Biomaterials*, 1996, **17**, 103.
31. J. P. Santerre, K. Woodhouse, G. Laroche and R. S. Labow, *Biomaterials*, 2005, **26**, 7457.
32. M. Martina and D. W. Hutmacher, *Polym. Int.*, 2007, **56**, 145.

33. C. W. Tabor and H. Tabor, *Annu. Rev. Biochem.*, 1984, **53**, 749.
34. M. Cooke, N. Leeves and C. White, *Arch. Oral. Biol.*, 2003, **48**, 323.
35. H. P. Til, H. E. Falke, M. K. Prinsen and M. I. Willems, *Food Chem. Toxicol.*, 1997, **35**(3–4), 337.
36. J. Y. Zhang, E. J. Beckman, J. Hu, G. G. Yang, S. Agarwal and J. O. Hollinger, *Tissue Eng.*, 2002, **8**, 771.
37. R. G. J. C. Heijkants, R. V. van Calck, T. G. van Tienen, J. H. de Groot, P. Buma, A. J. Pennings, R. P. H. Veth and A. J. Schouten, *Biomaterials*, 2005, **26**, 4219.
38. T. G. Tienen, R. G. J. C. Heijkants, J. H. de Groot, A. J. Schouten, A. J. Pennings, R. P. H. Veth and P. Buma, *J. Biomed. Mater. Res. Part B: Appl. Biomater.*, 2006, **76B**, 389.
39. J. J. Guan, M. S. Sacks, E. J. Beckman and W. R. Wagner, *J. Biomed. Mater. Res.*, 2002, **61**, 493.
40. J. Guan, J. J. Stankus and W. R. Wagner, *J. Control. Release*, 2007, **120**(1–2), 70.
41. J. J Guan and W. R. Wagner, *Biomacromolecules*, 2005, **6**, 2833.
42. J. J. Guan, M. S. Sacks, E. J. Beckman and W. R. Wagner, *Biomaterials*, 2004, **25**, 85.
43. K. D. Kavlock, T. W. Pechar, J. O. Hollinger, S. A. Guelcher and A. S. Goldstein, *Acta Biomaterialia*, 2007, **3**, 475.
44. W. S. Wang, P. Ping, H. J. Yu, X. S Chen and X. B. Jing, *J. Polym. Sci. Part A: Polym. Chem.*, 2006, **44**, 5505.
45. M. K. Hassan, K. A. Mauritz, R. F. Storey and J. S. Wiggins, *J. Polym. Sci. Part A: Polym. Chem.*, 2006, **44**, 2990.
46. G. R. Saad, Y. J. Lee and H. Seliger, *Macromol. Biosci.*, 2001, **1**, 91.
47. M. M. Ding, J. H. Li, X. T. Fu, J. Zhou, H. Tan, Q. Gu and Q. Fu, *Biomacromolecules*, 2009, **10**, 2857.
48. X. Jiang, J. H. Li, M. M. Ding, H. Tan, Q. Y. Ling, Y. P. Zhong and Q. Fu, *Eur. Polym. J.*, 2007, **43**, 1838.
49. J. Xie, M. Ihara, Y. Jung, I. K. Kwon, S. H. Kim, Y. H. Kim and T. Matsuda, *Tissue Eng.*, 2006, **12**, 449.
50. S. I. Jeong, S. H. Kim, Y. H. Kim, Y. Jung, J. H. Kwon, B. S. Kim and Y. M. Lee, *J. Biomater. Sci. Polym. Ed.* 2004, **15**, 645.
51. M. Nagata and Y. Sato, *Polymer*, 2004, **45**, 87.
52. M. Nagata, K. Kato, W. Sakai and N. Tsutsumi, *Macromol. Biosci.*, 2006, **6**, 333.
53. J. Y. Shen, X. Y. Pan, C. H. Lim, M. B. Chan-Park, X. Zhu and R. W. Beuerman, *Biomacromolecules*, 2007, **8**, 376.
54. O. Jeon, S. H. Lee, S. H. Kim, Y. M. Lee and Y. H. Kim, *Macromolecules*, 2003, **36**, 5585.
55. D. Cohn and A. F. Salomon, *Biomaterials*, 2005, **26**, 2297.
56. H. M. Younes, E. Bravo-Grimaldo and B. G. Amsden, *Biomaterials*, 2004, **25**, 5261.
57. B. Amsden, S. Wang and U. Wyss, *Biomacromolecules*, 2004, **5**, 1399.

58. B. G. Amsden, G. Misra, F. Gu and H. M. Younes, *Biomacromolecules*, 2004, **5**, 2479.
59. A. C. Albertsson and I. K. Varma, *Biomacromolecules*, 2003, **4**, 1466.
60. Y. Ikada and H. Tsuji, *Macromol. Rapid Commun.*, 2000, **21**, 117.
61. O. Dechy-Cabaret, B. Martin-Vaca and D. Bourissou, *Chem. Rev.*, 2004, **104**, 6147.
62. D. Cohn and A. Hotovely-Salomon, *Polymer*, 2005, **46**, 2068.
63. C. L. Wanamaker, L. E. O'Leary, N. A. Lynd, M. A. Hillmyer and W. B. Tolman, *Biomacromolecules*, 2007, **8**, 3634.
64. C. L. Wanamaker, W. B. Tolman and M. A. Hillmyer, *Biomacromolecules*, 2009, **10**, 443.
65. S. Hiki, M. Miyamoto and Y. Kimura, *Polymer*, 2000, **41**, 7369.
66. Z. Zhang, D. W. Grijpma and J. Feijen, *Macromol. Chem. Phys.*, 2004, **205**, 867.
67. J. H. Kim and J. H. Lee, *Polym J.*, 2002, **34**, 203.
68. E. M. Frick and M. A. Hillmyer, *Macromol. Rapid Commun.*, 2000, **21**, 1317.
69. E. M. Frick, A. S. Zalusky and M. A. Hillmyer, *Biomacromolecules*, 2003, **4**, 216.
70. C. Y. Ba, J. Yang, Q. H. Hao, X. Y. Liu and A. Cao, *Biomacromolecules*, 2003, **4**, 1827.
71. L. Jia, L. Z. Yin, Y. Li, Q. B. Li, J. Yang, J. Y. Yu, Z. Shi, Q. Fang and A. Cao, *Macromol. Biosci.*, 2005, **5**, 526.
72. A. P. Pego, C. L. A. M. Vleggeert-Lankamp, M. Deenen, E. A. J. F. Lakke, D. W. Grijpma, A. A. Poot, E. Marani and J. Feijen, *J. Biomed. Mater. Res. Part A*, 2003, **67A**, 876.
73. A. P. Pego, B. Siebum, M. J. A. Van Luyn, X. J. G. Y. Van Seijen, A. A. Poot, D. W. Grijpma and J. Feijen, *Tissue Eng.*, 2003, **9**, 981.
74. K. S. Bisht, Y. Y. Svirkin, L. A. Henderson, R. A. Gross, D. L. Kaplan and G. Swift, *Macromolecules*, 1997, **30**, 7735.
75. Y. Takahashi and R. Kojima, *Macromolecules*, 2003, **36**, 5139.
76. J. Yang, Q. H. Hao, X. Y. Liu, C. Y. Ba and A. Cao, *Biomacromolecules*, 2004, **5**, 209.
77. M. Nagata, T. Tanabe, W. Sakai and N. Tsutsumi, *Polymer*, 2008, **49**, 1506.
78. Q. P. Hou, D. W. Grijpma and J. Feijen, *Acta Biomaterialia*, 2009, **5**, 1543.
79. A. V. Grego and G. Mingrone, *Clin. Nutr.*, 1995, **14**, 143.
80. P. B. Mortensen, *Biochim. Biophys. Acta*, 1981, **664**, 349.
81. J. Fu, J. Fiegel, E. Krauland and J. Hanes, *Biomaterials*, 2002, **23**, 4425.
82. Y. D. Wang, G. A. Ameer, B. J. Sheppard and R. Langer, *Nat. Biotech.*, 2002, **20**, 602.
83. D. Motlagh, J. Yang, K. Y. Lui, A. R. Webb and G. A. Ameer, *Biomaterials*, 2006, **27**, 4315.
84. C. J. Bettinger, E. J. Weinberg, K. M. Kulig, J. P. Vacanti, Y. D. Wang, J. T. Borenstein and R. Langer, *Adv. Mater.*, 2006, **18**, 165.

85. J. Gao, A. E. Ensley, R. M. Nerem and Y. D. Wang, *J. Biomed. Mater. Res.*, 2007, **83A**, 1070.
86. Q. Z. Chen, A. Bismarck, U. Hansen, S. Junaid, M. Q. Tran, S. E. Harding, N. N. Ali and A. R. Boccaccini, *Biomaterials*, 2008, **29**, 47.
87. C. J. Bettinger, B. Orrick, A. Misra, R. Langer and J. T. Borenstein, *Biomaterials*, 2006, **27**, 2558.
88. W. L. Neeley, S. Redenti, H. Klassen, S. Tao, T. Desai, M. J. Young and R. Langer, *Biomaterials*, 2008, **29**, 418.
89. S. Redenti, W. L. Neeley, S. Rompani, S. Saigal, J. Yang, H. Klassen, R. Langer and M. J. Young, *Biomaterials*, 2009, **30**, 3405.
90. Z. J. Sun, C. Chen, M. Z. Sun, C. H. Ai, X. L. Lu, Y. F. Zheng, B. F. Yang and D. L. Dong, *Biomaterials*, 2009, **30**, 5209.
91. Y. D. Wang, Y. M. Kim and R. Langer, *J. Biomed. Mater. Res. Part A*, 2003, **66A**, 192.
92. Q. Y. Liu, M. Tian, T. Ding, R. Shi and L. Q. Zhang, *J. Appl. Polym. Sci.*, 2005, **98**, 2033.
93. Q. Y. Liu, M. Tian, T. Ding, R. Shi, Y. X. Feng, L. Q. Zhang, D. F. Chen and W. Tian, *J. Appl. Polym. Sci.*, 2007, **103**, 1412.
94. Q. Y. Liu, M. Tian, R. Shi, L. Q. Zhang, D. F. Chen and W. Tian, *J. Appl. Polym. Sci.*, 2007, **104**, 1131.
95. C. L. E. Nijst, J. P. Bruggeman, J. M. Karp, L. Ferreira, A. Zumbuehl, C. J. Bettinger and R. Langer, *Biomacromolecules*, 2007, **8**, 3067.
96. J. L. Ifkovits, R. F. Padera and J. A. Burdick, *Biomed. Mater.*, 2008, **3**, 034104.
97. J. P. Bruggeman, B. J. de Bruin, C. J. Bettinger and R. Langer, *Biomaterials*, 2008, **29**, 4726.
98. Z. J Sun, L. Wu, X. L. Lu, Z. X. Meng, Y. F. Zheng and D. L. Dong, *Appl. Surface Sci.*, 2008, **255**, 350.
99. Z. J. Sun, L. Wu, W. Huang, X. L. Zhang, X. L. Lu, Y. F. Zheng, B. F. Yang and D. L. Dong, *Mater. Sci. Eng. C*, 2009, **29**, 178.
100. J. Yang, A. R. Webb and G. A. Ameer, *Adv. Mater.*, 2004, **16**, 511.
101. J. Yang, D. Motlagh, J. B. Allen, A. R. Webb, M. R. Kibbe, O. Aalami, M. Kapadia, T. J. Carroll and G. A. Ameer, *Adv. Mater.*, 2006, **18**, 1493.
102. D. Motlagh, J. Allen, R. Hoshi, J. Yang, K. Lui and G. Ameer, *J. Biomed. Mater. Res. Part A*, 2007, **82A**, 907.
103. Y. Kang, J. Yang, S. Khan, L. Anissian and G. A. Ameer, *J. Biomed. Mater. Res. Part A*, 2006, **77A**, 331.
104. L. A. Hidalgo-Bastida, J. J. A. Barry, N. M. Everitt, F. R. A. J. Rose, L. D. Buttery, I. P. Hall, W. C. Claycomb and K. M. Shakesheff, *Acta Biomaterialia*, 2007, **3**, 457.
105. X. Q. Zhang, H. H. Tang, R. Hoshi, L. De Laporte, H. J. Qiu, X. Y. Xu, L. D. Shea and G. A. Ameer, *Biomaterials*, 2009, **30**, 2632.
106. H. J. Qiu, J. Yang, P. Kodali, J. Koh and G. A. Ameer, *Biomaterials*, 2006, **27**, 5845.
107. J. Yang, A. R. Webb, S. J. Pickerill, G. Hageman and G. A. Ameer, *Biomaterials*, 2006, **27**, 1889.

108. H. C. Zhao and G. A. Ameer, *J. Appl. Polym. Sci.*, 2009, **114**, 1464.
109. T. Ding, Q. Y. Liu, R. Shi, M. Tian, J. Yang and L. Q. Zhang, *Polym. Degrad. Stab.*, 2006, **91**, 733.
110. L. J. Lei, T. Ding, R. Shi, Q. Y. Liu, L. Q. Zhang, D. F. Chen and W. Tian, *Polym. Degrad. Stab.*, 2007, **92**, 389.
111. L. J. Lei, L. Li, L. Q. Zhang, D. F. Chen and W. Tian, *Polym. Degrad. Stab.*, 2009, **94**, 1494.
112. Q. Y. Liu, T. W. Tan, J. Y. Weng and L. Q. Zhang, *Biomed. Mater.*, 2009, **4**, 025015.
113. Q. Y. Liu, S. Z. Wu, T. W. Tan, J. Y. Weng, L. Q. Zhang, L. Liu, W. Tian and D. F. Chen, *J. Biomater. Sci. Polym. Ed.*, 2009, **20**, 1567.
114. Q. Y. Liu, J. Y. Wu, T. W. Tan, L. Q. Zhang, D. F. Chen and W. Tian, *Polym. Degrad. Stab.*, 2009, **94**, 1427.
115. I. Djordjevic, N. R. Choudhury, N. K. Dutta and S. Kumar, *Polymer*, 2009, **50**, 1682.
116. J. Dey, H. Xu, J. H. Shen, P. Thevenot, S. R. Gondi, K. T. Nguyen, B. S. Sumerlin, L. P. Tang and J. Yang, *Biomaterials*, 2008, **29**, 4637.
117. S. Fakirov and T. Gogeva, *Makromol. Chem.*, 1990, **191**, 603.
118. S. Fakirov, C. Fakirov, E. W. Fischer and M. Stamm, *Polymer*, 1991, **32**, 1173.
119. S. Fakirov, C. Fakirov, E. W. Fischer and M. Stamm, *Polymer*, 1992, **33**, 3818.
120. R. J. B. Sakkers, R. A. J. Dalmeyer, J. R. Wijn and C. A. Blitterswijk, *J. Biomed. Mater. Res.*, 2000, **49**, 312.
121. H. Hayen, A. A. Deschamps, D. W. Grijpma, J. Feijen and U. Karst, *J. Chromatogr. A*, 2004, **1029**(1–2), 29.
122. G. J. Beumer, C. A. Van Blitterswijk and M. Ponec, *J. Biomed. Mater. Res.*, 1994, **28**, 545.
123. A. M. Radder, H. Leenders and C. A. van Blitterswijk, *Biomaterials*, 1994, **16**, 507.
124. E. A. Bakkum, J. B. Trimbos, R. A. J. Dalmeyer and C. A. Blitterswijk, *J. Mater. Sci. Mater. Med.* 1995, **6**, 41.
125. A. M. Radder, H. Leenders and C. A. van Blitterswijk, *J. Biomed. Mater. Res.*, 1996, **30**, 341.
126. J. M. Bezemer, R. Radersama, D. W. Grijpma, P. J. Dijkstra, J. Feijen and C. A. van Blitterswijk, *J. Control. Release*, 2000, **64**(1–3), 179.
127. S. Wang, A. C. A. Wan, X. Y. Xu, S. J. Gao, H. Q. Mao, K. W. Leong and H. Yu, *Biomaterials*, 2001, **22**, 1157.
128. Z. Zhao, J. Wang, H. Q. Mao and K. W. Leong, *Adv. Drug Deliv. Rev.*, 2003, **55**, 483.
129. X. Huang, C. C. Li, L. C. Zheng, D. Zhang, G. H. Guan and Y. N. Xiao, *Polym. Int.*, 2009, **58**, 893.
130. N. Arabuli, G. Tsitlanadze, L. Edilashvili, D. Kharadze, T. Goguadze and V. Beridze, *Macromol. Chem. Phys.*, 1994, **195**, 2279.
131. M. X. Li, R. X. Zhuo and F. Q. Qu, *J. Polym. Sci. Part A: Polym. Chem.*, 2002, **40**, 4550.

132. H. L. Guan, C. Deng, X. Y. Xu, Q. Z. Liang, X. S. Chen and X. B. Jing, *J. Polym. Sci. Part A: Polym. Chem.*, 2005, **43**, 1144.
133. L. Y. Wang, Y. J. Wang and L. Ren, *J. Appl. Polym. Sci.*, 2008, **109**, 1310.
134. M. X. Deng, J. Wu, C. A. Reinhart-King and C. C. Chu, *Biomacromolecules*, 2009, **10**, 3037.
135. K. Guo and C. C. Chu, *J. Appl. Polym. Sci.*, 2010, **117**, 3386.
136. K. Hemmrich, J. Salber, M. Meersch, U. Wiesemann, T. Gries, N. Pallua and D. Klee, *J. Mater. Sci. Mater. Med.*, 2008, **19**, 257.
137. D. Yamanouchi, J. Wu, A. N. Lazar, K. C. Kent, C. C. Chu and B. Liu, *Biomaterials*, 2008, **29**, 3269.
138. J. M. Bezemer, P. O. Weme, D. W. Grijpma, P. J. Dijkstra, C. A. van Blitterswijk and J. Feijen, *J. Biomed. Mater. Res.*, 2000, **51**, 8.
139. C. J. Bettinger, J. P. Bruggeman, J. T. Borenstein and R. S. Langer, *Biomaterials*, 2008, **29**, 2315.
140. V. Janout and S. L. Regen, *J. Am. Chem. Soc.*, 2005, **127**, 22.
141. J. H. Park, S. Kwon, J. O. Nam, R. W. Park, H. Chung, S. B. Seo, I. S. Kim, I. C. Kwon and S. Y. Jeong, *J. Control. Release*, 2004, **95**, 579.
142. J. E. Gautrot and X. X. Zhu, *Angew. Chem. Int. Ed.*, 2006, **45**, 6872.
143. L. Y. Lee, S. C. Wu, S. S. Fu, S. Y. Zeng, W. S. Leong and L. P. Tan, *Eur. Polym. J.*, 2009, **45**, 3249.
144. M. A. Shaker, J. J. E. Dore and H. M. Younes, *J. Biomater. Sci.*, 2010, **21**, 507.
145. R. T. Tran, P. Thevenot, D. Gyawali, J. C. Chiao, L. P. Tang and J. Yang, *Soft Matter*, 2010, **6**, 2449.
146. D. G. Barrett and M. N. Yousaf, *Macromolecules*, 2008, **41**, 6347.

CHAPTER 9
Functionalization of Poly(L-lactide) and Applications of the Functionalized Poly(L-lactide)

XIULI HU AND XIABIN JING

State Key Laboratory of Polymer Physics and Chemistry, Changchun Institute of Applied Chemistry, Chinese Academy of Sciences, Changchun 130022, P. R. China

9.1 Introduction

The term 'biodegradable polymer' refers to a class of polymer materials that can be degraded or hydrolysed *in vivo* into small molecules, which can be absorbed through metabolism or excreted by the body, or destroyed by microorganisms in the environment. With the increasing severity of environmental problems and awareness of environmental protection, the disposal problem of the non-degradable petroleum-based plastics has raised the demand for biodegradable polymers as means of reducing the environmental impact. Biodegradable materials and environmentally friendly products are becoming one of research hotspots nowadays.

Among the biodegradable polymers under investigation, poly(lactic acid) (PLA) has received the most attention. Its prominent role is based on the following reasons: (1) its raw material, L-lactic acid, can be efficiently produced by fermentation from renewable resources such as starch and other polysaccharides, which are easily available from corn, sugar beet, sugar cane, potatoes and other biomasses; (2) PLA is biodegradable and biocompatible and has represented its potential applications in a number of growing technologies such as drug delivery, sutures, orthopaedics and temporary matrixes or tissue engineering scaffolds; (3) PLA has high strength and has excellent shaping and moulding ability so that it can be easily processed by conventional processing techniques like injection moulding, blow moulding, thermoforming and extrusion; (3) additionally, PLA can be obtained in large-scale efficiently.

Since the first preparation of PLA by Carothers in 1932,[1] people have done a lot of research to improve the synthesis of PLA and the following four technologies have been developed: direct polycondensation, azeotropic condensation polymerization, solid state polymerization and ring-opening polymerization (ROP). Among them, ROP is the most commonly studied one due to the possibility of accurate control of chemistry and thus varying the properties of the resulting polymers in a more controlled manner. It is usually employed for the synthesis of polymers with high molecular weight. ROP is completed generally in two steps (Scheme 9.1): firstly, lactic acid is condensed to PLA oligomers, which are allowed to undergo cracking and cyclization at high temperature and low pressure in the presence of catalyst to give lactides (LAs); secondly, high molecular weight PLA is obtained by ROP of LA in the presence of catalyst.

Although PLA has been increasingly used in biomedical applications, including degradable sutures, drug carriers, implant materials and tissue engineering scaffolds,[2] its hydrophobic and semicrystalline properties and absence of functional or reactive groups along the polymer backbone limit its medical applications: (1) PLA is built up by ester bonds and its poor hydrophilic properties reduce its biocompatibility, which may lead to rejection, inflammation, infection and other body tissue necrosis and thrombosis *in vivo*;[2,3] (2) PLA is a kind of linear polyester, and in some applications it does not have the required strength, it is brittle, its thermal deformation temperature is low and it has poor impact resistance; (3) its degradation is difficult to control; (4) PLA is comparatively expensive. All these disadvantages urge scientists to do a lot of research on the modification of PLA. In addition to blending with other polymers and adding modifying agents, functionalization of PLA *via* polymerization and chemical reactions is an important approach to PLA modification. It involves incorporating monomer units with appropriate functionality into the PLA backbones, or reacting its terminal hydroxyl groups

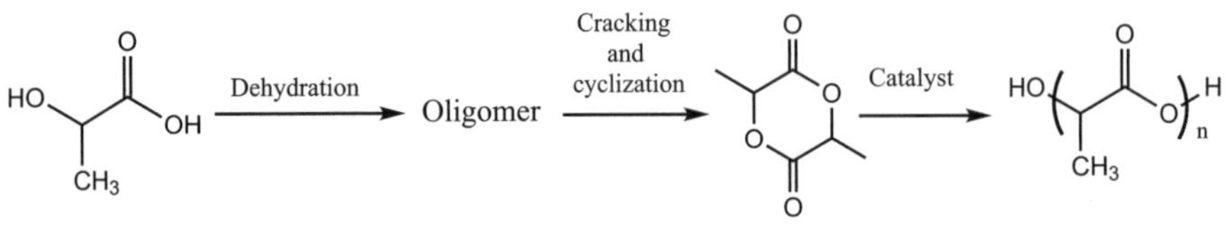

Scheme 9.1 Synthetic route of PLA by ROP.

Functionalization of Poly(L-lactide)

with functionalized moieties. In the following, recent progresses in PLA functionalization are reviewed.

9.2 PLA Functionalization

Over the past few decades, many functionalized monomers for chemical modification of PLA were designed and synthesized, including (1) morpholine-2,5-dione derivatives, (2) α-amino acid N-carboxyanhydrides (NCA), (3) cyclic carbonates, and (4) lactones. The monomer units are introduced into the PLA backbones *via* ROP with lactide and the obtained copolymers are usually random copolymers, block copolymers or block copolymers with a third hydrophilic block – poly(ethylene glycol) (PEG).

9.2.1 Morpholine Diones

Morpholine dione cyclic monomers are obtained from α-amino acids ($H_2NC(R)HCOOH$). The synthetic route of morpholine-2,5-dione derivatives with functional substituents is outlined in Scheme 9.2. To avoid side reactions at the pendant functional groups of the amino acid residues, these groups have to be temporarily blocked by protective groups which are stable to the subsequent reactions, including the polymerization reaction (Scheme 9.3) of the morpholine-2,5-dione derivatives. Finally the protective groups must be selectively removed without cleavage of ester and/or amide bonds of the polymer main chains. According to the different R groups, versatile functionalized morpholine-dione derivatives have been reported, such as those containing carboxyl, amine, thiol and hydroxyl groups, *etc.* (Table 9.1). The ε-amine group of L-lysine was protected by the benzyloxycarbonyl group (1a).

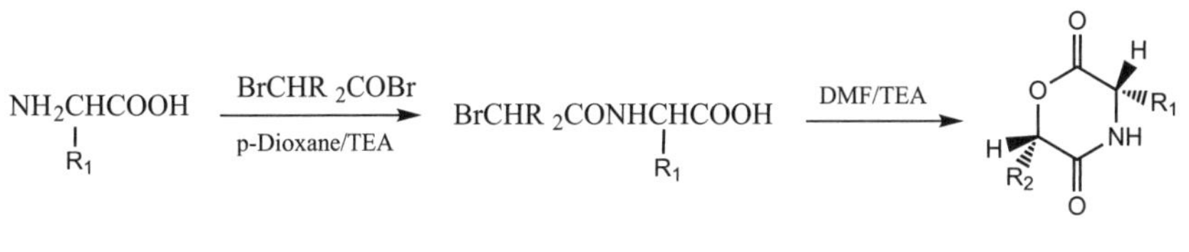

Scheme 9.2 Synthetic route of morpholine-2,5-dione cyclic monomer.

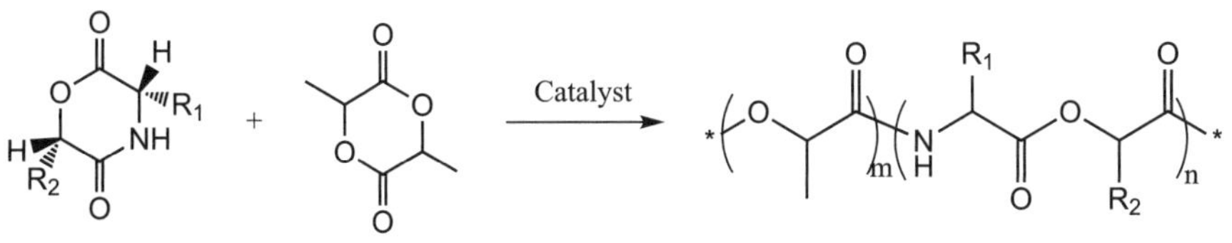

Scheme 9.3 Ring opening copolymerization of LA and morpholine-2,5-dione derivatives.

Table 9.1 Structure of morpholine-2,5-dione derivatives and polymers therefrom.

Entry	Monomer (a)	Polymer (b)	Polymer after deprotection (c)	Ref.
1	$R_1 = CH_3$ $R_2 = (CH_2)_4$ NHCOOBn (Bn = benzyl)	$R_1 = CH_3$ $R_2 = (CH_2)_4$ NHCOOBn (Bn = benzyl)	$R_1' = CH_3$ $R_2' = (CH_2)_4 NH_2$	5,6
2	$R_1 = H$ or CH_3 $R_2 = $ alkyl	$R_1 = H$ or CH_3 $R_2 = $ alkyl		7,8
3	$R_1 = (CH_2)_2 COOBn$ $R_2 = H$	$R_1 = (CH_2)_2 COOBn$ $R_2 = H$	$R_1' = (CH_2)_2 COOH$ $R_2' = H$	11
4	$R_1 = CH_2 OBn$ $R_2 = H$	$R_1 = CH_2 OBn$ $R_2 = H$	$R_1' = CH_2 OH$ $R_2' = H$	12–15
5	$R_1 = CH_2 COOBn$ $R_2 = H$	$R_1 = CH_2 COOBn$ $R_2 = H$	$R_1' = CH_2 COOH$ $R_2' = H$	16
6	$R_1 = CH_2 S$-MBz MBz = p-methoxybenzyl $R_2 = CH_3$	$R_1 = CH_2 S$-MBz MBz = p-methoxybenzyl $R_2 = CH_3$	$R_1' = CH_2 SH$ $R_2' = CH_3$	17

The benzyloxycarbonyl group could be selectively removed by catalytic hydrogenation (1c). The β-carboxylic acid group was protected by benzyl group (3a, 5a) and it could also be removed by catalytic hydrogenation (3c, 5c). The protective group for the thiol group was the p-methoxybenzyl group (6a) and it could be removed by trifluoromethanesulfonic acid (6c). In the following, we will give some specific examples.

Lou et al.[4] reviewed a variety of poly(ester amide)s synthesized from morpholine-2,5-dione derivatives. Their copolymerization with LA was extensively investigated. Langer[5,6] investigated the copolymerization of LA and 3-[N-(carbonylbenzoxy)-L-lysyl]-6-L-methyl-2,5-morpholinedione. And the degradability of these copolymers was enhanced regardless the deprotection of lysine residue. Feng et al.[7,8] have reported on the enzyme-catalysed ROP of several morpholine-2,5-dione derivatives. Shirahama et al.[9,10] studied the biodegradation of copolymers of LA with optically active 3,6-dimethylmorpholine-2,5-dione. In recent years, many studies focused on the amphiphilic block copolymers of PEG-polyesters in the expectation of achieving unique properties and corresponding applications. Guan et al.[11] reported an ABA-type triblock copolymer PLGBG-PEG-PLGBG consisting of PEG and poly{(lactic acid)-co-[(glycolic acid)-alt-(γ-benzyl-L-glutamic acid)]} (PLGBG) which was synthesized by ROP of LA and morpholine-2,5-dione derived from γ-benzyl-L-glutamic acid (BEMD) using dihydroxyl PEG as a macroinitiator. After catalytic hydrogenation, the pendant ester groups were converted to carboxyl groups, leading to the formation of PLGG-PEG-PLGG with pendant carboxyl groups. The copolymers could form micelles in aqueous solution with the cmc dependent on the composition of the copolymer. Presence of the pendant carboxyl groups on PLGG-PEG-PLGG was expected to facilitate further modifications of the polymer, such as attaching drug molecules, short peptides and oligosaccharides onto the carboxyl groups. Poly(ester amide)s with carboxylic acid, hydroxyl, amine and thiol groups by ROP of suitable 2,5-morpholinedione derivatives and deprotection of the functional substituents have been reported.[12–17]

9.2.2 α-Amino acid N-Carboxyanhydride (NCA)

NCA monomers were one of the most widely studied functionalized monomers. In 1906, Leuchs et al.[18–20] synthesized the first α-amino acid NCA monomer. Since then, preparation of poly(amino acid)s has been widely investigated by many groups. This interest stems from the wide variety of polypeptides. However, it was not until 1997, when Deming[21] reported the first living initiating system for the ROP of NCAs, that high molecular weight and structurally homogeneous polypeptides were obtained. Until now, triphosgene became the general material for the synthesis of NCA monomer and the synthetic route is shown in Scheme 9.4.

Several excellent reviews have been dedicated to the ROP of NCAs,[22–24] elucidating the mechanistic aspects of this polymerization. Glutamic acid, lysine, phenylalanine and cysteine NCA monomers have all been reported and their copolymerization with polyester was studied[25–31] (Table 9.2). With glutamic acid or aspartic acid NCA monomer, carboxyl-containing polymers can be obtained, and using lysine NCA monomer, amino-containing polymers are available. Deng[27] reported the synthesis of triblock copolymer polyester-b-PEG-b-poly(amino acid) by the polyester-b-PEG-NH_2 initiated ROP of NCA. Later on, triblock copolymers, PEG-b-poly(L-lactide)-b-poly(γ-benzyl-L-glutamic acid) (PEG-b-PLLA-b-PBLG) and PEG-b-poly(L-lactide)-b-poly(Z-L-lysine) (PEG-b-PLLA-b-PZLL) were synthesized. The side carboxyl groups and amino groups could be readily deprotected by catalytic hydrogenation with Pd/C to give corresponding copolymers PEG-b-PLLA-b-PLGA and PEG-b-PLLA-b-PLL. Langer et al.[5] synthesized lysine-b-LA copolymers through ROP of lactide and Cbz protected lysine cyclic monomer with stannous octoate as

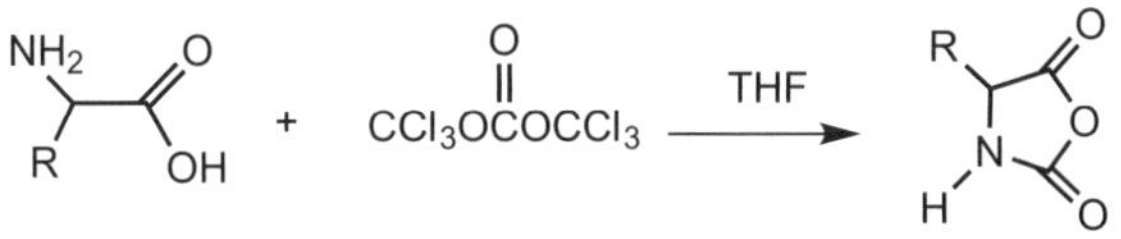

Scheme 9.4 Synthetic route of NCA monomer.

Table 9.2 Structures of α-amino acid NCA derivatives.

Entry	Monomer (a)	Polymer (b)	Polymer after deprotection	Ref.
1	R=(CH$_2$)$_2$COOCH$_2$C$_6$H$_5$	R=(CH$_2$)$_2$COOCH$_2$C$_6$H$_5$	R'=(CH$_2$)$_2$COOH	24, 27–29
2	R=(CH$_2$)$_4$NHCOOCH$_2$C$_6$H$_5$	R=(CH$_2$)$_4$NHCOOCH$_2$C$_6$H$_5$	R'=(CH$_2$)$_4$NH$_2$	25, 26
3	R=H, CH$_3$, CH$_2$C$_6$H$_5$	R=H, CH$_3$, CH$_2$C$_6$H$_5$		25
4	R=CH$_2$SCOOCH$_2$C$_6$H$_5$	R=CH$_2$SCOOCH$_2$C$_6$H$_5$	R'=CH$_2$SH	30
5	R=CH$_2$O-(CH$_2$CH$_2$O)$_n$-H	R=CH$_2$O-(CH$_2$CH$_2$O)$_n$-H		15
6	R=CH$_2$CH=CH$_2$	R=CH$_2$CH=CH$_2$		35
7	R=CH$_2$CCH	R=CH$_2$C=CH		36

catalyst. The pendant amine groups can be used to attach drug and protein molecules. Disadvantages connected with these polypeptides are the inevitable use of protecting group chemistry during monomer synthesis. The most often used α-amino-protecting groups are the 9-fluorenylmethoxycarbonyl (Fmoc),[32,33] the benzyloxycarbonyl (Z) and the *tert*-butyloxycarbonyl (Boc) groups.[34] The protecting groups should confer solubility in the most common solvents and prevent or minimize epimerization during the coupling, and their removal should be fast, efficient and free of side reactions and should render easily eliminatable by-products. Albericio et al.[35] have recently written an excellent review on the protecting groups of amino acids. NCA monomers based on DL-allylglycine[36] and γ-propargyl L-glutamate[37] were synthesized and post-functionalization was carried out *via* thiol-ene chemistry and azide-alkyne click chemistry, in which protection and deprotection processes were not needed. Besides linear polypeptides, other architectures such as cyclic, star shaped and brushed polypeptides have been extensively investigated. Hadjichristidis[22] described these structures in a related review. These polypeptides possess the ability to form α-helix and β-sheet motifs. These secondary structures contribute significantly to the self-assembling character of polypeptide chains, leading to novel supramolecular structure with potential biomedical and pharmaceutical applications. For example, Sun[26] reported the direct formation of giant vesicles from poly(L-lysine)-block-poly(L-phenylalanine) (PLL-b-PPA) block copolymers in water solution. Deming et al.[38,39] synthesized a series of diblock copolypeptides *via* a living polymerization of several NCAs. These copolypeptides were used in the synthesis of ordered silica structures and inorganic hollow spheres. Polypeptide and its copolymers have shown good mechanical properties, film forming and spinnability and they have great potential use in tissue engineering scaffolds. In comparison to natural proteins, synthetic polypeptides offer more advantages in stability and processability. The only limitation is that NCA monomers can not randomly copolymerize with aliphatic cyclic ester monomers.

9.2.3 Cyclic Carbonates

Polycarbonates are a class of polymers that contain carbonate in the main chain. The chemistry of cyclic carbonates has been explored since 1930s by Carothers *et al.*[1] and became a rich area of research within the past 20 years. Two main approaches to cyclic carbonates have been investigated. Brunelle[40,41] has focused on the synthesis of aromatic cyclic carbonate oligomers and their applications in the preparation of bisphenol A polycarbonates. At the same time, Kricheldorf,[42–44] Hocker[45] from Germany, Endo[46–48] from Japan, Zhuo[49,50] and Jing[51–53] from China have been exploring aliphatic cyclic carbonates as useful monomers for the preparation of polycarbonates. Functionalized cyclic carbonates and their polymerization have been attracting special attention because of their comparably easier synthetic method and higher polymerization activity. On the other hand, the degradation products of polycarbonate are carbon dioxide and neutral alcohol and they do not have side effects *in vivo*. Therefore, biodegradable polycarbonates have been widely used in drug delivery, implant materials, tissue engineering, *etc.* According to monomer ring size, the reported cyclic carbonates contain five-membered,[54,55] six-membered, seven-membered[56,57] or larger rings. Among them, six-membered cyclic carbonate was the most investigated due to its structural stability and ring-opening ability. The universal synthetic route to six-membered cyclic carbonate is shown in Scheme 9.5. The ability of cyclic carbonate monomers to undergo ROP depends on both thermodynamic and kinetic factors. Different kinds of initiators and catalysts have been investigated for ROP of various cyclic carbonate monomers according to cationic, anionic, coordination and enzymatic mechanisms.[58] For example, Endo *et al.*[59] reported the synthesis of cyclic carbonate using propane-1,3-diols and ethyl chloroformate in the presence of a stoichiometric amount of triethylamine. Formation of a specific ammonium salt as an intermediate seems to be the driving force for the reaction proceeding at low temperatures favouring six-membered rings and the yields are up to 60%. Extensive efforts have been devoted to the synthesis of functionalized monomers. By changing the chemical structure, various types of functional groups such as hydroxy, amino, ester and carboxyl groups have been synthesized[60] (Table 9.3).

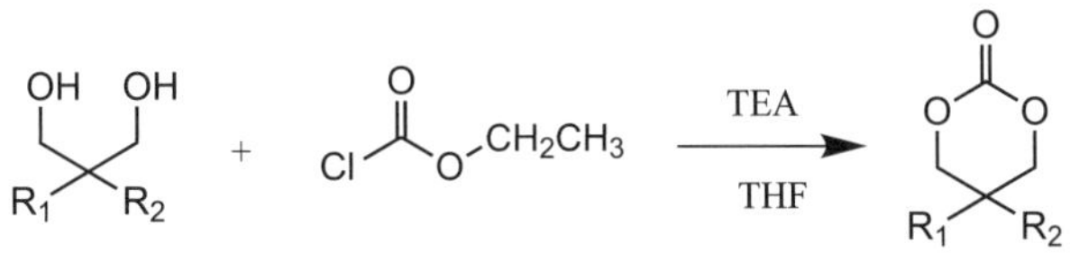

Scheme 9.5 Synthetic route of six-membered cyclic carbonate.

Table 9.3 Structures of six-membered cyclic carbonate derivatives.

Entry	Monomer (a)	Polymer (b)	Polymer after deprotection	Ref.
1	$R_1 = H$ $R_2 = OCH_2C_6H_5$	$R_1 = H$ $R_2 = OCH_2C_6H_5$	$R_1' = H$ $R_2' = OH$	61, 62
2	$R_1, R_2 = (CH_2O)_2CHC_6H_5$	$R_1, R_2 = (CH_2O)_2CHC_6H_5$	$R_1', R_2' = (CH_2OH)_2$	51
3	$R_1, R_2 = (CH_2O)_2C(CH_3)_2$	$R_1, R_2 = (CH_2O)_2C(CH_3)_2$	$R_1', R_2' = (CH_2OH)_2$	66
4	$R_1 = CH_3$ $R_2 = COOCH_2C_6H_5$	$R_1 = CH_3$ $R_2 = COOCH_2C_6H_5$	$R_1' = CH_3$ $R_2' = COOH$	70–72, 73
5	$R_1 = H$ $R_2 = NHCOOCH_2C_6H_5$	$R_1 = H$ $R_2 = NHCOOCH_2C_6H_5$	$R_1' = H$ $R_2' = NH_2$	46, 53
6	$R_1 = CH_3$ $R_2 = CH_2OOCCH=CHC_6H_5$	$R_1 = CH_3$ $R_2 = CH_2OOCCH=CHC_6H_5$		67
7	$R_1 = CH_3$ $R_2 = COOCH_2CH=CH_2$	$R_1 = CH_3$ $R_2 = COOCH_2CH=CH_2$		68
8	$R_1 = CH_3$ $R_2 = COOCH_2C{\equiv}CH$	$R_1 = CH_3$ $R_2 = COOCH_2C{\equiv}CH$		69

Take hydroxyl groups as an example: 5-ethyl-5-hydroxymethyl-1,3-dioxan-2-one is obtained from 1,1,1-trimethylolpropane. Protection and deprotection processes are necessary to avoid the reaction of –OH groups with an initiator or growing chain end, leading to branched or cross-linked structure of the resultant polycarbonates. The synthesis of cyclic carbonate 5-benzyloxy-trimethylene carbonate from glycerol was reported by Grinstaff.[61,62] Gross et al.[63–65] reported polycarbonates bearing vicinal diol pendant groups by ROP of trimethylene carbonate (TMC) with 1,2-O-isopropylidene-D-xylofuranose-3,5-cyclic carbonate, and by subsequent deprotection of the ketal protecting groups. Two hydroxyl-carrying cyclic carbonate 9,9-dimethyl-2,4,8,10-tetraoxaspiro[5,5]undecan-3-one was obtained from the monoacetal diol of pentaerythritol.[51,66] These hydroxy functional groups can be used to chemically bind drugs, proteins or carbohydrate polymers. Other functional groups have also been extensively investigated.[37,52,53,67–69] Bisht and his co-workers[70–72] synthesized carbonate monomer 5-methyl-5-benzyloxycarbonyl-1,3-dioxan-2-one (MBC) and moreover, prepared homopolymer of MBC and copolymers of MBC and TMC under catalysis of lipases, and then polymers containing pendant carboxyl groups were obtained by debenzylation. Lee et al.[73] reported functional poly(ester-carbonate)s by copolymerization of trans-4-hydroxy-L-proline with cyclic carbonate bearing a pendent carboxylic group. The existence of functional groups extends its application by changing the physical, chemical and biological properties to meet different needs.

9.2.4 Lactones

Functionalized aliphatic polyesters based on lactones have been extensively investigated. The synthesized functionalized lactones include four-membered, six-membered and seven-membered lactones, and among them, substituted caprolactone (CL) is the most commonly used. Lou[4] has given a detailed review on these functionalized lactones and so we will not discuss them here.

9.2.5 Cyclic Diesters

Cyclic diesters are another class of extensively studied monomers, and they have similar structures to lactides. Fujino and Ouchi[74] reported the synthesis of poly(malic acid) by ring-opening polymerization of the corresponding monomer, and hydrogenation of the pendant carboxyl protecting groups. Kimura et al.[12,75] synthesized functional aspartic acid monomer and its homopolymer and copolymers.

9.3 Applications of the Functionalized PLAs

Versatile functional aliphatic polyesters have been prepared by copolymerization with the above mentioned functional monomers. The introduction of reactive groups along the chains is highly desirable for fine tuning the polymer properties, such as chemical reactivity, biocompatibility, biodegradability and hydrophilicity. By attaching a number of biologically active compounds, such as drugs, peptides, sugars and other substances, these polyesters were endowed with specific biological activities, making them applicable to separation, diagnosis, treatment and other functions. With the development of modern biomedicine, these intelligent polymers are in higher demand. In the following, the applications of these functionalized PLAs are reviewed.

9.3.1 Drug Delivery Systems

Controlled drug delivery technology was first used in agriculture, mainly for chemical fertilizers, pesticides and herbicides release. Not until the 1960s did they begin to expand to the medical field. In the mid 1970s, the controlled release of large molecular weight drugs (*e.g.* peptides) began to be designed. In 1980s, various release mechanisms were extensively studied. In the 1990s people began to work on smart drug delivery systems. At present, controlled release has become an important content of polymer engineering as well as pharmaceutical chemistry.

According to dosage form, drug delivery systems are generally divided into microspheres, liposomes, micelles, gels and electrospinning fibres. Traditionally, the carrier polymers are considered as auxiliary materials in medicine formulations. The initial carrier was mostly based on non-degradable polymers, such as polyethylene, polypropylene, ethylene-vinyl acetate copolymer, *etc*. With the development of biotechnology in the medical field and the lessons learnt from the preclinical and clinical trials, biodegradable materials have become the prominent part of drug carriers due to the fact that they eliminate the need for second surgery for removal of them from the body. Among them, PLA has been approved by the US Food and Drug Administration (FDA) for use in bio-medicine because of its good biocompatibility, biodegradation and absorption capacity.

In the past 30 years, PLA and its copolymers were widely used as controlled release systems for a number of drugs that have short half-life, poor stability, easy degradability and toxic side effects and the controlled release systems effectively widened the route of administration, reduced the administration frequency and improved the bioavailability of the drugs. For example, microspheres based on PLA and its copolymers have been widely investigated in encapsulation of bioactive agents for therapeutic applications. Anderson[76] gave a good review on using biodegradable PLA and PLGA microspheres incorporating bone morphogenetic protein (BMP) and leuprorelin acetate as well as interactions with the eye, central nervous system and lymphoid tissue and their relevance to vaccine development.

In recent years, electrospinning has gained widespread interest as a potential polymer processing technique for applications in drug delivery and has afforded great flexibility in selecting materials. This can be attributed to electrospinning's relative ease of use, adaptability and the ability to fabricate fibres with diameters in the nanometer scale. A number of drugs, such as antibiotics, anticancer drugs, proteins and DNA, have been investigated for delivery using these nanofibres. A number of different drug loading methods have also been utilized, such as coating, embedding and encapsulation (coaxial and emulsion electrospinning). Kenawy[77] explored electrospun fibre mats made either from PLA, poly(ethylene-co-vinyl acetate) (PEVA), or from a 50:50 blend of the two as drug delivery vehicles using tetracycline hydrochloride as a model drug. Their first results indicated that the release profiles from the electrospun mats were comparable to a commercially available drug delivery system, Actisite, as well as to cast films of various formulations. Kim et al.[78] reported the successful incorporation and sustained release of a hydrophilic antibiotic drug (Mefoxin) from electrospun poly(lactide-co-glycolide) (PLGA)-based nanofibrous scaffolds without the loss of bioactivity. They also found that the usage of amphiphilic block copolymer (PEG-b-PLA) gave a more sustained release than the PLGA mat did. Xu et al.[79] reported the core-sheath nanofibres prepared by electrospinning a water-in-oil emulsion in which the aqueous phase consists of a poly(ethylene oxide) (PEO) solution in water and the oily phase is a chloroform solution of an amphiphilic poly(ethylene glycol)-poly(L-lactic acid) (PEG-PLA) diblock copolymer. The obtained fibres are composed of a PEO core and a PEG-PLA sheath with a sharp boundary in between. The overall fibre size and the relative diameters of the core and the sheath can be changed by adjusting the emulsion composition and the emulsification parameters. This technology can be applied in place of concentric electrospinning and is especially suitable for fabricating nanofibres containing water-soluble drugs. They also incorporated the anticancer drug BCNU into the electrospun PEG-PLA mats.[80] The effect of BCNU released from electrospun PEG-PLA mats on the growth of rat Glioma C6 cells was also examined. By embedding the drug in the polymer fibres, the authors were able to protect it from degradation and preserve its anticancer activity. Additionally, they were able to effectively control the release rate of BCNU from the electrospun mats by altering the drug loading. Liang et al.[81] elaborated a review on these functional nanofibres use in various biomedical applications, such as tissue engineering, wound dressing, enzyme immobilization and drug (gene) delivery.

In the past few decades, polymer micelles have emerged as promising drug carriers and attracted special attention due to their attractive features and advantages over other types of carriers, such as enhanced water solubility, prolonged blood circulation time, higher curing efficacy, reduced side reaction and toxicity. Even more, the passive accumulation of the micelles in a solid tumour via the enhanced permeability and retention (EPR) effect of the vascular endothelia at the tumour was observed by Maeda[82,83] and many other researchers.[84,85] Along with the specific interaction and recognition between antibody and antigen or between ligand and receptor and cellular uptake of the drug by receptor- or antigen-mediated endocytosis were discovered and became the main mechanism of active targeting, more and more researchers engage themselves in developing 'magic bullets', i.e. targeting drugs.[86] In order to solve the problem of burst release, 'polymeric prodrug' was introduced[87] and proved to be effective.[86,88] For example, Li[89] synthesized poly(ethylene glycol) – poly(glutamic acid) block copolymer and the terminus of poly(ethylene glycol) was functionalized with vinyl sulfone groups for attaching monoclonal antibody targeting groups. The anticancer drug adriamycin was conjugated to glutamic acid units through an amide bond. Their results suggested that conjugation of a receptor-homing ligand to the PEG-terminal of the polymer-drug conjugate enhanced the targeted delivery of anticancer agents. Zhu[90] conjugated the anthracyclin antineoplastic agent doxorubicin (adriamycin) to DalB02, an IgG1 kappa murine monoclonal antibody (mAb) against surface-associated antigens on human chronic lymphocytic leukaemia (CLL) B cells. Their *in vivo* results indicated that the percentages of the injected dose (%ID) of both I-131-DalB02 and the I-131-DalB02-containing conjugate that were localized in the tumour were much higher than the %ID of the respective preparations that were localized in normal tissues of D-10-1-xenografted mice. And the conjugate was a more effective inhibitor of established D-10-1 xenografts than the free drug. Nasongkla *et al.*[91] synthesized PEG-PCL with maleic acid at the PEG terminal for coupling short-peptide cRGD-SH and the obtained copolymer was used as targeted drug carrier. Attachment of the cyclic RGD ligand greatly enhanced internalization of the micelles in tumour endothelial cells that over-express $\alpha_v\beta_3$ integrins. Czech Ulbrich *et al.*[92] connected both doxorubicin and antibodies to PHPMA polymers *via* hydrazone and disulfide bonds to obtain targeted carriers. Farokhzad *et al.*[93] designed and synthesized A10 2-fluoro-pyrimidine nucleotide aptamer (Aptamer) modified PEG-PLGA nanoparticles for docetaxel loading and achieved prostate cancer specific targeting *in vivo*. The bioconjugates exhibited remarkable efficacy and reduced toxicity as measured by mean body weight loss (BWL) *in vivo*. After a single intratumoural injection of the bioconjugates, complete tumour reduction was observed in five of seven LNCaP xenograft nude mice and 100% of these animals survived the 109-day study. Dhar *et al.*[94] constructed PLGA-b-PEG nanoparticles with PSMA targeting aptamers (Apt) on the surface as a vehicle for the platinum(IV). The cytotoxic activities of Pt(IV)-encapsulated PLGA-b-PEG NPs with the PSMA aptamer on the surface (Pt-NP-Apt), cisplatin and the non-targeted Pt(IV)-encapsulated NPs (Pt-NP) against human prostate PSMA-over-expressing LNCaP and PSMA(–)PC3 cancer cells revealed that the effectiveness of PSMA targeted Pt-NP-Apt nanoparticles is approximately an order of magnitude greater than that of free cisplatin. Cheng *et al.*[95] reported a similar approach, using A10 RNA aptamer (Apt) modified PEG-PLGA nanoparticles to load docetaxel, and they studied nanoparticle distribution in various organs. Their results showed richer accumulation of drugs in the spleen comparing to the nanoparticles without targeting groups. Feng's group has carried out a lot of work on folic acid targeting nanomicelles.[96] Ma *et al.*[97] observed lactose mediated liver-targeting effect by *ex vivo* imaging technology using lactose functionalized PEG-PLA nanomicelles. Several excellent reviews[98–101] have more detailed information on these functionalized polymer conjugates.

9.3.2 Artificial Oxygen Carriers

Blood continuously delivers oxygen to and takes carbon dioxide away from all tissues and cells and is of the most essential importance for human lives.

The use of blood products in medical practice often meets various difficulties, such as shortage of blood supply, blood type incompatibility, short preservation period, immunological responses and potential infections. An alternative to blood transfusion, based on oxygen-carrying solutions, has been sought for over a century and rapid progress has been made in this field.

In the 1960s, Bunn and Jandl[102] demonstrated that when haemoglobin (Hb) is free in plasma, it dissociates into αβ dimmers, which are rapidly filtered out in the kidney. The 'first-generation' haemoglobin-products are based on the observations that cross-linking Hb can overcome subunit dissociation and renal toxicity,[103] including polymerized Hbs, intramolecularly cross-linked Hbs and perfluorocarbon emulsions. The 'second-generation' products are based on a better understanding of the mechanisms of vasoconstriction.[104] But clinical trials have been disappointing because of unexpected toxicity. From the early 1990s to the present, researchers have been dedicated to the development of the third generation of red blood cell substitutes. Haemoglobin vesicles (HbV) prepared by encapsulating Hb molecules into lipid or polymeric vesicles[105,106] exhibited much better performance than chemically modified Hbs. The HbVs, with a diameter of 100–250 nm, have a layer of membrane which prevents the Hb molecules from contacting directly with the immunological system[107] or penetrating vascular wall.[108] It is expected to be the most promising artificial oxygen carrier. The matrix used for HbVs should combine good biocompatibility, non-toxicity and non-immunogenicity. In 1957, Chang[109] first prepared nylon and colloidin microcapsules. Many studies were reported on liposome-encapsulated Hb (LEH) since Djordjevich and Miller established the method in 1977. Recently, biodegradable polymer PLA was more and more studied as the HbV matrix due to its favourable properties such as: (1) its biocompatibility and biodegradability, (2) its chemical structure diversity by copolymerization with other monomers, and (3) good mechanical properties compared with traditional HbV matrixes. Chang's group first tried to prepare artificial blood cells with PLA materials in 1976.[110] Recently, they reported some approaches to nanoscale Hb encapsulation with PEG-PLA.[107,111] Rameez et al.[105] described polymersome-encapsulated haemoglobin (PEH). The polymersome was self-assembled from biodegradable and biocompatible amphiphilic diblock copolymers composed of poly(ethylene oxide) (PEO), poly(caprolactone) (PCL) and PLA and their results indicated that Hb encapsulation did not affect the oxygen binding properties of Hb. Sun et al.[106] constructed an oxygen carrier by encapsulating carbonylated haemoglobin (CO-Hb) molecules into polypeptide vesicles made from poly(L-lysine)-block-poly(L-phenylalanine) (PLL-b-PPA) diblock copolymers in aqueous medium. The CO-Hb encapsulated in the PLL-b-PPA vesicles was more stable than free CO-Hb under ambient conditions. In the presence of an O_2 atmosphere, the CO-Hb in the vesicle could be converted into oxygen-binding haemoglobin (O_2-Hb) under irradiation of visible light. In another paper,[69] azidized haemoglobin molecules were conjugated onto polymer micelles containing propargyl groups via click reaction between the propargyl groups in the micelles and the azido groups in the haemoglobin molecules. The propargyl groups were attached to the core surface of the micelles and expected to react with the azido groups without difficulty and the haemoglobin molecules were hidden in the PEG corona of the micelles, being protected against the immunological systems. Results indicated that the conjugated haemoglobins retained their O_2-binding ability.

9.3.3 Protein Separation and Purification

Many potential and interesting applications can be obtained when bioactive macromolecules are properly immobilized on solid surfaces with their activity preserved, such as biosensor chips, microarrays, implantable medical devices and purification of biochemical agents.[112,113] To create biomaterials that allow for the attachment of proteins and peptides, the frequently used strategy is to introduce functional groups into already established biomaterials to which the desired bioactive substances are to be attached. The functional groups can be introduced during polymer synthesis or *via* surface reactions after the polymer has been processed into a medical device. A number of protein-immobilizing substrates have been developed, such as gold-coated surfaces,[114] glass slides treated with organosilanes,[115] titanium and titanium oxide,[116,117] polyelectrolyte multilayer films[118] and hydrogels.[119] But with the development in techniques and devices, more and more new systems are based on polymeric materials.[120] PEO brushes have been especially recognized as biocompatible and resistant to protein adsorption due to the hydrophilic but uncharged nature of the polymer. PEG-modified surfaces have been developed in order to avoid non-specific protein adsorption.[121,122] Current immobilization methods are mainly based on the following three mechanisms: physical adsorption, covalent and bioaffinity immobilization. Rusmini *et al.*[120] reviewed current immobilization strategies in a recent paper. No matter what mechanism is used, the immobilized protein should at least meet the following requirements: (1) full retention of protein conformation and activity, (2) selective and specific, in other words, non-specific protein adsorption needs to be avoided or at least minimized. Physical adsorption has always been accompanied with weak attachment, which leads to proteins being removed by buffers when performing the assays. Bioaffinity immobilization, such as biotin and avidin, ligand and receptor, antigen and antibodies, has been extensively investigated due to its selectivity and specificity. The interaction between biotin and avidin or streptavidin is one of the strongest non-covalent affinity interactions in nature ($K_{aff} = 10^{13-15}$ M^{-1} in solution), and as such it has been used for a wide range of applications including immunoassays, cytochemistry, protein purification and diagnostics.[123] For example, Lu *et al.*[124] developed bioactive fibrous mats based on biotin and streptavidin interaction for protein immobilization and protein separation/purification. The fibrous mat was obtained by co-electrospinning biotinylated poly(ethylene glycol)-b-poly(L-lactide)-b-poly(L-lysine) and PLGA. The biotin species retained their ability to specifically recognize and bind streptavidin, and the immobilized streptavidin could further combine with biotinylated antibodies, antigens and other biological moieties. Li *et al.*[125] successfully incorporated biotin into PLA nanofibres through electrospinning to prepare membrane substrates for biosensors based on biotin-streptavidin specific binding. Pre-blocking the membranes effectively eliminated non-specific binding between streptavidin and PLA. Their preliminary biosensor assays confirmed that streptavidin immobilized on the membrane surface could capture a biotinylated DNA probe. Other biocompatible surfaces using avidin-biotin technology have also been reported.[126–129] Besides biotin and avidin, ligand/receptor and antigen/antibodies have also been investigated. Shi[52] immobilized testis-specific protease 50 (TSP50) onto functionalized PLA fibres via click chemistry. The TSP50-immobilized fibres can resist non-specific protein adsorption but preserve specific recognition and combination with anti-TSP50. ELISA results showed that anti-TSP50 can be selectively adsorbed from its solution onto the TSP50-immobilized fibres in the presence of BSA of as high as 10^4 times concentration. And the anti-TSP50 can also be eluted off from the fibre when the pH changes. The eluted fibre can re-combine anti-TSP50 at an efficiency of 75% compared to the original TSP50-immobilized fibre. This method can be used in the detection, separation and purification of anti-TSP50. Lu et al.[130] used poly(L-lactic acid)-b-poly(L-cysteine) (PLA-b-PCys) ultrafine fibres in glutathione S-transferase (GST) purification. The reduced glutathione (GSH) was first conjugated to PLA-b-PCys fibre surfaces via disulfide bonds. Their results showed that the GSH moieties on the fibre surface retain the bioactivity of the free GSH and thus can bind specifically with GST. In addition, the bound GSH is not as active as free GSH so that the captured GST can be eluted off from the fibre. In this way, GST can be separated and purified easily. The preliminary purification efficiency was reported to be 6.5 mg $(gPCys)^{-1}$.

9.4 Conclusions

PLA and its copolymers offer prospective applications in a number of fields, such as tissue engineering scaffolding, drug delivery systems, implant materials, gene delivery, protein separation and purification and disease diagnosis. We have reviewed concisely the methods of PLA functionalization and applications of the functionalized PLAs in drug delivery, artificial oxygen carriers and protein separation and purification that our group has worked on in the past decade. More research is needed to get tailored properties with respect to degradability and strength for specific applications. Moreover, there is a great potential to use PLA in a number of unexplored applications by replacing the conventional polymers, where it can contribute a significant role in the form of composites, copolymers and blends to obtain the required properties for different applications.

References

1. W. H. Carothers and F. J. Van Natta, *J. Am. Chem. Soc.*, 1930, **52**, 314.
2. A. Duda and S. Penczek, *Polimery*, 2003, **48**, 16.
3. K. A. Athanasiou, *et al., Biomaterials*, 1996, **17**, 93.

4. X. D. Lou, C. Detrembleur and R. Jerome, *Macromol. Rapid. Commun.*, 2003, **24**, 161.
5. D. A. Barrera, E. Zylstra, P. T. Lansbury and R. Langer, *J. Am. Chem. Soc.*, 1993, **115**, 11010.
6. D. A. Barrera, E. Zylstra, P. T. Lansbury and R. Langer, *Macromolecules*, 1995, **28**, 425.
7. Y. Feng, D. Klee and H. Hocker, *Macromol. Chem. Phys.*, 2002, **203**, 819.
8. Y. K. Feng, D. Klee, H. Keul and H. Hocker, *Macromol. Chem. Phys.*, 2000, **201**, 2670.
9. H. Shirahama, M. Shiomi, M. Sakane and H. Yasuda, *Macromolecules*, 1996, **29**, 4821.
10. H. Shirahama, A. Tanaka and H. Yasuda, *J. Polym. Sci. Pol. Chem.*, 2002, **40**, 302.
11. H. L. Guan, Z. G. Xie, P. B. Zhang, C. Deng, X. S. Chen and X. B. Jing, *Biomacromolecules*, 2005, **6**, 1954.
12. Y. Kimura, K. Shirotani, H. Yamane and T. Kitao, *Polymer*, 1993, **34**, 1741.
13. G. John and M. Morita, *Macromol. Rapid. Commun.*, 1999, **20**, 265.
14. G. John and M. Morita, *Macromolecules*, 1999, **32**, 1853.
15. G. John, S. Tsuda and M. Morita, *J. Polym. Sci. Polym. Chem.*, 1997, **35**, 1901.
16. D. Wang and X. D. Feng, *Macromolecules*, 1997, **30**, 5688.
17. P. Veld, P. J. Dijkstra and J. Feijen, *Macromol. Chem. Phys.*, 1992, **193**, 2713.
18. H. Leuchs, *Berichte Der Deutschen Chemischen Gesellschaft*, 1906, **39**, 857.
19. H. Leuchs and W. Geiger, *Berichte Der Deutschen Chemischen Gesellschaft*, 1908, **41**, 1721.
20. H. Leuchs and W. Manasse, *Berichte Der Deutschen Chemischen Gesellschaft*, 1907, **40**, 3235.
21. T. J. Deming, *Nature*, 1997, **390**, 386.
22. N. Hadjichristidis, H. Iatrou, M. Pitsikalis and G. Sakellariou, *Chem. Rev.*, 2009, **109**, 5528.
23. H. Sekiguchi, *Pure Appl. Chem.*, 1981, **53**, 1689.
24. H. R. Kricheldorf, *Angew. Chem. Int. Ed.*, 2006, **45**, 5752.
25. C. Deng, X. S. Chen, J. Sun, T. C. Lu, W. S. Wang and X. B. Jing, *J. Polym. Sci. Polym. Chem.*, 2007, **45**, 3218.
26. J. Sun, X. S. Chen, C. Deng, H. J. Yu, Z. G. Xie and X. B. Jing, *Langmuir*, 2007, **23**, 8308.
27. C. Deng, X. S. Chen, H. J. Yu, J. Sun, T. C. Lu and X. B. Jing, *Polymer*, 2007, **48**, 139.
28. H. Y. Tian, X. S. Chen, H. Lin, C. Deng, P. B. Zhang, Y. Wei and X. B. Jing, *Chemistry – a European Journal*, 2006, **12**, 4305.
29. J. Sun, C. Deng, X. S. Chen, H. J. Yu, H. Y. Tian, J. R. Sun and X. B. Jing, *Biomacromolecules*, 2007, **8**, 1013.

30. C. Deng, H. Y. Tian, P. B. Zhang, J. Sun, X. S. Chen and X. B. Jing, *Biomacromolecules*, 2006, **7**, 590.
31. J. Sun, X. S. Chen, T. C. Lu, S. Liu, H. Y. Tian, Z. P. Guo and X. B. Jing, *Langmuir*, 2008, **24**, 10099.
32. L. A. Carpinoa and G. Y. Han, *J. Org. Chem.*, 1972, **37**, 3404.
33. L. A. Carpino and G. Y. Han, *J. Am. Chem. Soc.*, 1970, **92**, 5748.
34. E. Hlebowicz, A. J. Andersen, L. Andersson and B. A. Moss, *J. Pept. Res.*, 2005, **65**, 90.
35. A. Isidro-Llobet, M. Alvarez and F. Albericio, *Chem. Rev.*, 2009, **109**, 2455.
36. J. Sun and H. Schlaad, *Macromolecules*, 2010, **43**, 4445.
37. C. Xiao, C. Zhao, P. He, Z. Tang, X. Chen and X. Jing, *Macromol. Rapid Commun.* 2010, DOI: 10. 1002/marc. 200900821.
38. J. N. Cha, G. D. Stucky, D. E. Morse and T. J. Deming, *Nature*, 2000, **403**, 289.
39. M. S. Wong, J. N. Cha, K. S. Choi, T. J. Deming and G. D. Stucky, *Nano Lett.*, 2002, **2**, 583.
40. D. J. Brunelle and T. G. Shannon, *Macromolecules*, 1991, **24**, 3035.
41. D. J. Brunelle, *Adv. Polycarbonates*, 2005, **898**, 1.
42. S. Kobayashi, H. Kikuchi and H. Uyama, *Macromol. Rapid Commun.*, 1997, **18**, 575.
43. H. R. Kricheldorf, *Angew. Chem. Int. Ed.*, 1979, **18**, 689.
44. S. Namekawa, H. Uyama, S. Kobayashi and H. R. Kricheldorf, *Macromol. Chem. Phys.*, 2000, **201**, 261.
45. F. Schmitz, H. Keul and H. Hocker, *Macromol. Rapid Commun.*, 1997, **18**, 699.
46. F. Sanda, J. Kamatani and T. Endo, *Macromolecules*, 2001, **34**, 1564.
47. T. Takata, F. Sanda, T. Ariga, H. Nemoto and T. Endo, *Macromol. Rapid Commun.*, 1997, **18**, 461.
48. T. Ariga, T. Takata and T. Endo, *Macromolecules*, 1997, **30**, 737.
49. Y. Zhou, R. X. Zhuo and Z. L. Liu, *Macromol. Rapid Commun.*, 2005, **26**, 1309.
50. X. J. Zhang, H. J. Mei, C. Hu, Z. L. Zhong and R. X. Zhuo, *Macromolecules*, 2009, **42**, 1010.
51. Z. G. Xie, C. H. Lu, X. S. Chen, L. Chen, Y. Wang, X. L. Hu, Q. Shi and X. B. Jing, *J. Polym. Sci. Polym. Chem.*, 2007, **45**, 1737.
52. Q. Shi, X. S. Chen, T. C. Lu and X. B. Jing, *Biomaterials*, 2008, **29**, 1118.
53. X. L. Hu, X. S. Chen, Z. G. Xie, H. B. Cheng and X. B. Jing, *J. Polym. Sci. Polym. Chem.*, 2008, **46**, 7022.
54. S. Jana, A. Parthiban and C. L. L. Chai, *Chem. Commun.*, 2010, **46**, 1488.
55. Haba N. Furuichi and Y. Akashika, *Polym. J.*, 2009, **41**, 702.
56. R. Wu, T. F. Al-Azemi and K. S. Bisht, *Biomacromolecules*, 2008, **9**, 2921.
57. J. Matsuo, F. Sanda and T. Endo, *Macromol. Chem. Phys.*, 2000, **201**, 585.
58. G. Rokicki, *Prog. Polym. Sci.*, 2000, **25**, 259.

59. J. Matsuo, S. L. Nakano, F. Sanda and T. Endo, *J. Polym. Sci. Polym. Chem.*, 1998, **36**, 2463.
60. J. Yang, Q. H. Hao, X. Y. Liu, C. Y. Ba and A. Cao, *Biomacromolecules*, 2004, **5**, 209.
61. W. C. Ray and M. W. Grinstaff, *Macromolecules*, 2003, **36**, 3557.
62. J. B. Wolinsky, W. C. Ray, Y. L. Colson and M. W. Grinstaff, *Macromolecules*, 2007, **40**, 7065.
63. Y. Q. Shen, X. H. Chen and R. A. Gross, *Macromolecules*, 1999, **32**, 3891.
64. R. Kumar, W. Gao and R. A. Gross, *Macromolecules*, 2002, **35**, 6835.
65. Y. Q. Shen, X. H. Chen and R. A. Gross, *Macromolecules*, 1999, **32**, 2799.
66. E. J. Vandenberg and D. Tian, *Macromolecules*, 1999, **32**, 3613.
67. X. L. Hu, X. S. Chen, H. B. Cheng and X. B. Jing, *J. Polym. Sci. Polym. Chem.*, 2009, **47**, 161.
68. X. L. Hu, X. S. Chen, S. Liu, Q. Shi and X. B. Jing, *J. Polym. Sci. Polym. Chem.*, 2008, **46**, 1852.
69. Q. Shi, Y. B. Huang, X. S. Chen, M. Wu, J. Sun and X. B. Jing, *Biomaterials*, 2009, **30**, 5077.
70. T. F. Al-Azemi and K. S. Bisht, *Macromolecules*, 1999, **32**, 6536.
71. T. F. Al-Azemi and K. S. Bisht, *J. Polym. Sci. Polym. Chem.*, 2002, **40**, 1267.
72. T. F. Al-Azemi, J. P. Harmon and K. S. Bisht, *Biomacromolecules*, 2000, **1**, 493.
73. R. S. Lee, J. M. Yang and T. F. Lin, *J. Polym. Sci. Polym. Chem.*, 2004, **42**, 2303.
74. T. Ouchi and A. Fujino, *Macromol. Chem. Phys.*, 1989, **190**, 1523.
75. Y. Kimura, K. Shirotani, H. Yamane and T. Kitao, *Macromolecules*, 1988, **21**, 3338.
76. J. M. Anderson and M. S. Shive, *Adv. Drug Deliver. Rev.*, 1997, **28**, 5.
77. E. R. Kenawy, G. L. Bowlin, K. Mansfield, J. Layman, D. G. Simpson, E. H. Sanders and G. E. Wnek, *J. Control. Release*, 2002, **81**, 57.
78. K. Kim, Y. K. Luu, C. Chang, D. F. Fang, B. S. Hsiao, B. Chu and M. Hadjiargyrou, *J. Control. Release*, 2004, **98**, 47.
79. X. L. Xu, X. L. Zhuang, X. S. Chen, X. R. Wang, L. X. Yang and X. B. Jing, *Macromol. Rapid. Commun.*, 2006, **27**, 1637.
80. X. L. Xu, X. S. Chen, X. Y. Xu, T. C. Lu, X. Wang, L. X. Yang and X. B. Jing, *J. Control. Release*, 2006, **114**, 307.
81. D. Liang, B. S. Hsiao and B. Chu, *Adv. Drug Deliver. Rev.*, 2007, **59**, 1392.
82. Y. Matsumura and H. Maeda, *Cancer Res.*, 1986, **46**, 6387.
83. H. Maeda, *Adv. Enzyme Regulation*, 2001, **41**, 189.
84. A. N. Lukyanov, W. C. Hartner and V. P. Torchilin, *J. Control. Release*, 2004, **94**, 187.
85. V. P. Torchilin, *Expert Opin. Ther. Pat.*, 2005, **15**, 63.
86. R. Haag and F. Kratz, *Angew. Chem. Int. Ed.*, 2006, **45**, 1198.
87. J. Bartulin, H. G. Batz, G. Franzman, V. Hofmann, M. Przybyls, H. Ringsdor and H. Ritter, *Angew. Chem. Int. Ed.*, 1974, **13**, 417.

88. J. W. Singer, B. Baker, P. De Vries, A. Kumar, S. Shaffer, E. Vawter, M. Bolton and P. Garzone, *Polymer Drugs in the Clinical Stage: Advantages and Prospects*, 2003, **519**, 81.
89. C. Li, *Adv. Drug Deliver. Rev.*, 2002, **54**, 695.
90. Z. P. Zhu, J. Kralovec, T. Ghose and M. Mammen, *Cancer Immunol. Immun.*, 1995, **40**, 257.
91. N. Nasongkla, X. Shuai, H. Ai, B. D. Weinberg, J. Pink, D. A. Boothman and J. M. Gao, *Angew. Chem. Int. Ed.*, 2004, **43**, 6323.
92. K. Ulbrich, T. Etrych, P. Chytil, M. Jelinkova and B. Rihova, *J. Control. Release*, 2003, **87**, 33.
93. O. C. Farokhzad, J. J. Cheng, B. A. Teply, I. Sherifi, S. Jon, P. W. Kantoff, J. P. Richie and R. Langer, *Proc. Natl. Acad. Sci. USA*, 2006, **103**, 6315.
94. S. Dhar, F. X. Gu, R. Langer, O. C. Farokhzad and S. J. Lippard, *Proc. Natl. Acad. Sci. USA*, 2008, **105**, 17356.
95. J. Cheng, B. A. Teply, I. Sherifi, J. Sung, G. Luther, F. X. Gu, E. Levy-Nissenbaum, A. F. Radovic-Moreno, R. Langer and O. C. Farokhzad, *Biomaterials*, 2007, **28**, 869.
96. J. Pan and S. S. Feng, *Biomaterials*, 2009, **30**, 1176.
97. P. A. Ma, S. Liu, Y. B. Huang, X. S. Chen, L. P. Zhang and X. B. Jing, *Biomaterials*, 2010, **31**, 2646.
98. M. J. Vicent and R. Duncan, *Trends Biotechnol.*, 2006, **24**, 39.
99. C. Li and S. Wallace, *Adv. Drug Deliver. Rev.*, 2008, **60**, 886.
100. R. Satchi-Fainaro, R. Duncan and C. M. Barnes, Polym. Therapeutics II: Polymers as Drugs, *Conjugates and Gene Delivery Systems*, 2006, **193**, 1.
101. X. L. Hu and X. B. Jing, *Expert Opin. Drug Delivery*, 2009, **6**, 1079.
102. H. F. Bunn and J. H. Jandl, *Trans. Assoc. Am. Physicians*, 1968, **81**, 147.
103. M. A. Marini, G. L. Moore, S. M. Christensen, R. M. Fishman, R. G. Jessee, F. Medina, S. M. Snell and A. I. Zegna, *Biopolymers*, 1990, **29**, 871.
104. Q. H. Gibson, *Prog. Biophys. Mol. Biol.*, 1959, **9**, 1.
105. S. Rameez, H. Alosta and A. F. Palmer, *Bioconjugate Chem.*, 2008, **19**, 1025.
106. J. Sun, Y. B. Huang, Q. Shi, X. S. Chen and X. B. Jing, *Langmuir*, 2009, **25**, 13726.
107. T. M. S. Chang, *Trends Biotechnol.*, 2006, **24**, 372.
108. T. Kyokane, S. Norimizu, H. Taniai, T. Yamaguchi, S. Takeoka, E. Tsuchida, M. Naito, Y. Nimura, Y. Ishimura and M. Suematsu, *Gastroenterology*, 2001, **120**, 1227.
109. T. M. S. Chang, *J. Intern. Med.*, 2003, **253**, 527.
110. T. M. S. Chang, *J. Bioengineering*, 1976, **1**, 25.
111. T. M. S. Chang and D. Powanda, *Artif. Cells Blood Substit. Immobil. Biotechnol.*, 2003, **31**, 231.
112. P. Angenendt, *Drug Discov. Today*, 2005, **10**, 503.
113. S. Bodovitz, T. Joos and J. Bachmann, *Drug Discov. Today*, 2005, **10**, 283.
114. Y. J. Ding, J. Liu, H. Wang, G. L. Shen and R. Q. Yu, *Biomaterials*, 2007, **28**, 2147.

115. X. L. Sun, C. L. Stabler, C. S. Cazalis and E. L. Chaikof, *Bioconjugate Chem.*, 2006, **17**, 52.
116. S. Tosatti, R. Michel, M. Textor and N. D. Spencer, *Langmuir*, 2002, **18**, 3537.
117. R. Michel, J. W. Lussi, G. Csucs, I. Reviakine, G. Danuser, B. Ketterer, J. A. Hubbell, M. Textor and N. D. Spencer, *Langmuir*, 2002, **18**, 3281.
118. D. L. Elbert, C. B. Herbert and J. A. Hubbell, *Langmuir*, 1999, **15**, 5355.
119. K. Y. Lee and D. J. Mooney, *Chem. Rev.*, 2001, **101**, 1869.
120. F. Rusmini, Z. Y. Zhong and J. Feijen, *Biomacromolecules*, 2007, **8**, 1775.
121. E. Ostuni, R. G. Chapman, R. E. Holmlin, S. Takayama and G. M. Whitesides, *Langmuir*, 2001, **17**, 5605.
122. R. G. Chapman, E. Ostuni, L. Yan and G. M. Whitesides, *Langmuir*, 2000, **16**, 6927.
123. M. Wilchek and E. A. Bayer, *Biomol. Eng.*, 1999, **16**, 1.
124. T. C. Lu, X. S. Chen, Q. Shi, Y. Wang, P. B. Zhang and X. B. Jing, *Acta Biomater.*, 2008, **4**, 1770.
125. D. P. Li, M. W. Frey and A. J. Baeumner, *J. Membrane Sci.*, 2006, **279**, 354.
126. C. L. Smith, J. S. Milea and G. H. Nguyen, *Immobilisation of DNA on Chips II*, 2005, **261**, 63.
127. R. N. Orth, T. G. Clark and H. G. Craighead, *Biomed. Microdevices*, 2003, **5**, 29.
128. X. L. Sun, K. M. Faucher, M. Houston, D. Grande and E. L. Chaikof, *J. Am. Chem. Soc.*, 2002, **124**, 7258.
129. N. P. Huang, J. Voros, S. M. De Paul, M. Textor and N. D. Spencer, *Langmuir*, 2002, **18**, 220.
130. T. C. Lu, J. Sun, X. Q. Dong, X. S. Chen, Y. Wang and X. B. Jing, *Sci. China Ser. B*, 2009, **52**, 2033.

CHAPTER 10
Biodegradation of Poly (3-hydroxyalkanoates)

RACHANA BHATT,[1,2] KAMLESH PATEL[1] AND
UJJVAL TRIVEDI[1]

[1] BRD School of Biosciences, Sardar Patel University, Vallabh Vidyanagar – 388 120, Gujarat, India; [2] Department of Microbiology, Shri A. N. Patel Postgraduate Institute, Charotar Education Society, Anand – 388 001, Gujarat, India

10.1 Introduction

In the second half of twentieth century there evolved a group of materials never seen before on this earth, plastics – a synthetic product whose production skyrocketed as soon as it was invented. They have low cost of production, ease of processing and outstanding mechanical as well as physical properties. They have heterogeneous properties with many practical uses and all of them possess a common property of high resistance to natural degradation processes. As a result, they have replaced wood, cotton, paper and other natural materials in many applications. They are widely used in packaging, building materials and commodity as well as hygienic products. The fact that plastics are mostly non-degradable is the primary cause of increasing amounts of solid waste, whose major content is synthetic polymers. This is the primary driving force for scientists trying to find a solution to the waste problem. Efforts have been made to design and develop biodegradable alternatives, sometimes also referred to as 'bioplastics'. The term bioplastics may refer to biodegradable plastics and/or plastics from a biological origin.

Many living systems (mainly plants and microorganisms) produce materials with great potential for replacing synthetic plastics. One of the typical groups of materials found in nature is polyhydroxyalkanoates (PHAs). These are natural polyesters that are produced by many microorganisms, especially bacteria, as a means to store carbon and energy. PHAs are synthesized and deposited in the form of intracellular granules and sometimes account for as much as 90% of cellular dry weight (Figure 10.1).

Bacteria require a carbon source and a nitrogen source as well as mineral salts for their growth and survival. Accumulation of PHAs in the bacterial cell usually occurs if a carbon source is provided in excess, and if at least one other nutrient, which is essential for growth, has been depleted. As early as 1926, Maurice Lemoigne isolated the first of the PHAs – polyhydroxybutyrate (PHB, a homopolymer whose building unit is 3-hydroxybutyric acid) from *Bacillus megaterium*. At the time of writing, as many as 190 different monomers are found making up a variety of PHAs. The type of bacterium and growth conditions determines the chemical composition of PHAs. The molecular weights of PHAs typically range from 2×10^5 to 3×10^6 Da. PHAs are synthesized from a variety of substrates and they are deposited intracellularly as insoluble spherical inclusions or PHA granules.[2–4] PHA granules are usually 50–500 nm in diameter and there are 5–10 granules per cell. These granules consist of an amorphous hydrophobic PHA polyester at the core surrounded by a phospholipid monolayer with attached or embedded proteins that include the PHA synthase (an enzyme responsible for the synthesis of PHA), PHA depolymerase (an enzyme responsible for the degradation of PHA) as well as some structural and regulatory proteins (Figure 10.2).[5]

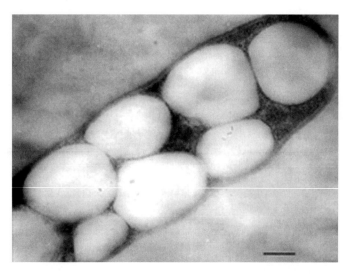

Figure 10.1 Electron micrograph of *Cupriavidus necator* accumulating PHB.[1] Bar – 0.2 μm (reproduced with permission).

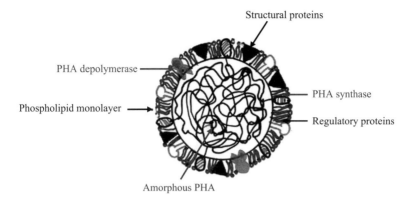

Figure 10.2 Structure of PHA granule.

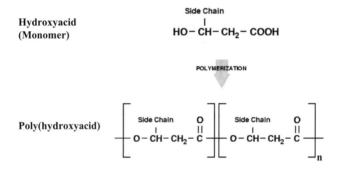

Figure 10.3 Chemical structure of PHAs. Side chain = CH_3 to C_2H_5 are short chain length PHAs. Side chain = C_3H_7 to $C_{11}H_{23}$ are medium chain length PHAs. Side chain > $C_{14}H_{29}$ are long chain length PHAs.

The large variety of microorganisms capable of synthesizing PHAs comprises heterotrophic (organisms requiring organic compound as carbon source), chemolithotrophic (organisms requiring reduced inorganic compounds as carbon source), phototrophic (utilizing light as carbon source), aerobic and anaerobic Eubacteria (a class of bacteria), as well as Archeaebacteria (bacteria living in extreme environmental conditions).[6]

PHAs are classified into three classes: short chain length PHA (PHA_{SCL}), medium chain length PHA (PHA_{MCL}) and long chain length PHA (PHA_{LCL}) (Figure 10.3). PHAs are thermoplastic and/or elastomeric, insoluble in water, enantiomerically pure, non-toxic, biocompatible, piezoelectric compounds that exhibit high degree of polymerization.

A question arises, why are the PHAs so interesting? The answer is given mainly by their properties. Considering physical properties such melting point, tensile strength, Young's modulus and elongation to break; polyhydroxybutyrate (PHB) closely resembles polypropylene (PP) – a widely used synthetic polymer. Contrary to PP, polyhydroxybutyrate is biodegradable in a reasonable time period. PHB is degraded under usual waste disposal conditions in, roughly estimated, several months. However, for polypropylene, we have to wait tens of years or centuries![7]

Thus, owing to the thermoplastic properties of PHAs and their biodegradability, they have attracted industrial interest, and bacteriologically produced PHB and its copolymers with 3-hydroxyvaleric acid are commercially available. Although PHAs are water-insoluble, hydrophobic and partially crystalline polymers, they can be degraded by a large variety of microorganisms.

10.2 Degradation of Plastics

Plastic degradation can be classified into two types: (i) abiotic degradation and (ii) biotic degradation.

10.2.1 Abiotic Degradation

In general, abiotic degradation of plastics in the environment results in a partial degradation which is due to photo-oxidation, hydrolysis, oxidation and photolysis. This will result in the formation of fragments which may remain in the environment for an indefinite period or may be further degraded biologically.

10.2.2 Biotic Degradation

Due to biodegradation, plastics may be partially degraded (into smaller fragments) or completely degraded (mineralized) into simple molecules such as methane (CH_4), carbon dioxide (CO_2) and water (H_2O). The degradation fate of a plastic depends upon the environmental conditions and the number and types of microorganisms involved in the degradation process. Biodegradation is enhanced if there is presence of ester, ether or amide bonds in the structure of plastic.

Several factors affecting biodegradation are: (1) microbiological parameters – the diversity and distribution of microbes, activity and adaptability of microbes; (2) physicochemical properties – temperature, pH, humidity, oxygen level, nutrients, activators, inhibitors, *etc.*; (3) material properties – plastic composition, its molecular weight, molecular weight distribution, crystallinity, glass transition temperature, porosity, hydrophobicity, steric configuration, *etc.*; (4) material processing – type of processing, surface characteristics, material thickness, additives, fillers and coatings.

For effective and efficient biodegradation following elements are essential:

- appropriate microorganisms
- environmental conditions conducive for the growth of microorganisms
- vulnerability of the plastic to microbial attack.

Biodegradation is essentially attributed to the role of enzymes produced and secreted by microorganisms which leads to mineralization of plastics to CO_2,

H_2O, *etc.* The enzymes are responsible for breakdown of these high molar mass products, for transport into cells followed by its utilization in the cell. The essential characteristics that an enzyme must possess include:

- accessibility to polymer for its activity
- ability to readily act on polymers at the specific site of action.

Microorganisms utilize the polymers as a carbon source and occasionally as nitrogen source, because of the enzymatic activity on these unusual substrates. The limited reports on microbial activity on polyethylene is attributed to no points of attack in its structure, except at the termination of the carbon chain, as the repeating C–C bonds are not amenable to attack.

10.2.3 Standard Methods for Plastic Biodegradation Studies

Biodegradable plastics are defined by the American Society for Testing and Materials (ASTM) as degradable plastics in which the degradation results from the action of naturally occurring microorganisms such as bacteria, fungi and algae.[8-12] Different parameters used to determine biodegradability of plastics are based on visual observations of plastic material, quantitative determination of microbial growth on plastics and determination of microbial activity in a given environment. These parameters are monitored by different indicators listed in Table 10.1. Several indicators assessing the biodegradability of polymeric materials include weight loss, molecular weight, CO_2 evolution, tensile strength, % elongation, *etc.*

Table 10.1 Practical test methods to assess biodegradability of plastic materials.

Test parameters	Measurement/ indicator	Microbial system	References
Microbial growth	Microscopic observation, determination of turbidity, protein or phospholipid content	Pure or mixed culture	13–15
Polymer utilization	Weight loss, turbidity, CO_2 evolution	Pure, mixed culture, natural ecosystem or cell-free enzyme extract	16, 17
Changes in polymer characteristics	Change in molecular weight by GPC, tensile strength and percentage elongation by NMR/FTIR	Pure culture	10, 18, 19
Microbial activities	Gas production or oxygen consumption	Sewage sludge, compost, pure/mixed culture or samples from natural ecosystem	8–11, 18
Clear zone test	Turbidimetric	Pure culture	20, 21

10.3 Biodegradation of Polyhydroxyalkanoates

As described earlier, PHAs are accumulated in the form of granules and serve as storage compounds for carbon and energy under growth-limiting conditions.[22] Thus, PHA can be degraded within the cell for its own growth and sustainability, *i.e.* intracellular degradation of PHA. Intracellular degradation is the active degradation (mobilization) of an endogenous storage PHA by the accumulating bacterium itself. Enzymes responsible for such intracellular degradation of PHA are intracellular PHA depolymerases (i-PHA depolymerases). Reports on biodegradation of PHA-based materials in the environment are also available. This is due to the enzymes secreted by microorganisms into the external environment, *i.e.* extracellular degradation. Thus, extracellular degradation is the utilization of an exogenous polymer using extracellular PHA depolymerases (e-PHA depolymerases) not necessarily produced by the accumulating microorganism.

Intracellular PHA depolymerases are unable to hydrolyse extracellular PHA, and vice versa. This is due to differences in the two biophysical conformations of PHA when present in different conditions. Within the cell, PHA exists in an amorphous (native) form and is covered by a monolayer of phospholipid embedded with several proteins. Within the cell, such granules are known as 'native PHA granules' or 'nPHA'.[5,23] After the release of these native granules from the cell either by cell lysis or solvent extraction processes or after the damage of the surface layer, the polymer denatures and becomes crystalline. This type of PHA granules is known as 'denatured PHA granules' or 'dPHA'.[24-27]

Biodegradation of PHAs can thus be viewed as two major aspects: (1) extracellular degradation of PHA and (2) intracellular degradation of PHAs.

10.3.1 Extracellular Degradation of PHA

10.3.1.1 Isolation of PHA-degrading Microorganisms

Several bacteria and fungi possess the ability to degrade extracellular PHA. Various natural ecological niches such as soil, compost, aerobic and anaerobic sewage sludge, freshwater, shallow as well as deep sea waters and estuarine sediments are reported to be rich sources of PHA-degrading microorganisms under both aerobic and anaerobic conditions. Isolation of these microorganisms can be done both by solid culture techniques as well as liquid culture techniques. In liquid culture techniques, PHA-degrading microorganisms can be enriched from samples by their inoculation in mineral salts medium (liquid enrichment) containing the desired PHA to be utilized as the sole source of carbon and energy by the organisms. Mineral salts are added to provide the basic necessities for the growth of microorganisms.

Alternatively, solid agar media containing PHA as the sole source of carbon can also be inoculated with dilutions of the samples of interest in order to obtain a wide variety of PHA degraders. The organism producing extracellular PHA depolymerase will utilize the PHA in its surrounding area and will form a clear zone of hydrolysis against the opaque agar medium containing PHA (Figure 10.4). This clear zone technique can be performed aerobically (in the

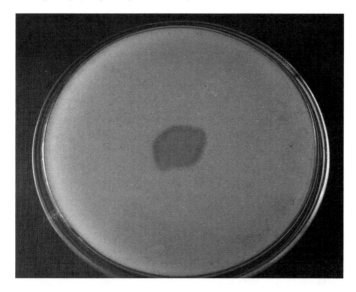

Figure 10.4 Growth of PHB depolymerase producing bacteria on solid opaque medium containing PHB showing zone of hydrolysis of PHB.

presence of oxygen), anaerobically (in the absence of oxygen), or in agar shake tubes for the selection of aerobic, anaerobic or microaerophilic microorganisms. The PHA-hydrolysing ability can be assessed by measuring the diameter of the colonies and the clear transparent zones surrounded by these colonies.

Only short chain length PHAs can be prepared as opaque suspension of granules. All medium chain length PHAs form rubber-like aggregations and cannot be used for the clear zone technique directly. For the isolation of microorganisms degrading such PHAs, thin solution cast films are layered over the solid agar media containing mineral salts.[28] A more sensitive clear zone technique for isolation of PHA degraders is one involving the use of PHA emulsions, which have a milky appearance.

Emulsions of PHA are prepared by dissolving the polymer in chloroform, adding a surfactant and emulsifying with water, followed by sonication[29] or by adding acetone-dissolved PHA to cold water[30] and removing the solvent by dialysis or evaporation. This yields an opaque, stable suspension of granules as an emulsion which can be added to the media for preparing the agar plates.[31]

10.3.1.2 Diversity of Extracellular PHA Degrading Microorganisms

PHA-degrading microorganisms, especially PHB-degrading microorganisms, are present in all terrestrial and aquatic ecosystems. It has been almost 50 years since the first PHA-degrading microorganisms belonging to the genera *Bacillus*, *Pseudomonas* and *Streptomyces* were isolated.[32] Most characterized PHA-degrading bacteria are specific for either PHA_{SCL} or PHA_{MCL} as their carbon source/substrate. However, some bacteria possess the ability to degrade both

Table 10.2 PHA-degrading bacteria reported for their ability to degrade different PHAs.

Group	Polymer degraded	Representative strains	References
I	P(3HB)	Alcaligenes faecalis T1	33
		Comamonas sp.	34
II	P(6HX)	Isolate SK850	30
III	P(3HO)	Pseudomonas fluorescens GK13	28
IV	P(3HB), P(3HV)	Pseudomonas lemoignei	35
V	P(3HB), P(6HX)	Isolae SK860	30
VI	P(3HO),P(3HD-co-3HO)	Pseudomonas fluorescens GK13	28
VII	P(6HX), P(3HO), P(3HD-co-3HO)	Isolate SK853	30
VIII	P(3HB), P(3HO), P(3HD-co-3HO)	Isolate SK827	30
IX	P(3HB), P(3HB-co-3HX-co-4HX), P(3HO), P(3HD-co-3HO)	Streptomyces exfoliatus K10	36
X	P(3HB-co-3HX-co-4HX), P96HX), P(3HO), P(3HD-co-3HO)	Isolate SK801	30
XI	P(3HB), P(3HB-co-3HX-co-4HX), P(6HX), P(3HO), P(3HD-co-3HO)	Isolate SK844	30

3HB, 3-hydroxybutyrate; 6HX, 6-hydroxyhexanoate; 3HO, 3-hydroxyoctanoate; 3HV, 3-hydroxyvalerate; 3HD, 3-hydroxydecanoate; 3HX, 3-hydroxyhexanoate; 4HX, 4-hydroxyhexanoate.

types of PHAs. This property of bacteria is due to their ability to synthesize more than one PHA depolymerase having different substrate specificities. Numerous PHA degrading bacteria, in particular PHB-degrading bacteria have been isolated and characterized since 1990. Table 10.2 lists PHA-degrading bacteria according to their polymer degrading ability.

The ability to degrade PHA is not restricted to bacteria; as many as 95 genera of PHA-degrading fungi have also been identified. Of the different genera of PHA-degrading fungi isolated from soil and marine environments, overwhelming majority (97%) were found to be members of the division Amastigomycota. Of these, the Basidiomycotina, Deuteromycotina, and Ascomycotina accounted for 48%, 27% and 19% of the total, respectively; and the Zygomycotina accounted for 2%, which indicates that higher fungi are the predominant degraders of PHAs in the environment.[37] Most studies on the environmental degradation of PHAs have shown that fungi belonging to the group Deuteromycota, including mainly *Aspergillus* spp. and *Penicillium* spp., contribute considerably to PHA breakdown in the ecological system (Table 10.3).

10.3.1.3 Biochemical Properties of Extracellular PHA Depolymerases

All the enzymes have specific properties which are determined by their structure. All the purified PHA depolymerases show specificity for either PHA_{SCL} or PHA_{MCL}. However, they have certain common characteristics.

1. The enzyme consists of single polypeptide chain with low molecular weight (<100 kDa).

Table 10.3 Some of the fungal strains reported to degrade PHA.

Strain	Type of PHA degraded	References
***Aspergillus* spp.**		
A. penicilloides	PHB	38
A. fumigatus	PHB, poly(3HB-co-3HV)	39, 40
A. fumigatus	PHB	41
A. fumigatus Pdf1	PHB, PHV, poly(3HB-co-3HV)	42
A. fumigatus 202	PHB	43
***Penicillium* spp.**		
P. funiculosum	PHB	44
P. adametzii	PHB	38
P. daleae	PHB	38
P. chermisinum	PHB	45
P. pinophilium	PHB	46
P. funiculosum	PHB	47
P. funiculosum LAR 18	PHB	48
P. minioluteum LAR 14	PHB	48
P. simplicissimum LAR13	PHB	49
Penicillium sp. DS9713a-01	PHB	50
Penicillium sp. DS9701-09a	PHB	51
***Paecilomyces* spp.**		
Paecilomyces marquandii	PHB	38
Paecilomyces lilacinus D218	PHB	52
Paecilomyces lilacinus F4-5	PHB, poly (3HB-co-3HV)	53
Other species		
Candida guilliermondii	PHB	54
Cephalosporium sp.	PHB	55
Cladosporium sp.	PHB	55
Debaromyces hansenii	PHB	54
Emericellopsis minima W2	PHB, poly (3HB-co-3HV)	56
Eupenicillium sp. IMI 300465	PHB	57
Mucor sp.	PHB	55

2. High stability at a wide range of pH, temperature, ionic strength and retain activity in solvents.
3. Gives maximum activity at alkaline pH (7.5 to 9.8).
4. Activity inhibited in presence of reducing agents like dithiothreitol (DTT). This indicates the importance of essential disulfide bonds to maintain the structure of enzyme in an active form.
5. Most of the PHA depolymerases are serine hydrolases. Their hydrolytic activity on polymers is due to the presence of serine (an amino acid) at the active site of the enzyme. This was proved by the inhibition of enzyme activity in presence of diisopropyl-fluoryl phosphate (DFP) or acylsulfonyl derivatives. This compounds bind to the active site serine and irreversibly inhibits enzyme activity.[58]
6. PHA depolymerases have high affinity for hydrophobic materials and hence their purification protocols include hydrophobic interaction chromatography.

Extracellular bacterial PHA depolymerases are non-glycosylated in nature, in contrast to all fungal PHA depolymerases which possess additional carbohydrate moieties at their N-terminal. Data indicates that this carbohydrate moiety becomes covalently attached to the depolymerase during secretion of the enzyme across the cell wall and the presence of carbohydrate is not essential for the activity of the enzyme.[59] It is possible that glycosylation enhances the resistance of the exoenzymes to elevated temperatures or to hydrolytic cleavage by proteases.[60] The synthesis of PHA_{SCL} depolymerase is generally repressed if easily utilizable carbon sources (*e.g.* glucose) are present. However, after the exhaustion of these soluble nutrients, synthesis of PHA depolymerases is de-repressed.[61] One outstanding property of PHB depolymerases is that it catalyses the reverse reaction, *i.e.* the synthesis of esters or transesterification if the reaction is performed in the absence of water (*e.g.* in solvents) showing its additional application along with PHA degradation.[62,63]

Most of the purified PHA depolymerases are of prokaryotic origin and are specific for only PHA_{SCL}. However, during last ten years, several bacterial PHA_{MCL} depolymerases exhibiting broad substrate specificities for structurally different aliphatic and aromatic PHA_{MCL} have been isolated and biochemically characterized.[28,64,65] Poly(3-hydroxyoctanoate) (PHO) depolymerase of *Pseudomonas fluorescens* GK13 is one of the most studied PHA_{MCL} depolymerases.[28] It is specific for PHA_{MCL} and for synthetic esters such as *p*-nitrophenylacyl esters with six or more carbon atoms in the fatty acid moiety. PHO depolymerase consists of two identical subunits having molecular mass of 25 kDa each. In contrast to PHA_{SCL} depolymerase, it is not inhibited by DTT or EDTA, proving that the disulfide bridges are not important in maintaining the active structure of the enzyme. No significant homologies were obtained when many sequences of PHA_{SCL} depolymerases were compared, except for small regions in the neighbourhood of a lipase box (Gly-Xaa-Ser-Xaa-Gly), an aspartate and histidine residue. Regulation of PHA_{MCL} depolymerase production is similar to that of PHA_{SCL} depolymerase. Higher production of the enzyme was observed in the presence of PHA_{MCL}. The presence of sugars or fatty acids repressed its synthesis.[28] To date there are no reports on the isolation of a fungal PHA_{MCL} depolymerase.

Brucato and Wong purified PHB depolymerase from *Penicillium funiculosum* as the first eukaryotic enzyme.[44] Fungal depolymerases show some characteristics, which are distinct from the common properties of bacterial PHB depolymerases. The following are the characteristics that are common among the fungal depolymerases.

(1) PHB depolymerases consist of one single polypeptide chain with a molecular mass ranging from 33 to 57 kDa.
(2) Fungal PHB depolymerases possess an acidic or neutral pI value.
(3) They are glycoproteins.
(4) They are susceptible to DFP or PMSF, which is a serine esterase inhibitor.
(5) They are very sensitive to dithiothreitol, indicating the essential role of disulfide bridges in the active sites.

10.3.1.4 Mechanism of PHA Hydrolysis

Studies on the purified PHB depolymerase of *Alcaligenes faecalis* on the hydrolysis of end-labelled 3-hydroxybutyrate oligomers showed both endohydrolase and exohydrolase activity.[66] Similar conclusions were drawn by Koning et al.[67] who showed that PHA depolymerase of *Pseudomonas fluorescens* GK 13 was able to hydrolyse covalently cross-linked dPHA$_{MCL}$. It is assumed that most, if not all, extracellular PHA depolymerases, including fungal enzymes have endohydrolase and exohydrolase activity.[68] Depending on the depolymerases, the end products are only monomers, monomers and dimers, or a mixture of oligomers. If oligomers are the end products of PHA depolymerases, microorganisms have oligomer hydrolases or dimer hydrolases that cleave the oligomer/dimers to the monomer. PHA depolymerase of *Comamonas* strains, *Pseudomonas stutzeri* and *Streptomyces exfoliatus* have high endogenous dimer-hydrolase activities.[69] Other bacteria, which have PHA depolymerase without dimer-hydrolase activity, have additional enzymes that are necessary for efficient hydrolysis of oligomers/dimers to monomers. Such hydrolases are located extracellularly in *A. faecalis* T1 and *Pseudomonas* sp.[70,71] or intracellularly in *P. lemoignei*,[72] *Rhodospirillum rubrum* and *Zoogloea ramigera*.[73,74] Bacteria which have intracellular hydrolases also have carrier systems for the uptake of oligomers/dimers. The polymer enantiomeric specificity of PHA depolymerases has been studied mainly with dPHB depolymerase of *A. faecalis* T1[75,76] and with nPHB depolymerase of *phaZ7* of *P. lemoignei*.[77] All PHA depolymerases analysed so far are specific for 3HA-esters in the natural (*R*) configuration. Those in the (*S*) configuration are not cleaved. However, copolymers of (*R*)- and (*S*)-3HA are hydrolysed between two adjacent (*R*)-3HA.

10.3.1.5 Structure of PHA$_{SCL}$ Depolymerases

PHA depolymerases are highly specific with respect to the length of the monomer carbon side chain of the PHA substrate. All extracellular PHA$_{SCL}$ depolymerase proteins have a complex domain structure typically comprised of four functional domains namely: signal peptide, catalytic domain (two types), linker region (three types) and substrate-binding domain (two types) (Figure 10.5).

Figure 10.5 A domain model of PHA$_{SCL}$ depolymerases[13] (modified image).

(1) The signal peptide (SP) is 22–58 amino acids long. It is responsible for transport of the enzyme across the cytoplasmic membrane of the cell. During the secretion process of the enzyme this signal peptide is cleaved off by signal peptidases.
(2) A large catalytic domain (320–400 amino acids) is situated at the N terminus of the PHA depolymerase. Serine, aspartate and histidine are found to be strictly conserved in the active centre of the catalytic domain of all PHA depolymerases. Serine is part of a lipase box pentapeptide Gly-Xaa1-Ser-Xaa2-Gly and is present in almost all known enzymes such as lipases, esterases and serine proteases. The region around the histidine resembles the oxyanion hole known from lipases.[78] The oxyanion amino acid stabilizes the transition state of the hydrolysis reaction by allowing the formation of a hydrogen bridge from the histidine peptide bond to the negatively charged enzyme-bound tetrahedral intermediate (oxyanion) (see Figure 10.6). Two types of catalytic domains are differentiated on the basis of arrangement of catalytic amino acids within the primary amino acid sequence: type I catalytic domain: histidine (oxyanion)-serine-aspartate-histidine; type II catalytic domain: serine-aspartate-histidine-histidine (oxyanion).
(3) A linker (40–100 amino acids) provides the flexible region between the catalytic and substrate-binding domains in order to increase the hydrolytic efficiency of the catalytic domain. Three different types of linkers are reported: Threonine-rich, fibronectin type 3-sequences and cadherine like.[79]
(4) Substrate-binding domain (SBD) of PHA depolymerases is situated at the C-terminal end. It binds to PHA granules and consists of 40–60 amino acids. It is highly specific in its binding to PHA and this property can be used for affinity purification of dPHB depolymerases.[80–82] PHA depolymerases lacking this substrate-binding domain do not show any hydrolysing activity, proving the importance of this domain. There are two types of SBD: namely, type 1 and type 2. Histidine (His-49), arginine (Arg-44) and a cysteine residue (Cys-2) are conserved in both types of binding domains. In type 1 additionally a second cysteine at position 57 or 59 is present.[83]

These three functional domains (namely catalytic domain, linker domain and substrate-binding domain) are essential for the enzymatic degradation of water-insoluble polymers. Hence it has been proposed that the degradation of the polymer should proceed in three steps: (1) adsorption of the enzyme to the polymer, (2) non-hydrolytic disruption of the structure of the polymer, and (3) hydrolysis.[84]

10.3.1.6 *Pseudomonas lemoignei: A Model Bacteria Producing PHB Depolymerase*

The most intensively studied PHA-degrading organism is *Paucimonas* (*Pseudomonas*) *lemoignei*.[85] It belongs to the β-subclass of the Proteobacteria. *P. lemoignei* is unique among PHA-degrading bacteria because it synthesizes at least seven PHB depolymerases (*phaZ1* to *phaZ7*). Microorganisms having the ability to degrade natural polymers such as cellulose or chitin have also shown the presence of several isoenzymes within single strain. Glycosylation in *P. lemoignei* dPHB depolymerase is not well understood. It might be that the carbohydrate moiety becomes non-covalently attached to the depolymerase during secretion of the enzyme across the cell wall. The presence of carbohydrate is not essential for activity of *P. lemoignei* depolymerases because recombinant *E. coli* or *B. subtilis* strains harbouring the corresponding depolymerase gene synthesized the non-glycosylated form of the enzyme but had almost the same specific activity. It is possible that glycosylation enhances the resistance of the exoenzyme to elevated temperature and/or to hydrolytic cleavage by proteases of competing microorganisms. PHB depolymerases of *P. lemoignei* apparently catalyse the reverse reaction, *i.e.* the synthesis of esters for transesterification, if the reaction is performed in the absence of water (*e.g.* in solvents).[62,63]

10.3.1.7 Catalytic Mechanism of PHB Depolymerase of Penicillium funiculosum

The crystal structure analysis of PHB depolymerase from *Penicillium funiculosum* has been studied recently.[84] According to the structure described, the spatial arrangement of the catalytic residues in the enzyme indicates that the mechanism of the depolymerase reaction may be similar to that of a lipase/serine esterase. The overall catalytic mechanism is as shown in Figure 10.6.

Asp$_{121}$-His$_{155}$-Ser$_{39}$ forms the active site of the enzyme. Ser$_{39}$ plays a central role in the catalytic reaction. Ser$_{39}$ does not hydrogen bond to His$_{155}$ in the ligand-free state, and it is likely that Ser$_{39}$ changes its conformation upon binding of a substrate, forming a hydrogen bond with His$_{155}$, and raising its nucleophilicity.

(1) The oxygen atom of Ser$_{39}$ makes a nucleophilic attack on the carbonyl carbon atom of a bound PHB chain. The His$_{155}$-Asp$_{121}$ hydrogen bonding system may enhance the nucleophilicity of the hydroxyl group of Ser$_{39}$.

(2) Oxygen atom of Ser$_{39}$ binds to the carbonyl carbon of PHB making a tetrahedral intermediate having a formal negative charge on the oxygen atom derived from the carbonyl group. This charge is stabilized by interactions with the –NH group from amino acids in a site termed as the oxyanion hole. These interactions also help to stabilize the transition state that precedes the formation of tetrahedral intermediate.

(3) A proton transfers from the positively charged His$_{155}$ to the O formed by the cleavage of the ester bond (acyl-enzyme intermediate).

(4) The first product is now free to depart from the enzyme, completing the first stage of hydrolytic reaction – acylation of the enzyme.

(5) The next stage is deacylation of enzyme and it begins with the entry of a water molecule. The ester group of the acyl-enzyme is now hydrolysed by a process that essentially repeats steps 2–4.

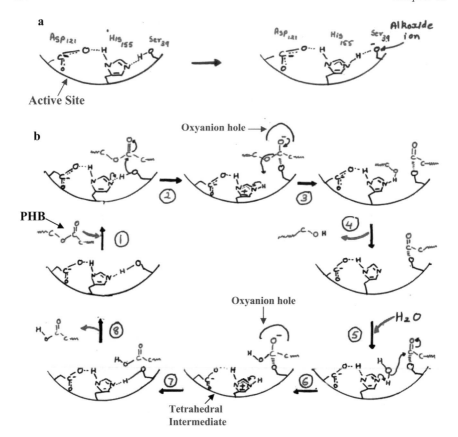

Figure 10.6 A proposed mechanism for the action of the PHB depolymerase from *P. funiculosum*[70] (Image modified). a: catalytic triad in active site of PHB depolymerase; b: catalytic mechanism.

(6) His_{155} draws a proton away from water. The resulting OH^- ion attacks carbonyl carbon atom of the acyl group, forming a tetrahedral intermediate.

(7) This structure breaks down to release the carboxylic acid product.

(8) Finally, the carboxylic acid product is released and the enzyme can enter into another round of catalysis.

10.3.1.8 Influence of Physicochemical Properties of the PHAs on Biodegradability

PHA depolymerases differ highly in their specificities towards various PHAs. The polymer itself also influences the biodegradability. The most important factors are the following.

(1) The stereoregularity of the PHA: PHA depolymerases hydrolyses only ester linkages in the (*R*) configuration.[86–88]

(2) The crystallinity of PHA: the more crystalline the PHA, the less its degradability will be.[89]
(3) The molecular mass of PHA: low molecular mass PHAs are generally degraded more rapidly than high molecular mass PHAs.
(4) The monomeric composition of PHA: The composition of PHA will definitely affect the degradation rates of the polymers. The more complex the polymeric composition, the more time it will take to degrade.

However, the degradation of PHA in complex ecosystems can not be predicted from laboratory experiments using pure culture and/or purified enzymes alone.

10.3.2 Intracellular Degradation of PHA

The major role of PHAs is of a beneficial reserve material, the ability to store large quantities of reduced carbon without significant effects on the osmotic pressure of the cell.[2] As discussed earlier, the proteins attached to PHA granules include PHA polymerase/synthase, PHA depolymerase, structural proteins known as phasins and some other proteins with unknown functions. PHA depolymerases present inside the cells are known as intracellular PHA depolymerases (i-PHA depolymerases). i-PHA depolymerase is an enzyme responsible for degradation/mobilization/hydrolysis of PHA for the survival of the cell in absence of an exogenous carbon source.[27,90] There are many studies on intracellular mobilization of PHB in PHB-accumulating organisms.

10.3.2.1 Diversity of Microbial Strains Studied for Intracellular PHA Degradation

Intracellular PHA degradation is studied from the point of view of understanding the cyclic metabolism of PHAs and to understand the physiological role of occurrence of PHA granules in the cell. Lemoigne (1925) was the first to observe this intracellular mobilization of PHB granules in *Bacillus*.[91] Later, several studies on the metabolism of PHA mobilization and the role of PHA depolymerase were carried out in *Bacillus megaterium*,[91,92] *Rhodospirillum rubrum*,[24,93] *Azotobacter beijerinckii*,[94] *Paracoccus denitrificans*,[95] *Zoogloea ramigera*,[96] *Spirillium*,[97] *Hydrogenomonas* H16,[98] *Pseudomonas oleovorans*,[99] *Micrococcus halodenitirificans*, *Legionella pneumophila*,[100] *Rhodobacter sphaeroids*,[101] *Bacillus thuringiensis* and[102] *Pseudomonas oleovorans*.[103,104] Detailed study on intracellular PHB mobilization has been carried out in *Ralstonia eutropha* (now better known as *Cupriavidus necator*).[105-111]

10.3.2.2 Mechanism of Intracellular PHA Hydrolysis

Intracellular degradation of PHA by depolymerases has not been studied in as much depth as compared to extracellular PHA degradation. *C. necator*, an organism able to accumulate PHB, was the organism of research to study the intracellular mobilization. This is mediated by a cyclic metabolic route: from acetyl-CoA via acetoacetyl-CoA and 3HB-CoA to PHB by β-ketothiolase, acetoacetyl CoA reductase and PHB synthase, respectively (PHB biosynthesis sequence) and from PHB via 3HB and acetoacetate to acetoacetyl-CoA and acetyl-CoA by intracellular PHB (i-PHB) depolymerase, 3HB dehydrogenase, acetocaetate:succinyl-CoA transferase and ketothiolase (PHB degradation sequence). However, the recent research efforts have led to a possibility that thiolysis of PHB would be a more appropriate metabolic route to be followed by the organism to degrade PHB in order to make the whole process energy efficient. But the major role for the thiolysis of PHB will be played by intracellular PHB depolymerase, leading to a degradation sequence of PHB to 3-hydroxybutryl CoA to acetoacetyl-CoA and acetyl-CoA. It has been reported that the intracellular PHB degradation due to PHB depolymerase led to synthesis of 3-hydroxybutyryl-CoA instead of 3-hydroxybutyric acid.[110] Thus the mode of action of intracellular PHB depolymerase on PHB is definitely different from that of extracellular PHB on crystalline PHB. Much of the details on the mechanism of action of intracellular PHB depolymerase is yet to be explored.

Intracellular PHB Depolymerase of *Cupriavidus necator*. Note that *C. necator* was previously known as *Ralstonia eutropha*, *Wautersia eutropha*, *Alcaligenes eutropha* and *Hydrogenomonas eutropha*. iPHB depolymerases of PHA_{SCL}-accumulating bacteria were not described before the year 2000 but since then not less than seven putative iPHB depolymerases and two 3-hydroxybutyrate oligomer hydrolases have been postulated for *C. necator*.[105,107,108,112–114] Depending on similarities of the primary amino acid sequences to each other, these hydrolases were named PhaZa1 to PhaZa5 and PhaZd1 and PhaZd2 for the iPHB depolymerases and PhaZb and PhaZc for the oligomer hydrolases, respectively. Alternative names for PhaZb and PhaZc are PhaY1 and PhaY2.[113]

Intracellular PHB Depolymerase of *Rhodospirillum rubrum*. *R. rubrum* possesses a putative intracellular PHB depolymerase system consisting of a soluble PHB depolymerase, a heat-stable inactivator and a 3-hydroxybutyrate dimer hydrolase. It consists of one polypeptide of around 35 kDa and has a pH optima of 9 and temperature optima of 55 °C. The purified enzyme was inactive with dPHB. It does not show lipase, protease or esterase activity with *p*-nitrophenyl fatty acid esters. It is highly substrate specific.[27]

Intracellular PHB Depolymerase of *Paracoccus denitrificans*. This enzyme does not have any similarity with i-PHB depolymerase of *R. rubrum* but has high similarities to i-PHB depolymerase of *C. necator*.

Intracellular PHA_{MCL} Depolymerase. iPHA depolymerases have been described first for PHA_{MCL}-accumulating *Pseudomonas putida* (*P. oleovorans*). pH optima of the enzyme with PHO as a substrate was 8.[104,115–118]

10.4 Conclusions

Microbial polyhydroxyalkanoates (PHAs) represent a promising class of thermoplastic polyesters. They are receiving much attention as biodegradable substitutes for non-degradable plastics. Numerous bacteria have been found to exhibit the capacity for intracellular PHA accumulation. As biodegradability is one of the most prominent characteristics of PHAs, their biotechnological exploitation should always be accompanied by a thorough investigation of environmental and biological factors responsible for the degradation of PHA-based products released in the environment. Detailed studies on intracellular as well as extracellular PHA degradation by several bacterial as well as fungal PHA depolymerases have been carried out. This knowledge of environmental and enzymatic degradability of PHAs will open up many possibilities for designing processes to produce tailor-made PHAs. Several blends and grafts of PHAs can be designed on the basis of this information. Further investigations and efforts must be undertaken by scientists in order to reduce the production costs of these PHAs and provide industrial sustainability and commercialization to PHAs.

References

1. M. Potter, M. H. Padkour, F. Mayer and A. Steinbuchel, *Microbiology*, 2002, **148**, 2413.
2. A. J. Anderson and E. A. Dawes, *Microbiol. Rev.*, 1990, **54**, 450.
3. L. L. Madison and G. W. Huisman, *Microbiol. Mol. Biol. Rev.*, 1999, **63**, 21.
4. M. Zinn, B. Witholt and T. Egli, *Adv. Drug Delivery*, 2001, **53**, 5.
5. K. Grage, A. C. Jahns, N. Parlane, R. Palanisamy, I. A. Rasiah, J. A. Atwood and B. H. A. Rehm, *Biomacromolecules*, 2009, **10**, 660.
6. A. Steinbuchel, Polyhydroxyalkanoic acid in *Biomaterials*, ed. D. Byrom, Macmillan Publishers Ltd, Basingstoke, 1991, p. 123.
7. B. Witholt and B. Kessler, *Curr. Opin. Biotechnol.*, 1999, **10**, 279.
8. American Society for Testing & Materials (D 5209-92) in *Annual Book of ASTM Standards, Vol 8.03*, ASTM, Philadelphia, 1994, p. 372.
9. American Society for Testing & Materials (D 5210-92) in *Annual Book of ASTM Standards, Vol 8.03*, ASTM, Philadelphia, 1994, p. 376.
10. American Society for Testing & Materials (D 52477-92) in *Annual Book of ASTM Standards, Vol 8.03*, ASTM, Philadelphia, 1994, p. 390.
11. American Society for Testing & Materials (D 5271-93) in *Annual Book of ASTM Standards, Vol 8.03*, ASTM, Philadelphia, 1994, p. 409.
12. American Society for Testing & Materials (D 5338-92) in *Annual Book of ASTM Standards, Vol 8.03*, ASTM, Philadelphia, 1994, p. 139.
13. American Society for Testing & Materials (G21-70) in *Annual Book of ASTM Standards, Vol 8.03*, ASTM, Philadelphia, 1986, p. 817.
14. American Society for Testing & Materials (G22-76) in *Annual Book of ASTM Standards, Vol 8.03*, ASTM, Philadelphia, 1986, p. 822.

15. T. M. Aminabhavi, R. H. Balundgi and P. E. Cassidy, *Polym. Plast. Technol. Eng.*, 1990, **29**, 235.
16. H. Brandl and P. Puchner, *Biodegradation*, 1992, **2**, 237.
17. J. Mergaert, C. Anderson, A. Wouters, J. Swings and K. Kersters, *FEMS Microbiol. Rev.*, 1992, **103**, 317.
18. B. Lee, A. L. Pometto, A. Fratzke and T. B. Bailley, *Appl. Environ. Microbiol.*, 1991, **57**, 678.
19. D. L. Kaplan, J. M. Mayer, M. Greenberger, R. A. Gross and S. McCarthy, *Polym. Degrad. Stab.*, 1994, **45**, 165.
20. H. Brandi, R. Bachofen, J. Mayer and E. Wintermantel, *Can. J. Microbiol.*, 1995, **41**, 143.
21. L. J. Foster, S. J. Zervas, R. W. Lenz and R. C. Fuller, *Biodegradation*, 1995, **6**, 67.
22. H. G. Schelgel, H. Kaltwasser and G. Gottschalk, *Arch. Mikrobiol.*, 1961, **38**, 209.
23. M. Potter and A. Steinbuchel, *Biomacromolecules*, 2005, **6**, 552.
24. J. Merrick and M. Doudoroff, *J. Bacteriol.*, 1964, **88**, 60.
25. R. J. Griebel and J. M. Merrick, *J. Bacteriol.*, 1971, **108**, 782.
26. P. J. Hocking, R. H. Marchessault, Biopolyesters in: *Chemistry and Technology of Biodegradable Polymers*, ed. G. J. L. Griffin, Chapman & Hall, London, 1994, p. 48.
27. D. Jendrossek and R. Handrick, *Ann. Rev. Microbiol.*, 2002, **56**, 403.
28. A. Schirmer, D. Jendrossek and H. G. Schelgel, *Appl. Environ. Microbiol.* 1993, **59**, 1220.
29. D. M. Horowitz and J. K. M. Sanders, *Can. J. Microbiol.*, 1995, **115**, 123.
30. A. Schirmer, C. Matz and D. Jendrossek, *Can. J. Microbiol.*, 1995, **41**, 170.
31. D. Jendrossek, *Appl. Microbiol. Biotechnol.*, 2007, **74**, 1186.
32. A. A. Chowdhury, *Arch. Mikrobiol.*, 1963, **47**, 167.
33. T. Tanio, T. Fukui, Y. Shirakura, T. Saito, K. Tomita and T. Kaiho, *Eur. J. Biochem.*, 1982, **124**, 71.
34. D. Jendrossek, I. Knoke, R. Habibian, A. Steinbuchel and H. Schelgel, *J. Environ. Polym. Degrad.*, 1993, **1**, 53.
35. B. Muller and D. Jendrossek, *Appl. Microbiol. Biotechnol.*, 1993, **38**, 487.
36. B. Klingbeil, R. Kroppenstedt and D. Jendrossek, *FEMS Microbiol. Lett.*, 1996, **142**, 215.
37. D. Y. Kim and Y. H. Rhee, *Appl. Microbiol. Biotechnol.*, 2003, **61**, 300.
38. J. Mergaert, C. Anderson, A. Wouters, J. Swings and K. Kersters, *FEMS Microbiol. Rev.*, 1992, **103**, 317.
39. J. Mergaert, A. Webb, C. Anderson, A. Wouters and J. Swings, *Appl. Environ. Microbiol.*, 1993, **59**, 3233.
40. J. Mergaert, C. Anderson, A. Wouters and J. Swings, *J. Environ. Polym. Degrad.*, 1994, **2**, 177.
41. T. M. Scherer, R. C. Fuller, R.W. Lenz and S. Goodwin, *Polym. Degrad. Stab.*, 1999, **64**, 267.
42. S. Iyer, R. Shah, A. Sharma, D. Jendrossek and A. Desai, *J. Polym. Environ.*, 2000, **8**, 197.

43. R. Bhatt, K. C. Patel and U. Trivedi, *J. Polym. Environ.*, 2010, **18**, 141.
44. C. Brucato and S. Wong, *Arch. Biochem. Biophys.*, 1991, **290**, 497.
45. J. Mergaert, A. Wouters, C. Anderson and J. Swings, *Can. J. Microbiol.*, 1995, **41**, 154.
46. J. S. Han, Y. J. Son, C. S. Chang and M. N. Kim, *J. Microbiol.*, 1998, **36**, 67.
47. S. Miyazaki, K. Takahashi, M. Shiraki, T. Saito, Y. Tezuka and K. Kasuya, *J. Polym. Environ.*, 2000, **8**, 175.
48. M. N. Kim, A. R. Lee, J. S. Yoon and J. J. Chin, *Eur. Polym. J.*, 2000, **36**, 1677.
49. J. S. Han and M. N. Kim, *J. Microbiol.*, 2002, **40**, 20.
50. C. Su-Qin, C. Shan, D. Liu and H. Xia, *World J. Microbiol, Biotechnol.*, 2006, **22**, 729.
51. H. Liu, H. Zhang, S. Chen, D. Liu and H. Xia, *J. Polym. Environ.*, 2006, **14**, 419.
52. Y. Oda, H. Osaka, T. Urakami and K. Tonomura, *Curr. Microbiol.*, 1997, **34**, 230.
53. B. I. Sang, W. Lee, K. Hori and H. Unno, *World J. Microbiol. Biotechnol.*, 2006, **22**, 51.
54. K. E. Gonda, D. Jendrossek and H. P. Molitoris, *Hydrobiologia*, 2000, **426**, 173.
55. M. Matavulj and H. Molitoris, *FEMS Microbiol. Rev.*, 1992, **103**, 323.
56. D. Y. Kim, J. H. Yun, H. W. Kim, K. S. Bae and Y. H. Rhee, *J. Microbiol.*, 2002, **40**, 129.
57. D. W. McLellan and P. J. Halling, *FEMS Microbiol. Rev.*, 1988, **52**, 215.
58. P. Gruchulski, F. Bouthillier, R. Kazlauskas, A. Serrequi, I. Schrag, E. Ziomek and M. Cygler, *Biochemistry*, 1994, **33**, 3494.
59. B. H. Briese, B. Schmidt and D. Jendrossek, *J. Environ. Polym. Degrad.*, 1994, **2**, 75.
60. D. Jendrossek, A. Frisse, A. Behrends, M. Andermann and H. D. Kratzin, *J. Bacteriol.*, 1995, **177**, 596.
61. D. Jendrossek, B. Muller and H. G. Schelgel, *Eur. J. Biochem.*, 1993, **218**, 701.
62. A. Kumar, R. A. Gross and D. Jendrossek, *J. Org. Chem.*, 2000, **65**, 7800.
63. Y. Suzuki, S. Taguchi, S. Saito, K. Toshima, S. Matsumura and Y. Doi, *Biomacromolecules*, 2001, **2**, 541.
64. H. Kim, H. S. Ju and J. Kim, *Appl. Microbiol. Biotechnol.*, 2000, **53**, 323.
65. H. M. Kim, K. E. Ryu, K. S. Bae and Y. H. Rhee, *Biosci. Bioeng.*, 2000, **89**, 196.
66. K. Nakayama, T. Saito, T. Fukui, Y. Shirakura and K. Tomita, *Biochim. Biophys. Acta*, 1985, **827**, 63.
67. G. M. Koning, H. M. van Bilsen, P. J. Lemstra, W. Hazenberg and B. Witholt, *Polymer*, 1994, **35**, 2090.
68. T. M. Scherer, C. Fulller, S. Goodwin and R. W. Lenz, *Biomacromolecules*, 2000, **1**, 577.
69. M. Shiraki, T. Shimada, M. Tatsumichi and T. Saito, *J. Environ. Polym. Degrad.*, 1995, **3**, 13.

70. Y. Shirakura, T. Fukui, T. Tanio, K. Nakayama, R. Matsuno and K. Tomita, *Biochim. Biophys. Acta*, 1983, **748**, 331.
71. Y. Shirakura, T. Fukui, T. Saito, Y. Okamoto, T. Narikawa, K. Kide, K. Tomita, T. Takemasa and S. Masamune, *Biochim. Biophys. Acta*, 1986, **880**, 46.
72. F. Delafield, M. Doudoroff, N. Palleroni, C. Lusty and R. Contopoulos, *J. Bacteriol.*, 1965, **90**, 1455.
73. J. Merrick and C. Yu, *Biochemistry*, 1966, **5**, 3563.
74. Y. Tanaka, T. Saito, T. Fukui, T. Tanio and K. Tomita, *Eur. J. Biochem.*, 1981, **118**, 177.
75. B. M. Bachmann and D. Seebach, *Macromolecules*, 1999, **32**, 1777.
76. Y. Doi, S. Kitamura and H. Abe, *Macromolecules*, 1995, **28**, 4822.
77. R. Handrick, S. Reinhadt, M. L. Focarete, M. Scandola and G. Adamus, *J. Biol. Chem.*, 2001, **276**, 215.
78. K. Jaeger, S. Ransac, B. Dijkstra, D. Colson, M. Heuvel and O. Misset, *FEMS Microbiol. Rev.*, 1994, **15**, 29.
79. M. Nojiri and T. Saito, *J. Bacteriol.*, 1997, **179**, 6965.
80. A. Behrends, B. Klingbell and D. Jendrossek, *FEMS Microbiol. Lett.*, 1996, **143**, 191.
81. K. Kasuya, T. Ohura, K. Masuda and Y. Doi, *Int. Biol. Macromol.*, 1999, **24**, 329.
82. T. Ohura, K. Kasuya and Y. Doi, *Appl. Environ. Microbiol.*, 1999, **65**, 189.
83. K. Yamashita, Y. Aoyagi, H. Abe and Y. Doi, *Biomacromolecules*, 2001, **2**, 25.
84. T. Hisano, K. Kasuya, Y. Tezuka, N. Ishii, T. Kobayashi, M. Shiraki, E. Oroudjev, H. Hansma, T. Iwata, Y. Doi, T. Saito and K. Miki, *J. Mol. Biol.*, 2006, **356**, 993.
85. D. Jendrossek, *Int. J. Syst. Evol. Microbiol.*, 2001, **51**, 905.
86. H. Abe, I. Matsubara, Y. Doi, Y. Hori and A. Yamaguchi, *Macromolecules*, 1994, **27**, 6018.
87. H. Abe, I. Matsubara and Y. Doi, *Macromolecules*, 1995, **28**, 844.
88. P. J. Hocking, M. R. Timmins, T. M. Scherer, R. C. Fuller, R. W. Lenz and R. H. Marchessault, *J. M. S. Pure Appl. Chem. A*, 1994, **32**, 889.
89. Y. Kumagai, Y. Kanasawa and Y. Doi, *Makromol. Chem.*, 1992, **193**, 53.
90. T. Saito and T. Kobayashi in *Biopolymers vol. 3b: Polyesters*, ed. Y. Doi and A. Steinbuchel, Wiley-VCH, 2002, p. 23.
91. M. Lemoigne, *Ann. Inst. Pastuer*, 1925, **39**, 144.
92. R. Macrae and J. Wilkinson, *J. Gen. Microbiol.*, 1958, **19**, 210.
93. R. Handrick, S. Reinhardt, P. Kimmig and D. Jendrossek, *J. Bacteriol.*, 2004, **186**, 7243.
94. P. J. Senior and E. A. Dawes, *Biochem. J.*, 1973, **134**, 225.
95. S. Gao, A. Maehar, T. Yamane and S. Ueda, *FEMS Microbiol. Lett.*, 2001, **196**, 159.
96. T. Saito, H. Saegusa, Y. Miyata and T. Fukui, *FEMS Microbiol. Rev.*, 1992, **103**, 333.
97. A. Hayward, W. Forsyth and J. Roberts, *J. Gen. Microbiol.*, 1959, **20**, 510.

98. H. Hippe and H. Schelgel, *Arch. Mikrobiol.*, 1967, **56**, 278.
99. L. Foster, E. Stuart, A. Tehrani, R. Lenz and R. Fuller, *Int, J. Biol. Macromol.*, 1996, **19**, 177.
100. B. James, W. S. Mauchline, P. J. Dennis, C. W. Keevil and R. Wait, *Appl. Environ. Microbiol.*, 1999, **65**, 822.
101. T. Kobayashi, K. Nishikori and T. Saito, *Curr. Microbiol.*, 2004, **49**, 199.
102. C. L. Tseng, H. J. Chen and G. C. Shaw, *J. Bacteriol.*, 2006, **188**, 7592.
103. E. S. Stuart, L. J. Foster, R. W. Lenz and R. C. Fuller, *Int. J. Biol. Macromol.*, 1996, **19**, 171.
104. G. W. Huisman, E. Wonink, R. Meima, B. Kazemier, P. Terpstra and B. Witholt, *J. Biol. Chem.*, 1991, **266**, 2191.
105. R. Handrick, S. Reinhardt and D. Jendrossek, *J. Bacteriol.*, 2000, **182**, 5916.
106. T. Kobayashi, M. Shiraki, T. Abe, A. Sugiyama and T. Saito, *J. Bacteriol.*, 2003, **185**, 3485.
107. G. M. York, J. Lupberger, J. Tian, A. G. Lawrence, J. Stubbe and A. J. Sinskey, *J. Bacteriol.*, 2003, **185**, 3788.
108. T. Abe, T. Kobayashi and T. Saito, *J. Bacteriol.*, 2005, **187**, 6982.
109. K. Uchino and T. Saito, *J. Biochem.*, 2006, **139**, 615.
110. K. Uchino, T. Saito, B. Gebauer and D. Jendrossek, *J. Bacteriol.*, 2007, **189**, 8250.
111. K. Uchino, T. Saito and D. Jendrossek, *Appl. Environ. Microbiol.*, 2008, **74**, 1058.
112. T. Kobayashi and T. Saito, *J. Biosci. Bioeng.*, 2003, **96**, 487.
113. A. Pohlmann, W. F. Fricke, F. Reinecke, B. Kusian, H. Liesegang, R. Cramm, T. Eitinger, C. Ewering, M. Potter, E. Schwartz, A. Strittmatter, I. Voss, G. Gottschalk, A. Steinbuchel, B. Friedrich and B. Bowien, *Nat. Biotechnol.*, 2006, **24**, 1257.
114. H. Saegusa, M. Shiraki, C. Kanai and T. Saito, *J. Bacteriol.*, 2001, **183**, 94.
115. L. J. Foster, R. W. Lenz and R. C. Fuller, *FEMS Microbiol. Lett.*, 1994, **118**, 279.
116. L. J. Foster, E. S. Stuart, A. Tehrani, R. W. Lenz and R. C. Fuller, *Int. J. Biol. Macromol.*, 1996, **19**, 177.
117. L. J. Foster, R. W. Lenz and R. C. Fuller, *Int. J. Biol. Macromol.*, 1999, **26**, 187.
118. L. I. de Eugenio, P. Garcia, J. M. Luengo, J. M. Sanz, J. S. Roman, J. L. Garcia and M. A. Prieto, *J. Biol. Chem.*, 2007, **282**, 4951.

CHAPTER 11
Degradation of Biodegradable and Green Polymers in the Composting Environment

ACKMEZ MUDHOO,[1] ROMEELA MOHEE,[1]
GEETA D. UNMAR[1] AND SANJAY K. SHARMA[2]

[1] University of Mauritius, Department of Chemical and Environmental Engineering, Faculty of Engineering, Réduit, Mauritius; [2] Jaipur Engineering College & Research Centre, JECRC Foundation, Jaipur, Rajasthan, India

11.1 Introduction

A significant majority of synthetic polymers are recalcitrant to any microbial attack[1–3] due to their excessive molecular mass, high number of aromatic rings, unusual bonds or halogen substitutions.[4] Once such a material enters the natural ecosystem, the harmful effects are long-lasting and give way to several other hygienic, health and environmental complications. This explains clearly why plastics then contribute considerably to the contamination of the environment.[2] Living conditions in the biosphere have consequently changed drastically and to such an extent that the presence of non-biodegradable residues is affecting the survival of many species.[5] With the development of technologies for physical, energetic and chemical recycling of polymeric waste progressing in parallel, an increasing interest through intensive research and development is now being devoted to the synthesis and characterization of polymers with enhanced sensitivity and susceptibility to biodegradation. Thence, the use of biodegradable polymers, namely in applications with a relatively short useful life cycle of the products such as packaging, would be an ecologically green and affordable alternative for reducing the accumulation of solid plastics waste and the related disposal issues.[2] In this line of thought, the need to dispose of plastics in an ecologically sound and environmentally safe manner has fostered the growth of biodegradable polymers industry.[6]

11.1.1 Biodegradable Polymers

Biodegradable polymers are designed to degrade upon disposal by the action of living microorganisms.[7] Promising and fruitful progress has been made in the development of practical processes and products from polymers such as starch, cellulose and lactic acid. The need to create alternative biodegradable water-soluble polymers for down-the-drain products such as detergents, pharmaceuticals and cosmetics has taken more importance[7] and fits very well with the maturing 'Green Chemistry' concept.[8,9] Consumers have, however, so far attached meagre value to the biodegradability of such commodities, urging the industries to compete frantically on a cost-performance basis with existing familiar products.[7] Biopolymers are polymers that are generated from renewable natural sources[10,11] are often biodegradable and nontoxic.[12] It is earnestly perceived that polymers based on biological resources are 'greener' than synthetic polymers even although the latter may also be biodegradable. The justification, which is all the more logical and verifiable, for using renewable resources is that the carbon dioxide (CO_2) burden in the environment is neutral for biologically based polymers but is positive for polymers based on mineral oil.[11]

According to the American Society for Testing and Materials (ASTM), biopolymers are degradable polymers in which degradation is brought about from the action of naturally occurring bacteria, fungi and algae.[12,13] Biopolymers can be produced by biological systems involving microorganisms, plants and animals, or chemically synthesized from biological materials such as sugars, starch, natural fats or oils.[12] Chemical synthetic pathways are designed, tested, validated and applied in converting these raw materials into biodegradable polymers through extraction of the native polymer from a plant or animal tissue, and a chemical or biotechnological route of monomer polymerization. An alternative to petroleum-based polymers, some biopolymers degrade in only a few weeks while the degradation of others may draw out to several months.[12]

11.1.2 Degradability through Composting

In principle, the properties relevant for the use of the polymer as well as for its biodegradability are determined by the molecular structure of the biodegradable polymer. In Europe, standards have been proposed for evaluating the biodegradability of biopolymers based on the composting technique. A number of standard methods have been developed to estimate the extent of biodegradability of biodegradable polymers under various conditions and with a variety of microorganisms.[14] They tend to be used mainly in the countries where they have been developed, although there is considerable overlap between the standard procedures of different countries. This commonness hence provides wide scope for the development of consistent and international standards.

The growth of composting, a proven sustainable bioremediation technique, as an ecologically sound waste management technique[6] supports the need for biodegradable polymers in the economic scene. Polyesters such as poly-ε-caprolactone (PCL), poly(lactic acid)-based, poly(hydroxybutyrate-co-hydroxyvalerate), thermoplastic starch and modified starch formulations, poly(vinyl alcohol) and protein polymers are examples of biodegradable polymeric materials being introduced into the marketplace. This review chapter will approach the coupled topic of biodegradation of biopolymers and composting through an outline of the mechanistic aspects of polymer biodegradation, a discussion on the essentials of the composting process and a reasonably comprehensive appraisal of the progress achieved in research on the degradation of polyhydroxyalkanoates (PHAs), poly(lactic acid)-based polymers, polyethylenes and PCLs in the composting environment. To achieve a balanced, approach, a variety of research conducted on biodegradable polymers and composting has been presented and discussed. However, this discussion is not claimed to be complete, but in the interim, the authors hope to foster further research and discussion so that the body of knowledge in this field is expanded, enriched and applied.

11.2 Degradation of Biodegradable Polymers

Biodegradation of a biopolymer implies the utilization of the polymer substrate as the sole carbon source.[6] The production of CO_2[15] under aerobic conditions[16] and methane (CH_4) under anaerobic environments and humic substances (compost) is the true measure of biodegradation.[6] The use of positive controls (naturally biodegradable polymers such as cellophane) and negative controls (recalcitrant polymers such as polystyrene) provides a relative and quantitative measure of biodegradability. Biodegradation is to be a very important property for water-soluble or water-immiscible polymers because they eventually enter streams which can neither be recycled nor incinerated[3] for economic, environmental and pollution prevention reasons and concerns.[17] It is important to consider the microbial degradation of natural and synthetic biodegradable polymers in order to understand what is necessary for the biodegradation and mechanisms involved.[3] This requires an understanding of the interactions between the polymeric materials and the microorganisms and the biochemical changes involved during the biodegradation process.[3]

11.2.1 Polymer Biodegradation Mechanisms

Because of a lack of water solubility and the size of the polymer molecules, microorganisms are unable to transport the polymeric material directly into the cells where most biochemical processes take place.[18] Extracellular enzymes must first be secreted which 'depolymerize' the polymers (*i.e.* initiate an initial breakdown of the polymer structure) outside the cells.[19] The end products of these initial metabolic processes include water, carbon dioxide and/or methane together with new biomass. The 'depolymerization' property of polymers by enzymes is of great interest in the synthesis, development and, more importantly, application of biodegradable plastics. Polyesters play the dominant role in biodegradable polymers.[19] The chain mobility of the polymer chains proved to be the most relevant factor controlling polyester biodegradability, usually excluding many aromatic polyesters such as polyethylene terephthalate (PET) from biodegradation. Recently a new thermophilic hydrolase having an optimum performance at 65 °C from *Thermobifida fusca* (TfH) has been isolated, characterized and expressed in recombinant *Escherichia coli*. This enzyme is especially active in degrading polyesters containing aromatic constituents, and exhibits a 65% sequence similarity to *Streptomyces albus* isolated lipase, and combines similarly to lipases and esterases.[19]

The extracellular enzymes being too large to penetrate deeply into the polymer material, have to act only on the polymer surface. In this respect, the biodegradation of polymers may be considered and is usually a surface erosion process.[18] Although the enzyme-catalysed polymer chain length reduction is in the primary process of biodegradation, non-biotic chemical and physical processes can co-act on the polymer, either simultaneously or in a first stage solely on the polymer. These non-biotic effects include chemical hydrolysis, thermal polymer degradation, and oxidation or scission of the polymer chains by irradiation.[20] For some materials, these effects are used directly to induce the biodegradation process, but they must also to be taken into account when biodegradation is caused mainly by extracellular enzymes. Because of the coexistence of biotic and non-biotic processes, the overall mechanism of polymer degradation can be referred to as 'environmental degradation'.[18] Biodegradation of polymeric bio-based materials involves cleavage of hydrolytically or enzymatically sensitive bonds in the polymer leading to polymer erosion.[21] Glycosidic bonds, but ester and peptide linkages as well, are subject to hydrolysis through nucleophilic attack on the carbonyl carbon atom.[22,23] Ether linkages are cleaved by aerobic bacteria through monooxygenase reactions which transform the comparably stable ether linkage through hydroxylation to a hemiacetal structure of low stability.[24]

Depending on the mode of degradation, polymeric biomaterials can be further classified into hydrolytically degradable polymers and enzymatically degradable polymers.[25] The biodegradability of microbial polymers was evaluated from enzymatic degradation by enzymes such as polyhydroxybutyrate depolymerase from *Ralstonia pickettii* T1 or lipase from *Rhizopus oryzae* in the laboratory and composting in natural environments by Hsieh *et al.*[26] It was thereupon reported that physical properties of biodegradable polymers are significantly influenced by their chemical structures and crystalline morphologies. In particular, the chemical structure of the polymer influences not only the crystal growth mechanism, but also its degradation mechanism.[27,28] A recent article in the '*The Chemical Engineer*' magazine (March 2009, Issue 813, p. 6, *Biodegradable PET bottles are "green breakthrough"*, United Kingdom) highlights the development of a form of polyethylene PET which is biodegradable in both aerobic and anaerobic conditions within five years (ENSO BOTTLES, a company based in Phoenix, Arizona). The researchers who worked on this PET project have modified the PET polymer chain with organic compounds and microbial nutrients thereby making the polymer structure weak and more susceptible to faster microbial breakdown.

11.2.2 Assessment of Biodegradable Polymers Degradability

There is a difference between a short-term degradation process, normally engineered by technologists, and the long-term deterioration, brought about by natural environment (*i.e.* microorganisms) in polymers.[29] Using microbial, chemical and photolytic modes of degradation, it is now possible to promote short-term degradation processes in a polymer. ASTM, British Standards (BS) and ISO (International Organization for Standardization) propose test methods that can be used to assess and/or simulate actual disposal conditions (*in situ* testing) for evaluating the degradability of polymers.[29]

11.2.2.1 *ASTM series*

- ASTM D6692-01 Standard Test Method for determining the biodegradability of radiolabelled polymeric plastic materials in seawater.
- ASTM D6340-98(2007) Standard Test Methods for determining aerobic biodegradation of radiolabelled plastic materials in an aqueous or compost environment.
- ASTM D5338-98(2003) Standard Test Method for determining aerobic biodegradation of plastic materials under controlled composting conditions.
- ASTM D-5526-94 Standard Test Method for determining anaerobic biodegradation of plastic materials under accelerated landfill conditions.
- ASTM E1720-01 Standard Test Method for determining ready, ultimate, biodegradability of organic chemicals in a sealed vessel CO_2 production test.

11.2.2.2 *BS series*

- BS 8472 Method for the determination of compostability (including biodegradability and eco-toxicity) of packaging materials based on oxo-biodegradable plastics.
- BS EN ISO 14855-2:2009 Determination of the ultimate aerobic biodegradability of plastic materials under controlled composting conditions.

Degradation of Biodegradable and Green Polymers 337

Method by analysis of evolved carbon dioxide. Gravimetric measurement of carbon dioxide evolved in a laboratory-scale test.
- BS EN 13432:2000 Packaging. Requirements for packaging recoverable through composting and biodegradation. Test scheme and evaluation criteria for the final acceptance of packaging.

11.2.2.3 ISO series

- ISO 14851:1999 Determination of the ultimate aerobic biodegradability of plastic materials in an aqueous medium – Method by measuring the oxygen demand in a closed respirometer.
- ISO 14852:1999 Determination of the ultimate aerobic biodegradability of plastic materials in an aqueous medium – Method by analysis of evolved carbon dioxide.
- ISO 14855-2:2007 Determination of the ultimate aerobic biodegradability of plastic materials under controlled composting conditions – Method by analysis of evolved carbon dioxide-Part 2: Gravimetric measurement of carbon dioxide evolved in a laboratory-scale test.

Besides the latter standards for determining the ultimate aerobic/anaerobic biodegradability[30] of polymeric materials, the ISO 14855-1 is a common test method that measures carbon dioxide evolution using continuous infrared analysis, gas chromatography or titration.[15,31] This method is a small-scale test for determining the ultimate aerobic biodegradability of polymeric materials, where the amounts of compost inoculum and test sample is one-tenth that of ISO 14855-1. This method is well versed in ISO/DIS 14855-2 in which the carbon dioxide evolved from test vessel is determined by gravimetric analysis of carbon dioxide absorbent. There are also some European Standards for assessing the biodegradability of polymeric materials.

11.2.2.4 European Standards

- European Standard EN 13432 (2000) Packaging – Requirements for packaging recoverable through composting and biodegradation – Test scheme and evaluation criteria for the final acceptance of packaging.
- European Standard prEN 14046 (2000) Packaging – Evaluation of the ultimate aerobic biodegradability and disintegration of packaging materials under controlled composting conditions – Methods by analysis of released carbon dioxide.
- European Standard prEN 14045 (2000) Packaging – Evaluation of the disintegration of packaging materials in practical oriented tests under defined composting conditions.

The biodegradation of a polymer is initially accompanied by a loss of chemical and/or physical changes which are followed by loss of mass (weight loss or mineralization of the polymer).[25,29] The carbon balance approach, weight loss analysis, microscopy, spectroscopy and thermal analyses are techniques that, when combined, deepen the understanding of the degradation of biodegradable polymers.[25] The degradability of biodegradable polymers is predetermined by their chemical and physical structure.[2] Of specific importance is the need of having hydrolysable or oxidizable bonds in the polymer backbone.[20] The accessibility of these bonds for enzymes secreted by the microorganisms and a sufficient flexibility of macromolecules are subsequent requirements for an effective biodegradation. Polymers with ester bonds in the macromolecular backbone are an attractive group of polymers in the family of synthetic biodegradable plastics. Several aliphatic polyesters or co-polyesters based on hydroxyacids, diacids and diols, lactones or lactides have been found to be completely biodegradable.[32–34]

11.2.3 Biodegradable Polymers Blends

For the application of biodegradable polymers, where the material has to disintegrate in an environment containing microbial populations of bacteria and fungi,[35] a well-defined degradation behaviour is generally required.[36] To adapt their degradation behaviour completely to these standards, further research and development is required. One possibility to influence the degradation is to incorporate naturally biodegradable components in the polymers. These promoters of biodegradation should normally accelerate the disintegration if they degrade much faster than the matrix polymer[36] since they increase the effective surface accessible to the microorganisms,[37] and consequently, tend to control the disintegration. Hence, the development of new biodegradable polymers may be addressed either alone or in blends[14] with materials like various forms of starch and products derived from it, cellulose/chitosan/chitin,[38–42] natural fibres,[43] biopolyesters, agroindustrial residues[44] and soy-based components.[12] Starch is rapidly metabolized and is an excellent base material for polymer blends or for infill of more environmentally inert polymers where it is metabolized to leave less residual polymer on biodegradation.[14] In a study to assess the biodegradability and mechanical properties of poly-(β-hydroxybutyrate-co-β-hydroxyvalerate) starch blends, it was concluded that the addition of starch to poly-(β-hydroxybutyrate-co-β-hydroxyvalerate) not only reduced the cost but also lead to a completely biodegradable material whose degradation rate could be tailored by adjusting the starch/PHA ratio.[45] A recent study also showed the beneficial improvements in the structural and thermal properties of PCL following the grafting of acrylic acid onto PCL in a PCL/chitosan blend.[41] A recent article in the '*The Chemical Engineer*' magazine (December 2008/January 2009, Issue 810, p. 9, *New additive makes plastic compostable*, United Kingdom) highlights the development of an additive (by Australia's Goody Environment) that can render any plastic (polymeric materials, in general) biocompostable with a shelf-life of up to five years. Unlike, starch-based technologies, the plastics remain suitable for combining with standard plastics for recycling. The additive has been tested to ISO 14851 and 16929 standards and has been demonstrated to meet the Australian (AS 4736) and European standards for biodegradability and compostability. According to the Director of Goody Environment, Nick Paech, the additive can make up to 1 to 10% of the formulation, and that a 26 g PET bottle will be reduced to 13 g of compost within 90 days. Some additional examples of studies assessing the biodegradability of blended biodegradable polymers are given in Table 11.1.

11.3 Composting Process Essentials

There are many well-established bioremediation technologies applied for environmental management and protection.[56] One such technology is the application of composting and the use of compost materials. Composting matrices and composts are rich sources of microorganisms,[57] which can degrade contaminants to harmless compounds such as carbon dioxide and water. Composting is applied in bioremediation as a means of degrading organic compounds and lessening their environmental toxicity.[56,58] In addition, composting stabilizes wastes for ultimate disposal in traditional manners in landfills or on farmland. Composting has been practised to reduce the volume and water content of feedstock, to destroy pathogens and to destroy odour-producing nitrogenous and sulfurous compounds.[59] Composting is also considered as a remediation method for handling contaminated soil, sediment and organic wastes.[60] Mechanical treatment by grinding, mixing and sieving out non-degradable or disturbing materials (metals, glass, stones and gravel) gives good conditions for biological treatment of compostable materials.[61] The biological treatment builds up stable organic compounds through humification[57] and reduces concentrations of organic pollutants. Composting can be used to lower the levels of chemical contaminants in residues or in soils to which polluted residues have been added.[56] The processes of remediation in compost are similar to those that occur biologically in soil.

11.3.1 Composting Chemistry

Composting can be defined as the self-heating controlled biological decomposition of organic substrates carried out by successive microbial populations combining both mesophilic and thermophilic activities, leading to the production of a final product sufficiently stable for storage and application to land without adverse environmental effects.[57,62] Under optimal conditions, composting proceeds from the initial ambient state through three phases namely the mesophilic or moderate-temperature phase, which lasts for a few days; the thermophilic or high-temperature phase, which can last from a few days to several months, and the cooling (and maturation) phase which lasts for several months.[63] The 'self-heating' is due to heat liberation from microbial

Table 11.1 Biodegradation of biodegradable polymers in blends with natural polymers.

Reference	Substrates/Additives	Experimental conditions	Biodegradation performance
Mohee and Unmar[46]	Willow Ridge Plastics – PDQ-H additive (Plastic A) and Ecosafe Plastic – Totally Degradable Plastic Additives (TDPA) additive (Plastic B).	Controlled and natural composting environments.	Cumulative CO_2 evolution for Plastic A was much higher than that for Plastic B. Plastic A therefore showed a higher level of biodegradation in terms of CO_2 evolution than Plastic B. Another experiment was undertaken to observe any physical change of Plastics A and B as compared to a reference plastic, namely, compostable plastic bag (Mater-Bi product – Plastic C). Thermophilic temperatures were obtained for about 3–5 days of composting. After 55 days of composting, Plastic C degraded completely while Plastic A and Plastic B did not undergo any significant degradation. It was concluded that naturally based plastic made of starch would degrade completely in a 60 days.
Mohee et al.[47]	Mater-Bi Novamont (MB) and Environmental Product Inc. (EPI). Cellulose filter papers (CFP) were used as a positive control.	Biodegradability was assessed under aerobic and anaerobic conditions. For aerobic conditions, organic fractions of municipal solid wastes were composted. For the anaerobic process, anaerobic inoculum from a wastewater treatment plant was used.	A biodegradation of 27.1% on a dry basis for MB plastic within a period of 72 days of composting. EPI plastic did not biodegrade under either condition.

Kim et al.[48]	Poly(butylene succinate) (PBS) and bioflour, PBS biocomposite filled with rice-husk flour (RHF) reinforcing.	Natural and aerobic compost soil environments.	Weight loss and the reduction in mechanical properties of PBS and the biocomposites in the compost soil burial test were significantly greater than those in the natural soil burial test. The biodegradability of the biocomposites was enhanced with increasing bioflour content because the bioflour is easily attacked by microorganisms.
Vikman et al.[49]	New starch-based biopolymers: commercial starch-based materials and thermoplastic starch films prepared by extrusion from glycerol and native potato starch, native barley starch, or cross-linked amylomaize starch.	Enzymatic hydrolysis was performed using excess *Bacillus licheniformis* α-amylase and *Aspergillus niger* glucoamylase at 37 °C and 80 °C.	The enzymatic method is a rapid means of obtaining preliminary information about the biodegradability of starch-based materials.
Gu et al.[50]	Cellulose acetate (CA) films with degree of substitution (DS) values of 1.7 and 2.5.	In-laboratory composting testing vessels conditions at 53 °C.	The CA 1.7- and 2.5- DS films completely disappeared by the end of 7- and 18-day exposure time periods in the biologically active bioreactors, respectively.
Rosa et al.[51]	Unripe coconut fibers used as fillers in a biodegradable polymer matrix of starch/ethylene vinyl alcohol (EVOH)/glycerol.	Effects of fiber content on the mechanical, thermal, and structural properties were evaluated.	Addition of coconut fiber into starch/EVOH/glycerol blends reduced the ductile behavior of the matrix by making the composites more brittle. At low fiber content, blends were more flexible, with higher tensile strength than at higher fiber levels. Temperature at the maximum degradation rate slightly shifted to lower values as fiber content increased. Comparing blends with and without fibers, there was no drastic change in

Table 11.1 (Continued)

Reference	Substrates/Additives	Experimental conditions	Biodegradation performance
			melt temperature of the matrix with increase of fiber content, indicating that fibers did not lead to significant changes in crystalline structure. Starch alone degraded readily, but the starch/EVOH/glycerol blends exhibited much slower degradation in compost.
Chen et al.[52]	Blends of poly(L-lactide) (PLLA) and chitosan with different compositions prepared by precipitating out PLLA/chitosan from acetic acid-dimethyl sulfoxide mixtures with acetone.	Blends were characterized by Fourier transform infrared analysis (FTIR), X-ray photoelectron spectroscopy (XPS), differential scanning calorimetry (DSC), 13C solid-state NMR and Wide-angle X-ray diffraction (WAXD).	FTIR and XPS results showed that intermolecular hydrogen bonds existed between two components in the blends, and the hydrogen bonds were mainly between carbonyls of PLLA and amino groups of chitosan. Blending chitosan with PLLA suppressed the crystallization of the PLLA component.
Xie et al.[53]	A series of phosphate-modified Konjac synthesized by esterification of natural polysaccharide Konjac.	Comparative studies conducted to examine the flocculation efficiency and biodegradability for the modified and unmodified products.	Both modified derivatives and the unmodified parent show good flocculation ability and biodegradability. Within the polymer concentration studied, the higher the phosphoric content, the better the flocculation

Alexy et al.[54]	Effect of collagen hydrolysate (CH) and glycerol on polyvinyl alcohol (PVA) water sensitivity has been investigated.	Laboratory analyses.	efficiency is, and the modified analogues show superior biodegradability to that of the parent polymer. P–O–P bond was cut during the aging process at 30 °C. CH content affected water penetration into the prepared blown films, affecting therefore their solubility. Increasing contents of CH in PVA-based blends shortens the time to the first disruption of the film after immersing in water. Pure PVA film presented limited biodegradation at low temperature (5 °C) while CH addition in the blend significantly increased the biodegradation rate at that temperature.
Zhao et al.[55]	Cornstarch was methylated and blend films were prepared by mixing methylated-cornstarch (MCS) with PVA.	Mechanical properties, water resistance and biodegradability of the MCS/PVA film were investigated.	Enzymatic, microbiological and soil burial biodegradation results indicated that the biodegradability of the MCS/PVA film strongly depended on the starch proportion in the film matrix. The degradation rate of starch in the starch/PVA film was hindered by blending starch with PVA.

metabolic activities taking place during the biodegradation, as depicted by Equation 11.1.[57]

$$C_6H_{12}O_6(aq) + 6O_2(g) \rightarrow 6CO_2(aq/g) + 6H_2O(l) \quad \Delta G = -2829.86 \text{ kJ mol}^{-1}$$
(11.1)

This thermophilic stage is governed by the basic principles of heat and mass transfer and by the biological constraints of living microorganisms.[64] Published data on the rate of heat production by organic material decomposition suggest that there are wide variations in the rate of heat released by composts. This is because the rate of heat production is a function of the chemical, physical and biological properties of the composting materials and substrates.[63,65,66] Decomposition by microorganisms during composting occurs predominantly in the thin liquid biofilms on the surface of the organic particles or other suitable substrates.[57] If the moisture content drops below a critical level, which is normally 25–30%, the microbial activity will decrease and the microorganisms will become dormant. On the other hand, if the moisture content goes above 65%, oxygen depletion and losses of nutrients through leaching may occur,[65] again lowering the biodegradation rates.

11.3.2 Physical Parameters in Composting

Composting and other aerobic solid-state fermentation processes rely on gas transfer through a biologically active porous medium with evolving physical and chemical properties. Two of the physical parameters of particular importance to process control and analysis are the air-filled porosity and the permeability of the matrix.[67] Air-filled porosity is the volume fraction of air (often reported on a percentage basis) in a porous matrix. Permeability is a measure of the ability of fluids to flow through a multi-phase material.[68] Composting is a solid-phase process which exploits the phenomenon of microbial self-heating.[69–71] The material being composted serves as its own matrix, permitting gas exchange, and provides its own source of nutrients, water and an indigenous, diverse inoculum. It also serves as its own waste sink and thermal insulation.[72] Hence, metabolically generated heat is conserved within the system, elevating its temperature.

The physical properties of compost materials play an important role in every stage of the biodegradation process and compost production as well as in the handling and utilization of the end product. From the mixing of various feedstocks in a conducive carbon to nitrogen ratio and process monitoring and maintenance,[73] to the packaging and shipping of the final product, parameters such as bulk density and porosity dictate the requirements for the optimum composting environment and the design of machinery and aeration equipment used in the system. The wet bulk density of compost is a measure of the mass of material (solids and water) within a given volume and is important in the determination of initial compost mixtures.[65] The wet bulk density determines how much material can be placed at a certain site or hauled in a truck of a given size. The density of compost also influences the mechanical properties such as strength, porosity and ease of compaction,[73] and ultimately affects the rate of biodegradation. Air-filled porosity and permeability relationships[67] are different for different substrates and mix of these substrates, as each substrate and mix will have a different distribution of particle sizes and shapes, and these mixtures can be compacted to various wet and dry bulk densities.[74]

11.3.3 Composting Systems

Composting systems fall into two main categories: the 'fully or partially open to air' systems and the 'in-vessel' systems.[75] In the first category are systems ranging from the ones used from prehistoric times to the windrows, aerated static pile and 'household' systems used in the present day. In the second category fall the 'tunnel' systems, the rotary drum composting system and other 'in-vessel' or 'reactor'/'bioreactor' systems of various designs. Depending on the physical location where the composting experiments are being carried out, the substrate, the scale of operation, and the skills and the machinery available, one or the other type of system is used. A more extensive review of the various composting systems may be read in Gajalakshmi and Abbasi.[75]

11.3.4 Vermicomposting

One variant of the composting process is vermicomposting whereby earthworms are used in the degradation of the substrates. Due to their biological, chemical and physical actions, earthworms can be directly employed within bioremediation strategies to promote biodegradation of organic substrates.[76] In the vermicomposting process, a major fraction of the nutrients contained in the organic matter is converted to more bioavailable forms. The first step in vermicomposting occurs when earthworms break the substrate down to small fragments as a prelude to ingesting the substrate.[75,77] This increases the surface area of the substrate, facilitating microbial action. The substrate is then ingested and goes through a process of 'digestion' brought about by numerous species of bacteria and enzymes present in the worms' gut.[75] During this process, important nutrients such as nitrogen present in the feed material are converted into forms that are much more water-soluble and bioavailable than those in the parent substrate.[78,79] The earthworms derive their nourishment from the microorganisms that grow upon the substrate particles. At the same time, they promote further microbial activity in the residuals so that the faecal material that they excrete is further fragmented and made more microbially active than what was ingested as the parent substrate.[75]

11.4 Biopolymer Degradation and Composting

For natural ecosystems to operate and self-sustain healthily, they have to cope with increasing volumes of anthropogenic waste. With the mix and types of solid wastes generated today world-wide, it would be environmentally unsound and unsafe to allow for an uncontrolled and unmonitored biodegradation.[6] In this respect, new bioremediation processes that can 'catalyse' the degradation of these wastes in a 'green' and controlled manner are needed.[6,80] By virtue of its biochemical reactions that bring about degradation of organic matter, composting is one such process which is ecologically sound for the degradation of polymeric materials, when these are designed to be biodegradable using renewable resources as the major raw material component.[6,81,82] By composting the biodegradable polymeric materials,[83] a much-sought after carbon-rich soil with essential humic matter may be produced. Compost-amended soil can have beneficial effects by increasing soil organic carbon, increasing water and nutrient retention, reducing chemical inputs and suppressing plant diseases. A key requirement for official regulations and the decision as to which polymeric materials may be composted are to carry out systematic investigations on their biodegradability and the quality of the compost produced.[84] In this sense, composting is more than just an ecologically sound waste disposal, it is equally resource recovery[85] and hence a preferred treatment strategy for biodegradable polymeric materials.[86] There are many different types of compostable biodegradable polymers currently in development, ranging from oil-based plastics (such as vegetable oil and soya bean oil) to starch-based compounds.[16] Under favourable conditions of light, moisture and temperature, these polymers start to break down as a result of optimum microbial metabolic action. The sections to follow present a comprehensive review of research progress accumulated in recent years on the degradation of PHAs, poly(lactic acid)-based polymers, polyethylenes and PCLs in the composting environment.

11.4.1 Polyhydroxyalkanoates

Polyhydroxyalkanoates (PHAs) are bacterial storage polyesters currently receiving much attention due to their synthesis from renewable resources and their applicability as biodegradable and biocompatible plastics.[87–89] PHAs are energy/carbon storage materials accumulated under unfavourable growth conditions in the presence of excess carbon source.[90] PHAs are attracting much attention as substitutes for non-degradable petrochemically derived plastics because of their similar material properties to conventional plastics and their complete biodegradability in the natural environment upon disposal.[90] Most aliphatic polyesters are readily mineralized by a number of aerobic and anaerobic microorganisms that are widely distributed in nature. However, aromatic polyesters are more resistant to microbial attack than aliphatic polyesters.[91] The fungal biomass in most soils generally exceeds the bacterial biomass[92] and thus it is likely that the fungi may play a considerable role in degrading polyesters, just as they predominantly perform the decomposition of organic matter in the soil ecosystem.[89]

PHAs can be rapidly and completely degraded in municipal anaerobic sludge by various microorganisms. PHAs have always been found to be biodegradable, but the degradation rate varies with their specific properties.[93–95] PHAs that are intracellularly accumulated can be degraded by intracellular PHA depolymerases.[96] PHAs and polymers made of PHAs disposed to the environment can be degraded by extracellular PHA depolymerases.[97,98] Extracellular PHA depolymerases are secreted by many bacteria and fungi (which may also be readily found in composts) for the utilization of PHAs left in the environment. Intracellular PHA depolymerases are unable to hydrolyse extracellular PHA, and extracellular depolymerases can not hydrolyse intracellular granules due to the differences in the physical structures of intracellular native granules and extracellular denatured PHA.[90] Although PHAs can be degraded completely to CO_2, water and microbial biomass, weeks (and at times months) are required for them to degrade in most natural environments.[45] The biodegradation rate depends on temperature, moisture, sample thickness and the number of PHA depolymerase-producing microorganisms present. There are instances, such as during municipal composting, when the overall rate of biodegradation is normally quicker.[45]

Poly-3-hydroxybutyric acid (P3HB) is the most well known member of the PHAs family.[99,100] P3HB is similar to polypropylene in its physical properties but has the advantage of being biodegradable.[89,101] P3HB and its copolymers have been shown to be biodegradable by bacteria into water and carbon dioxide (and methane under anaerobic conditions)[89,102] in different natural environments such as soils,[103] composts and natural waters.[104-106] A monitoring system based on an aerobic biodegradation technique (Sturm test)[107] to investigate the biodegradation of poly-β-(hydroxybutyrate), poly-β-(hydroxybutyrate-co-β-valerate) and poly(ε-caprolactone) in compost derived from municipal solid waste has been developed by Rosa et al.[108] After thermal analysis of these polymers using DSC, melting temperature and crystallinity determination, it was found that the poly-β-(hydroxybutyrate) degraded faster than the other two polymers, most plausibly accounted for by the chemical structure of this polymer making it more susceptible to microbial attack.[108]

A series of miscible blends consisting of cellulose acetate propionate (CAP) and poly(ethylene glutarate) (PEG) or poly(tetramethylene glutarate) (PTG) were evaluated in a static bench-scale simulated municipal compost environment.[109] Samples were removed from the compost at different intervals, and the weight loss was determined before evaluation by gel permeation chromatography (GPC), scanning electron microscopy (SEM), and ^{1}H NMR. Buchanan et al.[109] found that at fixed CAP DS, when the content of polyester in the blend was increased, the rate of composting and the weight loss due to composting increased, while when the CAP was highly substituted, little degradation was observed within 30 days and almost all of the weight loss was attributable to the loss of polyester. The overall set of data suggested that the initial degradation of the polyester was by chemical hydrolysis and the rate of hydrolysis was very dependent upon the temperature profile of the compost, and hence on the succession in predominance of the various microbial populations, and upon the DS of the CAP. The biodegradability of injection moulded specimens prepared by blending poly(hydroxybutyrate-co-valerate) (PHBV) with cornstarch in the presence or absence of poly(ethylene oxide) (PEO) was assessed in natural compost by measuring changes in physical and chemical properties over a period of 125 days.[110] It was found that the incorporation of PEO into starch-PHBV blends had a negligible effect on the rate of weight loss, while starch in blends degraded faster than PHBV and it accelerated PHBV degradation. PHB and PHV moieties within the copolymer degraded at similar rates, regardless of the presence of starch, as determined by ^{1}H NMR spectroscopy.[110] SEM showed homogeneously distributed starch granules embedded in a PHBV matrix, typical of a filler material, and the starch granules had been rapidly depleted as a result of exposure to compost. In another study, it has also been found that natural aliphatic copolyester, 3-hydroxybutyrate-co-3-hydroxyvalerate, and its blends with the synthetic aliphatic-aromatic copolyester of 1,4-butanediol with adipic and terephthalic acids degrade faster in compost than in seawater.[111] Quite recently, the biodegradation of poly(butylene succinate) (PBS) was studied under controlled composting conditions.[112] The biodegradation process of PBS under controlled composting conditions was found to proceed in three distinct phases. The biodegradation in the first phase was slow, then accelerated in the second phase and showed stabilization in the third phase.[112] Out of the four bacterial strains isolated from the compost and identified, *Aspergillus versicolor* was found to be the best PBS-degrading microorganism.

11.4.2 Poly(lactic acid)-based Polymers

Poly(lactic acid)-based polymers or polylactides (PLAs) are biopolyesters derived from lactic acid, which is generally produced by the fermentation of corn sugars.[113,114] The synthesis of polylactide is shown in Scheme 11.1. PLA possesses a panoply of favourable material properties that enable its penetration into diverse markets. Their different properties arise chemically, either from the length of the pendant groups that extend from the polymer backbones, or from the distance between the ester linkages in the polymer backbones.[115] The structure variation has offered a wide range of mechanical properties. Its clarity and physical properties enable its use in food and product packaging applications

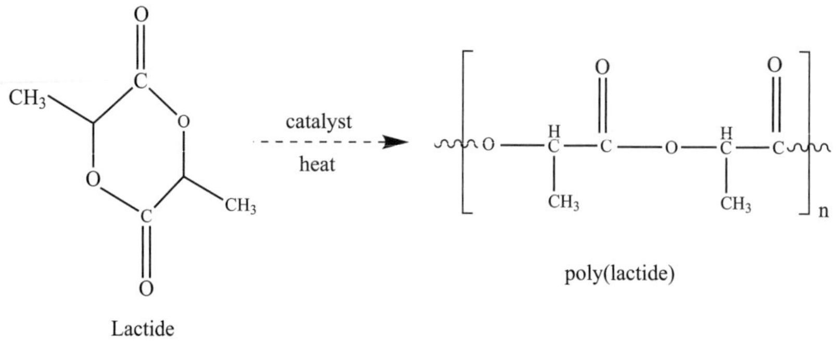

Scheme 11.1 Synthesis of poly(lactide) (PLA).

and it is currently being commercially produced at a large scale as and for a bio-based packaging material.[113] Life cycle assessment (LCA) shows that the production of PLA polymers consumes around two times less energy than conventionally petroleum based polymers.[116] Vink et al.[117] have demonstrated using LCA that it has the lowest non-renewable fossil resource content compared to a variety of other common polymers. Therefore, the first issue that needs to be addressed is the environmental impacts associated with PLAs as regards its biodegradability.

The disposal of PLA polymeric packaging residues in composting facilities can be an important method of reducing the amount of packaging materials that are disposed as municipal solid waste. PLAs are also being engineered to degrade in composting environments. Real composting conditions differ from the simulated ones because of factors such as weather, microbial growth and pH.[116] Kale et al.[116] assessed the compostability of two commercially available PLA packages (a bottle and a tray) in real composting conditions. The degradation of a PLA bottle composed of 96% L-lactide and 4% D-lactide with bluetone additive and a tray composed of 94% L-lactide and 6% D-lactide were evaluated in a composting pile having temperatures around 65 °C, a relative humidity of 65% wet weight moisture content and a pH of 8.5. The packages were placed in compost in duplicate sets and were taken out on 1, 2, 4, 6, 9, 15, and 30 days. The molecular weight and the glass transition, melting and decomposition temperatures were monitored to assess the changes in the packages' physical properties. After 4 days of being in the compost pile, an initial fragmentation of the packages was observed. At 15 days of the composting process, the trays started to become a part of the compost whereas the bottles showed slower degradation and started breaking apart. After 4 days for the trays and at 6 days for the bottles, a molecular weight reduction of 77% and 85% were observed, respectively. The difference between the degradation times between the PLA bottles and trays was attributed to their initial difference in crystallinity. Kale et al.[116] reported that the PLA bottles and trays had however degraded much faster under the 'real' composting conditions than in previous studies of simulated composting conditions.

The composting of extruded PLA in combination with pre-composted yard wastes in a laboratory composting system was studied by Ghorpade et al.[118] Yard waste and PLA mixtures containing 0%, 10% or 30% PLA (dry weight basis) were placed in composting vessels for four weeks. The data sets for the monitoring indicated that microbial degradation of PLA had occurred and there was no significant difference ($P > 0.05$) in CO_2 emission between the 0% and 30% PLA mixtures. Based on the observed pH variations, Ghorpade et al.[118] deduced that for 30% PLA, substantial chemical hydrolysis and lactic acid generation had occurred. The then lowered pH likely suppressed any further microbial activity. Supported by the GPC results showing a significant decrease in PLA molecular weight as a result of composting, Ghorpade et al.[118] concluded that PLA can be efficiently composted when added in small amounts of less than 30% by weight to pre-composted yard waste. Gattin et al.[119] also studied the biodegradation of a co-extruded starch/PLA polymeric film in liquid, inert solid and composting media. It was found that, irrespective of the biodegradation medium used, the percentage of mineralization was better than the required 60% value for the definition of a biodegradable material. The presence of starch was found to have facilitated the biodegradation of the PLA component, especially in liquid media.

More recent studies on the degradation of PLA by composting have been reported by Vargas et al.,[120] Kijchavengkul et al.[121] and Iovino et al.[122] The effects of electron beam (EB) irradiation on the backyard composting behaviour of PLA polymer were evaluated by Vargas et al.[120] Samples from thermoformed PLA drinking cups were exposed to 10 MeV EB irradiation at doses of 0, 72, 144 and 216 kGy. The irradiated PLA samples were then placed in a heat-sealed, plastic screen and added to an organic feedstock in a rotating composter within a computer-controlled environmental chamber for 10 weeks at 35 °C. The PLA molecular weight decreased as irradiation dose and composting time increased. After 1 week in the backyard composter, the molecular weight D values had increased to about 560 kGy and then fell to about 380 kGy after 2 weeks of composting. Kijchavengkul et al.[121] also studied the biodegradation of PLA polymers under real composting conditions and compared the latter results with the biodegradation values obtained in three simulated laboratory conditions. It was concluded that the laboratory conditions underrepresented the biodegradation that took place during the real composting conditions. The PLA bottles completely biodegraded in 30 days in real composting. Iovino et al.[122] investigated the aerobic biodegradation of a PLA composite (with and without the addition of maleic anhydride (MA), acting as coupling agent, thermoplastic starch (TPS) and short natural coir fibre) under controlled composting conditions using standard test methods. For comparison, TPS and matrix (containing 75 wt% of PLA and 25 wt% of TPS) were also tested. At the end of the incubation period, TPS appeared to be the most bio-susceptible material, being totally biodegraded, and the matrix showed a higher level of biodegradation through higher amounts of evolved CO_2 than PLA. The fibres seemed to have played a secondary role in the process as confirmed by the slight differences in CO_2 produced. SEM micrographs of the aged compost samples showed the formation of patterns and cracks on the surface of the tested materials aged in the compost, evidencing a profound loss of structure following biodegradation. Table 11.2 summarizes some additional studies conducted to assess the biodegradability of PLA during composting.

11.4.3 Polyethylenes

Polyethylene (PE) is a thermoplastic commodity made by the chemical industry through the addition polymerization of ethene and which heavily used in a variety of consumer products. PE is classified into several different categories, based mostly on its density and branching. The mechanical properties of PE depend significantly on variables such as the extent and type of branching, the crystal structure and the molecular weight. The following are types of PE:

Table 11.2 Summary of research on the biodegradability of PLA polymeric materials during composting.

Reference	Substrates	Experimental conditions	Main results
Ganjyal et al.[123]	Extruded starch acetate foams PLA with pre-conditioned yard waste. Extruded foams of high amylose starch used as control.	Laboratory composting system	Control material degraded completely in 15 days. Rate of degradation was faster for foams with higher PLA contents. Starch acetate foams took 130 days to degrade completely.
Kale et al.[124]	PLA bottles	Simulated composting conditions according to ASTM and ISO standards. Results compared with a novel method of evaluating package biodegradation in real composting conditions.	Both cumulative measurement respirometric (CMR) and the gravimetric measurement respirometric (GMR) systems showed similar trends of biodegradation at day 58. Mineralization was 84.2% and 77.8%, respectively.
Kale et al.[125]	PLLA biodegradable packages	Real composting conditions and under ambient exposure. The packages were subjected to composting for 30 days, and the degradation of the physical properties was measured at 1, 2, 4, 6, 9, 15 and 30 days.	PLLA packages made of 96% L-lactide exhibited lower degradation than PLA packages made of 94% L-lactide.
Tuominen et al.[126]	Lactic acid-based polymers	Controlled composting conditions (European Standard prEN 14046).	All polymers biodegraded to over 90% of the positive control in 6 months.
Itävaara et al.[127]	PLA polymers	Different elevated temperatures in both aerobic and anaerobic, aquatic and solid state conditions.	In the aerobic aquatic headspace test, mineralization of PLA was very slow at room temperature, but faster under thermophilic conditions. At similar elevated temperatures, the biodegradation of PLA was much faster in anaerobic solid state conditions than in aerobic aquatic

Table 11.2 (*Continued*)

Reference	Substrates	Experimental conditions	Main results
			conditions. Degradation behaviour of PLA in natural composting process was similar to that in the aquatic biodegradation tests, biodegradation starting only after the beginning of the thermophilic phase.
Alauzet et al.[128]	Two lactic acid-based stereocopolymers, namely 50/50 and 96/4 L/D poly(L-lactic-co-D-lactic acids) and corresponding oligomers.	Degradation characteristics in various worm-free and worm-containing media were investigated under various conditions including direct feeding using impregnated paper or coated tree leaves, model composting and vermicomposting. Data were compared with abiotic degradation in sterile neutral phosphate buffer.	High molar mass PLA can be ingested by earthworms provided they are disintegrated first. However, they cannot be bio-assimilated before hydrolytic degradation generates oligomers.
Hakkarainen et al.[129]	PLLA polymeric films	Microorganisms isolated from compost.	After 5 weeks in biotic environment, the films had fragmented to fine powder, while the films in corresponding abiotic medium still looked intact. SEM micrographs showed the formation of patterns and cracks on the surface of the films aged in biotic medium, while the surface of the sterile films remained smooth.

- Ultra high molecular weight PE (UHMWPE) has outstanding toughness, cut, wear and excellent chemical resistance. UHMWPE is used in a wide diversity of applications. These include can and bottle handling machine parts, moving parts on weaving machines, bearings, gears, artificial joints, edge protection on ice rinks and butchers' chopping boards.

- High molecular weight PE (HMWPE) and high density PE (HDPE) are used in products and packaging such as milk jugs, detergent bottles, margarine tubs, garbage containers and water pipes.
- High density cross-linked PE (HDXLPE), cross-linked PE (PEX) and medium density PE (MDPE) are typically used in gas pipes and fittings, sacks, shrink film, packaging film, carrier bags and screw closures.
- Low density PE (LDPE) and linear low density PE (LLDPE) are used predominantly in film applications due to their toughness, flexibility, and relative transparency.
- Very low density PE (VLDPE) is used for hose and tubing, ice and frozen food bags, food packaging and stretch wrap.

To enhance the degradation of the PE,[130] chemical or photo initiators or both are added to the degradable polymeric material.[131] For PE films containing photo-oxidants and pro-oxidants,[132] the primary initiators of oxidation are light and temperature, respectively. Both the pro-oxidant and the photo-oxidant produce free radicals on the long PE chain, causing the material to lose some of its physical properties, to become oxidized, and to become more accessible to microbial biodegradation.[131] The environmental degradation of PE proceeds by a synergistic action of photo- and thermo-oxidative degradation and biological activity.[133] Since biodegradation of commercial high molecular weight PE proceeds slowly, abiotic oxidation is the initial and rate-determining step. More than 200 different degradation products, including alkanes, alkenes, ketones, aldehydes, alcohols, carboxylic acid, keto-acids, dicarboxylic acids, lactones and esters, have been identified in thermo- and photo-oxidized PE. In biotic environment, these abiotic oxidation products and oxidized low molecular weight polymer can be thereafter assimilated by microorganisms.[133] In a view to promote the green biodegradation of polyethylene and its polymeric derivatives, composting has also been relatively widely studied as a matrix providing the necessary biodegradability conditions. A discussion of a selection of such studies follows.

The degradation performance of 11 types of commercially produced degradable starch-polyethylene plastic compost bags was evaluated in municipal yard waste compost sites.[134] The bags differed in starch content (5 to 9%) and pro-oxidant additives. Each compost site was seeded with test strips (200 to 800 of each type) taped together, which were recovered periodically over an 8- to 12-month period. The first 8 months indicated that the materials recovered from the interior of the compost row demonstrated very little degradation, whereas materials recovered from the exterior had degraded well. In the second-year study, degradation was however observed in several of the plastic materials recovered from the interior of the compost row. In the same year, Chiellini et al.[135] analysed the degradation of starch-filled PE films, consisting of HDPE and LDPE and 0–20% starch, exposed for 60 days to controlled composting conditions. It was concluded that the oxidation of the PE matrix was dependent upon the PE type and starch content. LDPE and HDPE polyethylene films filled with starch up to a maximum level of 20% by weight were tested for biodegradation under windrow composting consisting of various putrescible wastes and assembled for controlled biostabilization management under static conditions were used.[136] The physical and chemical deterioration of the polyethylene-starch films exposed to the controlled composting environment were recorded and analysed with respect to the different composting evolution and were compared with the data collected in pure culture systems and in bench scale tests simulating an aerobic biostabilization process. It was deduced that partial removal of starch from the different films had occurred as a result of massive surface colonization by various microorganisms. For the composting trials experiencing prolonged severe temperature conditions, a negligibly detectable oxidation of the polyethylene matrix was also observed. From the study of Vallini et al.,[136] the efficiency of controlled composting systems can be maintained in assessing reproducible conditions for an accelerated biostabilization of putrescible matter and hence versatility in the evaluation of the biodegradation of polyethylene-based materials. The thermo-oxidative degradation of polyethylene films containing pro-oxidant had also been studied at three temperatures under controlled composting and oxygen concentration variation.[137] While it was demonstrated that temperature played a most influential part in determining the rate of the thermo-oxidative degradation of the polyethylene materials, the oxygen concentration however had very little significance for the same. The rate of aerobic biodegradation evaluated under the controlled composting conditions using measurements of produced CO_2, showed that the degree of bioassimilation was around 60%, and continued to increase even after 180 days.[137] In the same year, Bonhomme et al.[138] investigated the degradation of a commercial environmentally degradable polyethylene in two stages: firstly in abiotic oxidation in an air oven to simulate the effects of a compost environment and secondly in the presence of selected microorganisms. It was observed that microbial growth had occurred in the presence of the PE samples that had been compression moulded to thick sections but had not been deliberately pre-oxidized. Measurable erosion of the film surface[20] was also observed in the vicinity of the microorganisms and the FTIR results showed that the decay of the oxidation products in the surface of the polymer film was due to the formation of protein and polysaccharides, attributable to the growth of the microorganisms.

The degradation of compost bags strips made of degradable PE and nondegradable LDPE and HDPE was evaluated in soil mixed with 50% (w/w) mature municipal solid waste compost by Orhan et al.[139] The plastic films were buried for 15 months at room temperature in 2 L desiccator jars containing soil adjusted to 40% of maximum water holding capacity. It was found that the tested polymeric films could be arranged in order of decreasing microbial susceptibility to degradation as follows: degradable PE >> LDPE > HDPE. Sedlarik et al.[140] prepared a lactose (L) filled composite of metallocene linear LDPE (mLLDPE) as a new environmentally friendly polymeric material. The effect of L on the material was characterized through its mechanical, physicochemical and rheological properties, and biodegradability in the composting environment (up to 4 months). The microorganisms present in the compost bed showed a significant influence on the properties of the new material. The biodegradation of the highest-filled composite (up to 40% wt starch) was found substantially higher than that of the others. This was in agreement with the surface morphology results from the SEM. Griffin[141] reported the degradation of buried LDPE for the composting garbage environment. Strong supporting evidence was found to explain the initial stage of the breakdown of the polymer through the transfer by diffusion of unsaturated lipids from the compost to the polymer associated with the generation of peroxides by autoxidation.

11.4.4 Poly-ε-caprolactones

Poly-ε-caprolactone (PCL) is a semicrystalline polyester and is of great interest as it can be obtained by the ring opening polymerization (ROP) of a relatively cheap monomeric unit 'ε-caprolactone'.[25] Scheme 11.2 shows the synthesis of PCL. PCL is also highly processible as it is soluble in a wide range of organic solvents, has a low melting point (55–60 °C) and a glass transition temperature of –60 °C while having the ability to form miscible blends with wide range of polymers.[25] PCL undergoes hydrolytic degradation due to the presence of hydrolytically labile aliphatic ester linkages[142] but the rate of degradation is rather slow and may vary from 2 to 3 years depending on several factors mentioned earlier in this chapters. PCL is susceptible to microbial degradation[143] and is degradable in several biotic environments,[144] including river and lake waters, sewage sludge, farm soil, paddy soil, creek sediment, roadside sediment, pond sediment and compost.[145] Table 11.3 presents some studies where the biodegradability of PCL has been studied. It has been reported that esterase and other kinds of lipase could degrade PCL.[146] The degradation times of PCL have also been shown to vary with molecular weight, crystallinity degree and morphology.[20,143,144]

Biodegradation of PCL was examined by measuring the release of CO_2 when the polymer was mixed with dog food (used as a model fresh waste) under controlled laboratory composting conditions.[154] It was found out that the quantity of PCL decomposition, evaluated as the difference in the quantity of CO_2 evolution in the presence and absence of PCL, was in proportion to the PCL mixing level. The percentage of PCL decomposition, which is calculated as a ratio of the quantity of PCL decomposition to the mixing level of PCL, was 84% after 11 days in the composting using dog food, but was 59% after the

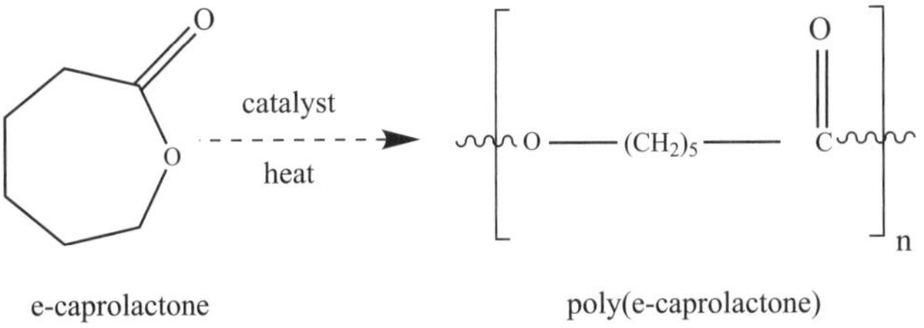

Scheme 11.2 Synthesis of poly-ε-caprolactone by ROP. Terminal –H and –OH have been added.

Table 11.3 Research studies on the biodegradability of PCL

Reference	Substrates	Biodegradation conditions
Lei et al.[147]	3D polycaprolactone–20% tricalcium phosphate composites	In vitro degradation and Von Kossa assays
Lefèvre et al.[148]	Polycaprolactone films	Biodegradation by pure strain of microorganisms isolated from an industrial composting unit
Kunioka et al.[149]	PCL powders	Controlled composting medium at 58 °C using the microbial oxidative degradation analyzer based on ISO 14855-2
Kulkarni et al.[150]	Multi-block copolymers consisting of PCL segments and poly(p–dioxanone) segments	Hydrolytic and Pseudomonas lipase catalysed enzymatic degradation
Deng et al.[151]	Water-resistant composite plastics prepared from soy protein isolate, PCL and toluene-2,4-diisocyanate as compatibilizer	Burial in two series of sheets in soil and culturing in a mineral salt medium containing microorganisms
Khatiwala et al.[152]	PCL films	Biodegradation in presence of bacterium *Alcaligenes faecalis*
Hiraishi et al.[153]	Resilon (PCL-based thermoplastic)	Biodegradation by cholesterol esterase using agar-well diffusion assay of serially diluted aqueous Resilon emulsions dispersed in agar

same period using matured compost.[154] Therefore, it was confirmed that a higher PCL decomposition rate was achieved by mixing PCL with fresh waste than by mixing it with the matured compost this being plausible attributable to a more active microbial flora in fresh waste. Ohtaki et al.[155] have also investigated the effects of temperature (40, 50, and 60 °C) on the biodegradability of PCL in a bench-scale composting reactor under controlled laboratory composting conditions. The quantity of CO_2 evolved in association with the decomposition of PCL was determined as the difference between the quantity of CO_2 evolved in the presence of the PCL and that evolved in the absence of the PCL. The degradability of PCL was calculated as a ratio of the quantity of PCL decomposition to the mixed quantity of PCL in the compost raw material. The optimum temperature for the PCL degradation was found to be approximately 50 °C, at which about 62% of the PCL was decomposed over 8 days of composting. The effect of the type of inoculum on the degradability of PCL when composting at 50 °C was then examined.

The biodegradability of PCL was studied in blends and composites of modified and granular starch. Four types of PCL-starch compositions were prepared: (i) PCL-granular starch blends; (ii) hydrophobic coating of starch particles by *n*-butylisocyanate (C_4 starch) and octadecyltrichlorosilane (C_{18} starch), followed by melt blending with PCL; (iii) PCL-starch blends compatibilized by PCL-g-dextran grafted copolymer (PGD); and (iv) PCL-grafted starch particles as obtained by *in situ* ring-opening polymerization of caprolactone (CL) initiated directly from hydroxyl functions at the granular starch surface by Singh *et al.*[156] The biodegradability of these polymeric materials was measured by monitoring the percentage weight loss in composting conditions and the rate of fungal colonization when samples were used as a sole carbon source for fungus (*A. niger*). It was found that the inherent biodegradability of host polymer was enhanced with surface compatibilization[156] during composting for longer incubation. It was observed that the weight loss during composting increased with the decrease in interfacial tension between filler and polymer. In general, Singh *et al.*[156] concluded that the inherent biodegradability of the PCL blends does not depend very significantly on the concentration of starch in the polyester matrix, but on the compatibilization efficiency.

The biodegradation of polypropylene (PP) (chosen as a non-degradable plastic), poly(L-lactic acid) (PLLA) and poly(butylene succinate) (PBS) (chosen as slowly degrading plastics) and polycaprolactone (PCL) and poly(butylene succinate-co-adipate) (PBSA) (chosen as easily degradable plastics) were tested in compost made with animal fodder.[157] It was observed that the biodegradation of PLLA and PBS depended on their shape all through the biodegradation tests while the shapes of PCL and PBSA exerted some deal of influence on their biodegradability only at the early stage of the biodegradation reactions, while at the late stage, the biodegradation proceeded almost independently of their shape in the compost matrix. PCL, poly(vinyl alcohol) (PVAl) and their blends had been incubated in the presence of a pure strain of microorganisms isolated from an industrial compost for household refuse.[158] In the conditions used, pure PCL films were completely assimilated over periods of 600–800 hours, whilst pure PVAl could not be degraded even over much longer exposure times. Unexpectedly, the blends, even PCL rich, were not altered in the presence of the microorganisms. De Kesel *et al.*[158] showed that the inactivation of the strain by PVAl did not occur and was hence not responsible for the lack of degradability. PVAl, even when present in small amounts in the incubation medium, was adsorbed on the surface of PCL or blend films thereby rendering the PCL is inaccessible to the microorganisms for degradation. Šašek *et al.*[2] have tested the biodegradation of poly(ester-amide)s (prepared by the anionic copolymerization of ε-capro-lactam and ε-caprolactone) and aromatic-aliphatic copolyesters based on glycolysed polyethylene terephthalate from used beverage bottles and ε-caprolactone during composting under controlled conditions and treatment with ligninolytic fungi. Both methods resulted in the degradation of the copolymers, composting being more robust.

11.5 Concluding Remarks

Composting and green polymers have long-term mutual interests.[159] In the future, compostable wastes other than yard trimmings, mixed organic fruit and food wastes may become a potentially significant feedstock for composting facilities. In the same area of green waste management, this diversification of feedstocks could be extended to include all sorts of compostable, and hence biodegradable, polymeric materials and items comprising packaging films, bottles and jars, foam packaging, food-service items and consumer products. This review has summarized and pondered on a variety of research studies with several biodegradable polymers and bio-based polymers (PHAs, PLAs, polyethylenes and PCLs) that have been tested for their degradability in the composting environment. The results of these studies collectively advocate the potential of composting as a sound bioremediation technique for biopolymer wastes. It has been made clear, based on the above discussions, that synthetic polymers designed to be biodegradable should be composed from units which are linked by bonds that can be readily cleaved extracellularly by hydrolytic enzymes, and that such enzymes must be excreted by the specific microbial stains into the compost biofilms to be effective in the depolymerization reaction.

Depending on the origin, different categories of biodegradable polymers have been identified. These may be broadly classified as agro-polymers such as starch or cellulose from agro-resources,[160–164] polymers obtained by microbial production (*e.g.* PHAs), chemically synthesized polymers from monomers derived from agro-resources[162–164] (*e.g.* PLAs) and chemically synthesized polymers from monomers obtained conventionally by chemical synthesis.[160] The development of biodegradable polymers is best thought of in the context of diminishing of crude oil reserves and green engineering. In coming years, it will be largely driven through the present skills of petrochemistry to produce suitable replacement of conventional polymers. The fermentation industry backed by advances in genetic engineering may yield microbes that will convert renewable materials to biopolymers more efficiently. Hence, such state-of-the-art technologies will allow designing industrially feasible and cost-effective[12] biological synthetic pathways to a wider range of biodegradable polymers. In conclusion, by virtue of their special degradability properties and broad biotechnological applications, biodegradable polymers have an exceptionally promising future.

Acknowledgements

The authors wish to express their gratitude to all the researchers whose valuable research findings and discussions have been used in writing this chapter. Our apologies are nevertheless due to others whose work could not be reported due to lack of space. Finally, the authors are thankful to other colleagues and the anonymous reviewers whose suggestions and criticisms have helped improve this chapter.

References

1. F. Kawai, *Adv. Biochem. Eng. Biotechnol.*, 1995, **52**, 151.
2. V. Šašek, J. Vitásek, D. Chromcová, I. Prokopová, J. Brožek and J. Náhlík, *Folia Microbiol.*, 2006, **51**, 425.

3. A. A. Shah, F. Hasan, A. Hameed and S. Ahmed, *Biotechnol. Adv.*, 2008, **26**, 246.
4. M. Alexander, *Science*, 1981, **211**, 132.
5. J. M. Luengo, B. García, A. Sandoval, G. Naharroy and E. R. Olivera, *Curr. Opin. Microbiol.*, 2003, **6**, 251.
6. R. Narayan, *Biodegradable Plastics*, Institute of Standards & Technology (NIST), 1993.
7. R. A. Gross and B. Kalra, *Science*, 2002, **297**, 803.
8. P. T. Anastas and M. M. Kirchhoff, *Acc. Chem. Res.*, 2002, **35**, 686.
9. P. T. Anastas and J. B. Zimmerman, *Environ. Sci. Technol.*, 2003, **37**, 94.
10. Y. W. Cho, S. S. Han and S. W. Ko, *Polymers*, 2000, **41**, 2033.
11. G. Scott, *Polym. Degrad. Stab.*, 2007, **68**, 1.
12. M. Flieger, M. Kantorová, A. Prell, T. Řezanka and J. Votruba, *Folia Microbiol.*, 2003, **48**, 27.
13. M. Shimao, *Curr. Opin. Biotechnol.*, 2001, **12**, 242.
14. R. Jayasekara, I. Harding, I. Bowater and G. Lonergan, *J. Polym. Environ.*, 2005, **13**, 231.
15. B. De Wilde and J. Boelens, *Polym. Degrad. Stab.*, 1998, **59**, 7.
16. S. T. Wellfair, *The Plymouth Student Scientist*, 2008, **1**, 243.
17. B. Singh and N. Sharma, *Polym. Degrad. Stab.*, 2008, **93**, 561.
18. R.-J. Müller, *Biodegradability of Polymers: Regulations and Methods for Testing*, ed. A. Steinbüchel, CPL Press, UK, 2003, p. 365.
19. R.-J. Müller, *Process Biochem.*, 2006, **41**, 2124.
20. A. Göpferich, *Biomaterials*, 1996, **17**, 103.
21. D. S. Katti, S. Lakshmi, R. Langer and C. T. Laurencin, *Adv. Drug Deliv. Rev.*, 2002, **54**, 61.
22. B. Schink, P. H. Janssen and J. Frings, *FEMS Microbiol. Rev.*, 1992, **103**, 311.
23. J. M. Wasikiewicz, F. Yoshii, N. Nagasawa, R. A. Wach and H. Mitomo, *Radiat. Phys. Chem.*, 2005, **73**, 287.
24. D. I. Stirling and H. Dalton, *J. Gen. Microbiol.*, 1980, **116**, 277.
25. L. S. Nair and C. T. Laurencin, *Prog. Polym. Sci.*, 2007, **32**, 762.
26. W.-C. Hsieh, H. Mitomo, K.-I. Kasuya and T. Komoto, *J. Polym. Environ.*, 2006, **1(4)**, 79.
27. M. Mochizuki, K. Mukai, K. Yamada, N. Ichise, S. Murase and Y. Iwaya, *Macromolecules*, 1997, **30**, 7403.
28. H. Abe, Y. Doi, H. Aoki and T. Akehata, *Macromolecules*, 1998, **31**, 1791.
29. D. Raghavan, *Polym.-Plast. Technol. Eng.*, 1995, **34**, 41.
30. A. Ohtaki and K. Nakasaki, *Waste Manag. Res.*, 2000, **18**, 184.
31. A. Hoshino, M. Tsuji, M. Momochi, A. Mizutani, H. Sawada, S. Kohnami, H. Nakagomi, M. Ito, H. Saida, M. Ohnishi, M. Hirata, M. Kunioka, M. Funabash and S. Uematsu, *J. Polym. Environ.*, 2007, **15**, 275.
32. R. Chandra and R. Rustgi, *Progr. Polym. Sci.*, 1998, **23**, 1273.
33. W. Amass, A. Amass and B. Tighe, *Polym. Int.*, 1998, **47**, 89.

34. L. Averous, *J. Macromol. Sci. C Polym. Rev.*, 2004, **44**, 231.
35. V. A. Alvarez, R. A. Ruseckaite and A. Vázquez, *Polym. Degrad. Stab.*, 2006, **91**, 3156.
36. A. Keller, D. Bruggmann, A. Neff, B. Müller and E. Wintermantel, *J. Polym. Environ.*, 2000, **8**, 91.
37. L. G. Carr, D. F. Parra, P. Ponce, A. B. Lugão and P. M. Buchler, *J. Polym. Environ.*, 2006, **14**, 179.
38. E. Psomiadou, I. Arvanitoyannis, C. G. Biliaderis, H. Ogawa and N. Kawasaki, *Carbohydr. Polym.*, 1997, **33**, 227.
39. J. Yang and K. Suliao, *Plast. Sci. Technol.*, 2009, **37**, 43.
40. S. Wu, S. M. Lai and H. T. M. Liao, *J. Appl. Polym. Sci.*, 2002, **85**, 2905.
41. C.-S. Wu, *Polymers*, 2005, **46**, 147.
42. A. R. Sarasam, R. K. Krishnaswamy and S. V. Madihally, *Biomacromolecules*, 2006, **7**, 1131.
43. I. Ammar, R. Ben Cheikh, A. R. Campos and A. M. Cunha, *Polym. Compos.*, 2006, **27**, 341.
44. K. Arevalo, E. Aleman, G. Rojas, L. Morales and L. J. Galan, *New Biotechnol.*, 2009, **25**, 287.
45. B. A. Ramsay, V. Langlade, T. Pierre, J. Carreau and J. A. Ramsay, *Appl. Environ. Microbiol.*, 1993, **1242**.
46. R. Mohee and G. Unmar, *Waste Manag.*, 2007, **27**, 1486.
47. R. Mohee, A. Mudhoo and G. D. Unmar, *Int. J. Environ. Waste Manag.*, 2008, **2**, 3.
48. H.-S. Kim, H.-J. Kim, J.-W. Lee and I.-G. Choi, *Polym. Degrad. Stab.*, 2006, **91**, 1117.
49. M. Vikman, M. Itävaara and K. Poutanen, *J. Polym. Environ.*, 1995, **3**, 23.
50. J.-D. Gu, D. T. Eberiel, S. P. McCarthy and R. A. Gross, *J. Polym. Environ.*, 1993, **1**, 143.
51. M. F. Rosa, B. Chiou, E. S. Medeiros, D. F. Wood, L. H. C. Mattoso, W. J. Orts and S. H. Imam, *J. Appl. Polym. Sci.*, 2009, **111**, 612.
52. C. Chen, L. Dong and M. K. Cheung, *Eur. Polym. J.*, 2005, **41**, 958.
53. C. Xie, Y. Feng, W. Cao, Y. Xia and Z. Lu, *Carbohydr. Polym.*, 2007, **67**, 566.
54. P. Alexy, D. Bakoš, G. Crkoňová, Z. Kramárová, J. Hoffmann, M. Julinová, E. Chiellini and P. Cinelli, *Polym. Test.*, 2003, **22**, 811.
55. G. Zhao, L. Ya, F. Cuilan, M. Zhang, C. Zhou and Z. Chen, *Polym. Degrad. Stab.*, 2006, **91**, 703.
56. Z. Khan and Y. Anjaneyulu, *Rem. J.*, 2006, **16**, 109.
57. J. Ryckeboer, J. Mergaert, K. Vaes, S. Klammer, D. De Clercq, J. Coosemans, H. Insam and J. Swings, *Ann. Microbiol.*, 2003, **53**, 349.
58. H. M. KeenerW. A. DickH. A. J. Hoitink*Land Application of Agricultural, Industrial, and Municipal By-products*, eds. J. F. Power W. A. Dick, Soil Science Society of America, Madison, 2001, p. 315.
59. A. M. H. Veeken, F. Adani, K. G. J. Nierop, P. A. de Jager and H. M. V. Hamelers, *J. Environ. Qual.*, 2001, **30**, 1675.

60. Z. C. Symons and N. C. Bruce, *Natural Prod. Rep.*, 2006, **23**, 845.
61. A. Zach, M. Latif, E. Binner and P. Lechner, *Compost Sci. Util.*, 1999, **7**, 25.
62. C. Petiot and A. De Guardia, *Compost Sci. Util.*, 2004, **12**, 69.
63. A. E. Ghaly, F. Alkoaik and A. Snow, *Can. Biosys. Eng.*, 2006, **48**, 1.
64. H. M. KeenerC. MaruggR. C. HansenH. A. J. Hoitink, Optimizing the efficiency of the composting process, In: *Proceedings of the International Composting Research Symposium*, eds. H. A. J. Hoitink, H. M. Keener, Renaissance Publications, Columbus, 1993, p. 59.
65. R. Mohee and A. Mudhoo, *Powder Technol.*, 2005, **155**, 92.
66. A. Mudhoo and R. Mohee, *J. Environ. Informatics*, 2006, **8**, 100.
67. T. L. Richard, A. Veeken, V. De Wilde and H. M. V. Hamelers, *Biotechnol. Prog.*, 2004, **20**, 1372.
68. J. L. Lange and B. V. Antohe, *ASME J. Fluids Eng.*, 2000, **122**, 619.
69. A. Mudhoo and R. Mohee, *J. Environ. Informatics*, 2007, **9**, 87.
70. A. Mudhoo and R. Mohee, *J. Environ. Informatics*, 2008, **11**, 74.
71. R. Mohee, G. D. Unmar, A. Mudhoo and P. Khadoo, *Waste Manag.*, 2008, **28**, 1624.
72. J. A. Hogan, F. C. Miller and M. S. Finstein, *Appl. Environ. Microbiol.*, 1989, **55**, 1082.
73. J. M. Agnew and J. J. Leonard, *Compost Sci.Util.*, 2003, **11**, 238.
74. J. Bear, *Dynamics of Fluids in Porous Media*, Elsevier, New York, 1972.
75. S. Gajalakshmi and S. A. Abbasi, *Crit. Rev. Environ. Sci. Technol.*, 2008, **38**, 311.
76. Z. A. Hickman and B. J. Reid, *Environ. Int.*, 2008, **34**, 1072.
77. V. K. Garg, R. Gupta and P. Kaushik, *Int. J. Environ. Pollut.*, 2009, **38**, 385.
78. J. Martín-Gil, L. M. Navas-Graci, E. Gómez-Sobrino, A. Correa-Guimaraes, S. Hernández-Navarro, M. Sánchez-Báscones and M. del Carmen Ramos-Sánchez, *Bioresour. Technol.*, 2008, **99**, 1821.
79. T. N. Ravikumar, N. A. Yeledhalli, M. V. Ravi and K. Narayana Rao, *Karnataka J. Agric. Sci.*, 2008, **21**, 222.
80. S. Chatterjee, P. Chattopadhyay, S. Roy and S. K. Sen, *J. Appl. Biosci.*, 2008, **11**, 594.
81. B. Ceccanti, G. Masciandaro, C. Garcia, C. Macci and S. Doni, *2006, Water Air Soil Pollut.*, 2006, **177**, 383.
82. J. A. Marín, J. L. Moreno, T. Hernández and C. García, *Biodegradation*, 2006, **17**, 251.
83. G. Kale, R. Auras, S. P. Singh and R. Narayan, *Polym. Test.*, 2007, **26**, 1049.
84. U. Pagga, D. B. Beimborn and M. Yamamoto, *J. Polym. Environ.*, 1996, **4**, 173.
85. T. Kijchavengkul and R. Auras, *Polym. Int.*, 2008, **57**, 793.
86. I. Körner, K. Redemann and R. Stegmann, *Waste Manag.*, 2005, **25**, 409.
87. A. P. Bonartsev, V. L. Myshkina, D. A. Nikolaeva, E. K. Furina, T. A. Makhina, V. A. Livshits, A. P. Boskhomdzhiev, E. A. Ivanov, A. L.

Iordansakii and G. A. Bonartseva, *Commun. Curr. Res. Educs. Topics Trends Appl. Microbiol.*, 2007, 295.
88. A. Steinbüchel, *Polyhydroxy Alkanoic Acids*, ed. D. Byrom, Stockton, New York, 1991, p. 124.
89. S. Reddy, M. Thirumala and S. Mahmood, *J. Microbiol.*, 2008, **4**.
90. Y. Lee and J. Choi, *Waste Manag.*, 1999, **19**, 133.
91. R.-J. Müller, I. Kleeberg and W.-D. Deckwer, *J. Biotechnol.*, 2001, **86**, 87.
92. D. Y. Kim and Y. H. Rhee, *J. Appl. Microbiol. Biotechnol.*, 2003, **61**, 300.
93. S. F. Williams, D. P. Martin, D. M. Horowitz and O. P. Peoples, *Int. J. Biol. Macromol.*, 1999, **25**, 111.
94. I. Janigova, I. Lacik and I. Chodak, *Polym. Degrad. Stab.*, 2002, **77**, 35.
95. Y. He, X. T. Shuai, A. Cao, K. Kasuya, Y. Doi and Y. Inoue, *Polym. Degrad. Stab.*, 2001, **73**, 193.
96. R. Griebel, Z. Smith and J. M. Merrick, *Biochemistry*, 1968, **7**, 3676.
97. F. P. Delafield, M. Doudoroff, N. J. Palleroni, C. J. Lusty and R. Contopoulos, *J. Bacteriol.*, 1965, **90**, 1455.
98. D. Jendrossek, A. Schirmer and H. G. Schlegel, *J. Appl. Microbiol. Biotechnol.*, 1996, **46**, 451.
99. K.-M. Lee, D. F. Gimore and M. J. Huss, *J. Polym. Environ.*, 2005, **13**, 213.
100. W. J. Page, *Can. J. Microbiol.*, 1995, **41**, 1.
101. H. Brandl and P. Puchner, *Biodegradation*, 1992, **2**, 237.
102. G. Swift, *Acc. Chem. Res.*, 1993, **26**, 105.
103. J. Mergaert, A. Webb, C. Anderson, A. Wouters and J. Swings, *Appl. Envi. Microbiol.*, 1993, **59**, 3233.
104. J. Mergaert, A. Wouters, C. Anderson and J. Swings, *Can. J. Microbiol.*, 1995, **41**, 154.
105. M. Matavulj and H. P. Molitoris, *FEMS Microbiol. Rev.*, 1992, **103**, 323.
106. K. Mukai, K. Yamada and Y. Doi, *Polym. Degrad. Stab.*, 1993, **41**, 85.
107. R. N. Sturm, *J. Am. Oil Chem. Soc.*, 1973, **50**, 159.
108. D. S. Rosa, R. P. Filho, Q. S. H. Chui, M. R. Calil and C. G. F. Guedes, *Eur. Polym. J.*, 2003, **39**, 233.
109. C. M. Buchanan, C. N. Boggs, D. D. Dorschel, R. M. Gardner, R. J. Komarek, T. L. Watterson and A. W. White, *J. Polym. Environ.*, 1995, **3**, 1.
110. S. H. Imam, L. Chen, S. H. Gordon, R. L. Shogren, D. Weisleder and R. V. Greene, *J. Polym. Environ.*, 1998, **6**, 91.
111. M. Rutkowska, K. Krasowska, A. Heimowska and M. Kowalczuk, *Macromol. Symp.*, 2003, **197**, 421.
112. J.-H. Zhao, X.-Q. Wang, J. Zeng, G. Yang, F.-H. Shi and Q. Yan, *J. Appl. Polym. Sci.*, 2005, **97**, 2273.
113. B. Braun and J. R. Dorgan, *Polylactides: A New Paradigm in Polymer Science*, 2004. Available at http://www.nt.ntnu.no/users/skoge/prost/proceedings/aiche-2004/pdffiles/papers/004a.pdf. Accessed on 4 November 2009.
114. Y.-W. Wang, W. Mo, H. Yao, Q. Wu, J. Chen and G. Chen, *Polym. Degrad. Stab.*, 2004, **85**, 815.

115. G. Q. Chen, Q. Wu and J. C. Chen, *Tsinghua J. Sci. Technol.*, 2001, **6**, 193.
116. G. Kale, S. P. Singh and R. Auras, *Evaluation of Compostability of Commercially Available Biodegradable Packages in Real Composting Conditions* 2005. Available at http://www.nt.ntnu.no/users/skoge/prost/ proceedings/aiche-2005/non-topical/Non%20topical/papers/448f.pdf. Accessed on 5 November 2009.
117. E. Vink, *et al.*, *Polym. Degrad. Stab.*, 2003, **80**, 403.
118. V. M. Ghorpade, A. Gennadios and M. A. Hanna, *Bioresour. Technol.*, 2001, **76**, 57.
119. R. Gattin, A. Copinet, C. Bertrand and Y. Couturier, *Int. Biodeterior. Biodegrad.*, 2002, **50**, 25.
120. L. Fernando Vargas, B. A. Welt, P. Pullammanappallil, A. A. Teixeira, M. O. Balaban and C. L. Beatty, *Packaging Technol. Sci.*, 2009, **22**, 97.
121. T. Kijchavengkul, G. Kale and R. Auras, *Polymer Degradation and Performance, Chapter 3*, 2009, p. 31.
122. R. Iovino, R. Zullo, M. A. Rao, L. Cassar and L. Gianfreda, *Polym. Degrad. Stab.*, 2008, **93**, 147.
123. G. M. Ganjyal, R. Weber and M. A. Hanna, *Bioresour. Technol.*, 2007, **98**, 3176.
124. G. Kale, T. Kijchavengkul, R. Auras, M. Rubino, S. E. Selke and S. P. Singh, *Macromol. Biosci.*, 2007, **7**, 255.
125. G. Kale, R. Auras and S. P. Singh, *J. Polym. Environ.*, 2006, **14**, 317.
126. J. Tuominen, J. Kylmä, A. Kapanen, O. Venelampi, M. Itävaara and J. Seppälä, *Biomacromolecules*, 2002, **3**, 445.
127. M. Itävaara, S. Karjomaa and J.-F. Selin, *Chemosphere*, 2002, **46**, 879.
128. N. Alauzet, H. Garreau, M. Bouché and M. Vert, *J. Polym. Environ.*, 2002, **10**, 53.
129. M. Hakkarainen, S. Karlsson and A.-C. Albertsson, *Polymer*, 2000, **41**, 2331.
130. B. Schink and M. Stieb, *Appl. Environ. Microbiol*, 1983, **45**, 1905.
131. B. Lee, A. L. Pometto Iii, A. Fratzke and T. B. Bailey, Jr., *Appl. Environ. Microbiol.*, 1991, **57**, 678.
132. M. Koutny, J. Lemaire and A. M. Delort, *Chemosphere*, 2006, **64**, 1243.
133. M. Hakkarainen and A.-C. Albertsson, *Adv. Polym. Sci.*, 2004, **16**(9), 177.
134. K. E. Johnson, A. L. Pometto III and Z. L. Nikolov, *Appl. Environ. Microbiol.*, 1993, **59**, 1155.
135. E. Chiellini, F. Cioni, R. Solaro, G. Vallini, A. Corti and A. Pera, *J. Polym. Environ.*, 1993, **1**, 167.
136. G. Vallini, A. Corti, A. Pera, R. Solaro, F. Cioni and E. Chiellini, *J. Gen. Appl. Microbiol.*, 1994, **40**, 445.
137. I. Jakubowicz, *Polym. Degrad. Stab.*, 2003, **80**, 39.
138. S. Bonhomme, A. Cuer, A.-M. Delort, J. Lemaire, M. Sancelme and G. Scott, *Polym. Degrad. Stab.*, 2003, **81**, 441.
139. Y. Orhan, J. Hrenović and H. Büyükgüngör, *Acta Chim. Slov.*, 2004, **51**, 579.
140. V. Sedlarik, N. Saha and P. Saha, *Polym. Degrad. Stab.*, 2006, **91**, 2039.

141. G. J. L. Griffin, *J. Polym. Sci.: Polym. Symp.*, **57**, 281.
142. J. C. Middleton and A. J. Tipton, *Biomaterials*, 2000, **21**, 2335.
143. P. Jarrett, C. V. Benedict, J. P. Bell, J. A. Cameron and S. J. Huang, *Polymers as Biomaterials*, ed. S. W. Shalaby, A. S. Hoffman, B. D. Ratner T. A. Horbett, Plenum Press, New York, 1991, p. 181.
144. M. Rutkowska, K. Krasowska, A. Heimowska, I. Steinka, H. Janik, J. Haponiuk and S. Karlsson, *Pol. J. Environ. Studies*, 2002, **11**, 413.
145. A. C. Albertsson, R. Renstad, B. Erlandsson, C. Eldsater and S. J. Karlsson, *Appl. Polym. Sci.*, 1998, **70**, 61.
146. D. Darwis, H. Mitomo, T. Enjoji, E. Yoshii and K. Makuuchi, *Polym. Degrad. Stab.*, 1998, **62**, 259.
147. Y. Lei, B. Rai, K. H. Ho and S. H. Teoh, *Mater. Sci. Eng: C*, 2007, **27**, 293.
148. C. Lefèvre, A. Tidjani, C. Vander Wauven and C. David, *J. Appl. Polym. Sci.*, 2002, **83**, 1334.
149. M. Kunioka, F. Ninomiya and M. Funabashi, *Polym. Degrad. Stab.*, 2007, **9**(2), 1279.
150. A. Kulkarni, J. Reiche, J. Hartmann, K. Kratz and A. Lendlein, *Eur. J. Pharm. Biopharm.*, 2008, **68**, 46.
151. R. Deng, Y. Chen, P. Chen, L. Zhang and B. Liao, *Polym. Degrad. Stab.*, 2006, **91**, 2189.
152. V. K. Khatiwala, N. Shekhar, S. Aggarwal and U. K. Mandal, *J. Polym. Environ.*, 2008, **16**, 61.
153. N. Hiraishi, F. T. Sadek, N. M. King, M. Ferrari, D. H. Pashley and F. R. Tay, *Am. J. Dent.*, 2008, **21**, 119.
154. A. Ohtaki, N. Sato and K. Nakasaki, *Polym. Degrad. Stab.*, 1998, **61**, 499.
155. A. Ohtaki, N. Akakura and K. Nakasaki, *Polym. Degrad. Stab.*, 1998, **62**, 279.
156. R. P. Singh, J. K. Pandey, D. Rutot, Ph. Degée and Ph. Dubois, *Carbohydr. Res.*, 2003, **338**, 1759.
157. H.-S. Yang, J.-S. Yoon and M.-N. Kim, *Polym. Degrad. Stab.*, 2005, **87**, 131.
158. C. De Kesel, C. Van der Wauven and C. David, *Polym. Degrad. Stab.*, 1997, **55**, 107.
159. E. S. Stevens, *Biocycle*, 2002, 42.
160. M. Gáspár, Zs. Benkõ, G. Dogossy, K. Réczey and T. Czigány, *Polym. Degrad. Stab.*, 2005, **90**, 563.
161. L. Avérous and P. J. Halley, *Biofuels Bioproducts Biorefining*, 2009, **3**, 329.
162. R. Ouhib, B. Renault, H. Mouaziz, C. Nouvel, E. Dellacherie and J.-L Six, *Carbohydr. Polym.*, 2009, **77**, 32.
163. A. Shaabani, A. Rahmati and Z. Badri, *Catal. Commun.*, 2008, **9**, 13.
164. P. Bordes, E. Pollet and L. Avérous, *Prog. Polym. Sci.*, 2009, **34**, 125.

CHAPTER 12
Biodegradable Polymers: Research and Applications

X. W. WEI, G. GUO, C. Y. GONG, M. L. GOU AND ZHI YONG QIAN

Sichuan University, West China Medical School, West China Hospital, State Key Laboratory of Biotherapy and Cancer Center, 610041, Chengdu, China

12.1 Introduction

As we enter the twenty-first century, research at the interface of polymer chemistry and materials science has given rise to a favourable shift from biostable materials to biodegradable polymers in various fields. Biodegradable polymers that are either hydrolytically or enzymatically degradable play an enormous role in industrial, biomedical and other related areas.[1] Biodegradation is a natural process by which organic chemicals in the environment are converted to simpler compounds, mineralized and redistributed through elemental cycles such as the carbon, nitrogen and sulfur cycles.[2] Biodegradation is a term that everyone understands, yet no definition has been coined that is universally acceptable for biodegradable polymers. One of the major reasons for this is the wide range of disciplines, represented by, namely, biologists, biochemists, polymer chemists, legislators, environmentalists and manufacturers *etc*. The ones who are involved directly in biodegradable polymer research or with an opinion on requirements have different opinions on what they expect a polymer to do in the environment in order to be called or defined as biodegradable. Mainly, biodegradation of polymeric biomaterials involves cleavage of hydrolytically or enzymatically sensitive bonds in the polymer leading to polymer erosion.[3] Here we define biodegradable polymers as polymers that are degraded and catabolized eventually to carbon dioxide and water under hydrolytic or enzymatic reaction.

12.1.1 Biodegradable Polymers and the Environment

What accounted for the promoting interest in biodegradable polymers was the growing concern raised by the recalcitrance and unknown environmental fate of many of the currently used synthetic polymers. Dating back to the last century, synthetic polymers were originally developed for their durability and resistance to all forms of degradation, the crucial characteristics of which are achieved through control of molecular weight and functionality. They are widely welcomed due to their special properties, inexpensive prices and capability to enhance the comfort and quality of life in our modern industrial society. However, these same properties that made the polymers widely accepted have contributed to a disposal problem. Some of the polymers received the brunt of media attention on this issue, exemplified by plastics. They are commonly visible in the environment as litter and their low density made them occupy a high volume fraction of buried waste, which resulted in an obvious depletion in landfill.[4] In the 1970s, work was started in the United States (US) and elsewhere to produce photodegradable and biodegradable plastics for the packaging industry. The ideal polymeric materials were required to be non-toxically degradable, and have suitable mechanical properties, economic viability and processability. Since then, such plastics, both made from natural and synthetic polymers, were brought into view. They were together with three main technologies, incineration, recycling and burial in landfill sites, playing an important role in the waste management of plastics.[5] Despite visible plastics, unseen water-soluble polymers as surfactants in waste water streams also caught public attention. And people in related industries were taking action, continuing to develop biodegradable polymers. Great improvement was made in the detergent industry for they completely switched from non-biodegradable (branched alkyl) benzenesulfonate surfactants to biodegradable linear analogues. As a result, the streams of foam previously visible in the effluent from wastewater treatment plants in the 1960s are no longer seen.[1,6] Moreover, some research was concerned with looking for the ideal surfactants which could both eliminate the possibility of eutrophication and prevent acid accumulation of the used polymer. The review by Paik et al.[7] traces the history of the biodegradable polymer options that have been considered and the state of the research in 1990s.

More studies around the application of biodegradable polymers in the environment were carried out, e.g. some polymers have been recognized their potential for the release of insecticide into soil, which could release the chemical at the appropriate time controlled by the time of degrading.[8] Others contributed to the preventing of oil pollution and purification of the denitrifying bacteria in water.[9] In addition, the use of biodegradable polymers for renewable resources was analysed.[10]

12.1.2 Biodegradable Polymers and Biomedical Uses

The biomedical application of biodegradable polymers, for example enzymatically degradable natural polymers such as collagen, dates back thousands of years. However, the application of synthetic biodegradable polymers started only in the later half of 1960s.[11] Last century saw the development of a wide variety of new generation synthetic biodegradable polymers and analogous natural polymers which were specifically designed for biomedical applications.

The current biomedical applications of biodegradable polymeric materials can be classified into five groups.

(1) Surgical fixation, such as sutures, clips, bone pins and plates.
(2) Controlled drug delivery (both localized and targeting systems), such as microspheres and nanoparticles.
(3) Plain membranes for guided tissue regeneration.
(4) Multifilament meshes or porous structures for tissue engineering.
(5) Specialty packaging, such as packaging materials for pharmaceutical products, drugs and wound dressings.[12,13]

Each application mentioned above demands materials with specific physical, chemical, biological, biomechanical and degradation properties to provide efficient therapy.[14] Above all, the vital prerequisite for a polymer to be used in the biomedical field is biocompatibility, which is the ability of a material to perform with an appropriate host response in a specific application. For instance, there were several important properties listed for materials used in implantation.

(1) The material should not evoke a sustained inflammatory or toxic response.
(2) The material should have acceptable shelf life.
(3) The degradation time of the material should be compatible with the healing or regeneration course.
(4) The material should have appropriate mechanical properties for specific application and the variation in mechanical properties with degradation should match the healing or regeneration process.
(5) The degradation products should be non-toxic, and able to be metabolized and cleared from the body.
(6) The material should have appropriate permeability and processability for the intended application.[14,15]

What affects the biocompatibility of a polymeric material are some inherent properties, such as material chemistry, molecular weight, solubility, hydrophilicity/hydrophobicity, absorption, degradation and erosion mechanism, *etc*. Consequently, given the complexity and the variety of biomedical applications for which biodegradable polymers are currently used, it underlines the need for developing a wide range of biodegradable materials available for requirements of each medical application.

The work conducted currently in developing various biodegradable materials is basically some investigation on polymer synthesis. To summarize, polymers with required properties for specific use are developed as follows.

(1) Focusing on novel synthetic polymers with unique chemistries to increase the diversity of polymer structure.
(2) Attempting to adopt biosynthetic processes to form biomimetic polymer structures.
(3) Making efforts on combinatorial and computational approaches in biomaterial design to accelerate the discovery process.[14,16]

Nowadays, biodegradable polymers can be briefly classified into two categories: natural polymers and synthetic polymers.

In this chapter, we give introductions for two categories of biodegradable copolymers and deal with the recent development and application for each material, particularly highlighting the biomedical uses. The reader will have a comprehensive and consolidated overview of the immense potential and ongoing research in biodegradable polymer science and engineering, which is earnestly attempting to make the world of tomorrow greener.

12.2 Natural Biodegradable Polymers and their Derivatives

12.2.1 Starch and Derivatives

For materials applications, starch has been one of the principal polysaccharides of interest. Starch is a mixture of linear amylose (poly-α-1,4-D-glucopyranoside) and branched amylopectin (poly-α-1,4-D-glucopyranoside and α-1,6-D-glucopyranoside) (Scheme 12.1). The crops generally used for its production include rice, corn and potatoes. However, the produced starch, in the form of granules, varies in size and composition with the source: amylose usually makes up about 20 wt% of the granule, and the branched polymer, amylopectin, the remainder. Amylose is crystalline and soluble in boiling water while amylopectin is insoluble. But while used in food, both fractions are readily hydrolysed at the acetal link by enzymes, such as amylases and glucosidases.

The stability of starch, as a polymer, under different temperatures needs to be improved. At temperatures around 150 °C the glucoside links start to break, and when raising the temperature up to 250 °C the starch grain endothermally collapses. In addition, at rather low temperature retrogradation is observed, which is a reorganization of the hydrogen bonds and an aligning of the molecular chains during cooling. The extreme case of this is that a temperature of 10 °C leads to precipitation. Therefore, starch is either physically mixed or melted and blended on a molecular level with certain materials for specific use as a biodegradable polymer. When it comes to the chemical modification, the starch molecule has two important functional groups, the –OH group for

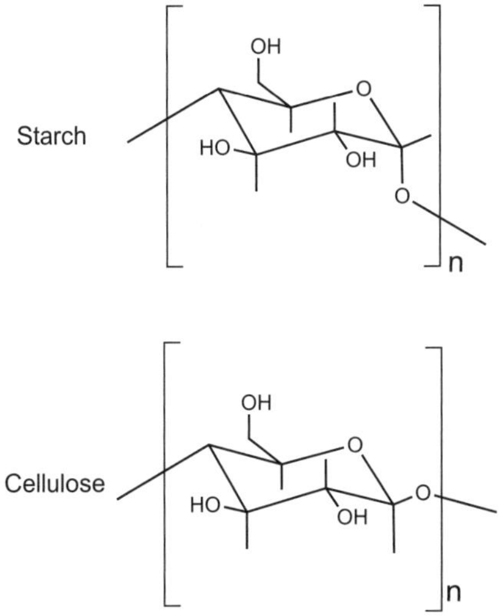

Biodegradable Polymers: Research and Applications

Scheme 12.1 Structures of starch and cellulose.

substitution reactions and the C–O–C bond for chain breakage. The nucleophilic hydroxyl group of glucose can achieve various properties by reacting with appropriate polymers or monomers.[17] The acetylation of starch is an easy and well-known example, the obtained starch acetate has several advantages over native starch, such as improved solubility and better retention of tensile properties in aqueous environments.[18,19]

Another field of starch converted to thermoplastic material has caught much attention. The so-called starch plastics offer an interesting alternative for synthetic polymers achieving rapid degradation instead of long-term durability.[20] Thermoplastic starch is generally processable as a traditional plastic; however, its sensitivity to humidity restricts it to some applications, including loose-fillers, expanded trays, shape moulded parts — mainly as a replacement for polystyrene.[21,22]

Starch and its derivatives are also favourable in pharmaceutics being an alternative for microcrystalline cellulose as an excipient intended for the production of pellets *via* extrusion-spheronization.[23] Dukić *et al.*[24] reported the application of a specific grade of modified starch for extrusion-spheronization purposes: a crystalline, high-amylose starch formed by gelatinization of amylose-rich starches, followed by enzymatic debranching of amylopectin molecules and retrogradation of linear amylose chains, produces a potential candidate.[24]

12.2.2 Cellulose and Derivatives

Cellulose was isolated for the first time some 150 years ago. Its structure is similar to amylose, composed of hundreds or thousands of D-glucopyranoside repeating units. These units are linked by acetal bonds formed between the hemiacetal carbon atom, C1, of the cyclic glucose structure in one unit and a hydroxyl group at the C3 atom in the adjacent unit. There exist two isomers due to the cyclic form, the α-isomer with an axial –OH group on the ring or the β-isomer with an equatorial –OH group. The only structural difference between amylose and cellulose is that starch forms the β-form while cellulose exists in the β-form (Scheme 12.1). And as a result, enzymes related to acetal hydrolysis reactions in the biodegradation of each of these two polysaccharides are different. Cellulose is mainly composed of very long repeating units, thus making it a highly crystalline, high molecular weight polymer. It is hardly fusible and soluble in any solvents, except the most aggressive, hydrogen bond-breaking ones. Due to these intrinsic characteristics, cellulose is usually converted into derivatives to make it more processable.

Cellulose acetate (CA) is one of the potentially useful cellulose esters, which is now commercially available. CA of various degree of substitution (DS) are now being widely used as films and coatings. CA films have a tensile strength comparable to polystyrene, which makes the polymer suitable for injection moulding.

Recently, cellulose nanofibres were prepared by fermentation, yielding a very pure cellulose product with unique physical properties, which distinguished it from plant-derived cellulose. Its fibres had a high aspect ratio and had a high surface area per unit mass. This property, when combined with its very hydrophilic nature, resulted in very high liquid loading capacity. It was an attractive candidate for a wide range of biomedical and biotechnology applications.[25]

12.2.3 Chitin and Chitosan

Chitin, poly(β-(1→4)-N-acetyl-D-glucosamine), is widely distributed in nature, is one of the most abundant organic compounds on Earth and is of great importance in different fields since it possesses unique structures and characteristics different from typical synthetic polymers. Chitin is obtained from lower animals, mainly extracted from crustaceans. The structure of chitin is similar to cellulose, with acetamide groups at the C2 positions in place of hydroxyl (Scheme 12.2).

Chitosan is the most important derivative of chitin, which can be obtained by deacetylation of chitin under alkaline conditions (Scheme 12.2). When the degree of deacetylation of chitin reaches about 50%, it becomes soluble in aqueous acidic media. Thanks to the –NH_2 functional group on the C2 position of the D-glucosamine repeating unit, protonation occurs so that polysaccharide is converted to a polyelectrolyte in acidic media. Chitosan is the only pseudo natural cationic polymer and thus has unique advantages while applied in different fields. The research concerned with chitosan is about various applications such as in the areas of biomedicine, membranes, drug delivery systems, hydrogels, water treatment, food packaging, $etc.$[26,27]

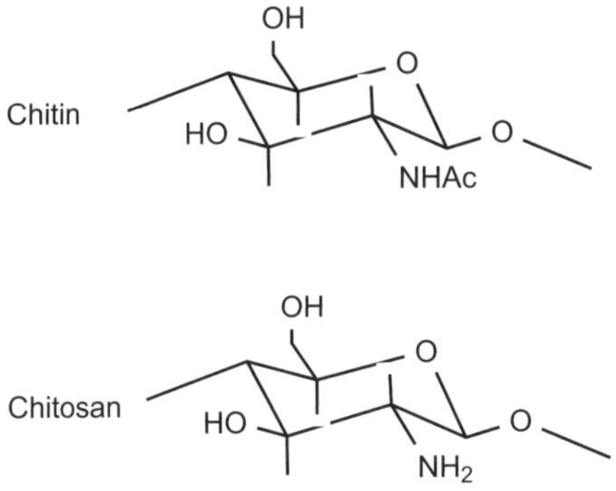

Scheme 12.2 Structures of chitin and chitosan.

Chitosan and its derivatives are suitable for tissue engineering applications due to their porous structure, gel-forming properties and ease of chemical modification. There is an excellent review on chitosan in responsive and *in situ*-forming scaffolds for bone tissue engineering applications by Martins *et al.*[28] And another new class of biocompatible and biodegradable composite hydrogels derived from chitosan and oxidized hyaluronic acid was reported.[29] The hydrogel was formed upon mixing without the addition of a chemical cross-linking agent and was able to support cell survival.

Chitosan is one of the most reported non-viral naturally derived polymers for gene delivery. As a polycation, it has strong affinity for DNA and can spontaneously form microspheric particles *via* complex coacervation.[30] One study on skin delivery of antisense oligonucleotides (AsODNs), which has exciting potential in the treatment of skin diseases, was recently achieved by AsODN-loaded chitosan nanoparticles. AsODN-loaded chitosan nanoparticles were topically applied to Sprague Dawley rats. Animal skin samples were taken for measurement of beta-galactosidase (beta-Gal) expression and histological control. After topical application of AsODN-loaded chitosan nanoparticles in different doses, beta-Gal expression reduced significantly. The results indicates that chitosan nanoparticles are a useful carrier for delivery of AsODNs into skin cells of rats and may be used for topical application on human skin.[31]

12.2.4 Alginic Acid

Alginic acid is found within the cell walls and intercellular spaces of brown algae and is a non-branched, binary copolymer of 1,4-linked β-D-mannuronic acid and α-L-guluronic acid monomers. Alginate is a block copolymer composed of two uronic acid with different block lengths and sequential arrangement. In solution, alginate acid form gels upon the introduction of counterions. The degree of cross-linking is dependent on various factors such as pH, type of counterion and the functional charge density of these polymers. The high functionality of gelling makes alginic acid a favourable biodegradable material in controlled release drug delivery systems.[11]

A study on dual-functional alginic acid hybrid nanospheres for cell imaging and drug delivery has been carried out.[32] Gold nanoparticle-encapsulated alginic acid poly(2-(diethylamino)ethyl methacrylate) monodisperse hybrid nanospheres (ALG-PDEA-Au) was prepared. These negatively charged ALG-PDEA-Au hybrid nanospheres could be internalized by human colorectal cancer cells and hence acted as novel optical-contrast reagents in tumour-cell imaging by optical microscopy. Moreover, these hybrid nanospheres could also serve as biocompatible carriers for the loading and delivery of an anti-cancer drug doxorubicin. The obtained hybrid nanospheres successfully combined two functions which might be of great potential in other biomedical-related areas.

There are also other biomedical applications of alginate acid. Alginate-based microcapsules for immunoisolation of pancreatic islets have been intensively investigated.[33] Also, research concerned with polyionic hydrocolloids for the intestinal delivery of protein drugs by alginate-based material has received much attention.[34]

12.2.5 Collagen

Collagen is considered to be a favourable natural polymer for biomedical use since it is a major natural constituent of connective tissue and a major structural protein of any organ. The current main sources of collagen are animal skin, mostly bovine or porcine, and the properties of collagen, such as mechanical strength, fluid absorption volume or haemostatic activity, differ depending on the species and locations of the organisms. Collagen is unique in its different levels of structural order: primary, secondary, tertiary and quaternary. And two types of covalent cross-linking, intramolecular and intermolecular, means that collagen molecules can form fibres *in vivo*. Biomaterials made of collagen are welcomed for several advantages: they are biocompatible and non-toxic and have well-documented structural, physical, chemical, biological and immunological properties.

Due to its biocompatibility and well-established safety profile, collagen represents a favourable matrix for on-site drug delivery. Ruszczak *et al.*[35] summarized some of the developments and applications of collagen as a biomaterial in drug delivery systems for antibiotics, where the efforts in the use of collagen and collagen-synthetic polymer composites for controlled drug delivery as well as collagen-based diffusion membranes for prolonged drug release have also been included.[35] They assumed that the next generation of collagen drug delivery system for antibiotics would be focused on both drug combination and different release profiles which would lead to better infection control.

Also, research on collagen in the field of tissue engineering was reported. Oliveira *et al.* created a transient cartilage template *in vitro*, which could serve as an intermediate for bone formation by the endochondral mechanism once implanted *in vivo*.[36] They found the prepared collagen sponges presented improved stiffness and supported chondrocyte attachment and proliferation. Another study on octacalcium phosphate (OCP) collagen composites enhancing bone healing in a dog tooth extraction socket model was also carried out.[37]

12.2.6 Gelatin

Gelatin, composed of 19 amino acids joined by peptide linkages, can be hydrolysed by numerous proteolytic enzymes to yield its constituent amino acids or peptide components.[38]

The present cancer gene therapy using gelatin is lacking in both efficiency and specificity in comparison with viral vectors. However, recent studies showed that complexes of therapeutic DNA with modified gelatin would possibly offer a safe and efficient strategy for systemic administration of therapeutic genes to solid tumours compared to injection of naked plasmid DNA. The *in vivo* and *in vitro* studies using gelatin for the delivery of therapeutic genes to cancerous cells have been summarized.[39]

Despite the application in drug delivery systems, the contribution of gelatin capsules to the control of soil-borne pests was also investigated.[40] The gelatin capsule (gel cap) formulation of 1,3-dichloropropene (1,3-D) has been a new concept to reduce the environmental release, transport and hazard potential of the use of 1,3-D to control soil-borne diseases and nematodes. The biological efficacy of the 1,3-D gel cap formulation under laboratory and greenhouse trial conditions has been evaluated. Results showed that 1,3-D gel cap application was as effective as 1,3-D liquid injection treatment on tomato and crops.

12.2.7 Other Biodegradable Natural Polymers

Some other biodegradable natural polymers such as dextran,[41] heparin,[42] xanthan,[43] pullulan,[44] elastin,[45] fibrin,[46] pectin[47] and hyaluronate,[48] were also investigated as biodegradable polymers in various fields.

Hyaluronate can be both obtained from natural sources and produced from microbial fermentation. Its application in drug delivery systems for either protein drug or gene delivery was reported. Hyaluronic acid (HA)-based, microscopic hydrogel particles (HGPs) with inherent nanopores and defined functional groups were synthesized by an inverse emulsion polymerization technique.[48] They have developed a biomimetic growth factor delivery system that effectively stimulated the chondrogenic differentiation of the cultured mesenchymal stem cells *via* the controlled presentation of bone morphogenetic protein-2 (BMP-2). The obtained hydrogel particles provided an improved BMP-2 delivery system for stimulating chondrogenic differentiation *in vitro*, with potential therapeutic application for cartilage repair and regeneration. In another study, hyaluronic acid and chitosan-DNA complex multilayered thin film was used in surface-mediated nonviral gene delivery.[49]

Dextran, a biocompatible, water-soluble polysaccharide, was modified for specific uses in biomedical research. The modification was performed on by making its hydroxyl groups into acetal moieties, such that it became insoluble in water but freely soluble in common organic solvents, thus enabling its use in the facile preparation of acid-sensitive microparticles.[41] Both hydrophobic and hydrophilic cargoes were successfully loaded into these particles. The protein ovalbumin (OVA) has been used in a model vaccine application. Also, work on hydrogel based on interpenetrating polymer networks of dextran and gelatin for vascular tissue engineering has been carried out.[50]

Fibrin has a particular use in suture-less surgery. Recent developments have established fibrin glue for tissue adhesives as attractive alternatives to sutures. The current available information on fibrin glue is discussed by Panda et al.[46]

12.3 Synthetic Polymers

12.3.1 Polyesters

12.3.1.1 Polyglycolide

Glycolic acid, also called hydroxyacetic acid, is produced in normal body metabolism. The polymer of glycolic acid, known as poly(glycolic acid) or polyglycolide (PGA), can be considered as one of the first biodegradable synthetic polymers investigated for biomedical applications (Scheme 12.3). The repeating units, $-(O-CO-CHR)-$ are derived from α-hydroxy acids, HO–CHR–COOH. Though PGA of low molecular weight was first synthesized and described about one hundred years ago, the synthesis of PGA was intensively investigated in the last century. The ring-opening polymerization method has gained most of attention in synthesis of higher molecular weight PGA.[51] Ring-opening polymerization of glycolide mainly occurs under the influence of an inorganic metal salt catalyst at a low concentration. The molecular weight of the final polymer is controlled by the temperature, time, concentration of catalyst and concentration of the chain-length determining agent.

Polyglycolide is a highly crystalline polymer (45–55% crystallinity) and therefore exhibits a high tensile modulus with very low solubility in organic solvents. Strong implants can be manufactured from this polymer with a self-reinforcing (SR) technique and can be used in the treatment of fractures and osteotomies.[52]

The scaffold structures of polyglycolide were reported: a study aiming at examining the proliferation behaviour of human gingival fibroblasts on polyglycolic acid fleeces with various structural characteristics was carried out, with the satisfactory result showing that textile PGA fleece seemed to be suitable as a scaffold structure for human gingival fibroblasts. And it was indicated that the structural parameters of the fleece had a significant influence on the proliferation of the cells.[53] Furthermore, another study using polyglycolide (PGA) as scaffold to explore the feasibility of using cartilage-derived morphogenetic protein-1 (CDMP1) induced dermal fibroblasts (DFs) as seed cells for fibrocartilage engineering was reported. Histological and immunohistochemical staining of the constructs after being *in vitro* cultured for 4 and 6 weeks were carried out to observe the fibrocartilage formation condition.[54]

Applications of PLA-based nanostructured materials in different areas have been summarized in a review by Singh *et al.*[55] A large variety of nanoparticles of different nature and size can be blended with PLA, thus generating a new class of nanostructured biomaterials or nanocomposites with interesting physical properties and applications. In addition biocompatible materials such as carbon nanotubes, cellulose nanowhiskers, hydroxyapitite, *etc.* could also be incorporated into the PLA matrix, which increase the potential of PLA for biomedical applications.

12.3.1.2 Poly(lactide)

The most commercially viable material to date is poly(lactide) (PLA). PLA is mostly synthesized by the ring-opening polymerization of lactide, which itself derives from renewable resources such as potato, corn and sugar beet (Scheme 12.3).[56] Lactide is a chiral molecule and exists in two optically active forms: L-lactide and D-lactide. The polymerization of both monomers leads to the formation of semi-crystalline polymers. However, the polymerization of racemic (D,L)-lactide and mesolactide results in the formation of amorphous polymers.[57] The particular advantage of PLA polymers is its hydrolysis to lactic acid, which is one of the metabolites in the carboxylic acid cycle.[58] For its biodegradability and biocompatibility, it was approved by FDA as far back as the 1970s. PLA was favourable as a material in different fields for many advantageous physical properties, such as good mechanical properties, transparency, thermal stability, oil resistance and gas impermeability, as well as easy processing.[59]

Copolymers of PLA and poly(ethylene glycol) (PEG) are attractive for application in drug delivery systems. Novel thymopentin (TP5) release systems prepared from bioresorbable PLA-PEG-PLA hydrogels were reported

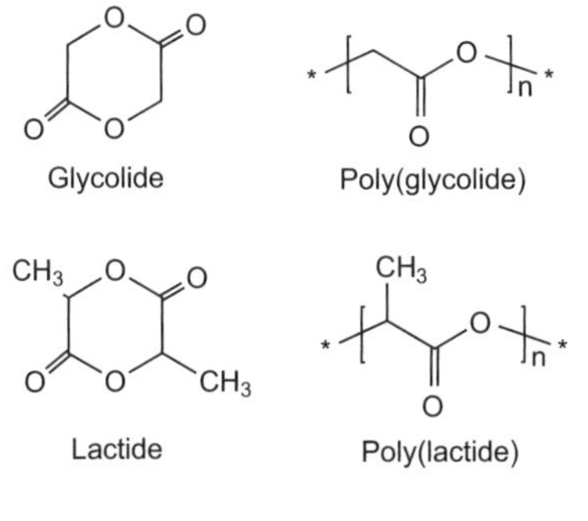

Scheme 12.3 Structures of polyesters.

recently.[60] Bioresorbable hydrogels were obtained by mixing PLLA-PEG-PLLA and PDLA-PEG-PDLA aqueous solutions due to stereocomplexation between PLLA and PDLA chains. Thymopentin was taken as a model drug to evaluate the potential of PLA-PEG-PLA hydrogels as carrier of hydrophilic drugs. The release profiles were characterized by an initial burst followed by slower release. Higher copolymer concentration led to a slower release rate and less burst effect due to more compact structure which disfavoured drug diffusion. Furthermore, *in vivo* studies proved the potential of TP5-containing hydrogels, especially those with a concentration of 25%, which combined with related results indicated the immunization efficacy of the TP5 release systems based on PLA/PEG hydrogels. In another study related to nanoparticles of PLA/PEG copolymer, three types of nanoparticles were formulated using different block copolymers AB, ABA and BAB (where 'A' is PLA and 'B' is PEG) encapsulating hepatitis B surface antigen (HBsAg) to evaluate their efficacy as an oral vaccine delivery system. Results indicated the efficacy of BAB nanoparticles as a promising carrier for oral immunization.[61]

12.3.1.3 Poly(lactide-co-glycolide)

Poly(lactide-co-glycolide) (PLGA) was obtained from copolymerization of lactide and glycolide, and its structure is shown in Scheme 12.4. The product exhibits better properties than PLA and PGA. Compared to PGA, PLGA usually exhibits lower crystallinities and T_m values.

Bulk erosion of PLGA occurred under hydrolysis of the ester bonds. The degradation rate of PLGA was more rapid than that of PLA and could be increased by increasing GA content.[62] The rate of degradation depended on a variety of parameters including the LA/GA ratio, molecular weight, and the shape and structure of the matrix. The popularity of the copolymer accounted to its good processability and controllable degradation rates. PLGA demonstrated good cell adhesion and biocompatibility, which made it a potential candidate in both tissue engineering and micron-/nano- drug delivery systems.[63]

Recently, investigation on poly(lactide-co-glycolide) implants was carried out, aiming at assessing the feasibility of hot-melt extrusion (HME) for preparing implants based on PLGA formulations with special emphasis on protein stability, burst release and release completeness. The results demonstrated that the release from all implants reached the 100% value in 60–80 days with nearly

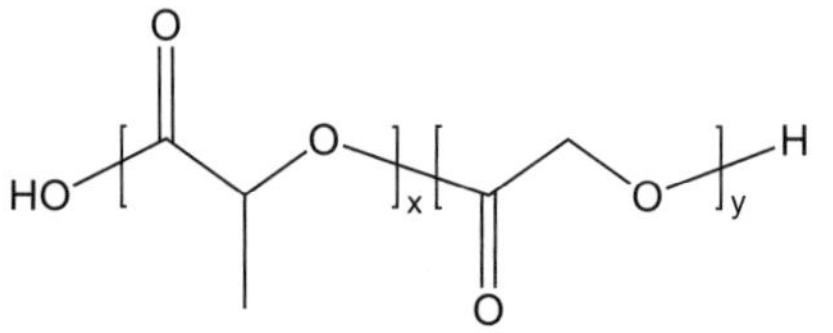

Scheme 12.4 Structure of poly(lactide-co-glycolide).

complete enzymatic activity of the last fraction of released lysozyme. Pure PLGA implants with up to 20% lysozyme loading could be formulated without an initial burst. A complete lysozyme recovery in active form with a burst-free and complete release from PLGA implants prepared by hot-melt extrusion was obtained.[64]

In another study, PLGA nanoparticles coated with poloxamer 188 or polysorbate 80 enabled an efficient brain delivery of the drugs after intravenous injection. Pharmacological tests that used rats could demonstrate that therapeutic amounts of the drugs were delivered to the sites of action in the brain and showed the high efficiency of the surfactant-coated PLGA nanoparticles for brain delivery.[65]

12.3.1.4 Polycaprolactone

Polycaprolactone (PCL) is a semi-crystalline, hydrophobic polymer with a relatively polar ester group and five non-polar methylene groups in its repeating unit. It is mostly synthesized by the ring-opening polymerization method from ε-caprolactone (ε-CL) monomers. The high olefin content imparts polyolefin-like properties to PCL. However, owing to the high degree of crystallinity and hydrophobicity, PCL degrades rather slowly and is less biocompatible with soft tissue, which restricts its further clinical application.[66] Therefore, the modification of PCL is proposed.[67–69] Poly(ethylene glycol) (PEG), chosen for its hydrophilicity, non-toxicity and absence of antigenicity and immunogenicity, can be attached to PCL, forming PCL/PEG copolymers (Scheme 12.5).[69–77] The PCL/PEG copolymers can be summarized into several types including PCL-PEG diblock copolymer, PCL-PEG-PCL (PCEC), or PEG-PCL-PEG (PECE) triblock copolymers and star-shaped polymers. For the synthesis and applications of each type one can refer to a recent review by Wei et al.[78] The PCL/PEG copolymers have various applications in biomedical uses, particularly, the triblock copolymers of PCEC and PECE found potential as *in situ* gel-forming hydrogel drug delivery system.

Gong et al. have made a series of investigations on PCL/PEG triblock copolymers.[69,75–77] Aqueous solution of PCEC copolymer displayed thermosensitive sol–gel–sol transition behaviour, which is a flowing sol at low temperature and turns into a non-flowing gel at body temperature. The cytotoxicity of PCEC copolymer was evaluated. *In vivo* gel-formation, degradation test, acute toxicity tests and histopathological study of PCEC hydrogels were performed in BALB/c mice by subcutaneous administration, and histopathologic study was done. The obtained PCEC hydrogel was non-toxic after

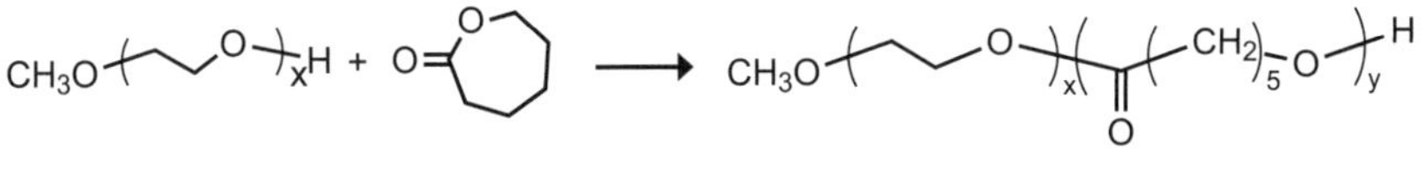

Scheme 12.5 Synthesis of MPEG-PCL by ring opening polymerization.

subcutaneous administration.[75] In their further study, sol–gel–sol transition and drug delivery behaviour was further investigated. The sol–gel phase transition mechanism was investigated using ^{13}C-nuclear magnetic resonance imaging and a laser diffraction particle size analyser. The *in vitro* release behaviors of several model drugs were studied. An anaesthesia assay was conducted using the tail flick latency (TFL) test to evaluate the *in vivo* controlled drug delivery. All results indicated that PCEC hydrogel is promising for use as an injectable local drug delivery system.[77]

12.3.1.5 Polyhydroxyalkanoates

The natural polyesters, which are synthesized and accumulated intracellularly during unbalanced growth by a wide variety of microorganisms, are receiving increased attention.[79] Microbial polyhydroxyalkanoates (PHAs) is one of the largest groups of these polyesters, and is the most appropriate candidate for biodegradable plastics.[80] The members of this family have the general structure given in Scheme 12.6. All of these polyesters contain units which are 100% optically pure at the β-position and their material properties varies from rigid brittle plastics, to flexible ones and to strong elastomers with the variation in size of pendant alkyl group and composition of the polymer. Poly(D-3-hydroxybutyrate) (PHB), with $R=CH_3$ is the most ubiquitous and most intensively studied PHA with high crystallinity and relatively high T_g. Considerable interest arose in the synthesize of PHB by controlled fermentation and chemical synthesis.[81] Also, both routes were suitable for the modification of PHB. Particularly, different copolymers with better mechanical properties than PHB were successfully obtained by feeding the bacteria with a variety of carbon sources in fermentation.[82]

An interesting investigation was made to indicate that PHB can form a new biodegradable adsorption material. Poly-3-hydroxybutyrate (PHB) was used as a new material to prepare a biomimetic adsorbent by a modified double emulsion solvent evaporation technique. The enrichment capacities of the adsorbent for toxic liposoluble organic compounds were evaluated by chlorobenzene (CB) and *o*-nitrochlorobenzene (*o*-NCB) with the adsorption isotherms, enrichment factor (EF) and enrichment kinetics.[83]

Ye *et al.* have looked into PHB scaffolds, and worked on the potential of PHB/poly(hydroxybutyrate-co-hydroxyhexanoate) (PHBHHx) (PHB/PHBHHx) to produce neocartilage upon seeding with differentiated human adipose-derived stem cells (hASCs). The study demonstrated that PHB/PHBHHx was a suitable material for cartilage tissue engineering.[84]

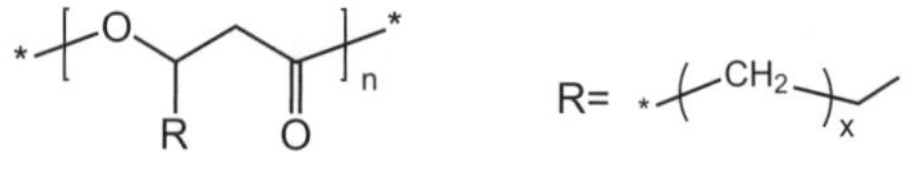

Scheme 12.6 Structure of polyhydroxyalkanoates.

12.3.2 Polyurethanes

Polyurethane is widely used in various fields, such as the manufacture of plastic foams, rubber goods, adhesives, paints and fibres. Polyurethane is commercially produced using polyols, such as alkanediols and glycerol, with toxic diisocyanate, which is derived from the even more toxic phosgene.[85] The produced polyurethane is generally resistant to biodegradation, whereas a low molecular weight urethane oligomer can be hydrolysed by some microorganisms.[86] Great efforts were paid in developing the next-generation polyurethane which ideally should be biodegradable and synthesized without using diisocyanate/phosgene.[87]

There have been many attempts to develop non-phosgene and non-diisocyanate routes for the synthesis of polyurethanes. Notably, enzymatically synthesized polyurethane may meet these requirements as a green polymer of all methods.[88] On the other side, design and synthesis of biodegradable and enzymatically cleavable polyurethane attracted much interest. The so-called biodegradable polyurethane could be produced by combining low molecular weight biodegradable hard segments of polyurethane with enzymatically hydrolysable linkages, such as an ester and carbonate (Scheme 12.7). A new class of poly(ester-urethane)s and poly(carbonate-urethane)s with their mechanical properties similar to those of conventional polyurethanes, are the most applicable substitutes for non-degradable polyurethanes.[89,90]

Polyurethanes have been widely used in drug delivery systems. Kim *et al.* investigated the feasibility of developing a temperature-responsive braided stent using shape memory polyurethane (SMPU) through finite element analysis.[91] The deployment process of the braided stents inside narrowed

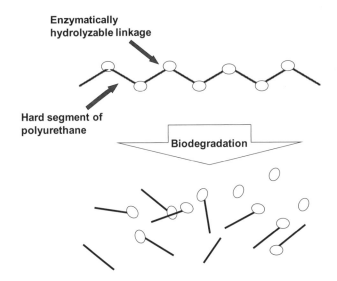

Scheme 12.7 Design of biodegradable polyurethane.

vessels was simulated, showing that the SMPU stents can be comfortably implanted while minimizing the overpressure onto the vessel walls, due to their thermo-responsive shape memory behaviour. More over, the work by Guo *et al.* reported that highly adjustable and precisely controllable drug release from a biodegradable stent coating was achieved using a unique family of nano-structured hybrid polyurethanes. The results showed that the studied hybrid polyurethane family allowed a drug release rate that was effectively manipulated through variation in polymer T_g, degradation rate and thickness increment rate.[92]

12.3.3 Polyamides

Although polyamides contain the same amide linkage that is found in polypeptides, aliphatic polyamides are generally resistant to microbial and enzymatic attacks.[2] What is responsible for their low biodegradability are the strong intermolecular interactions caused by hydrogen bonding. Thus, attempts were made to enhance the biodegradability of aliphatic polyamides mainly in two ways: (1) the introduction of ester linkages;[93] (2) the incorporation of α-amino acid residues. Poly(ester amide)s is a class of polymers with both advantages of polyesters and polyamides, and which appears to have ideal material and processing properties and good biodegradability.[94] It can either be synthesized by polycondensation of ester-containing diamines with dicarboxylic acids or by ring-opening polymerization of depsipeptides.[95] The presence of hydrolytically readily cleavable ester bonds in the backbone makes poly(ester-amide)s a promising material for their use in biomedical fields.

Studies on the blood and tissue compatibility of poly(ester amide) co-polymers were carried out. A family of biodegradable poly(ester amide) (PEA) co-polymers based on naturally occurring α-amino acids has been developed for applications ranging from biomedical device coatings to delivery of therapeutic biologics. To gain insight into this process, representative elastomeric PEAs designed for a cardiovascular stent coating were compared to non-degradable and biodegradable polymers in a series of *in vitro* assays to examine blood and cellular responses. These *in vitro* studies of the blood and tissue compatibility of these biodegradable, α-amino acid-based PEAs suggested that they might support a more natural healing response by attenuating the pro-inflammatory reaction to the implant and promoting growth of appropriate cells for repair of the tissue architecture.[96]

12.3.4 Polyanhydrides

Polyanhydrides have been intensively investigated as important biomaterials in medical fields due to their excellent biodegradability and biocompatibility. Over the decades, numerous studies have been carried out in academia and industry which were mainly concerned with chemical and physical characterization of these polymers, degradation properties, toxicity studies and various applications in medical fields. Polyanhydrides can be classified into several types including aliphatic polyanhydrides, unsaturated polyanhydrides, aromatic polyanhydrides, aliphatic-aromatic homopolyanhydrides, poly(ester-anhydrides) and poly(ether-anhydrides).[97] They are most commonly synthesized from diacid monomers by polycondensation, though other methods were also reported such as melt condensation, ring-opening polymerization, interfacial condensation, dehydro-chlorination, *etc.* For the specific syntheses, the excellent review by Kumar *et al.* provides more details.[97] Biodegradation of these polymers involves cleavages of hydrolytically sensitive bonds that lead eventually to polymer erosion. The anhydride bond in polyanhydrides is hydrolytically very labile and readily splits in the presence of water into two carboxylic acids. Biodegradation and elimination of polyanhydrides *in vivo* were systematically summarized in a review by Katti *et al.*, which also highlighted biocompatibility and toxicological aspects of the polymers.[3]

12.3.5 Polyphosphoesters

Polyphosphoesters (PPEs) refers to a wide range of biodegradable polymers with repeating phosphoester linkages in the backbone whose general structure is shown in Scheme 12.8. The initial application of polyphosphoesters in biomedical fields dates back to the 1980s.[98] Since then, polyphosphoesters have been attractive for extensive uses and have been investigated thoroughly looking at the versatility of the PPE's pendant groups. The introduction of bioactive molecules and various modifications can be achieved thanks to the pentavalent phosphorus atom, thus obtaining PPE with modified physical and chemical properties. When it comes to the syntheses of PPEs, polycondensation has been one of the most often used methods. The reactions are usually carried out between diols and phosphonic dihalides or diallyl phosphoesters.[99] There are also other ways of synthesizing reported, such as polyaddition, transesterification and ring-opening polymerization by enzymes and by other catalysts. For the specific reaction, we can refer to a review by Wang *et al.*[100]

Recently, novel thermo-responsive block copolymers of poly(ethylene glycol) and polyphosphoester were synthesized, and the thermo-induced self-assembly, biocompatibility and hydrolytic degradation behaviour were studied and its biocompatibility and thermo-responsiveness were ensured.[101] Similarly, aqueous dispersions of thermo-sensitive gold nanoparticles protected by diblock copolymers of poly(ethylene glycol) and polyphosphoester were prepared and studied.[102] These gold nanoparticles protected by thermo-sensitive diblock

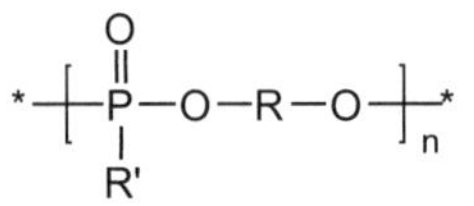

Scheme 12.8 Structure of phosphoester.

copolymers with tuneable collapse temperature are expected to be useful for biomedical applications.

12.3.6 Others

Some other synthetic biodegradable polymers such as poly(alkyl cyanoacrylates),[103] polyphosphazenes,[104] poly(ortho esters),[105] poly(ethylene terephthalate)-poly(ethylene glycol) copolymer[106] and poly(ether ester amide)s[107] have been of great interest in many fields.

Poly(alkyl cyanoacrylates) is one of the most favourable classes of biodegradable acrylate polymers used in medical fields (Scheme 12.9A). These are neutral polymers prepared by the anionic polymerization of alkyl cyanoacrylic monomers with a trace amount of moisture as the initiator. High inductive activation of methylene hydrogen atoms by the electron-withdrawing neighbouring groups was responsible for the hydrolytic sensitivity of the backbone. These polymers are one of the fastest degrading polymers, having degradation times ranging from few hours to few days. Among the alkyl esters, poly(methyl-2-cyanoacrylate) tend to be the most degradable and the degradability decreases as the alkyl size increases.[108] Due to these unique properties, poly(alkyl cyanoacrylates) have found applications in biomedical fields, particularly in drug delivery systems.[103]

The synthetic flexibility and versatile adaptability has made polyphosphazenes quite distinct from the polymers synthesized so far (Scheme 12.9B). These are high molecular weight, essentially linear polymers with an inorganic backbone of alternating phosphorous and nitrogen atoms bearing two side groups attached to each phosphorous atom. Biodegradable polyphosphazenes, due to their hydrolytic instability, non-toxic degradation products, ease of fabrication and matrix permeability, are an excellent platform for controlled drug delivery applications. The mode of degradation and drug delivery applications was discussed by Lakshmi et al.[109] A study on novel micelles from graft polyphosphazenes as potential anti-cancer drug delivery systems was also carried out.[110]

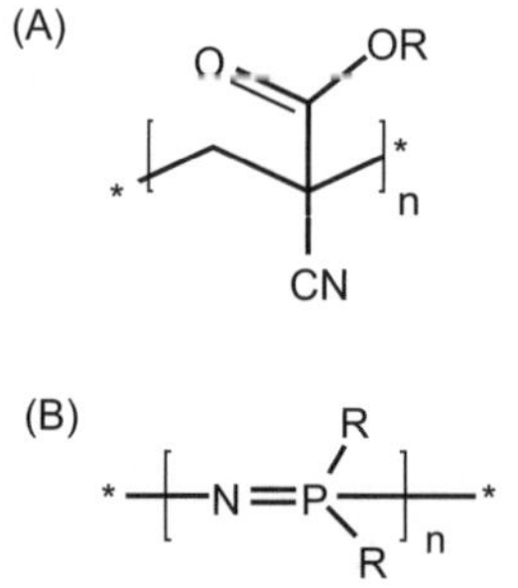

Scheme 12.9 Structures of poly(alkyl cyanoacrylates) and phosphogene.

Polyacetal and poly(ortho ester)-poly(ethylene glycol) graft copolymer thermogels was reported by Schacht *et al.* A series of materials having lower critical solution temperature (LCST) between 25 and 60 °C were prepared, but no erosion or drug release studies have as yet been completed.[111]

12.4 Conclusions

Tremendous progress has been attained in seeking and optimizing biodegradable polymers with appropriated properties in modern industrial applications and biomedical use. Reported studies on biodegradable polymers currently involved natural polymers such as cellulose and synthetic polymers such as polyesters. Owing to the presence of hydrolytic bonds (hydrolytically or enzymatically), these polymers can both be promising candidates as biomaterials with good biocompatibility and an attractive approach to environmental waste management. Further studies concerned with biodegradable materials might be facilitated by advances in synthetic organic chemistry and novel bioprocesses. Although many advances have been made, much work still remains before a wide range of biodegradable materials will be used commonly throughout the world, which is earnestly making the world of tomorrow greener.

References

1. G. Wift, *Acc. Chem. Res.*, 1993, **26**, 105.
2. R. Chandra and R. Rustgi, *Prog. Polym. Sci.*, 1998, **23**, 1273.
3. D. Katti, S. Lakshmi, R. Langer and C. Laurencin, *Adv. Drug Deliv. Rev.*, 2002, **54**, 933.
4. A. Thayer, *Chem. Eng. News*, 1989, **67**, 7.
5. W. Amass, A. Amass and B. Tighe, *Polym. Int.*, 1998, **47**, 89.
6. M. L. Hunter, M. da Moth and J. N. Lester, *Environ. Technol. Lett.*, 1988, **9**, 1.
7. Y. H. Paik, E. S. Simon and G. Swift, *Adv. Chem. Ser.*, 1996, **248**, 79.
8. P. A. Holmes, *Phys. Technol.*, 1985, **16**, 32.
9. D. M. Dohse and L. W. Lion, *Environ. Sci. Technol.*, 1994, **28**, 541.
10. E. D. Beach, R. Boyd and N. D. Uri, *Appl. Math. Model.*, 1996, **20**, 388.
11. E. Chiellini and R. Solaro, *Adv. Mater.*, 1996, **8**, 305.
12. P. A. Gunatillake, R. Adhikari, *Eur. Cell. Mater.*, 2003, **5**, 1.
13. M. Vert, *Biomacromolecules*, 2005, **6**, 538.
14. L. S. Naira and C. T. Laurencin, *Prog. Polym. Sci.*, 2007, **32**, 762.
15. A. Lloyd, *Med. Device Technol.*, 2002, **13**, 18.
16. M. Okada, *Prog. Polym. Sci.*, 2002, **27**, 87.
17. R. Chandra and R. Rustgi, *Prog. Polym. Sci.*, 1998, **23**, 1273.
18. R. L. Whistler, J. Bemiller and E. Paschall, eds., In *Starch: Chemistry and Technology*, Academic Press, New York, 1984.
19. S. Parandoosh and S. Hudson, *J. Appl. Polym. Sci.*, 1993, **48**, 787.

20. C. L. Swanson, R. L. Shogren, G. F. Fanta and S. H. Imam, *J. Polym. Environ.*, 1993, **1**, 155.
21. P. Colonna and C. Mercier, *Phytochemistry*, 1985, **24**, 1667.
22. L. Wang, R. L. Shogren and C. Carriere, *Polym. Eng. Sci.*, 2000, **40**, 499.
23. A. Dukić, M. Thommes, J. P. Remon, P. Kleinebudde and C. Vervaet, *Eur. J. Pharm. Biopharm.*, 2009, **71**, 38.
24. A. Dukić, R. Mens, P. A. Adriaensens, P. Foreman, J. Gelan, J. P. Remon and C. Vervaet, *Eur. J. Pharm. Biopharm.*, 2007, **66**, 83.
25. Y. Dahman, *J. Nanosci. Nanotechnol.*, 2009, **9**, 5105.
26. H. Honarkar and M. Barikani, *Monatsh Chem.*, 2009, **140**, 1403.
27. A. K. Mohanty, M. Misraa and G. Hinrichsen, *Macromol. Mater. Eng.*, 2000, **276**, 1.
28. A. M. Martins, C. M. Alves, F. K. Kasper, A. G. Mikos and R. L. Reis, *J. Mater. Chem.*, 2010, **20**, 1638.
29. H. Tan, C. Chu, K. Payne and R. Marra, *Biomaterials*, 2009, **30**, 2499.
30. J. M. Dang and K. W. Leong, *Adv. Drug Deliv. Rev.*, 2006, **58**, 487.
31. S. Ozbaş-Turan, J. Akbuğa and A. Sezer, *Oligonucleotides*, 2010, **20**, 147.
32. R. Guo, R. Li, X. Li, L. Zhang, X. Jiang and B. Liu, *Small*, 2009, **5**, 709.
33. P. Vos, M. Faas, B. Strand and R. Calafiore, *Biomaterials*, 2006, **27**, 5603.
34. M. George and T. Abraham, *J. Control. Release*, 2006, **114**, 1.
35. Z. Ruszczak and W. Friess, *Adv. Drug Deliv. Rev.*, 2003, **55**, 1679.
36. S. Oliveir, R. Ringshi, R. Legeros, E. Clark, M. Yost, L. Terracio and C. Teixeir, *J. Biomed. Mater. Res. A*, 2010, **94**, 371.
37. S. Iibuchi, K. Matsui, T. Kawai, K. Sasaki, O. Suzuki, S. Kamakura and S. Echigo, *Int. J. Oral Maxillofac. Surg.*, 2010, **39**, 161.
38. R. Langer and D. A. Tirrell, *Nature*, 2004, **428**, 487.
39. S. Nezhadi, P. Choong, F. Lotfipour and C. Dass, *J. Drug Target.*, 2009, **17**, 731.
40. Q. Wang, Z. Song, J. Tang, D. Yan, F. Wang, H. Zhang, M. Guo and A. Cao, *J. Agric. Food Chem.*, 2009, **57**, 8414.
41. E. Bachelder, T. Beaudette, K. Broaders, J. Dashe and J. Fréchet, *J. Am. Chem. Soc.*, 2008, **130**, 494.
42. S. Nelson and I. Greer, *Hum. Reprod. Update*, 2008, **14**, 623.
43. S. Comba and R. Sethi, *Water Res.*, 2009, **43**, 3717.
44. H. Kobayashi, O. Katakura, N. Morimoto, K. Akiyoshi and S. Kasugai, *J. Biomed. Mater. Res. B Appl. Biomater.*, 2009, **91**, 55.
45. I. Massodi, E. Thomas and D. Raucher, *Molecules*, 2009, **14**, 1999.
46. A. Panda, S. Kumar, A. Kumar, R. Bansal and S. Bhartiya, *Indian J. Ophthalmol.*, 2009, **57**, 371.
47. X. P. Tang, X. Y. Ding, L. Y. Chu, J. P. Hou, J. L. Yang, X. Song and Y. M. Xie, *Bioorg. Med. Chem.*, 2010, **18**, 1599.
48. A. Jha, W. Yang, C. Kirn-Safran, M. Farach-Carson and X. Jia, *Biomaterials*, 2009, **30**, 6964.
49. Q. Lin, K. Ren and J. Ji, *Colloids Surf. B*, 2009, **74**, 298.
50. Y. Liu and M. Chan-Park, *Biomaterials*, 2009, **30**, 196.
51. C. Jérôme and P. Lecomte, *Adv. Drug Deliv. Rev.*, 2008, **60**, 1056.

52. N. Ashammakhi and P. Rokkanen, *Biomaterials*, 1997, **18**, 3.
53. F. Bäumchen, R. Smeets, D. Koch and H. Gräber, *Oral Surg. Oral Med. Oral Pathol. Oral Radiol. Endod.*, 2009, **108**, 505.
54. G. Zhao, S. Yin, G. Liu, L. Cen, J. Sun, H. Zhou, W. Liu, L. Cui and Y. Cao, *Biomaterials*, 2009, **30**, 3241.
55. S. Singh and S. Ray, *J. Nanosci. Nanotechnol.*, 2007, **7**, 2596.
56. S. Wang, W. Cui and J. Bei, *Anal. Bioanal. Chem.*, 2005, **381**, 547.
57. J. Middleton and A. Tipton, *Biomaterials*, 2000, **21**, 2335.
58. A. J. Ragauskas, C. K. Williams, B. H. B. Davison, G. J. Cairney, C. A. Eckert, W. J. Fredrick, J. P. Hallett, D. J. Leak, C. L. Liotta, J. R. Mielenz, R. Murphy, R. Templer and T. Tschaplinski, *Science*, 2006, **311**, 484.
59. C. Williams, *Chem. Soc. Rev.*, 2007, **36**, 1573.
60. Y. Zhang, X. Wu, Y. Han, F. Mo, Y. Duan and S. Li, *Int. J. Pharm.*, 2010, **386**, 15.
61. A. Jain, A. Goyal, N. Mishra, B. Vaidya, S. Mangal and S. Vyas, *Int. J. Pharm.*, 2010, **387**, 253.
62. D. Mooney, M. Breuer, K. Mcnamara, J. Vacanti and R. Langer, *Tissue Eng.*, 1995, **1**, 107.
63. H. Kim, M. Kwon, J. Choi, S. Yang, J. Yoon, K. Kim and J. Park, *Biomaterials*, 2006, **27**, 2292.
64. Z. Ghalanbor, M. Körber and R. Bodmeier, *Pharm. Res.*, 2010, **27**, 371.
65. S. Gelperina, O. Maksimenko, A. Khalansky, L. Vanchugova, E. Shipulo, K. Abbasova, R. Berdiev, S. Wohlfart, N. Chepurnova and J. Kreuter, *Eur. J. Pharm. Biopharm.*, 2010, **74**, 157.
66. Z. Y. Qian, S. Li, Y. He, H. Zhang and X. Liu, *Biomaterials*, 2004, **25**, 1975.
67. M. L. Gou, H. Wang, M. J. Huang, Y. B. Tang, Z. Y. Qian, Y. J. Wen, A. L. Huang, K. Lei, C. Y. Gong, J. Li, W. J. Jia, C. B. Liu, H. X. Deng, M. J. Tu and G. T. Chao, *J. Mater. Sci. Mater. Med.*, 2008, **19**, 1033.
68. M. L. Gou, M. J. Huang, Z. Y. Qian, L. Yang, M. Dai, X. Y. Li, K. Wang, Y. J. Wen, J. Li, X. Zhao and Y. Q. Wei, *Growth Factors*, 2007, **25**, 202.
69. C. Y. Gong, Z. Y. Qian, C. B. Liu, M. J. Huang, Y. C. Gu, Y. J. Wen, B. Kan, K. Wang, M. Dai, X. Y. Li, M. L. Gou, M. J. Tu and Y. Q. Wei, *Smart Mater. Struct.*, 2007, **16**, 927.
70. X. Wei, C. Gong, S. Shi, S. Fu, K. Men, S. Zeng, X. Zheng, M. Gou, L. Chen, L. Qiu and Z. Qian, *Int. J. Pharm.*, 2009, **369**, 170.
71. M. L. Gou, M. Dai, X. Y. Li, X. H. Wang, C. Y. Gong, Y. Xie, K. Wang, X. Zhao, Z. Y. Qian and Y. Q. Wei, *J. Mater. Sci. Mater Med.*, 2008, **19**, 2605.
72. M. Gou, X. Li, M. Dai, C. Gong, X. Wang, Y. Xie, H. Deng, L. Chen, X. Zhao, Z. Qian and Y. Wei, *Int. J. Pharm.*, 2008, **359**, 228.
73. M. Gou, M. Dai, X. Li, L. Yang, M. Huang, Y. Wang, B. Kan, Y. Lu, Y. Wei and Z. Qian, *Colloids Surf. B. Biointerfaces*, 2008, **64**, 135.
74. M. Gou, M. Dai, Y. Gu, X. Li, Y. Wen, L. Yang, K. Wang, Y. Wei and Z. Qian, *J. Nanosci. Nanotechnol.*, 2008, **8**, 2357.

75. C. Y. Gong, S. Shi, P. W. Dong, B. Yang, X. R. Qi, G. Guo, Y. C. Gu, X. Zhao, Y. Q. Wei and Z. Y. Qian, *J. Pharm. Sci.*, 2009, **98**, 4684.
76. C. Gong, S. Shia, P. Dong, B. Kan, M. Gou, X. Wang, X. LiF. Luo, X. Zhao, Y. Wei and Z. Qian, *Int. J. Pharm.*, 2008, in press.
77. C. Gong, S. Shi, L. Wu, M. Gou, Q. Yin, Q. Guo, P. Dong, F. Zhang, F. Luo, X. Zhao, Y. Wei and Z. Qian, *Acta Biomaterialia*, 2009, **5**, 3358.
78. X. Wei, C. Gong, M. Gou, S. Fu, Q. Guo, S. Shi, F. Luo, G. Guo, L. Qiu and Z. Qian, *Int. J. Pharm.*, 2009, **381**, 1.
79. S. Philip, T. Keshavarz and I. Roy, *J. Chem. Technol. Biotechnol.*, 2007, **82**, 233.
80. Y. Tokiwa and B. Calabia, *Biotechnol. Lett.*, 2004, **26**, 1181.
81. J. Shelton, J. Lando and D. Agostini, *J. Polym. Sci. Polym. Lett. B*, 1971, **9**, 173.
82. M. Mejía, D. Segura, G. Espín, E. Galindo and C. Peña, *J. Appl. Microbiol.*, 2010, **108**, 55.
83. X. Zhang, C. Wei, Q. He and Y. Ren, *J. Hazard. Mater.*, 2010, **177**, 508.
84. C. Ye, P. Hu, M. Ma, Y. Xiang, R. Liu and X. Shang, *Biomaterials*, 2009, **30**, 401.
85. G. Howard, *Biodeterior. Biodegrad.*, 2002, **49**, 245.
86. S. Owen, T. Otani, S. Masaoka and T. Ohe, *Biosci. Biotechnol. Biochem.*, 1996, **60**, 244.
87. S. Matsumura and K. Toshima, *Appl. Microbiol. Biotechnol.*, 2006, **70**, 12.
88. G. Rokicki and A. Piotrowska, *Polymer*, 2002, **43**, 2927.
89. T. Takamoto, H. Shirasaka, H. Uyama and S. Kobayashi, *Chem. Lett.*, 2001, **6**, 492.
90. A. Hung, *J. Polym. Sci. Part B Polym. Phys.*, 2003, **41**, 679.
91. J. Kim, T. Kang and W. Yu, *J. Biomech.*, 2010, **43**, 632.
92. Q. Guo, P. Knight and P. Mather, *J. Control. Release*, 2009, **137**, 224.
93. Z. Qian, S. Li, Y. He, H. Zhang and X. Liu, *Colloid. Polym. Sci.*, 2003, **281**, 869.
94. Z. Qian, S. Li, Y. He, H. Zhang and X. Liu, *Polym. Degrad. Stabil.*, 2003, **81**, 279.
95. Z. Qian, S. Li, Y. He, H. Zhang and X. Liu, *Colloid. Polym. Sci.*, 2004, **282**, 1083.
96. K. DeFife, K. Grako, G. Cruz-Aranda, S. Price, R. Chantung, K. Macpherson, R. Khoshabeh, S. Gopalan and W. Turnell, *J. Biomater. Sci. Polym.*, 2009, **20**, 1495.
97. N. Kumar, S. Robert, B. Langer and A. Domb, *Adv. Drug. Deliv. Rev.*, 2002, **54**, 889.
98. W. Song, J. Du, N. Liu, S. Dou, J. Cheng and J. Wang, *Macromolecules*, 2008, **41**, 6935.
99. J. Pretula, K. Kaluzynski, B. Wisniewski, R. Szymanski, T. Loontjens and S. Penczek, *J. Polym. Sci., Part A: Polym. Chem.*, 2008, **46**, 830.
100. Y. Wang, Y. Yuan, J. Du, X. Yang and J. Wang, *Macromol. Biosci.*, 2009, **9**, 1154.
101. Y. Wang, L. Tang, Y. Li and J. Wang, *Biomacromolecules*, 2009, **12**, 66.

102. Y. Yuan, X. Liu, Y. Wang and J. Wang, *Langmuir*, 2009, **25**, 298.
103. C. Vauthier, C. Dubernet, C. Chauvierre, I. Brigger and P. Couvreur, *J. Control. Release*, 2003, **93**, 151.
104. A. Ambrossio, H. Allcock, D. Katti and C. Laurencin, *Biomaterials*, 2002, **23**, 1667.
105. J. Heller, J. Barr, S. Ng, K. Abdellauoi and R. Gurny, *Adv. Drug Deliv. Rev.*, 2002, **54**, 1015.
106. Z. Qian, S. Li, Y. He and X. Liu, *Polym. Degrad. Stabil.*, 2004, **83**, 93.
107. Z. Qian, S. Li, Y. He, X. Zhang and X. Liu, *Colloid Polym. Sci.*, 2003, **282**, 133.
108. T. Chirila, P. Rakoczy, K. Garrett, X. Lou and I. Constable, *Biomaterials*, 2002, **23**, 321.
109. S. Lakshmi, D. Katti and C. Laurencin, *Adv. Drug Deliv. Rev.*, 2003, **55**, 467.
110. C. Zheng, L. Qiu, X. Yao and K. Zhu, *Int. J. Pharm.*, 2009, **373**, 133.
111. E. Schacht, V. Toncheva, K. Vandertaelen and J. Heller, *J. Control. Release*, 2006, **116**, 219.

CHAPTER 13
Impacts of Biodegradable Polymers: Towards Biomedical Applications

Y. OMIDI AND S. DAVARAN

Research Center for Pharmaceutical Nanotechnology, Faculty of Pharmacy, Tabriz University of Medical Sciences, Tabriz, Iran

13.1 Introduction

In the third millennium of mankind's history, it is now deeply believed that mimicking the natural process of life at the cellular/molecular level is a key for the advancement of novel medicines to optimize quality of life and control diseases. We are now moving fast to develop new generation of nano-biomedicines with maximal efficiency and minimal inadvertent consequences. To pursue such an aim, smart and specific approaches are necessary to control aberrant biofunctionalities of cells/tissues. Of these, one important matter is to avoid the undesired bioimpacts of materials used for medical interventions. In general, biocompatibility (*i.e.* cyto- and/or geno-compatibility) of advanced materials looms as a key factor. In fact, pharmacokinetically, a xenobiotic is subjected to ADME (adsorption, distribution, metabolism and elimination) processes. Thus, to avoid or to lessen toxic impacts, natural 'biodegradation' is a keystone of such a process. Biodegradation is the chemical breakdown of materials by the physiological environment. The word is often used in relation to ecology, waste management and environmental remediation, termed as 'bioremediation'.

RSC Green Chemistry No. 12
A Handbook of Applied Biopolymer Technology: Synthesis, Degradation and Applications
Edited by Sanjay K. Sharma and Ackmez Mudhoo
© Royal Society of Chemistry 2011
Published by the Royal Society of Chemistry, www.rsc.org

Basically, the evolution of biodegradable polymers appeared to be through chain modification of existing materials or by modulating the chemical composition of synthetic polymers using several polymerization techniques. Biodegradable polymers with controlled degradation characteristics can be used as an important part of biomedical sciences such as drug delivery, cell and tissue engineering, tissue repair, stem cell therapy and molecular detection and/or therapy.[1–3] To date, many types of natural and synthetic biodegradable polymers, for example albumin, chitosan, and hyaluronic acid and derivatives, have been investigated for medical and pharmaceutical applications. Use of natural polymers (*e.g.* cellulose and starches) appears to be still common in biomedical research; nevertheless, synthetic biodegradable polymers are increasingly exploited in pharmaceutical and tissue-engineering products. Basically, synthetic polymers can be prepared with chemical structures tailored towards optimized physical properties, as result of which they can be considered as biomedical materials with well-defined purities and compositions superior to those attainable when using natural polymers. There exist various established chemical classes of synthetic biodegradable polymers which offer good enough biocompatibility. Of note, the rate of biodegradation and mechanical strength of such advanced biodegradable polymers can be tuned.

From biocompatibility view points, the biological degradation of polymers is deemed to be very important in biomedicine for many reasons. For instance, the biologically degradation of a polymeric implant within the body clearly indicates that surgical intervention may not be required for removal at the end of its functional life, thus excluding need for any further surgical operation. Further, in tissue engineering and regenerative medicine, implementation of biodegradable polymers confers great potential to design such an approach to approximate tissues through providing a polymer scaffold that can withstand mechanical stresses and provide an appropriate surface as an extracellular matrix for attachment and growth of cells. And ideally, it should be degraded at an appropriate rate, allowing the load to be transferred to the new tissue.

Principally, organic material can be degraded aerobically with oxygen, or anaerobically without oxygen. Of the biodegradable polymers, bioabsorbable polymers can be applied for wound closure (sutures, staples), osteosynthetic materials (orthopedic fixation devices, pins, screw, rods, bone plates), cardiovascular surgery (stents, grafts) and intestinal surgery (anastomosis rings).[4] Bioabsorbable polymers also find applications as matrix materials for implanted drug release devices or drug-containing microspheres or microcapsules, and drug delivery vehicles, such as micro/nano-particles, micelles, hydrogels and injectable delivery systems.

In general, the major biomedical applications of biodegradable polymers[5,6] are as follows:

- bone replacement
- drug delivery using synthetic biodegradable polymers
- green biodegradable packaging materials
- tissue engineering with synthetic biodegradable polymers

- green biodegradable sutures
- vascular stents
- general and reconstructive surgery.

13.2 Classification of Biodegradable Polymers

Biodegradable polymers can be either natural or synthetic. In general, synthetic polymers are deemed to offer greater advantages than natural ones since they can be simply designed and tailored to provide the intended desired properties. Now, various families of synthetic biodegradable polymers are available for selection for a specific application.

As shown in Figure 13.1, depending upon evolution of the synthesis process, several classifications of the different biodegradable polymers have been proposed as follows:

- polymers from biomass such as the agro-polymers from agro-resources such as starch and cellulose
- polymers from microbial production such as poly hydroxyalkanoates (PHAs)
- conventionally and chemically synthesized polymers such as polylactic acid (PLA)
- polymers obtained from petrochemical resources such as polycaprolactones (PCL).

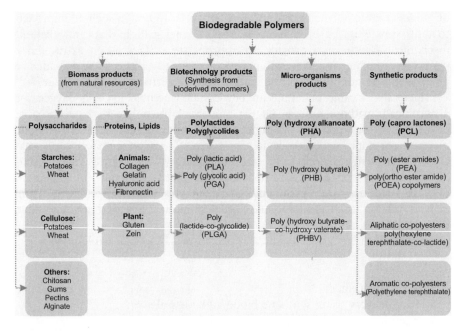

Figure 13.1 Classifications of the different biodegradable polymers.

13.3 Biodegradable Polyesters

Over the past few decades, several biodegradable polyesters have been extensively studied for a wide variety of pharmaceutical and biomedical applications such as drug delivery and tissue engineering.[7–10] Among these polymers, α-hydroxy acid polymers and copolymers are the most widely investigated. These polymers continue to be the most widely used synthetic biodegradable polymers in clinical applications. The ester linkages of aliphatic and aliphatic-aromatic co-polyesters can easily be cleaved by hydrolysis under alkaline, acid or enzymatic catalysis. This feature makes polyesters suitable for several applications, such as bioresorbable, bioabsorbable, environmentally degradable and recyclable polymers. Current applications of these polymers include surgical sutures and resorbable implants, with significant interest to further expand the use of these materials to drug encapsulation and delivery applications.[11,12]

Poly(glycolic acid) (PGA), poly(lactic acid) (PLA) and their copolymers (Figure 13.2) have been investigated for a wider range of applications than any other types of biodegradable α-hydroxy acid polymers. Several PLA-based polymers have been used significantly in pharmaceutical and biomedical applications. These polymers are used in a range of clinical applications as bone fixatives in suture reinforcement, as drug delivery systems and as scaffolds for cell transplantation and guided regeneration.[13–16]

A simple hydrolysis of the ester backbone to non-harmful and non-toxic compounds is the main mechanism for degradation of PLA and PGA within the body. The degradation products are either excreted by the kidneys or eliminated as carbon dioxide and water through well-known biochemical pathways. In fact, various applications of PLA/PGA polymers will soon be brought to market since these materials are considered safe, non-toxic and biocompatible by regulatory agencies in virtually all developed countries. They are deemed to be more cost effective than those utilizing novel polymers with unproven biocompatibility. They also have the potential for tailoring the

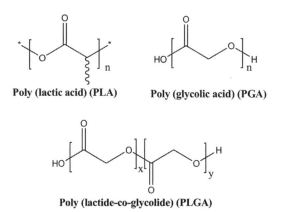

Figure 13.2 Polyester polymers and copolymers.

structure to alter degradation rates and mechanical properties by using various copolymerization strategies.

13.3.1 Properties of PLA/PGA Polymers

PGA, PLA and their copolymers 'poly (lactide-co-glycolide)' (PLGA) are believed to be some of the most commonly used biodegradable polymer materials, in part because of their properties that can be merely tuned by changing the polymer composition within the basic PLA/PGA theme.

PGA is a highly crystalline material, which possesses a high melting point (225–230 °C) and variable solubility in organic solvents, but which is generally low and depends on polymer molecular weight. Because of the ester bond in the polymer backbone, it is still susceptible to hydrolysis. Despite its low solubility, this polymer has been fabricated into a variety of forms and structures, for which different techniques (*i.e.* extrusion, injection, compression moulding, and particulate leaching and solvent casting) are used to develop PGA-based structures.[17] Fibres of PGA have a high strength, rigidity and modulus (7 GPa). This has promoted investigation to their use in bone internal fixation devices.[18,19] However, lower solubility and inelastic nature of PGA materials can limit their utility in some applications.

Unlike glycolide, lactide is a chiral molecule and exists in two distinct optically active forms (*i.e.* L-lactide and D-lactide). A semi-crystalline polymer can be the result of polymerization of these monomers, while polymerization of a racemic mixture of L- and D-lactides forms poly-D,L-lactide (PDLLA), which is amorphous and has a glass transition temperature of 55–60 °C. Tuning of the degree of crystallinity can be performed through alteration of the D to L enantiomer ratio within the polymer. It should be highlighted that the selection of the PLA stereochemistry can directly influence the polymer properties such as processability and biodegradability. Poly(L-lactide) (PLLA) is often the polymer of choice for cast/extruded biomedical devices because it breaks down into L(+)-lactic acid units, which is the naturally occurring stereoisomer and is therefore excreted with minimal toxicity.[13]

Of these aforementioned co-polyesters, wide research has been performed in developing a full range of PLGA polymers. Many researchers have exploited both L- and DL-lactides for co-polymerization, in which the ratio of glycolide to lactide at different compositions allows control of the degree of crystallinity of the polymers. Basically, the degree of crystallinity of the PLGA is reduced upon copolymerization of the crystalline PGA with PLA, and as a result the rates of hydration and hydrolysis can be increased. This clearly means that the degradation time of the copolymer is largely dependent upon the ratio of monomers used in synthesis. The degree of crystallinity and the melting point of the polymers are directly related to the molecular weight of the polymer, while swelling behaviour and accordingly the biodegradation rate are directly influenced by the crystallinity of the PLGA polymer. The T_g (glass transition temperature) of the PLGA copolymers are greater than the physiological temperature, therefore these copolymers are glassy in nature and have a fairly rigid chain structure.[20] These characteristics grant them significant mechanical strength to be formulated as drug delivery devices. Although the PLGAs are crystalline copolymers, those from D,L-PLA and PGA appear to be amorphous in nature. Thus, in short, it can be said that the higher the content of glycolide, the quicker the rate of degradation, even though the 50:50 ratio of PGA:PLA exhibits the fastest degradation as an exception. Based on our cytotoxicity examinations, there exists a correlation between degree of biodegradation and cytotoxicity (unpublished data produced by Omidi *et al.*).

13.3.2 Pharmaceutical Application of Biodegradable Polyesters

PLGA copolymers are semi-crystalline biocompatible copolymers that have been approved for certain clinical applications by the FDA. Of the biodegradable polyesters, PLA and PLGA are generally used as preformed scaffolds.[13] The pattern of drug release from these biodegradable structures appears to be dependent upon their composition. For example, multi-layer PLGA was shown to undergo pseudo surface degradation, while multi-layer PLLA degraded to a lesser extent over the same study period even though the two dominating release phases for both hydrophilic and lipophilic drugs were diffusion and zero-order release and PLGA had a shorter diffusion phase and a longer zero-order phase and the contrary was true for PLLA, perhaps due to the faster degradation of PLGA.[21] This is an important issue since in conventional methods of drug delivery (*e.g.* tablet or injection), the plasma level is raised as a dose is applied but it will be rapidly decreased as the drug is metabolized and eliminated. Thus to keep up the therapeutic levels, a controlled drug delivery systems is used, from which the drug is released at a constant, predetermined rate, and possibly targeted to a particular site. One of the most prominent approaches is exploitation of an appropriate polymer membrane to entrap or encapsulate drugs in a polymer matrix and as a result control its release where the drug diffuses out into the tissues following implantation. In some cases, erosion or dissolution of the polymer contributes to the release mechanism. To advance drug delivery systems for utilization in tissue engineering and regenerative medicine, researchers have focused on developing various copolymer systems such as incorporation of poly(ethylene glycol) (PEG) onto biodegradable nanoparticles.[22]

Poly(L-lactic acid) (PLLA) and its copolymers with D-lactic acid and/or glycolic acid have been extensively explored as biodegradable drug delivery carriers due to their superior biocompatibility and versatile process-abilities.[15,18,23–25] Biodegradable PLGA polymers have also been used as scaffolds for tissue engineering applications with several cell types including chondrocytes, hepatocytes and most recently, bone marrow-derived cells.[26,27]

In order to increase hydrophilicity of PLGA polymers, copolymerization of lactide and glycolide with poly (ethylene glycol) (PEG) has been investigated. The choice of PEG as a precursor is largely due to its good biocompatibility and hydrophilicity. For example, PLGA-PEG-PLGA thermo-sensitive triblock copolymers were shown to have great potential for delivery of macromolecules such as antibodies[28] and growth hormone.[29] The solubility of PLGA-PEG-PLGA triblock copolymers can be tuned by using appropriate molecular weight PEG.

In 2007, Chen et al. reported synthesis of biologically active forms of biodegradable and thermo-sensitive triblock copolymers with different block lengths (PLGA-PEG-PLGA) by ring-opening polymerization of D,L-lactide and glycolide with PEG.[30] They showed successful delivery of protein (e.g. lysozyme) in biologically active form for longer duration by varying block lengths and concentrations of triblock copolymers.

The usefulness of 'PLGA-PEG copolymers' or 'PEGylated PLGA' has been reported for the improved delivery of anticancer agents such as doxorubicin[31] and paclitaxel[32] and safe delivery of genes.[33]

In an attempt to improve cell attachment to PLGA copolymers, Yoon et al. (2004) modified PLGA by attaching cell-adhesive peptides such as Gly-Arg-Gly-Asp-Tyr (GRGDY) peptide and demonstrated significant improvement in cell attachment compared to unmodified polymer.[34] The same group also investigated the effect of immobilization of hyaluronic acid (HA) to promote the regeneration of cartilage tissue.[35] Interestingly, PEG-PLGA-PEG as a biodegradable non-ionic triblock copolymer has been exploited to enhance gene delivery efficiency in skeletal muscle showing a relatively lower toxicity.[36]

Furthermore, cationic PLGA has been formulated using poly(D,L-lacticcoglycolic acid) (PLGA), 1,2-dioleoyl-3-(trimethylammonium)propane (DOTAP) and asialofetuin (AF), and shown to be a highly effective delivery system for shuttling of the therapeutic gene IL-12 to liver tumour cells,[37] even though we have already shown that the cationic lipid-based gene delivery systems may induce non-specific toxicogenomics in target cells.[38]

Davaran et al. have synthesized cyclodextrin-based star-shaped polyesters bearing PLGA side chains, which are covalently attached via the hydroxyl groups in cyclodextrin. They have demonstrated that such copolymers can be used for sustained delivery of peptides such as insulin and anti-cancer drugs.[39,40]

Besides, dextran-graft-poly(L-lactide) (Dex-g-PLLA) and dextran-graft-poly(D-lactide) (Dex-g-PDLA) amphiphilic copolymers have been synthesized with well-defined compositions in a dilute aqueous solution.[41] Such a system displayed tuneable degradation properties by varying the number of grafted PLA chains as well as applying stereocomplexation.

13.3.3 Impacts of Micro and Nano Fabrication of PLGA-based Copolymers

Polymeric nanoparticles (NPs) and microparticles (MPs) are classified based on their sizes, composition and structure. The MPs range in diameter from 1 to 250 µm, while the size of NPs range between 1 and 100 nm, even though in biological applications particles less than 300 nm also can be used. Different types of biodegradable NPs are shown in Figure 13.3.

Basically, NPs and nano-structured delivery systems can be prepared mainly by different methods including dispersion of the preformed polymers, polymerization of monomers and conjugation. And accordingly several methods have been suggested to prepare biodegradable NPs from PLA, PLG, PLGA and poly(ε-caprolactone) by dispersing the preformed polymers.[42] However, unfortunately, these conventional methods often lack uniform characteristics, and as a result many investigations have been conducted to resolve such problem, *e.g.* using fluidic nano precipitation system (FNPS) to fabricate highly uniform PLGA particles.[43] Besides, huge attention has been devoted to advance such nanoparticles with functionalized moieties such as targeting agents.

Among enormous investigations on implementation of targeting nanosystems to test the hypothesis that NP enhance antigen presentation, Prasad *et al.* (2010) loaded dendritic cells (DCs) of patients' blood monocyte precursors with

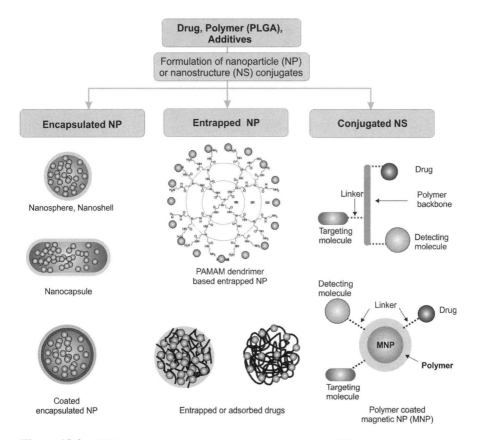

Figure 13.3 Different types of biodegradable nanoparticles (NPs).

NP-encapsulated tumour lysates to stimulate freshly isolated autologous CD8+ T cells. Their results revealed that these nanosystems can be clinically translatable, highly efficient and personalized polymer-based immunotherapy for solid organ malignancies.[44] Recently, plasma polymerized allylamine-treated PLGA scaffold particles have been reported to act as a structural support for neural stem cells injected directly through a needle into the lesion cavity using magnetic resonance imaging-derived co-ordinates.[45] Upon such an implantation, the neuro scaffolds integrated efficiently within the host tissue, forming a primitive neural tissue. Thus, these neuro scaffolds could be considered as a more advanced method for enhanced brain repair. Besides, ethionamide-loaded PLGA nanoparticles were shown to exhibit significant improvement in pharmacokinetic parameters such as C(max), t(max), AUC (0–infinity), AUMC(0–infinity) and mean residence time (MRT) of encapsulated ethionamide compared to free drug. Thus, PLGA nanoparticles of ethionamide have great potential in reducing dosing frequency of ethionamide in treatment of multi-drug resistant tuberculosis.[46] In short, these studies, along with many others, highlight the importance of the nanoscale biodegradable polymers in biomedical sciences. The wet chemical methods for nanoparticle preparation mainly include several classical methods: nanoprecipitation, emulsion-diffusion, double emulsification, emulsion-coacervation, polymer-coating, interfacial deposition and layer-by-layer.[47,48] The most commonly used methods are illustrated in Figure 13.4. In the following sections, we briefly discuss some methods of nanoparticle fabrication.

13.3.3.1 Solvent Evaporation Method

In this method, the polymer is dissolved in an organic solvent such as dichloromethane, chloroform or ethyl acetate. The drug is dissolved or dispersed into the preformed polymer solution, and this mixture is then emulsified into an aqueous solution to make an oil (O) in water (W) (*i.e.* O/W) emulsion by using a surfactant/emulsifying agent (*e.g.* poly(vinyl alcohol) (PVA), polysorbate-80, poloxamer-188, *etc.*). After the formation of a stable emulsion, the organic solvent is evaporated either by increasing the temperature under pressure or by continuous stirring under vacuum. The W/O/W method has also been used to prepare the water-soluble drug-loaded NPs. In both these methods, high-speed homogenization or sonication is used.[49]

13.3.3.2 Spontaneous Emulsification/Solvent Diffusion Method

In a modified version of the solvent evaporation method, the water-soluble solvent (such as acetone or methanol) along with the water-insoluble organic solvent (such as dichloromethane or chloroform) were used as an oil phase. An interfacial turbulence is created between two phases leading to the formation of smaller particles, which appears to be due to the spontaneous diffusion of water-soluble solvent (*e.g.* acetone or methanol). As the concentration of

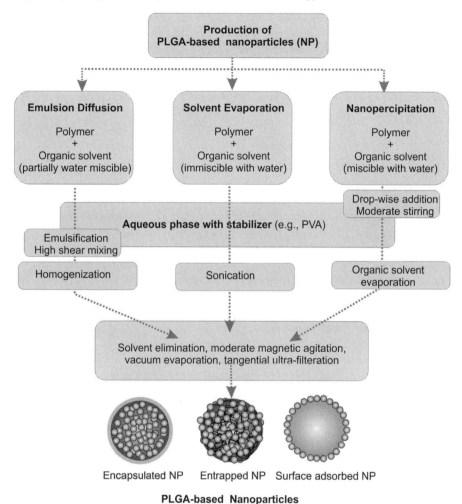

Figure 13.4 Most commonly used methods for preparation of biodegradable nanoparticles (NPs).

water-soluble solvent (*e.g.* acetone) increases, a considerable decrease in particle size can be achieved.[50,51]

13.3.3.3 Salting Out/Emulsification Diffusion Method

The aforementioned methods, basically, require the use of organic solvents – which are hazardous to the environment as well as to the physiological system. Such concern has been highlighted by many authorities. Accordingly, the FDA has specified the residual amount of organic solvents in injectable colloidal systems. Thus, in order to meet these requirements and to avoid surfactants and chlorinated solvents, researchers have developed safer methods of preparation of NPs, *e.g.* the salting out method and the emulsification-solvent diffusion technique for different drugs.[52–54] Basically, the emulsification-diffusion method consists of emulsifying an organic solution containing an oil, a polymer and a drug in an aqueous solution of a stabilizing agent. The subsequent addition of water to the system induces solvent diffusion into the external phase, and as a result the colloidal particles are formed.[47,55]

13.3.3.4 Nanoprecipitation

The nanoprecipitation technique (the so-called solvent displacement method) for preparation of NPs was first developed by Fessi and co-workers.[48] In this method, NP formation is instantaneous and the entire procedure is carried out in only one step. This method requires two solvents that are miscible, and ideally both the polymer and the drug must dissolve in the solvent (first system), but not in the non-solvent (the second system). Nanoprecipitation occurs by a rapid desolvation of the polymer when the polymer solution is added to the non-solvent. Once the polymer-containing solvent has diffused into the dispersing medium, the polymer precipitates – resulting in immediate drug entrapment. The rapid nanoparticle formation is due to interfacial turbulence that takes place at the interface of the solvent and the non-solvent and results from complex and cumulated phenomena such as flow, diffusion and surface tension variations.[48,56] Nanoprecipitation often enables the production of small nanoparticles (100–300 nm) with narrow unimodal distribution and a wide range of preformed polymers can be used, such as poly(D,L-lactic-co-glycolic acids), cellulose derivatives or poly ε-caprolactones. This method does not require extended shearing/stirring rates, sonication or very high temperatures, and is characterized by the absence of oily-aqueous interfaces, all conditions that might damage a protein structure. Moreover, surfactants are not always needed and unacceptable toxic organic solvents are generally excluded from this procedure.[48]

13.3.3.5 Interfacial Deposition

In 1984, Uno *et al.* reported a new method of preparing monocored water-loaded microcapsules through use of the process of interfacial polymer deposition. A solution of ethylcellulose or polystyrene in dichloromethane was added drop-wise to an O/W emulsion and n-hexane was dispersed as fine droplets in an aqueous gelatin solution. Successive evaporation of dichloromethane (at 40 °C) and n-hexane (at 80 °C) resulted in monocored water-loaded ethylcellulose or polystyrene microcapsules.[57] This led researchers to improve the methodology towards nano-sized particles. In 2002, Platt *et al.* utilized the interface between two immiscible liquids for spontaneous synthesis of NPs.[58] The major hurdle associated with this method is extraction of the nanoparticles away from the interface and into one of the liquid phases. NPs preferentially adsorb at the interface of two immiscible liquids to reduce the interfacial tension. In general, the immiscible liquids used are 1,2-dichloroethane and water. In addition, as shown in Figures 13.3 and 13.4, there exist several methods for incorporating drugs into the particles. For example, drugs can be entrapped in the polymer matrix, encapsulated in a NP core, surrounded by a shell-like polymer membrane, chemically conjugated to the polymer, or bound to the particle's surface by adsorption. In a very comprehensive review, Mora-Huertas *et al.* (2010) have discussed various methodologies, the types of polymers, solvent and stabilizer particle size for preparation of nanoparticles.[48]

13.3.4 Biocompatible Magnetite-PLGA Composite Nanoparticles

One particularly interesting material in targeted and controlled drug delivery systems is magnetic nanoparticles (MNPs), especially in the form of magnetite nanocrystals, which are already used in diagnostic applications as contrast agent in magnetic resonance imaging (MRI). Among various polymers used for such purposes, PLGA (alone or as PEGylated) have been shown to confer high potential as a carrier for imaging agents (*e.g.* for fluorescence, magnetic resonance imaging, and combined photoacoustic and ultrasound imaging). The use of PLGA for encapsulating both hydrophobic and hydrophilic agents has so far been successfully demonstrated since the utility of PLGA for imaging is broad as its size can be tuned for the application (*e.g.* as intratumoral, intravascular, intraperitoneal).[59] Interestingly, PLGA-based core-shell structure nanoparticles of superparamagnetic iron oxide (CSNP-SPIO) have been developed for MRI using a hydrophobic PLGA core and a positively charged glycol chitosan shell. The CSNP-SPIO were able to be internalized by cells and accumulated in lysosomes. Shortly after intravenous administration of the (99m)Tc-labelled CSNP-SPIO, a high level of radioactivity was observed in the liver. These findings clearly indicate that these biodegradable imaging nanosystems can serve as an efficient MRI contrast imaging agent.[60] Successful use of PLGA-based CSNP-SPIO as carrier for delivery of dexamethasone for the local treatment of arthritis[61] and gadolinium-DTPA contrast agent for enhanced MRI[62] clearly emphasizes its great potential as an advanced delivery system for a wide range of applications.

Magnetic NPs (MNPs) appear to be a class of advanced nanosystems that can be manipulated under the influence of an external magnetic field. They are commonly composed of magnetic elements (*e.g.* iron, nickel, cobalt and their oxides) that can be coated with biodegradable polymers and functionalized through conjugation with targeting molecules (*e.g.* peptides and nucleic acids) for organ-specific therapeutic and diagnostic modalities. The unique ability of MNPs to be guided by an external magnetic field has been utilized for MRI, targeted drug and gene delivery, tissue engineering, cell tracking and bioseparation.[63,64]

Iron oxide MNPs (*e.g.* magnetite, Fe_3O_4, or its oxidized and more stable form of maghemite, γ-Fe_2O_3) are deemed to be superior to other metal oxide nanoparticles for their biocompatibility and stability. Thus, they are the most commonly used MNPs for biomedical applications. In fact, the magnetic nanocrystals can be accumulated in a specific region of a body by applying an external magnetic field; also they can be recruited for local treatment of aberrant tissue by increasing the temperature of the target cells. Principally, uncoated MNPs have hydrophobic surfaces with large surface area to volume ratios and a propensity to agglomerate, but an appropriate surface coating with biocompatible materials improve the dispersibility and stability of iron oxide MNPs.[65] To control the uniformity and size of MNPs, they have been recently fabricated by means of the supercritical fluid method.[66]

13.3.5 PLGA-based Carriers for Macromolecule Delivery

Biodegradable PLGA-based nanoparticles have been extensively investigated for sustained and targeted/localized delivery of not only low molecular weight compounds but also different macromolecules such as antisense and plasmid DNA[67,68] as well as peptides and proteins.[69,70] Depending on the nature of the PLGA used, the drug has been shown to have varying extents of interactions with the base polymer resulting in rapid or prolonged release profiles. However, due to the bulk degradation of the polymers, the achievement of zero-order release kinetics from these polymer matrices appeared to be difficult. Another concern about PLGA-based protein delivery systems seems to be the possibility of protein denaturation during the degradation mechanism of the polymer, when acidic degradation products are produced and may inevitably alter the integrity of the protein/peptide. To tackle such concern, researchers used surface eroding polymers that have a greater ability to achieve zero-order release kinetics for molecules delivered from the matrix and are able to protect hydrolytically sensitive molecules by encapsulation.[71,72]

The important factors in developing biodegradable micro/nanoparticles for protein drug delivery are the protein release profile (including burst release, duration of release and extent of release), size of particles, protein loading rate or encapsulation efficiency, and bioactivity of the released protein. Albumin, as a model protein, has been used in many studies;[73] nonetheless the bioactivity of the released protein from the carrier has not been investigated. Some researchers have utilized PLGA-based NPs for delivery of vaccines,[74,75] insulin,[76] erythropoietin,[77] antigens[78] and growth factors.[79] Most of these investigations have highlighted the importance of formulation composition and processes to preserve bioactivity of the loaded protein.

In fact, an ideal loading rate and release profile of a designated macromolecule entrapped/encapsulated in NP requires the implementation of an appropriate preparation method. Among various parameters that may influence the final objectives, the protein loading capacity (LC), the protein release profile and the bioactivity of the protein seem to be the most important factors. The properties of micro/nanoparticles (*e.g.* LC, release kinetics and particle size) largely depend on some factors related to the protein (*e.g.* type and concentration), the polymer (*e.g.* composition, MW and concentration), volume ratio between protein and polymer solutions, emulsification method (time and intensity) and surfactant (type and concentration).[47] Often, analysis of the available information in the literature suggests that the properties of the micro/nanoparticles are not easy to control. For example, the removal of solvent from W/O/W double emulsion appears to be a critical step during preparation process, since such a process takes time and it may result in lower loading capacity and encapsulation efficiency or even initial burst release phenomenon. Figure 13.5 represents the architectural aspects of single and double emulsions. The sequence of fabrication is primary emulsion, secondary emulsion, solvent extraction/evaporation, freeze-drying and drug release test. During such a process, there is a possible partitioning of protein into the oil phase (Figure 13.5a,b), or even into the water phase (Figure 13.5c,d). Besides, during generation of the secondary emulsion, water channels connecting internal (W1) and external (W2) aqueous phases allow proteins to escape from droplets resulting in more chances of protein denaturation by increased surface area of the oil–water interface (Figure 13.5c,d). In fact, upon removal of solvent from the emulsion, protein molecules can diffuse out from the emulsion into the aqueous bath and also can accumulate on the surface of micro/nanoparticles, resulting in the high initial burst release.

Among the various copolymers investigated, poly(lactide-co-glycolide)-grafted dextran (Dex-g-PLGA) was shown to be a suitable candidate in terms of the encapsulation efficiency and *in vitro* release profile for human growth hormone (hGH).[80] Using such a system, these researchers showed a near zero-order release kinetic with no significant initial burst, while it was possible

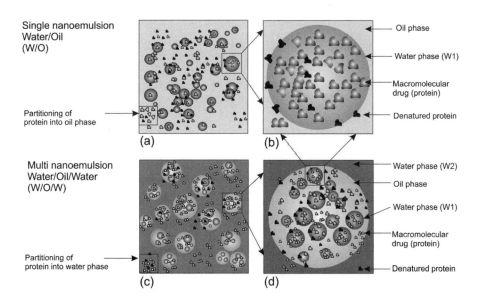

Figure 13.5 Schematic representation of single (a, b) and double (c, d) emulsions.

to control the size of the particles from 270 nm to 59 μm by changing the parameters such as the O/W ratio of the S/O/W emulsion and the stirring speed. The release rate of hGH appeared to be dependent upon the composition of Dex-g-PLGA. *In vivo* studies in mice revealed that the plasma concentration of hGH was maintained for 1 week without a significant initial burst. Figure 13.6 represents scanning electron microscopy (SEM) and confocal laser scanning microscopy (CLSM) of the particles. Morphologically, particles obtained by solvent evaporation of the solid-in-oil-in-water (S/O/W) emulsion appeared to be spherical with a smooth surface (Figure 13.6a,b), implying that the free polymer that existed in the organic phase may form a polymer matrix in which the nanoparticles are dispersed. The S/O/W emulsion method is deemed to be effective for the stable encapsulation of proteins in particles. In fact, the solidified protein particles in the S/O/W emulsion seem to suppress the contact between the protein and the water/oil interface – the phenomenon which is considered as one of the major causes of denaturation of protein-based drugs. Using CLSM, Kakizawa *et al.* showed the distribution of the encapsulated molecules within the particles (Figure 13.6c), which was confirmed by differential interference micrographs (Figure 13.6d).[80]

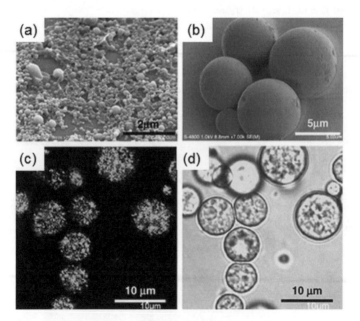

Figure 13.6 Scanning electron micrographs of (a) nanoparticles in the S/O suspension and (b) microparticles of Dex-g-PLGA. Confocal microscopy image (c) and differential interference micrograph (d) of Dex-g-PLGA microparticles containing a model compound (FD 40). Dex-g-PLGA: poly (lactide-co-glycolide)-grafted dextran. (Reproduced with Permission from Prof. N. Ida. Original work of citation for figure is Kakizawa Y., Nishio R., Hirano T., Koshi Y., Nukiwa M., Koiwa M., Michizoe J. and Ida N., *J Control. Release*, 2010, **142**(1), 8.)

Recently, Cui et al. (2010) investigated the effect of pigment epithelial-derived factor (PEDF) gene loaded in PLGA nanoparticles (PEDF-PLGANPs) on mouse colon carcinoma cells (CT26s) *in vitro* and *in vivo*.[81] These researchers showed lower cytotoxicity for blank PLGANPs (bPLGANPs) in comparison with cationic polymers such as polyethylenimine (PEI) in CT26s cells, while the PEDF-PLGANPs directly induced apoptosis in CT26 cells and inhibited proliferation in human umbilical vein endothelial cell (HUVEC). *In vivo* investigation revealed PEDF-PLGANPs inhibition in CT26 tumour growth by inducing CT26 apoptosis, decreasing MVD and inhibiting angiogenesis. These results indicate that PLGANP-mediated PEDF gene could provide an innovative strategy for the therapy of colon carcinoma.

Furthermore, PLGA micro/nanospheres have been utilized for sustained/controlled release of various genetic drugs (*e.g.* plasmid DNA and oligonucleotides). These biopolymers have been shown to improve delivery of DNA to antigen-presenting cells (APC) by efficient trafficking through local lymphoid tissue and uptake by dendritic cells.[82] Basically, the water-in-oil-in-water (W/O/W) double emulsion/solvent evaporation method has been successfully exploited for encapsulation/entrapment of plasmid DNA within PLGA microspheres. Since plasmid DNA is a highly charged macromolecule, DNA should be first dissolved in a small volume of an aqueous phase, emulsified in an oil phase, and then re-emulsified in an aqueous phase under high shear stress conditions to prepare PLGA microspheres. Nevertheless, such formulation process may structurally damage the encapsulated plasmid DNA, altering its structural integrity and biological activity. Besides, polymer degradation can induce an acidic microenvironment inside eroding PLGA microspheres, upon which the encapsulated DNA may undergo further structural damage during the release period.

In 2008, Mok and Park encapsulated plasmid DNA within PLGA nanospheres by using the polyethylene glycol (PEG) assisted solubilization technique on plasmid DNA in organic solvents.[83] Plasmid DNA was solubilized in an organic solvent mixture composed of 80%dichloromethane and 20%DMSO by producing PEG/DNA nano-complexes (with diameter $\sim$100 nm). DNA could be solubilized in the organic solvent mixture to a greater extent by increasing the weight ratio of PEG/DNA. DNA from the O/W PLGA nanospheres remained intact; however, DNA from the W/O/W PLGA nanospheres appeared to be completely degraded. *In vitro* release experiments revealed intact release of plasmid DNA from the O/W PLGA nanospheres with an initial burst profile followed by the sustained release – perhaps due to the non-uniform distribution of PEG/DNA nano-complexes within PLGA nanospheres. For the rapid release of DNA within 24 h, these researchers discussed that hydrophilic PEG/DNA nano-complexes entrapped within PLGA nanospheres were preferentially hydrated and swollen upon incubating in aqueous medium and then quickly released out by osmotic rupture of surrounding PLGA polymer phase. Based on microscopic analysis, many PLGA nanospheres were found in the intracellular area. The cells certainly expressed an exogenous GFP gene, nonetheless the GFP expression level in the macrophage cells was very low to quantify, compared to those transfected with commercially available polyplexes and lipoplexes. In fact, DNA-loaded PLGA nanospheres could not efficiently escape from the endosome and lysosome compartments after cellular uptake, in which incorporation of fusogenic peptides in the PLGA nanospheres seems to be desirable for an enhanced gene expression efficiency.

In addition to protein- and gene-based macromolecules, PLGA-based micro/nanosystems have been successfully used for delivery of vaccine.[84–86] Ingestion of intruding pathogens in the body results in migration of dendritic cells (DCs) from the periphery to the lymph nodes, where processed pathogen-derived antigens are presented to T cells. Such an interaction between DCs and T cells controls the type and magnitude of the resulting immune response, as a result of which some researchers in preclinical and clinical studies have exploited DCs in an attempt to improve vaccine efficacy. In fact, the efficacy of vaccines is strongly enhanced by antibody-mediated targeting of vaccine components to the antigen-presenting DCs.[78] However, the options to link a single antibody to multiple vaccine components (*e.g.* multiple antigens and immune modulators) are limited. To overcome such limitation, many researchers have focused on development and/or advancement of nano- and micrometer-sized slow-release vaccine delivery systems that specifically target human DCs. For example, in an interesting study, Cruz *et al.* (2010) coated the PLGA-based nano- and microparticles (with diameters ∼ 100–1000 nm) with a polyethylene glycol-lipid layer carrying the humanized targeting antibody hD1, which does not interact with complement or Fc receptors and recognizes the human C-type lectin receptor DC-SIGN on DCs.[87] The encapsulation of antigen resulted in almost 38% degradation for both nanoparticles and microparticles 6 days after particle ingestion by DCs, nevertheless the NPs effectively targeted human DCs. Consequently, targeted delivery only improved antigen presentation of NPs and induced antigen-dependent T cell responses at 10–100 fold lower concentrations than nontargeted NPs. The PLGA-PEG particles displayed approximately 20–30 µg antibody per mg PLGA. Analysis by fluorescence-activated cell sorting (FACS) flow cytometry and CLSM revealed that antibodies can present on the PLGA surface.

Recently, we have developed a star-shaped PLGA-β-cyclodextrin (PLGA-β-CD) copolymer by reacting L-lactide, glycolide and β-cyclodextrin in the presence of stannous octoate as a catalyst.[40] The structure of PLGA-β-CD copolymer was confirmed with ^{1}H-NMR, ^{13}C-NMR, and FT-IR spectra. Adriamycin (ADR) as an anticancer agent was encapsulated within the PLGA-β-CD nanoparticles and effects of the experimental parameters (*e.g.* copolymer composition, ADR concentration, copolymer concentration and poly(vinyl alcohol) concentration) on particular size and encapsulation efficiency were investigated. Figure 13.7 represents the SEM micrographs particles of PLGA-β-CD. We found that an increase in the internal aqueous phase volume can lead to a decrease in particle average size, and a decrease in the polymer concentration can cause an increase in particle average size (*i.e.* from 135.5 to 325.6 nm). All of the release profiles indicated a close relationship between each formulation variable and the amount of ADR released.

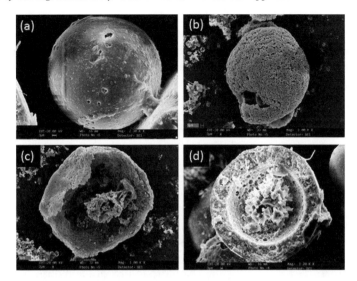

Figure 13.7 Scanning electron microscopy of intact (a and b) and cross-sectioned (c and d) particles of the star-shaped poly(lactide-co-glycolide)-β-cyclodextrin (PLGA-β-CD). PLGA with molar composition LA: GA: β-CD = 0.67: 0.23, 0.1 (PLGA-β-CD$_{10}$) (a and c), PLGA with molar composition LA: GA: β-CD = 0.6:0.2:0.2 (PLGA-β-CD$_{20}$) (b), and PLGA with molar composition LA: GA: β-CD = 00.3, 0.1, 0.6 (PLGA-β-CD$_{60}$) (d). (Reproduced from Mooguee M., Omidi Y. and Davaran S., *J. Pharm. Sci.*, 2010, **99**(8), 3389.)

13.3.6 Application of Polyester Polymers in Tissue Engineering

Biomaterials in general are used for the following cell/tissue-based purposes:

(1) to replace tissues that are diseased or otherwise nonfunctional (*e.g.* joint replacements, artificial heart valves and arteries, tooth reconstruction and intraocular lenses)
(2) to assist in the repair of tissue (*e.g.* sutures, bone fracture plates, ligament and tendon repair devices)
(3) to replace all or part of the function of the major organs (*e.g.* hemodialysis, oxygenation, left ventricular or whole heart assistance, liver perfusion and pancreas insulin delivery)
(4) to deliver drugs to the body, either to targeted sites (*e.g.* directly to a tumour) or sustained drug delivery systems.[88–90]

Basically, materials used to fabricate scaffolds for soft-tissue engineering are usually selected to provide transient structures, exhibiting adequate biological and mechanical properties.[91] Further, based on emergence of the biodegradable scaffolds, tissue engineering and regenerative medicine fields negate an absolute need for a permanent implant made of an engineered material remaining in the tissue. In fact, bioresorbable materials are deemed to grant a platform for further advancement of the field. For instance, cardiovascular disease is the leading cause of mortality in most of the developed countries. Of the cardiovascular diseases, one growing demand is for healthy autologous vessels for bypass grafting; however, availability of such tissue is limited and thereby fabricated prosthetic vascular conduits are often used. Although synthetic polymers have been extensively investigated as substitutes in vascular engineering, they often fail to meet all the biological requirements in terms of blood–material interfaces.[90] Of various tissue engineering strategies, bioresorbable materials are scaffolds not found in nature, but which over a period of time degrade in the body so that ultimately they are no longer part of the graft.[92] Vascular cell seeding of scaffolds and the design of bioactive polymers for *in situ* arterial regeneration have yielded promising results.

Huang *et al.* (2010) established a scaffold by embedding poly (β-hydroxybutyrate-co-β-hydroxyvalerate) (PHBV) microspheres into PLGA matrix with the aim of repairing bone defects. PLGA/PHBV scaffolds displayed good pore parameters (*e.g.* the porosity of PLGA/30%PHBV scaffold can reach to $81.273 +/- 2.192\%$). The morphology of the hybrid scaffold was rougher than that of pure PLGA scaffold, which had no significant effect on the cell behaviour. The *in vitro* evaluation suggested that the PLGA/PHBV is suitable as a scaffold for engineering bone tissue.[93]

In 2010, Kawazoe *et al.* developed a cell-leakproof porous PLGA-collagen hybrid scaffold through wrapping the surfaces of a collagen sponge (except the top surface used for cell seeding) with a bi-layered PLGA mesh. This hybrid scaffold displayed a structure consisting of a central collagen sponge formed inside a bi-layered PLGA mesh cup, showing high mechanical strength and cell seeding efficiency up to 90.0% with human mesenchymal stem cells (MSCs). The MSCs in the hybrid scaffolds showed round cell morphology after 4 weeks culture in chondrogenic induction medium and the immunostaining revealed that type II collagen and cartilaginous proteoglycan were detected in the extracellular matrices, that also showed up-regulation of genes encoding type II collagen, aggrecan and SOX9. Based on such findings, these researchers suggest that the cell-leakproof PLGA-collagen hybrid system is suitable as a scaffold for cartilage tissue engineering.[94]

Li *et al.* (2010), reported a three-dimensional (3D) bioscaffold for differentiation of rat bone marrow mesenchymal stem cells (BMSCs) into hepatocytes. Practically, for hepatocyte differentiation, they seeded the third passage BMSCs isolated from normal adult F344 rats into collagen-coated PLGA (C-PLGA) 3D scaffolds with hepatocyte differentiation medium for 3 weeks and characterized the hepatogenesis in scaffolds using various molecular techniques. Biocompatibility assessment indicated that the differentiated hepatocyte-like cells grew more stably in C-PLGA scaffolds than in controls during a 3-week differentiation period, upon which they suggested the C-PLGA bioscaffold for liver tissue engineering.[27]

Sahoo *et al.* (2010) discussed that an ideal scaffold should provide a combination of suitable mechanical properties along with biological signals for successful ligament/tendon regeneration in a mesenchymal stem cell-based tissue engineering process. To pursue such an aim, they developed a biohybrid fibrous scaffold system by coating bioactive bFGF-releasing ultrafine PLGA fibres over mechanically robust slowly degrading degummed knitted microfibrous silk scaffolds. On the ECM-like biomimetic architecture of ultrafine fibres, sustained release of bFGF mimicked the ECM in function, initially stimulating mesenchymal progenitor cell (MPC) proliferation, and subsequently, their tenogeneic differentiation. The biohybrid scaffold system facilitated MPC attachment, promoted cell proliferation and stimulated tenogeneic differentiation of seeded MPCs, while showing marked gene over-expression of ligament/tendon-specific ECM proteins and increased collagen production. Based on such findings, they proposed that such biohybrid fibrous scaffold is likely to confer appropriate mechanical properties for generation of a ligament/tendon analogue that has the potential to be used to repair injured ligaments/tendons.[95]

To explore the therapeutic potential of engineered neural tissue, in 2009 Xiong et al. combined genetically modified neural stem cells (NSCs) and PLGA polymers to generate an artificial neural network in vitro. In practice, the NSCs were transfected with either NT-3 or its receptor TrkC gene and seeded into a PLGA scaffold. Based upon immunoreactivity assessment, a high rate of differentiation toward neurons was suggested, in which immunostaining of synapsin-I and PSD95 revealed formation of synaptic structures. Furthermore, using FM1-43 dynamic imaging, they showed synapses in the differentiated neurons, which were found to be excitable and capable of releasing synaptic vesicles. Taken together, these researchers concluded that such an artificial PLGA construct is able to allow the NSCs to differentiate toward neurons, establish connections and exhibit synaptic activities, thus suggested great potential for construction in neural repair.[96]

Since the engineered biomaterials combined with growth factors have emerged as a new treatment alternative in bone repair and regeneration, Fu et al. (2008) encapsulated bone morphogenetic protein-2 (BMP-2) into a polymeric matrix (i.e. PLGA/hydroxylapatite (PLGA/HAp)) and characterized their individual performance in a nude mouse model. They examined the PLGA/HAp composite fibrous scaffolds loaded with BMP-2 through electrospinning and found that the PLGA/HAp composite scaffolds exhibited good morphology/mechanical strength and HAp nanoparticles were homogeneously dispersed inside PLGA matrix. Results from the animal experiments indicated that the bioactivity of BMP-2 released from the fibrous PLGA/HAp composite scaffolds was well maintained. They suggested BMP-2 loaded PLGA/HAp composite scaffolds as a promising bioscaffold for bone healing.[97]

Given these promising results, however, there exist some pitfalls for successful application of biodegradable scaffolds. For instance, there are few suitable techniques available to sterilize biodegradable polyester three-dimensional tissue engineering scaffolds because they are susceptible to degradation and/or morphological degeneration by high temperature and pressure.[98] To determine the optimal sterilization procedure (i.e. a sterile product with minimal degradation and deformation), Holy, et al. (2001) found that an argon plasma created at 100 W for 4 min was optimal for sterilizing PLGA scaffolds without affecting their morphology. They also compared the radio-frequency glow discharge (RFGD) plasma sterilization method with two well-established techniques (*i.e.* ethylene oxide (ETO) and gamma irradiation (gamma)) and found that both ETO and gamma irradiation posed immediate problems as sterilization techniques for 3D biodegradable polyester scaffolds. ETO, RFGD plasma sterilized and EtOH disinfected samples showed similar changes in M_w and mass over the 8-week time frame. They concluded that, of the three sterilization techniques studied, RFGD plasma was the best. Such sterilization methods have been used by others too.[99]

13.4 Functional Polymers: Cellular Toxicity

The polymer-based nanosystems can interact with cellular components and are transported within cells *via* vesicular machineries, where they interfere with normal function of the cell by interacting with intracellular molecular targets. Because they possess nano-scaled size and accordingly have a huge surface area, they can exhibit profound impacts on cell function even though their interactions with the cell membrane are very poorly understood. However, there is an increasing body of evidence that the surface properties of nanoparticles can lead to considerable inadvertent cytotoxicity.[100,101] Fischer *et al.* (2003) monitored cytotoxicity of various polycationic gene delivery systems in L929 mouse fibroblasts using the MTT assay and the release of the cytosolic enzyme lactate dehydrogenase (LDH). They showed a pattern for cellular toxicity as follow: poly(ethylenimine) = poly(L-lysine) > poly(diallyl-dimethylammonium chloride) > diethylaminoethyl-dextran > poly(vinyl pyridinium bromide) > Starburst dendrimer > cationized albumin > native albumin.[102] These researchers, interestingly, confirmed the molecular weight and the cationic charge density of the polycations as key parameters for the interaction with the cell membranes and accordingly the cell damage.

Despite these facts, the application of nanoparticles on a subcellular level necessitates an in depth study of their biocompatibility. To evaluate the possible toxicity of chitosan-modified PLGA nanoparticles on different cell lines, Nafee *et al.* (2009) examined impacts of different factors (*e.g.* nanoparticle concentration, exposure time, chitosan content in the particles and pH fluctuations) on cell viability and concluded that there is an undeniable impact of cell type, medium, presence/absence of serum on the colloidal state of the particles that consequently influence their interaction with the cells,[103] which may induce undesired intrinsic toxicity. In conclusion, based on potential of biopolymers to interact with target cells, biodegradable polymers impose significantly lower cytotoxicity compare to non-biodegradable polymers even though their genomic impact is yet to be fully understood. In following section, we will provide some important information upon this interesting subject.

13.5 Genocompatibility and Toxicogenomics of Polymers

So far, numerous polymers, including linear, branched, dendrimer, block and graft copolymers, have been examined as carriers for delivery of drugs/genes and as scaffolds for tissue engineering. However, surprisingly, little attention has been devoted for elucidating how a particular polymer *per se* is able to affect the normal function of the target cells/tissue perhaps by altering the gene expression profile and imposing changes in cellular phenotype. Now it seems we are faced with a new concept as 'functional excipients', in which the important field (*i.e.* 'polymer genocompatibly' or 'polymer toxicogenomics') has been increasingly highlighted. Figure 13.8 shows the structural architecture of some of these dendrimers.

We have previously reported that starburst polyamidoamine (PAMAM) dendrimers (*i.e.* PolyFect™ and SupeFect™) as well as polypropylenimine

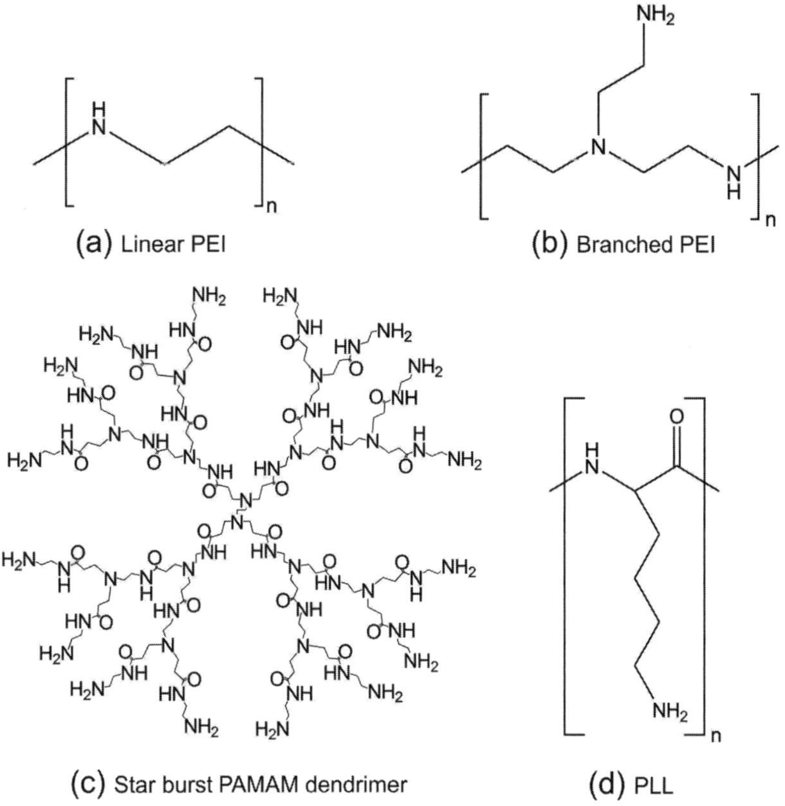

Figure 13.8 Selected examples of commonly used synthetic polycationic polymers for gene delivery. PEI: Poly(ethylenimine); PLL: Poly(L-lysine); PAMAM: Polyamidoaminedendrimer.

(*e.g.* DAB8 and DAB16) can inadvertently induce alterations in gene expression.[104,105] Of these dendrimers, we have previously shown dramatic alteration in gene expression induced by DAB16 dendrimer in A431 and A549 cells.[105] Figure 13.9 represents selected examples of DNA microarray data as the scatter plots of gene expression profiles induced by various cationic polymers such as polyamidoamine and linear and branched polyethylenimine in A431 cells.

Importantly, some of the genes altered by these polymers are related to cell defence and response to stress (*e.g.* ALOX5, TNFRSF7) and apoptosis (*e.g.* TNFRSF7). For example, in A549 cells, some of the altered genes were in association with cell defence, DNA repaire/damage and apoptosis (*e.g.* CCNH; ERCC1; PCNAM, CD14). In A549 cells treated with complexed DAB dendrimer with DNA (DBA16:DNA nanoparticles), we found expression changes for some important genes (*i.e.* TGFβ1, BCL2α1, IL5, CXCR4 AND PCKα). Of these, the BCL2 protein family is involved in a wide variety of cellular activities that also act, in particular, as anti- and pro-apoptotic regulators. The protein encoded by this gene is able to reduce the release of pro-apoptotic cytochrome c from mitochondria and block caspase activation which is the main apoptosis pathway. The up-regulation of TGFβ1 and BCL2α conceivably imply occurrence of apoptosis upon treatments with DAB16:DNApolyplexes; nonetheless, cells can also undergo a compensatory stage during apoptosis phenomena. We found that the altered genes induced by PolyFect™ and DAB16 dendrimers in A431 cells shows some commonalities and differences in gene expression pattern, presumably due to their positive charge and structural

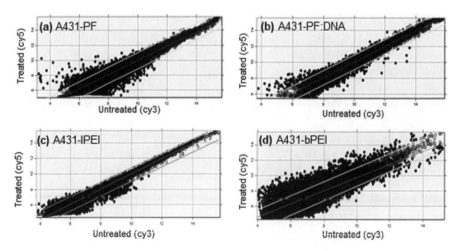

Figure 13.9 Selected examples of scatter plots of gene expression changes induced by cationic polymers in A431 cells. Data represent Log_2 transformed gene expression values. Above 2-fold change in expression of treated to untreated is indicated by reference lines, showing over-expressed genes (bold circles) and unchanged genes (unfilled circles). PF: Polyfect™; bPEI: branched polyethylenimine; lPEI: linear polyethylenimine (our unpublished data produced by Omidi *et al.*).

architecture. In A431 cells, cells treated with either DAB8 or DAB16 resulted in ~13% and ~7% similar and opposite patterns of gene expression changes, respectively. It should be remembered that the identity of the genes whose expression was significantly altered (*i.e.* the 'gene signature' of the delivery system) was markedly different in the two cell lines, despite the similar expression of the majority of the genes (80%) that remained unaffected.[106]

Likewise, Pluronic™ block copolymers were shown to cause various functional alterations in cells through interacting with cellular biomolecules and thus affecting various cellular functions such as mitochondrial respiration, ATP synthesis, activity of drug efflux transporters, apoptotic signal transduction and transcriptional activation of gene expression both *in vitro* and *in vivo*.[107] This polymer is able to enhance expression of reporter genes under the control of cytomegalovirus promoter and NF-KB response element in stably and transiently transfected mouse fibroblasts and myoblasts *in vitro*. These block copolymers acted as biological response modifying agents by up-regulating the transcription of genes through the activation of selected signalling pathways, in particular, NF-KB.[108] Furthermore, Plutonic P85 (P85) promoted transport of the pDNA to the nucleus in cells transiently transfected with DNA/PEI polyplex.[109] P85 has also been used successfully for DNA vaccines; however, it was reported that P85 simultaneously increased transgene expression and activated immunity, in which P85 alone and P85:DNA complexes were shown to increase the systemic expansion of CD11c+ (DC), and local expansion of CD11c+, CD14+ (macrophages) and CD49b+ (natural killer) cell populations. DNA/P85 polyplex can also increase maturation of local DC (CD11c+ CD86+, CD11c+ CD80+, and CD11c+ CD40+).[110] Thus, the activation of immunogenes in the antigen-presenting cells by P85:DNA complexes can highlight new insights for these kinds of polymers.

It has been shown that Pluronic™ can cause some alterations in hsp68 expression, suggesting that this polymer may affect stress-related pathways or there is a cross-talk between the stress and other pathways activated by the copolymer.[108] This finding is in accord with what we have observed for some other cationic polymers or lipids. Pluronic™ (Pluronic™ L61 and F127; also called as SP1017) has been reported to deliver plasmid DNA in skeletal and cardiac muscle, as well as in solid tumours. Unlike other polycations, Pluronic™ does not bind and condense the nucleic acids, it does not protect DNA from degradation or facilitate transport of the DNA into the cell and its effects involve transcriptional activation of gene expression.[109] The effect of Pluronic™ was reported to be related to the activation of gene expression by activating the NF-KB and p53 signalling pathways, in which pro-apoptotic AP-1 gene that is frequently regulated by the NF-KB system, was not responsive. This, perhaps, indicates that Pluronic-mediated effect on transcription is selective and not a result of a general non-specific activation of immune defence system such as NO-mediated burst. Kabanov's group has reported that Pluronic-block copolymers interact with biomembranes and induce gene expression through mechanisms that differ from the delivery of DNA into the cell. They also questioned whether up-regulation of expression of genes delivered into cells can also take place by other non-viral polymer-based gene delivery systems. We have observed that various polymers, in particular polycations, are able to affect gene expression related to immune response and cell defence.[38,111]

PLGA copolymers have been reported as a biocompatible polymer with no or little inadvertent cellular impacts. To assess such a concept, we have examined the genocompatibility and toxicogenomics of PLGA in A431 cells using microarray technology (our unpublished data). Briefly, the A431 cells at 40–50%confluency were exposed to routinely recommended concentration of PLGA for 4 h. After 24 h, 10μg total RNA was extracted and examined for yield and purity, then using a reverse transcription reaction it was converted to cDNA, i.e. aminoallyl-labelled cDNA (aa-cDNA). The aa-cDNA from untreated and treated cells were post-labelled with cyanine (Cy3 and Cy5) dyes and after purification were hybridized on a slide microarray housing 200 gene spots (Ocimum Biosolutions, India). After washing, the slide arrays were scanned (TECAN, Switzerland) and data were analysed using ArrayPro software. Table 13.1 represents some of the up- and down-regulated genes altered

Table 13.1 Gene expression changes induced by biodegradable poly(lactic-co-glycolic acid) (PLGA) in A431. T: treated; UT: untreated. (Unpublished data produced by Omidi et al.).

Gene ID	Description	T:UT ratio
NM_001020	ribosomal protein s16; rps16	2.72
NM_003292	translocated promoter region (to activated met oncogene); tpr	2.51
U41742	nucleophosmin-retinoic acid receptor alpha fusion protein npm-rar long form	2.32
NM_000879	interleukin 5 (colony-stimulating factor, eosinophil); il5	2.28
NM_001675	activating transcription factor 4; atf4	2.26
NM_003236	transforming growth factor, alpha; tgfa	2.26
NM_004794	rab33a, member ras oncogene family; rab33a	2.26
NM_001911	cathepsin g; ctsg	2.24
AF040965	unknown protein it12	2.10
NM_002965	s100 calcium-binding protein a9; s100a9	2.09
NM_004635	mitogen-activated protein kinase-activated protein kinase 3; mapkapk3	2.04
NM_002698	pou domain, class 2, transcription factor 2; pou2f2	1.68
NM_005190	cyclin c; ccnc	1.53
NM_002737	protein kinase c, alpha; prkca	1.39
NM_000858	guanylate kinase 1; guk1	0.96
NM_002064	glutaredoxin (thioltransferase); glrx	0.79
NM_001728	basigin; bsg	0.70
NM_001679	atpase, Na^+/K^+ transporting, beta 3 polypeptide; atp1b3	0.68
NM_021065	h2a histone family, member g; h2afg	0.49
NM_001237	cyclin a; ccna2	0.47
NM_000075	cyclin-dependent kinase 4, isoform 1; cdk4	0.40
NM_004417	dual specificity phosphatase 1; dusp1	0.33
NM_005319	h1 histone family, member 2; h1f2	0.32
NM_033301	ribosomal protein l8; rpl8	0.20

by PLGA in A431 cells. Based on our findings, in consensus with other researchers we also agree that the PLGA can be considered as a relatively safe polymer in comparison with s polycationic polymers such as PAMAM dendrimers and polyethylenimine polymers.

13.6 Final Remarks

Over the past several decades, the genesis of biodegradable and absorbable polymers was driven by demands for safe biomaterials to be implemented as delivery systems and bioscaffolds with minimal inadvertent biological responses. Of the many biopolymeric systems, it appears that the ester-based polymers still maintain an almost absolute dominance among clinically used systems. In fact, an ideal biopolymer should possess the highest biocompatibility (*i.e.* cytocompatibility and genocompatibility) and accordingly an ideal synthetic bioscaffold for gene delivery and tissue engineering should also incorporate methods of fabrication that allow for control of the structure and morphology of the biomaterial from the nanoscopic to macroscopic and molecular to macromolecular levels. So far, a great number of bioresorbable materials have been investigated as scaffolds for tissue engineering and tissue repair, including naturally occurring and synthetic polymers as well as porous bioactive ceramics and glasses and polymer/ceramic composites. The ideal scaffold for tissue engineering should have good cell affinity and enough mechanical strength to serve as an initial support. Synthetic scaffolds are advantageous, as material composition, micro- and macro-structure can be precisely engineered and scaffold properties tailored for specific applications, thus enabling the attainment of optimal conditions for cell survival, proliferation and subsequent tissue formation. Synthetic bioresorbable polymers have been increasingly applied as tissue engineering scaffolds during the past 10 years, particularly polyesteric polymers such as PLA, PGA and their copolymers. In fact, these FDA-approved polymers (*e.g.* Depot®, ProLease® and Trelstar®) are considered as the principle structural components of the biodegradable delivery systems and tissue engineering scaffolds since they degrade into lactic acid and glycolic acid through hydrolysis of the ester bond, and the material bulk erodes as water diffuses into the matrix more quickly than the polymer degrades. Likewise, PLGA is biocompatible due to the uptake of lactide and glycolide post-degradation into the citric acid cycle, and this allows for the use of PLGA without need for removal of the material after the intervention. In particular, biodegradable three-dimensional porous scaffolds are necessary for accommodation of the transplanted cells as a supportive matrix and to guide the formation of new tissue. Although a PLGA scaffold seems to be suitable for tissue engineering, its mechanical strength, small pore size and hydrophobic surface properties have limited its usage. The surface treatment of a scaffold seems to be an important approach for achieving good surface characteristics for bone-cell attachment and proliferation. Thus, in the near future, we will be faced with different advanced functionalized bioscaffolds to be used for implant drug delivery systems and tissue engineering. However, to translate these advanced materials towards clinical applications, we need to carefully examine their possible non-specific intrinsic bioimpacts.

Acknowledgements

The authors are grateful to the Research Center for Pharmaceutical Nanotechnology, Tabriz University of Medical Sciences for financial support.

References

1. M. Sokolsky-Papkov, K. Agashi, A. Olaye, K. Shakesheff and A. J. Domb, *Adv. Drug Deliv. Rev.*, 2007, **59**, 187.
2. L. E. Freed, G. Vunjak-Novakovic, R. J. Biron, D. B. Eagles, D. C. Lesnoy, S. K. Barlow and R. Langer, *Biotechnology (NY)*, 1994, **12**, 689.
3. M. H. Sheridan, L. D. Shea, M. C. Peters and D. J. Mooney, *J. Control. Release*, 2000, **64**, 91.
4. W. S. Pietrzak, D. R. Sarver and M. L. Verstynen, *J. Craniofac. Surg.*, 1997, **8**, 87.
5. S. J. Peter, M. J. Miller, A. W. Yasko, M. J. Yaszemski and A. G. Mikos, *J. Biomed. Mater. Res.*, 1998, **43**, 422.
6. M. Navarro, A. Michiardi, O. Castano and J. A. Planell, *J. R. Soc. Interface*, 2008, **5**, 1137.
7. A. R. Webb, J. Yang and G. A. Ameer, *Expert Opin. Biol. Ther.*, 2004, **4**, 801.
8. Y. Lemmouchi, E. Schacht, P. Kageruka, R. De Deken, B. Diarra, O. Diall and S. Geerts, *Biomaterials*, 1998, **19**, 1827.
9. C. Sasikala and C. V. Ramana, *Adv. Appl. Microbiol.*, 1996, **42**, 97.
10. T. M. Fahmy, R. M. Samstein, C. C. Harness and W. M. Saltzman, *Biomaterials*, 2005, **26**, 5727.
11. S. S. Lee, P. Hughes, A. D. Ross and M. R. Robinson, *Pharm. Res.*, 2010, **27**, 2043.
12. J. W. Lee and J. A. Gardella, Jr., *Anal. Bioanal. Chem.*, 2002, **373**, 526.
13. J. M. Lu, X. Wang, C. Marin-Muller, H. Wang, P. H. Lin, Q. Yao and C. Chen, *Expert Rev. Mol. Diagn.*, 2009, **9**, 325.
14. A. ChampaJayasuriya and N. A. Ebraheim, *J. Mater. Sci. Mater. Med.*, 2009, **20**, 1637.
15. F. Mohamed and C. F. van der Walle, *J. Pharm. Sci.*, 2008, **97**, 71.
16. E. J. Kim, S. J. Yoon, G. D. Yeo, C. M. Pai and I. K. Kang, *Biomed. Mater.*, 2009, **4**, 055001.
17. P. Ferruti, M. Penco, P. D'Addato, E. Ranucci and R. Deghenghi, *Biomaterials*, 1995, **16**, 1423.
18. X. Shi, Y. Wang, L. Ren, W. Huang and D. A. Wang, *Int. J. Pharm.*, 2009, **373**, 85.

19. M. J. Cuddihy and N. A. Kotov, *Tissue Eng. Part A*, 2008, **14**, 1639.
20. V. Hasirci, K. Lewandrowski, J. D. Gresser, D. L. Wise and D. J. Trantolo, *J. Biotechnol.*, 2001, **86**, 135.
21. S. C. Loo, Z. Y. Tan, Y. J. Chow and S. L. Lin, *J. Pharm. Sci.*, 2010, **99**, 3060.
22. T. Betancourt, J. D. Byrne, N. Sunaryo, S. W. Crowder, M. Kadapakkam, S. Patel, S. Casciato and L. Brannon-Peppas, *J. Biomed. Mater. Res. A.*, 2009, **91**, 263.
23. S. A. Burns, R. Hard, W. L. Hicks, Jr., F. V. Bright, D. Cohan, L. Sigurdson and J. A. Gardella, Jr., *J. Biomed. Mater. Res. A*, 2010, **94**, 27.
24. Y. Tang and J. Singh, *Int. J. Pharm.*, 2009, **365**, 34.
25. X. Yuan, L. Li, A. Rathinavelu, J. Hao, M. Narasimhan, M. He, V. Heitlage, L. Tam, S. Viqar and M. Salehi, *J. Nanosci. Nanotechnol.*, 2006, **6**, 2821.
26. K. Uematsu, K. Hattori, Y. Ishimoto, J. Yamauchi, T. Habata, Y. Takakura, H. Ohgushi, T. Fukuchi and M. Sato, *Biomaterials*, 2005, **26**, 4273.
27. J. Li, R. Tao, W. Wu, H. Cao, J. Xin, J. Guo, L. Jiang, C. Gao, A. A. Demetriou, D. L. Farkas and L. Li, *Stem Cells Dev.*, 2010, **19**, 1427.
28. S. Moffatt and R. J. Cristiano, *Int. J. Pharm.*, 2006, **317**, 10.
29. S. Chen and J. Singh, *Int. J. Pharm.*, 2008, **352**, 58.
30. S. Chen, R. Pieper, D. C. Webster and J. Singh, *Int. J. Pharm.*, 2005, **288**, 207.
31. J. Park, P. M. Fong, J. Lu, K. S. Russell, C. J. Booth, W. M. Saltzman and T. M. Fahmy, *Nanomedicine*, 2009, **5**, 410.
32. F. Danhier, N. Lecouturier, B. Vroman, C. Jerome, J. Marchand-Brynaert, O. Feron and V. Preat, *J. Control. Release*, 2009, **133**, 11.
33. H. Mok, J. W. Park and T. G. Park, *Pharm. Res.*, 2007, **24**, 2263.
34. J. J. Yoon, S. H. Song, D. S. Lee and T. G. Park, *Biomaterials*, 2004, **25**, 5613.
35. H. S. Yoo, E. A. Lee, J. J. Yoon and T. G. Park, *Biomaterials*, 2005, **26**, 1925.
36. C. W. Chang, D. Choi, W. J. Kim, J. W. Yockman, L. V. Christensen, Y. H. Kim and S. W. Kim, *J. Control. Release*, 2007, **118**, 245.
37. S. Diez, I. Migueliz and C. Tros de Ilarduya, *Cell. Mol. Biol. Lett.*, 2009, **14**, 347.
38. Y. Omidi, J. Barar and S. Akhtar, *Curr. Drug Deliv.*, 2005, **2**, 429.
39. S. Davaran, Y. Omidi, R. Mohammad Reza, M. Anzabi, A. Shayanfar, S. Ghyasvand, N. Vesal and F. Davaran, *J. Bioactive Compatible Polym.*, 2008, **23**, 115.
40. M. Mooguee, Y. Omidi and S. Davaran, *J. Pharm. Sci.*, 2010, **99**, 3389.
41. K. Nagahama, Y. Mori, Y. Ohya and T. Ouchi, *Biomacromolecules*, 2007, **8**, 2135.
42. J. M. Chan, P. M. Valencia, L. Zhang, R. Langer and O. C. Farokhzad, *Methods Mol. Biol.*, 2010, **624**, 163.
43. H. Xie and J. W. Smith, *J. Nanobiotechnol.*, 2010, **8**, 18.

44. S. Prasad, V. Cody, J. K. Saucier-Sawyer, W. M. Saltzman, C. T. Sasaki, R. L. Edelson, M. A. Birchall and D. J. Hanlon, *Nanomedicine*, 2011, **7**, 1.
45. E. Bible, D. Y. Chau, M. R. Alexander, J. Price, K. M. Shakesheff and M. Modo, *Biomaterials*, 2009, **30**, 2985.
46. G. Kumar, S. Sharma, N. Shafiq, P. Pandhi, G. K. Khuller and S. Malhotra, *Drug Deliv.*, 2011, **18**, 65.
47. D. Quintanar-Guerrero, E. Allemann, H. Fessi and E. Doelker, *Drug Dev. Ind. Pharm.*, 1998, **24**, 1113.
48. C. E. Mora-Huertas, H. Fessi and A. Elaissari, *Int. J. Pharm.*, 2010, **385**, 113.
49. I. D. Rosca, F. Watari and M. Uo, *J. Control. Release*, 2004, **99**, 271.
50. E. Cohen-Sela, M. Chorny, N. Koroukhov, H. D. Danenberg and G. Golomb, *J. Control. Release*, 2009, **133**, 90.
51. H. Murakami, M. Kobayashi, H. Takeuchi and Y. Kawashima, *Int. J. Pharm.*, 1999, **187**, 143.
52. M. A. Wheatley and J. Lewandowski, *Mol. Imaging*, 2010, **9**, 96.
53. S. Galindo-Rodriguez, E. Allemann, H. Fessi and E. Doelker, *Pharm. Res.*, 2004, **21**, 1428.
54. E. Allemann, J. C. Leroux, R. Gurny and E. Doelker, *Pharm. Res.*, 1993, **10**, 1732.
55. D. Quintanar-Guerrero, E. Allemann, E. Doelker and H. Fessi, *Pharm. Res.*, 1998, **15**, 1056.
56. E. J. Lee, S. A. Khan and K. H. Lim, *J. Biomater. Sci. Polym. Ed.*, 2011, **22**, 753.
57. K. Uno, Y. Ohara, M. Arakawa and T. Kondo, *J. Microencapsul.*, 1984, **1**, 3.
58. M. Platt, R. A. Dryfe and E. P. Roberts, *Chem. Commun.*, 2002, **20**, 2324.
59. A. L. Doiron, K. A. Homan, S. Emelianov and L. Brannon-Peppas, *Pharm. Res.*, 2009, **26**, 674.
60. P. W. Lee, S. H. Hsu, J. J. Wang, J. S. Tsai, K. J. Lin, S. P. Wey, F. R. Chen, C. H. Lai, T. C. Yen and H. W. Sung, *Biomaterials*, 2010, **31**, 1316.
61. N. Butoescu, C. A. Seemayer, M. Foti, O. Jordan and E. Doelker, *Biomaterials*, 2009, **30**, 1772.
62. A. L. Doiron, K. Chu, A. Ali and L. Brannon-Peppas, *Proc. Natl. Acad. Sci. USA*, 2008, **105**, 17232.
63. P. Pouponneau, J. C. Leroux and S. Martel, *Biomaterials*, 2009, **30**, 6327.
64. H. L. Liu, M. Y. Hua, H. W. Yang, C. Y. Huang, P. C. Chu, J. S. Wu, I. C. Tseng, J. J. Wang, T. C. Yen, P. Y. Chen and K. C. Wei, *Proc. Natl. Acad. Sci. USA*, 2010, **107**, 15205.
65. G. Liu, H. Yang, J. Zhou, S. J. Law, Q. Jiang and G. Yang, *Biomacromolecules*, 2005, **6**, 1280.
66. K. Byrappa, S. Ohara and T. Adschiri, *Adv. Drug Deliv. Rev.*, 2008, **60**, 299.
67. N. Nafee, S. Taetz, M. Schneider, U. F. Schaefer and C. M. Lehr, *Nanomedicine*, 2007, **3**, 173.
68. A. C. Kilic, Y. Capan, I. Vural, R. N. Gursoy, T. Dalkara, A. Cuine and A. A. Hincal, *J. Microencapsul.*, 2005, **22**, 633.

69. X. Wang, E. Wenk, X. Hu, G. R. Castro, L. Meinel, C. Li, H. Merkle and D. L. Kaplan, *Biomaterials*, 2007, **28**, 4161.
70. C. Wischke and H. H. Borchert, *J. Microencapsul.*, 2006, **23**, 435.
71. X. J. Xu, J. C. Sy and V. Prasad Shastri, *Biomaterials*, 2006, **27**, 3021.
72. F. von Burkersroda, L. Schedl and A. Gopferich, *Biomaterials*, 2002, **23**, 4221.
73. F. Kang and J. Singh, *Int. J. Pharm.*, 2003, **260**, 149.
74. B. Slutter, S. Bal, C. Keijzer, R. Mallants, N. Hagenaars, I. Que, E. Kaijzel, W. van Eden, P. Augustijns, C. Lowik, J. Bouwstra, F. Broere and W. Jiskoot, *Vaccine*, 2010, **28**, 6282.
75. B. Nayak, A. K. Panda, P. Ray and A. R. Ray, *J. Microencapsul.*, 2009, **26**, 154.
76. M. J. Santander-Ortega, D. Bastos-Gonzalez, J. L. Ortega-Vinuesa and M. J. Alonso, *J. Biomed. Nanotechnol.*, 2009, **5**, 45.
77. Y. Geng, W. Yuan, F. Wu, J. Chen, M. He and T. Jin, *J. Control. Release*, 2008, **130**, 259.
78. Y. Waeckerle-Men and M. Groettrup, *Adv. Drug Deliv. Rev.*, 2005, **57**, 475.
79. H. J. Chung, H. K. Kim, J. J. Yoon and T. G. Park, *Pharm. Res.*, 2006, **23**, 1835.
80. Y. Kakizawa, R. Nishio, T. Hirano, Y. Koshi, M. Nukiwa, M. Koiwa, J. Michizoe and N. Ida, *J. Control. Release*, 2010, **142**, 8.
81. F. Y. Cui, X. R. Song, Z. Y. Li, S. Z. Li, B. Mu, Y. Q. Mao, Y. Q. Wei and L. Yang, *Oncol. Rep.*, 2010, **24**, 661.
82. K. D. Newman, P. Elamanchili, G. S. Kwon and J. Samuel, *J. Biomed. Mater. Res.*, 2002, **60**, 480.
83. H. Mok and T. G. Park, *Eur. J. Pharm. Biopharm.*, 2008, **68**, 105.
84. S. M. Sivakumar, N. Sukumaran, R. Murugesan, T. S. Shanmugarajan, J. Anbu, L. Sivakumar, B. Anilbabu, G. Srinivasarao and V. Ravichandran, *Indian J. Pharm. Sci.*, 2008, **70**, 487.
85. D. J. Kirby, I. Rosenkrands, E. M. Agger, P. Andersen, A. G. Coombes and Y. Perrie, *J. Drug Target.*, 2008, **16**, 282.
86. P. Johansen, F. Estevez, R. Zurbriggen, H. P. Merkle, R. Gluck, G. Corradin and B. Gander, *Vaccine*, 2000, **19**, 1047.
87. L. J. Cruz, P. J. Tacken, R. Fokkink, B. Joosten, M. C. Stuart, F. Albericio, R. Torensma and C. G. Figdor, *J. Control. Release*, 2010, **144**, 118.
88. M. E. Furth, A. Atala and M. E. Van Dyke, *Biomaterials*, 2007, **28**, 5068.
89. J. Velema and D. Kaplan, *Adv. Biochem. Eng. Biotechnol.*, 2006, **102**, 187.
90. S. Ravi and E. L. Chaikof, *Regen. Med.*, 2010, **5**, 107.
91. A. Gloria, R. De Santis and L. Ambrosio, *J. Appl. Biomater. Biomech.*, 2010, **8**, 57.
92. S. T. Rashid, H. J. Salacinski, B. J. Fuller, G. Hamilton and A. M. Seifalian, *Cell. Prolif.*, 2004, **37**, 351.
93. W. Huang, X. Shi, L. Ren, C. Du and Y. Wang, *Biomaterials*, 2010, **31**, 4278.

94. N. Kawazoe, C. Inoue, T. Tateishi and G. Chen, *Biotechnol. Prog.*, 2010, **26**, 819.
95. S. Sahoo, S. L. Toh and J. C. Goh, *Biomaterials*, 2010, **31**, 2990.
96. Y. Xiong, Y. S. Zeng, C. G. Zeng, B. L. Du, L. M. He, D. P. Quan, W. Zhang, J. M. Wang, J. L. Wu, Y. Li and J. Li, *Biomaterials*, 2009, **30**, 3711.
97. Y. C. Fu, H. Nie, M. L. Ho, C. K. Wang and C. H. Wang, *Biotechnol. Bioeng.*, 2008, **99**, 996.
98. C. E. Holy, C. Cheng, J. E. Davies and M. S. Shoichet, *Biomaterials*, 2001, **22**, 25.
99. E. K. Moioli, L. Hong, J. Guardado, P. A. Clark and J. J. Mao, *Tissue Eng.*, 2006, **12**, 537.
100. B. J. Marquis, S. A. Love, K. L. Braun and C. L. Haynes, *Analyst*, 2009, **134**, 425.
101. C. Kirchner, A. M. Javier, A. S. Susha, A. L. Rogach, O. Kreft, G. B. Sukhorukov and W. J. Parak, *Talanta*, 2005, **67**, 486.
102. D. Fischer, Y. Li, B. Ahlemeyer, J. Krieglstein and T. Kissel, *Biomaterials*, 2003, **24**, 1121.
103. N. Nafee, M. Schneider, U. F. Schaefer and C. M. Lehr, *Int. J. Pharm.*, 2009, **381**, 130.
104. A. J. Hollins, Y. Omidi, I. F. Benter and S. Akhtar, *J. Drug Target.*, 2007, **15**, 83.
105. Y. Omidi, A. J. Hollins, R. M. Drayton and S. Akhtar, *J. Drug Target.*, 2005, **13**, 431.
106. S. Akhtar and I. Benter, *Adv. Drug Deliv. Rev.*, 2007, **59**, 164.
107. E. V. Batrakova and A. V. Kabanov, *J. Control. Release*, 2008, **130**, 98.
108. S. Sriadibhatla, Z. Yang, C. Gebhart, V. Y. Alakhov and A. Kabanov, *Mol. Ther.*, 2006, **13**, 804.
109. V. A. Kabanov, *Adv. Drug Deliv. Rev.*, 2006, **58**, 1597.
110. Z. Z. Gaymalov, Z. Yang, V. M. Pisarev, V. Y. Alakhov and A. V. Kabanov, *Biomaterials*, 2009, **30**, 1232.
111. J. Barar, H. Hamzeiy, S. A. Mortazavi-Tabatabaei, S. E. Hashemi-Aghdam and Y. Omidi, *Daru*, 2009, **17**, 139.

CHAPTER 14
Biodegradable Injectable Systems for Bone Tissue Engineering

RICHARD T. TRAN, DIPENDRA GYAWALI, PARVATHI NAIR AND JIAN YANG

The University of Texas at Arlington, Department of Bioengineering, 500 UTA Blvd, Arlington, Texas 76109, USA

14.1 Introduction

Natural bone has an intrinsic ability to heal itself, and this regenerative capacity is diminished with age, illness, or injury.[1] The current approaches for the treatment of critical-sized bone defects include the use of natural bone grafts or metallic prosthetic implants. Natural grafts can be categorized based upon their tissue source as autografts, allografts, and xenografts. Autografts currently serve as the gold standard for bone implantation, and are harvested from the patient's own body, which reduces the risk of graft rejection. Although, autografts are widely used in clinical applications, problems such as donor site morbidity, the invasive nature of surgery, long recovery times, and bone graft availability of a desired size and shape have restricted the use of autografts in orthopedic applications.[2,3] In the United States (US), there have been over a million surgical procedures involving large bone defects due to trauma, non-union healing fractures, or resection requiring the use of bone grafts. As the US population ages, there has been an increase in demand for bone grafts, and these surgical procedures have placed a large burden on the healthcare industry totaling over 5 billion dollars annually.[4,5] Thus, a clinical need exists for the development of alternative methods to regenerate bone to meet the shortage in bone grafts and address the limitations of the current available treatment choices.

Tissue engineering is a multidisciplinary field, which applies principles from material science, chemical engineering, and biological life sciences to develop alternatives to restore, improve, or maintain diseased or damaged tissue.[6] The foundation of tissue engineering relies on four key elements: cells, scaffolds, signals, and bioreactors.[7,8] In the general scheme for tissue engineering, cells are seeded onto a three-dimensional (3D) scaffold, a tissue is cultivated *in vitro*, proper signals are supplemented to the system, and finally the construct is implanted into the body as a prosthesis.[8] The general scheme for the key elements involved in the tissue engineering paradigm is illustrated in Figure 14.1.

The cells used in tissue engineering applications can be isolated from autologous, allogenic, or xenogenic sources, and may be tissue specific, stem cells, or progenitor cells. The harvested cells are then isolated and expanded *in vitro*. Scaffolds, which mimic the native extracellular matrix and provide a substrate for cell growth, can be composed of either a natural or synthetic material, and fabricated into a fibrous, foam, hydrogel, or capsule architectures. Signals are also introduced to the system for enhanced cell adhesion, proliferation, and differentiation within the construct. Bioreactors are often utilized to mimic the dynamic conditions inside the body, and provide many

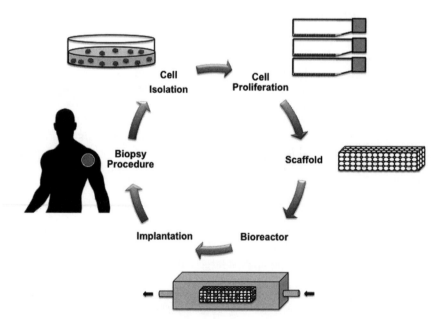

Figure 14.1 A representative schematic describing key elements of the traditional tissue-engineering paradigm.

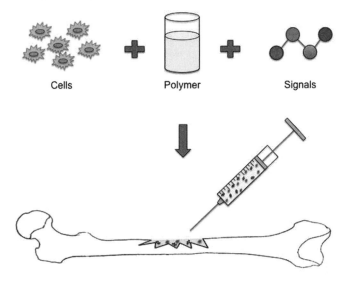

Figure 14.2 Schematic representing the concept behind injectable bone tissue engineering.

benefits such as improved mass transport and the application of mechanical stimuli to the developing tissue.[8,9]

Recently, many tissue engineering designs using injectable, *in situ* forming systems have been reported for orthopedic applications with many advantages over previous methods. Unlike tissue engineering approaches that utilize pre-fabricated scaffolds, injectable systems have garnered great interest within the field as a unique therapeutic method for difficult to reach areas of the body using minimally invasive procedures, and show the ability to conform to any shape irrespective of the defect geometry. Furthermore, injectable systems can be used as fillers to reinforce the mechanical properties of diseased/ injured bone and as a competent carrier of cells and therapeutic agents such as drugs and growth factors (Figure 14.2).[10–12]

14.2 Rationale and Requirements for Injectable Bone Tissue Engineering

Native bone is a complex tissue to engineer and understanding the structural, physical, chemical, biological, and cellular properties of natural bone is imperative to design a new approach that will fulfill not only the basic principles of tissue engineering, but also address the unique challenges of designing a tissue engineered bone. The following section will highlight the major criteria necessary in the design and development an ideal biodegradable injectable tissue engineered bone substitute.

14.2.1 Injectability and *In Situ* Cross-linking

The basic requirements for the selection of scaffold material for tissue engineering applications are challenging and multi-faceted. For injectable-based systems, the precursor solution of the polymer should be easily injectable through the required delivery system in order to facilitate the minimally invasive surgical procedure. The polymer viscosity prior to cross-linking plays a critical role in determining the injectability of the system. To ensure the injectability of the system, many *in situ* cross-linkable polymers are dissolved in biocompatible solvents such as water, *N*-methyl-2-pyrrolidone, and dimethyl sulfoxide. In addition to solvent solubility, the time required for polymer cross-linking should be within an appropriate time window to avoid surgical difficulties and failure of the implant. For example, slower cross-linking times may encourage unwanted migration of the material from the defect area, whereas extremely rapid cross-linking times may hinder the surgical ease of implantation.[13] Many factors such as the intensity of cross-linking stimuli, concentration of the initiating system, type of cross-linker, and material functionality play a major role in the cross-linking time of any injectable system.

14.2.2 Mechanical Properties

Biomaterials implanted inside the body are often subjected to a mechanically dynamic environment, and must be able to sustain and recover from repetitive deformations while allowing material/ tissue integration without irritating surrounding tissues.[14] To complicate matters, the mechanical properties of natural bone vary greatly depending on the location and function of the native bone tissue (Table 14.1). Thus, the mechanical properties of tissue engineered bone constructs should be given special attention in that the given materials are used for multiple purposes such as load bearing and tissue regeneration, and closely match the mechanical properties of natural bone as closely as possible to avoid problems associated with compliance mismatch and load shielding, which may lead to implant failure.[15,16]

In addition to preventing graft failure, many groups have shown that the mechanical properties of scaffolds may also play an important role in the biological outcome of the implanted device.[18,19] Scaffold mechanical properties have been shown to have an influence on the severity of the host inflammatory response, angiogenesis, and wound-healing properties.[20] Injectable materials should withstand and transfer the mechanostimulation needed to induce

Table 14.1 Mechanical properties for the various sections of native bone.[17]

Bone tissue	Mechanical strength		Young's modulus
	Tensile	*Compressive*	
Cortical bone (longitudinal)	78–151 MPa	131–244 MPa	17–20 GPa
Cortical bone (transverse)	55–66 MPa	106–131 MPa	6–13 GPA
Cancellous bone	5–10 MPa		50–100 MPa

metabolic activities of all types of bone cells including osteoblasts, osteocytes, osteoclasts, and osteoprogenitors. For example, osteocytes under mechanical stimulation can release osteoblastic factors to accelerate bone formation, whereas in the absence of mechanical cues, osteocytes have been shown to undergo apoptosis and recruit osteoclasts, resulting in bone resorption.[21] Furthermore, it has been reported that under fluid flow stimulation, osteoblasts have been shown to reorganize their cytoskeleton, up-regulate transmembrane focal adhesion proteins, and express osteoblast-specific proteins such as osteopontin involved in extracellular matrix adhesion.[17,22,23] Thus, in order to simulate the dynamic mechanical physiological environment, injectable materials should possess similar mechanical properties compared to native bone tissue.

14.2.3 Porosity

The porosity of the scaffold architecture is another important design requirement, which plays a critical role in the outcome of a tissue engineered graft. Cells seeded inside the scaffold rely heavily on the void spaces within the construct for cellular in-growth, vascularization, and the exchange of nutrients and waste products.[11] Thus, the implanted graft should have a highly organized, porous, and interconnected structure. Porosities of more than 90% and pore sizes greater than 300 μm are preferred for cellular penetration and vascularization while maintaining scaffold mechanical integrity.[24–28] Recent research has shown that nanoscale architectures of the scaffold can also influence the cellular behavior of the underlying material.[17] In particular to injectable materials, *in situ* porous generating techniques have been limited to gas foaming[29] and particulate leaching.[30] The foremost challenge for the success of injectable systems still lies in the creation of injectable scaffolds with an interconnected and homogeneous porous network.

14.2.4 Biodegradation

Tissue engineered devices should be fully resorbed by the body to match the rate of neotissue formation, and cleared from the body through normal physiological functions through a process known as biodegradation. Biodegradation is a process by which polymers are chemically reversed into their precursor monomers, which can be accomplished through dissolution, hydrolysis, and enzymatic activities.[31] Degradation mechanisms for polymers used in tissue engineering are categorized as either bulk degradation or surface erosion. Most polyesters, polyether-esters, and polyester-amides rely on bulk degradation and follow first-order profiles, whereas polyanhydrides and polyorthoesters degrade through surface erosion and yield zero-order profiles.[17]

Injectable polymers used as bone scaffolds are composed of cross-linked networks, which degrade through several mechanisms. Degradation of these networks primarily relies on the nature and location of the degradable groups and cross-linkable moieties. In the case of poly(propylene fumarate) (PPF) and poly(ethylene glycol maleate citrates) (PEGMC), the cross-linkable and degradable moieties are found alternating along the polymer backbone chain. As the polymer network degrades, it is broken down into multiple kinetic chains and the starting monomers. Lactic acid-based injectable materials are degraded into the original core molecules connected by the degradable units and kinetic chains formed during the free-radical polymerization of the photoreactive groups.[12] Finally, enzymatically cross-linked polymers are degraded through the pendant reactive groups along the proteolytic degradable polymeric backbone.[32]

In addition to chemical structure, the polymer crystallinity, crystal structures, molecular orientation, melting temperature (T_m), glass transition temperature (T_g), cross-linking density, external particulates present in the polymer network, and micro- and nanoscale structure of the scaffold also influence the degradation rate.[33]

14.2.5 Cellular Behavior

Bone is made up of 60% inorganic minerals (calcium phosphate), 30% organic (collagen type I, osteonectin, osteocalcin, osteopontin, proteoglycans, and glycoproteins), and 10% of cellular components (osteoblasts, osteocytes, and osteoclasts). Every component has a critical role for the healthy regeneration and function of bone. For all bone tissue engineering applications, the implanted graft material should be tightly integrated with the surrounding bone tissue to provide a suitable cellular environment for the production extracellular matrix proteins. As mentioned above, materials utilized in tissue engineering applications should mimic the natural extracellular matrix in order to provide mechanical support and regulate cell behavior including cell anchorage, segregation, communication, and differentiation.[34] In general, polymers used in injectable bone tissue engineering should provide suitable functionalities (carboxylic and hydroxyl groups) for facile modification of biomolecules onto the surface and into the bulk of the material to guide cellular behavior.

For example, arginine-glycine-aspartic acid (RGD), a short peptide sequence, has been shown to mediate cell attachment to various extracellular matrix proteins such as fibronectin, vitronectin, bone sialoprotein, and osteopontin.[35] More recently, a collagen-mimetic peptide sequence, glycine-phenlalanine-hydroxyproline-glycine-glutamate-arganine (GFOGER), has been reported to enhance osteoblast functionality and osseointegration *in vivo*.[36] Injectable polymers such as poly(propylene fumarate-co-ethylene glycol), photo-cross-linked poly(ethylene glycol), alginate and poly(N-isopropylacrylamideacrylic acid) hydrogels were functionalized with RGD peptide, and demonstrated an improvement in osteoblast attachment and spreading within the materials.[37–40]

Apart from these primary cellular responses, the materials should also contain cellular integrin-binding sites and growth factor-binding sites to promote osteoconductivity, osteoinductivity, and osseointegration. The inorganic component of bone not only promotes osteoblast adhesion and migration/

infiltration (osteoconduction), but also provides strength to the bone tissue. In case of neotissue formation, osteoinduction and angiogenesis are two critical processes of the tissue regeneration. Hence, numerous polylactide- and glycolide-based polymer composites are under active investigation as scaffolding materials for bone tissue engineering.[41–44] Transcriptional growth factors, such as transforming growth factor beta (TGF-b), bone morphogenic proteins (BMPs), platelet derived growth factors (PDGFs), and insulin-like growth factors (IGFs), are reported to have major role on the osteoinductivity of the MSCs.[45–48] Similarly, growth factors such as vascular endothelial growth factor (VEGF) and basic fibroblast growth factor (bFGF) are major cues for neovascularization.[49–51] To introduce these mobile cues to the polymer matrix, various drug delivery systems are constructed in the injectable system. Later sections of this chapter will discuss in detail the existing injectable delivery systems for these key factors.

14.2.6 Biocompatibility

Injectable systems must be compatible with cells, tissues, and bodily fluids in order to function properly and avoid future complications. Any leachable compounds and degradation products should not hinder the process of tissue formation.[52] Since injectable materials are often used to deliver cells and sensitive compounds, the cross-linking process after injection should occur under physiologically accepted conditions. While designing any injectable scaffolds, careful consideration towards the toxicity, carcinogenicity, and chronic inflammatory response induced by the material, cross-linking system, solvents, and the degradation products should be given special attention.

14.3 Network Formation

Numerous injectable biomaterials have been recently reported in the scientific community as a non-invasive or minimally invasive technique for the regeneration and healing of defective tissues. To induce *in situ* cross-linkability in the polymeric system, monomers liable to mild cross-linking strategies should be incorporated within the polymer chains. In general, these systems consist of injectable and monodisperse polymeric chains with sites for three-dimensional network expansion under specific stimuli. The ability of biomaterials to behave as suitable tissue-specific injectable materials depends entirely upon the underlying cross-linking mechanism of solidification.[11] Biomaterial scientists are actively exploring cross-linking mechanisms based upon chemical cross-linking and physical gelation strategies. Chemical cross-linking is a network formation method achieved by covalently bonding two or more monomers. On the other hand, physical gelation can be achieved when two or more polymeric chains come to assemble due to the physical interaction of the chains such as ionic, hydrophobicity, or self-assembly in response to the surrounding environmental stimuli such as pH, temperature, and polymer precipitation.

Chemically cross-linked networks provide higher cross-linking densities to the polymer network, are more favorable for the sustained release of therapeutics, and allow for the fabrication of scaffolds with enhanced mechanical properties. However, the toxicity of the chemical cross-linking agents used may adversely affect cell behavior and the incorporated bioactive molecules. On the other hand, physical gelation of the network may avoid the use of cross-linking agents, but shows a limited performance in their physical properties. In the next sections, we will discuss the mechanisms involved in the solidification of injectable materials.

14.3.1 Free Radical Polymerization (FRP)

FRP consists of two components: (1) a radical-generating initiator system and (2) radical-liable monomers or oligomers. The most common radical-generating initiator systems used in biomaterials are high-energy gamma rays, ultraviolet light sensitive photo initiators,[53] and redox initiators.[54] Due to their ability to cross-link under physiological conditions, the latter two have received increased attention in the field of biomaterials. However, all these initiating systems rely on the same principle in that the generation of free radicals to initiate the cross-linking process is required. Once the free radicals are generated, the monomers or oligomers containing radical-liable moieties (usually vinyl and thiol groups) will go under further propagation steps. During propagation, these radical initiators homolytically cleave the radical-liable moieties to induce the cross-linking propagation. Finally, two radicals in the propagating polymer chains bond covalently to terminate the cross-linking process (Figure 14.3).

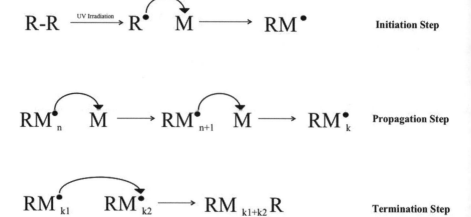

Figure 14.3 Representative schematic depicting the steps involved in free radical polymerization network formations.

Thus, the concentration of initiator, radical-liable moieties, and nature of solvent system all have a collective influence over the rate, stability, and kinetics of network formation. These parameters, in turn, reflect on the overall performance of the material in terms of cross-linking density, mechanical properties, and degradation profiles. Distinct differences in the application of systems cross-linked through photo-initiated and redox-initiated mechanisms have been reported in various literatures.[13,55,56] For example, redox-initiated systems are more favored in areas of limited light penetration and where homogeneous cross-linking of the network is preferred, whereas, photo-initiated systems are more favored where temporal and spatial control is required to develop complex structures such as patterned surfaces.

14.3.2 Chemical Cross-linking Systems (CCS)

CCS can be introduced into injectable materials by separately modifying soluble polymer chains with a pair of molecules that have specific affinity to each other. When these modified polymeric chains are injected simultaneously to the cross-linking site, they undergo rapid cross-linking to give rise to a covalent cross-linked network. The most commonly used pairs of molecules with specific affinities towards each other are N-hydroxysuccinimide (-NHS) to amine (-NH$_2$),[57] 1,4-addition of a doubly stabilized carbon nucleophile to an α,β-unsaturated carbonyl compound (Michael-type addition reaction),[58] and alkyne to azide (click chemistry).[59] In these types of reactions, the rate, stability, and kinetics of network formation are solely dependent upon the strength of the affinity of one molecule to its counterpart.

Another class of injectable materials using CCS is enzymatically cross-linked polymers. These polymers contain pendant phenol groups that undergo self-cross-linking in the presence of hydrogen peroxide (H_2O_2) and horseradish peroxidase (HRP). Phenol groups can be introduced to polymer chains such as chitosan,[60] gelatin,[61] hyluronic acid,[62] and dextran[63] by reacting their pendent carboxylic groups or amine groups with tyramine or 3,4-hydroxyphenylpropionic acid (HPA). This cross-linking strategy is inspired by the fact that treatment of proteins with peroxidase and H_2O_2 causes oxidation of phenol groups of tyrosine residues, resulting in cross-linking of protein molecules to form dityrosine and tertyrosine.[32] It has been reported that the cross-linking rate of the system is dependent on the concentration of HRP, H_2O_2, and phenol functionality. Gel formation is decreased with increasing HRP concentrations and decreasing H_2O_2 concentrations. An excess amount of H_2O_2 can oxidize HRP that results in deactivation of the cross-linking ability.[32] Besides HRP, another typical example of CCS is transglutaminase (TG) mediated glutaminamide-functionalized poly(ethylene glycol) (PEG) and poly(lysine-co-phenylalanine), which utilize calcium ions as cofactors.[64]

14.3.3 Thermally Induced Gelation Systems (TGS)

TGS are widely famous for their use in polymers, which have a unique ability to undergo sol-to-gel and gel-to-sol phase transitions as a function of temperature. These polymers are amphiphilic in nature (composed of both hydrophilic and hydrophobic segments), and display a sol-to-gel transition, which is attributed from the balance between intermolecular forces and hydrophobic section aggregation. The molecular weight of these segments is solely responsible for the sol-gel performance of these copolymers. For example, copolymers of methoxy poly(ethylene glycol) (MPEG)-poly(ε-caprolactone) (PCL) shows both gel-to-sol and sol-to-gel transition at elevated temperatures. Copolymers composed of MPEG ($M_n = 2000$) are in a gel state at 10 °C, and transition from gel to sol with increasing temperature.[65] When the MPEG segment molecular weight is reduced ($M_n = 750$), the material is liquid at room temperature, and transitions from sol to gel with increasing temperatures. In the case of copolymers with MPEG ($M_n = 750$), the sol-to-gel phase transition temperature decreased substantially with increasing molecular weight of PCL.[66] Various other PEG-based biodegradable copolymers have been reported to exhibit thermoresponsive properties. Linear or star-shaped copolymers such as poly(ethylene glycol-L-lactic acid) and poly(ethylene glycol-DL-lactic acid-co-glycolic acid)[69–70] have shown a sol-to-gel transition as the temperature is decreased. A more recently reported copolymer MPEG-b-(PCL-ran-PLLA) has been shown to display a liquid state at room temperature, and transitions to a gel state precisely at body temperature (37 °C), which is a major advantage for use in injectable therapies.[70]

Another class of TGS polymers is *N*-isopropylacrylamide-based copolymers that display a sol-to-gel transition as the temperature is elevated above their lower critical solution temperature (LCST). The drastic difference in solubility above the LCST is due to the entanglement and the gradual collapse of polymeric chain. Various copolymers have been reported to induce biodegradation to this polymeric system. The factors determining the gelation process include polymer concentration, molecular weight, and chemical structure of the copolymer. Other typical examples of thermosensitive polymers are poly-(ethylene oxide) and poly(propylene oxide) copolymers (poloxamers or pluronics),[71] cellulose derivatives,[72] chitosan,[73] and gelatin[74] have also been widely explored as thermoresponsive injectable materials in tissue engineering and drug delivery.

14.3.4 Self-assembly Systems (SAS)

Most of the SAS have been reported in amphiphilic polymers, which show the ability to self-assemble due to the desolvation, collapse, and intermolecular association of the hydrophobic portions of monomers.[75] In the case of charged (anionic, cationic, or zwitterionic) amphiphiles, further stability and structural specificity can be designed using intermolecular polar interactions, such as electrostatic and hydrogen bonding.[76] Various biologically inspired materials such as peptide- and protein-based systems have been reported with distinct amphiphilic properties for use in tissue regeneration and growth factor delivery. For example a self-assembling peptide-based hydrogel has been reported with the property of low viscosity at certain shear stress demonstrating the ability to be injectable. Following injection, the hydrogel recovers to its gel form to solidify in the defect cavity.[77] Others have demonstrated the three-dimensional encapsulation of biologically active molecules such as bone morphogenetic protein-2 (BMP-2)[78] and fibroblast growth factor (bFGF)[79] by mixing the suspension of these molecules in aqueous solution of amphiphilic peptide that undergo self-assembly to form a mechanically stable hydrogel.

Water-insoluble biodegradable polymers have also been injected in solutions with water-miscible, physiologically compatible solvents to show self-assembly *via* phase segregation. Following injection, the solvent diffuses into the tissue space and water diffuses to the polymer matrix. This results in the precipitation of water-insoluble polymer into a matrix at the injection site. The solvent systems that have been reported based on this approach are propylene glycol, acetone, 2-pyrrolidone, tetrahydrofuran, N-methyl-2-pyrrolidone, and dimethyl sulfoxide.[80] The rate of precipitation of the polymer depends upon many factors such as the concentration of the polymer in solvent, the molecular weight of the polymer, the solvent used, and the addition of a surfactant.[81]

14.3.5 Ion-mediated Gelation Systems (IGS)

IGS rely on the ability of di- or trivalent cations to form ionic interchain bridges between the polymeric chains. Alginate is the most wildly used polymer that has ability to cross-link *via* calcium or zinc cations. Structurally, alginate is a linear polysaccharide composed of homopolymeric blocks of 1,4-linked b-D-mannuronic (M) acid and a-L-guluronic (G) acid residues in various proportions and sequential arrangements.[82] These di- or trivalent cations have been reported to cross-link through different sites of the alginate chain.[83] However, calcium ions are more selective in the cross-linking ability through the polyguluronic acid block (GG) in a planar geometry. On the other hand, zinc cations are reported to be less selective for the cross-linking sites resulting in more extensive cross-linked alginate hydrogels.[84] The rate and kinetics of cross-linking are highly influenced by the concentration of multivalent cations and G-block segment sequences. However, higher concentrations of the alginate polymer chains also lead to a decreased cross-linking rate.[85] Table 14.2 provides a summary.

14.4 Injectable Ceramics

Natural bone is made of 60% of inorganic calcium phosphate minerals.[15,24] To this end, many researchers have developed synthetic bone substitutes based upon ceramics to better mimic the natural composition of bone. Ceramics have been widely used for orthopedic and dental applications, and have been used

Table 14.2 Biodegradable injectable system properties.

Network Formation	Material	Application	Reference
Ceramic setting	Calcium phosphate	Bone substitute	86–89
Free radical polymerization	Alginate	Cell encapsulation	90–92
	Chitosan	Ligament tissue	93–95
	Hyaluronic acid	Cartilage tissue	96–98
	Poly(ethylene glycol) based	Bone tissue	40,99–101
	Poly(L-lactide) based	Bone substitute	16,102,103
	Poly(vinyl alcohol)	Wound tissue	104–105
	Poly(propylene fumarates)	Bone tissue	29,31,39,52, 54,106–109
	Poly(alkylene maleate citrates)	Cell delivery Drug delivery Bone substitute	113,110
Chemical cross-linked systems	Chitosan	Cartilage tissue	60
	Dextran	Tissue engineering Protein delivery	63
	Gelatin	Cell delivery	61
	Hyaluronic acid	Protein delivery	62
	Poly(ethylene glycol) based	Bone tissue	50,58
Thermally induced gelation systems	Chitosan	Cardiac tissue Neural tissue Bone tissue	61,73,111,112
	Poloxamers or Pluronics	Lung tissue Bone tissue	71,113,114
	MPEG-PCL		66
	MPEG-b-(PCL-ran-PLLA)		70
	PEG-co-poly(a-hydroxy acid)	Drug delivery Protein delivery Cell delivery	67–69
Self-assembly systems	RGD-based fibers	Bone tissue	115
	Poly(DL-lactide-co-caprolactone)		80
	Peptide-based hydrogel	Bone tissue	116,117
Ion-mediated gelation systems	Alginate	Bone tissue	38,118 120

dating as far back as 1892 in the development of plaster of Paris ($CaSO_4$). The currently available bone cements, which are both injectable and biodegradable, can be organized into three categories: calcium phosphate cements (CPCs), bioglass, and bioactive glass cements.[121,122]

CPCs are widely used as bone substitutes and for augmentation in orthopedic applications due to their close resemblance to the mineral component of natural bone.[90] CPCs are a powder phase of calcium and/or phosphate salts, which set at body temperature when mixed in an aqueous phase. CPCs such as apatites including hydroxyapatite (HA), carbonated apatite (CA), and calcium-deficient hydroxyapatite (CDHA) can be further categorized as apatite or brushite cements depending on their rate of resorption.[121,123] Apatite cements have higher mechanical strength but slower degradation rate than the brushite-based cements.[121] In order to use CPCs in bone tissue engineering applications, it is imperative to incorporate macroporosity in the material to allow for tissue and new blood vessel formation. Recently, many groups have used mannitol,[124] sodium bicarbonate,[125] and albumin[126] as porogens to incorporate interconnected macropores in CPCs.

Bioglass is a member of the family of bioactive glass composed of silica (SiO_2), sodium oxide (Na_2O), calcium phosphate (CaP), and phosphoric anhydride (P_2O_5). Bioglass-like CPCs have excellent biocompatibility and have shown the ability to grow apatite layers on the surface resulting in better bone-implant integration. In addition, bioglass also promotes osteoblast attachment, proliferation, differentiation, enzymatic activity, and angiogenesis. Studies have shown that the degradation products of 45S5® Bioglass up-regulates the gene expression that controls osteogenesis and production of growth factors.[127] The degradation rates of bioglass are tailored to meet specific bone tissue engineering applications by changing the composition and processing environment. However, the mechanical properties of bioglass are plagued by low fracture toughness and poor mechanical strength, which makes it unsuitable for use in load-bearing applications.[122,123] 45S5® Bioglass is clinically available as Perioglas™ for treating periodontal disease, and Novabone™ as a filler for treating bone defects.[24] Bioglass ceramics such as apatite/wollastonite (A/W) glass exhibit better mechanical properties than the parent glass and sintered crystalline ceramics.[24,122] A/W glass have higher mechanical strength, excellent biocompatibility, and are used as fillers for bone defects caused by iliac grafts and as artificial vertebrae. A/W glass ceramics also show better bioactivity than sintered hydroxyapatite (HA).[128,129] Table 14.3 is a summary.

14.5 Injectable Cell Vehicles

The incorporation of cells into tissue engineered scaffolds has been based upon the use of two different approaches: (1) surface seeding of cells on prefabricated scaffolds and (2) encapsulation of cells in a 3D scaffold. While the use of surface seeding onto prefabricated scaffolds allows for the design of a precisely controlled porous network using different scaffold fabrication methods such as particulate leaching, gas foaming, and thermally phase-induced separation, scaffolds fabricated through this route cannot be implanted in a minimally invasive manner. On the other hand, the encapsulation of relevant cells into a 3D matrix is an inherently mild process, and can be used for injectable applications where the cells are mixed with the liquid cross-linkable solution and administered in a minimally invasive manner to the desired site *in vivo*. Since the liquid will diffuse and conform to the shape of the defect, the adhesion of scaffold to the tissue is improved when compared to prefabricated scaffolding

Table 14.3 Cement-based injectable system mechanical properties and applications.

Cement material	Compressive strength	Application	Reference
Hydroxyapatite	>400 MPa	Bone filler and prostheses coating materials	24,130
Biosorb (β-TCP)	15–150 MPa	Bone filler	124
Calcibon®	4–7 MPa	Bone substitute material	124
ChronOS Inject	7.5 MPa	Bone remodeling and cyst treatment	124
Bone Source™	26 MPa	Craniotomy cuts and cranial defects	121,131
Norian® SRS	50 MPa	Bone fractures	121,132,133
Cementek	20 MPa	Bone substitute material	123
45S5®Bioglass	~500 MPa	Bone filler, Middle ear prostheses, Periodontal disease	24,134
A/W glass ceramic	1080 MPa	Artificial vertebrae, Bone fillers, Intervertebral discs	129,130

approaches.[111,135] Polymeric networks that have the ability to uptake large quantities of water and demonstrate elastic properties are termed hydrogels, and have been used extensively in tissue engineering applications due to their close resemblance to native tissues. Hydrogels can be broadly classified based upon their nature of origin such as natural hydrogels and synthetic hydrogels. The following section summarizes the ongoing research on cell encapsulation for bone tissue engineering using injectable and biodegradable hydrogels.

14.5.1 Naturally Derived Hydrogels

Chitosan, alginate, and hyaluronic acid are all polysaccharide-based hydrogels similar to native ECM. Chitosan is a natural biopolymer with a striking resemblance to mammalian glycosaminoglycans. Alginate is non-mammalian polysaccharide that can be cross-linked under mild conditions with low toxicity. Alginate beads have been used to encapsulate MC3T-E1 osteoblasts, which were then mixed with calcium phosphate cement and chitosan-calcium phosphate cements. The alginate beads were found to improve cell viability significantly by protecting the cells from the cement hardening reaction.[136,137] Alginate gels encapsulated with murine embryonic stem cells (mESCs) were cultured in a rotary cell culture microgravity bioreactor. The resultant 3D mineralized constructs were found to have attributes of osteogenic lineage as well as the mechanical strength and mineralized Ca/P deposition.[138] Calcium alginate core of mineralized alginate/chitosan capsules were used to encapsulate human osteoprogenitor cells (STRO-1$^+$) and at the end of 7 days of *in vitro* subculture indicated the maintenance of osteoblastic phenotype. New bone formation with type I collagen matrix was seen when these polysaccharide capsules encapsulated with human bone marrow cells and rhBMP-2 were implanted in nude mice.[139] Thermoresponsive hydroxybutyl chitosan (HBC) was evaluated as an injectable therapeutic treatment of degenerated intervertebral discs (IVD). HBC gels were encapsulated with human mesenchymal stem cells (hMSCs), human annulus fibrosus cells (hAFC) and human nucleus pulposus cells (hNPs) and provided a suitable environment for the survival of the disk cells and hMSCs in a metabolically active and proliferative state.[140] Recent strategies involving hyaluronic acid for cell encapsulation for bone tissue engineering applications have involved modifying hyaluronic acid with methacrylates and thiols. *In vivo* studies using acrylated hyaluronic acid injected into rat calvarial (skull) defects showed that human mesenchymal stem cells (hMSCs) in the presence of bone morphogenetic factor-2 (BMP-2) demonstrated the ability to differentiate into specific cells such as endothelial and osteoblast cells.[101,137]

14.5.2 Synthetic-based Hydrogels

Poly(ethylene glycol) (PEG)-based hydrogels have been extensively studied as cell encapsulating networks for bone tissue engineering applications. Although PEG based hydrogels are highly hydrophilic, and have shown to resist protein adsorption and cell adhesion, studies have shown that osteoblasts and chondrocytes can survive in such hydrophilic conditions without any added biological cues.[111,141,142] Poly(ethylene glycol) diacrylate) (PEGDA) was used to encapsulate hMSCs, and facilitate differentiation into osteoblasts in the presence of osteogenic differentiation media consisting of ascorbic acid, dexamethasone, and β-glycerophosphate.[40,143,144] PEG-based hydrogels were modified by using RGD peptide sequence to improve osteoblast attachment, proliferation, and differentiation.[145] When RGD-modified PEGDA hydrogels were encapsulated with bone marrow stromal stem cells (bMSCs), the osteogenic activity of the cells improved and peaked at an optimum peptide concentration.[146] The addition of phosphate-containing molecule methacrylate phosphate (EGMP) improved hMSCs adhesion by promoting the mineralization within the hydrogel.[100]

The degradation profile of PEG hydrogels can be controlled by the addition of degradable linkages such as poly(α-hydroxy esters) or peptides that can be cleaved enzymatically.[101–141] Osteoblasts encapsulated in hydrogels through the copolymerization of poly(lactic acid)-b-poly(lactic acid) with PEGDA showed elevated ECM production in terms of osteopontin, type I collagen, and calcium phosphatase deposition.[101,142,147]

A new class of synthetic injectable and biodegradable hydrogels using fumaric acid has been developed, which includes poly(propylene fumarate) (PPF), poly(propylene fumarate-co-ethylene glycol) (Poly(PF-co-EG)), and oligo(poly(ethylene glycol) fumarate) (OPF).[29,148–152] Poly(alkyl fumarates) can be easily cross-linked with itself or in the presence of a cross-linking agent *in situ* to form a degradable polymeric network. The *in vitro* osteogenic differentiation of MSCs encapsulated in OPF gels (PEGDA mol. wt. 3 K and 10 K) in the presence of dexamethasone showed increased calcium content and

Table 14.4 Biodegradable injectable cell delivery vehicles for bone tissue engineering.

Material	Cell	Reference
Alginate beads	MC3T-E1 osteoblasts	136,137
Alginate gels	Murine embryonic stem cells	138
Alginate/chitosan capsules	Human osteoprogenitor cells (STRO-1$^+$)	139
	C2C12 myoblasts	139
	Bone marrow stromal cells	139
	Adipocytes	100
EGMP-containing PEGDA	hMSCs	100
Hydroxylbutyl chitosan gels	hMSCs	140
	hNPCs	140
	hAFCs	140
Hyaluronic acid gels	hMSCs	101,137
OPF/PEGDA	Rat MSCs	152
OPF	Rat bMSCs	153
PEGDA	hMSCs	143
PEGDA modified with RGD	Osteoblasts	145
PEGDA modified with poly(α-hydroxy ester)	Osteoblasts	101,142, 147
PPF	MSCs	154,155
PhosPEG	MSCs	156–159

osteopontin production with increased swelling behavior.[60,160] Rat bMSCs encapsulated in RGD-modified OPF gels supported osteogenic differentiation in the absence of any supplements (dexamethasone and β-glycerol phosphate).[153] PPF has also been investigated as a cell carrier in bone regeneration applications. Initial cell viability was increased when MSCs were encapsulated in gelatin microcapsules before adding them to the cross-linking PPF.[154,155]

Phosphoester hydrogels are photopolymerizable phosphate-containing PEG hydrogels (PhosPEG) that were designed to undergo degradation *via* hydrolysis at the phosphoester linkage. PhosPEG hydrogels were photo-cross-linked from the macromer precursor of poly(ethylene glycol)-di-[ethylphosphatidyl (ethylene glycol) methacrylate]. The presence of alkaline phosphatase (ALP) enhanced degradation by cleaving the phosphoester groups in the PhosPEG network, demonstrating enzymatic degradation. The by-products of the enzymatic degradation react with the calcium ion in the media promoting autocalcification and promoted osteogenic differentiation of encapsulated MSCs.[101,156–158] The incorporation of hydroxyapatite (HA) in thermosensitive poly(isopropylacrylamide-co-acrylic acid) (p(NiPAAm) with rabbit MSCs and bone morphogenic factor (BMP-2) showed increased osteogenic differentiation and extracellular matrix production.[159] Table 14.4 is a summary.

14.6 Injectable Drug Delivery Systems

Many of the previous injectable systems for bone tissue engineering have mainly relied on the material chemistry and physical properties to control cellular adhesion, proliferation, and differentiation.[5,159,161] However, bone regeneration is a complex process governed by an intricate interplay between various growth factors and cytokines to guide the healing process.[162] As this complex cascade of biological events for bone regeneration is further understood, new therapeutics and therapeutic release strategies are continually emerging towards the development of an ideal injectable system. Preferably, the administration of bioactive molecules or drugs from the delivery system should be precisely controlled to provide the appropriate dose over the therapeutic time frame to match the dynamic physiological needs of the regenerating tissue. Due to the hydrolytically unstable ester,[163,164] ether-ester,[165,166] anhydride,[167] or amide functional groups,[162,168] biodegradable polymers have been extensively researched in the area of controlled delivery of bioactive molecules and drugs.[169] Detailed reviews highlighting the concepts behind drug delivery for bone tissue engineering are available,[170–172] and the following section will briefly discuss the localized delivery of antibiotics and growth factors from injectable delivery systems.

14.6.1 Antibiotic Delivery

Infections associated with implanted devices are a significant challenge in the field of orthopedic tissue engineering.[173] Osteomyelitis is a deep bone infection caused by staphylococci and it often leads to bone loss and the spread of bacterial infection to the surrounding tissues. Treatment of this type of infection has proven difficult due to the short half-life of antibiotics, inadequate blood circulation to the infected area, and the systemic toxicity of the antibiotic, which limits the use of high systemic dosages.[174] Since the overall success of implanted materials is largely dependent on the prevention of bacterial in-growth to the defect site, many groups have focused on the localized delivery of antibiotics through injectable cement and polymer delivery systems. Early strategies to treat osteomyelitis have been researched since the mid-1990s, and relied on local antibiotic treatment through the release of antibiotics from non-biodegradable poly(methyl-methacrylate) (PMMA) cement carriers.[175] However, a major disadvantage to this approach is the non-biodegradable nature of the drug delivery vehicle, which requires a second surgery to remove the PMMA beads. Recent efforts to treat orthopedic infections have now moved to the use of injectable biodegradable cements and polymer systems.

Many of the injectable cement-based systems used in the treatment of orthopedic infections, such as calcium phosphate cements (CPCs),[176] β-dicalcium silicate (β-Ca_2SiO_4),[177] and hydroxyapatite cements (HACs),[178] have relied on the local delivery of antibiotics such as gentamicin or cephalexin monohydrate to increase the antibacterial activity against *E. coli* and *S. aureus* strains *in vitro*. A study led by Joosten *et al.*[178] evaluated the effects of gentamicin release from a HAC both *in vitro* and *in vivo*. Bone infections were induced into the right tibia of 'New Zealand' rabbits, and treated with HAC-loaded gentamicin. The *in vivo* results confirmed that no histopathological evidence of infection was found for the HAC/gentamicin-treated animals, whereas different stages of chronic osteomyelitis were found in all control groups.

Although antibiotics have been incorporated into many commercially available types of cement, the high curing temperatures required and poor release kinetics of the antibiotic have driven researchers to develop other delivery vehicles.[172] Peng et al.[179] have recently developed a novel thermosensitive implant composed of poly(ethylene glycol) monomethyl ether (mPEG) and poly(lactide-co-glycolic acid) (PLGA) copolymer (mPEG-PLGA) drug delivery system. The thermosensitive behavior of this system allows for the efficient loading of the drug or bioactive molecule without the use of harsh conditions, which can cause denaturing, aggregation, and undesirable chemical reactions. Similar to the control teicoplanin-loaded PMMA bone cements, the mPEG-PLGA hydrogel containing teicoplanin was effective in treating osteomyelitis in rabbits *in vivo*. Other strategies to treat osteomyelitis have been evaluated using poly(sebacic-co-ricinoleic-ester-anhydride) containing gentamicin, which increases in viscosity and becomes a semisolid gel when exposed to an aqueous environment.[175,180] Published studies have indicated a positive effect on established osteomyelitis in a rat model; however, these studies did not show complete eradication of the infection.[180]

14.6.2 Growth Factor Delivery for Osteogenesis

Growth factors are signaling polypeptides that bind to specific receptors of the target cell and are crucial in controlling important cellular functions.[162] After a bone fracture, the locally produced cytokines and growth factors direct the migration, adhesion, proliferation, and differentiation of osteoprogenitor cells into specific lineages as well as extracellular matrix production at the defect site.[181,182] Unfortunately, growth factors are plagued with a relatively short biological half-life, and a major challenge has been to design a delivery system which can administer a prolonged sustained release to maintain the activity of the growth factors.[170] Many osteoinductive growth factors have been identified such as fibroblast growth factors (FGFs),[183] insulin-like growth factors (IGFs),[184] epidermal growth factors (EGFs),[185] and platelet-derived growth factors (PDGFs). However, a majority of the growth factors utilized in injectable systems for orthopedic applications have been primarily been devoted towards the development of new bone formation through the use of the transforming growth factor beta superfamily.[163,186–188]

A number of delivery systems have been developed for the controlled release of bone morphogenic protein-2 (BMP-2). For example, calcium phosphate cement-based materials have shown the ability to deliver recombinant human bone morphogenic protein-2 (rhBMP-2) to increase alkaline phosphatase (ALP) activity in MC3T3-E1 cells *in vitro*,[189] and enhance bone formation both ectopically[190] and in an ulna osteotomy model.[88] In addition to cement-based systems, polymeric materials have also been heavily researched as potential delivery vehicles of BMP-2. Saito et al. have developed a temperature sensitive poly(D,L-lactic acid-polyethylene glycol) (PLA-PEG) block copolymer as an injectable delivery system for rhBMP-2, and was able to form new bone on the surface of murine femur 3 weeks after injection.[191] Hosseinkhani et al. have reported on the controlled release of BMP-2 using a novel injectable peptide amphiphile (PA) system, which has the ability to form a three-dimensional nanofibrous scaffold by mixing the PA aqueous solution with the BMP-2 suspension.[192] This system was able to induce significant increase in homogenous ectopic bone formation subcutaneously in the back of rats when compared to BMP-2 injection alone.

Delivery methods based on microparticle and nanoparticle designs have gained increased attention in the delivery of growth factors to induce osteogenesis with smaller amounts of BMP-2 and with improved release over more sustained times. Magnetic liposomes,[193] collagen minipellets,[194] cationic nanoparticles,[195] and poly-ε-caprolactone microparticles[196] have all been reported as vehicles with a more uniform release of growth factors to increase osteogenesis. Research groups have also investigated the controlled release of an osteogenic peptide, TP508, loaded poly(D,L-lactic-co-glycolic acid) (PLGA) microparticles added to a mixture of poly(propylene fumarate) (PPF). The PLGA TP508 loaded microparticles showed release of the osteogenic factor for up to 28 days, and also serve as a sacrificial porogen in the PPF matrix once degraded.[197] Radiographic, microtomograph, and histological results all confirmed that the PLGA/PPF system was shown to enhance the bone consolidation process in a rabbit model of distraction osteogenesis when compared to TP508 saline solutions and dextran only.[198] Table 14.5 shows selective experimental results of orthopedic therapeutic carrier systems.

Table 14.5 Selective experimental results of orthopedic therapeutic carrier systems.

Carrier material	Therapeutic agent	Matrix type	Cell/animal model	Reference
β-Dicalcium silicate	Gentamicin	Cement	L929 cells	178
Calcium alginate	hIGF-1	Hydrogel	Goat (meniscus)	199
Calcium phosphate	Bisphosphonate	Cement	Rat (femur)	200
	bFGF	Cement		201
	BMP-2	Cement		201
	Cephalexin	Cement	S. aureus	176
	Gentamicin	Cement	S. aureus	178,202
	rhBMP-2	Cement	Primate (fibular)	203
	rhBMP-2	Cement	Primate (vertabrate)	204
	rhBMP-2	Cement	Rabbit (radius)	205
	rhBMP-2	Cement	Rabbit (subcutaneous)	190
	rhBMP-2	Cement	Rabbit (ulnar)	88
	Salmon-calcitonin	Cement	Rat (abdomen)	206
	TGF-beta 1	Cement		201,207

Table 14.5 (*Continued*)

Carrier material	Therapeutic agent	Matrix type	Cell/animal model	Reference
Cationic nanoparticles	OP-1	Nanoparticle	Rat (intramuscular)	195
Chitosan/alginate	BMP-2	Hydrogel	Mouse (subcutaneous)	208
Chitosan/inorganic phosphates	BMP-2	Composite	Rat (calvarial)	209
Cholesterol-bearing pullulan	W9-peptide	Nanogel	Mouse (subcutaneous)	210
E-matrix	rhBMP-2	Scaffold	Rat (spinal)	211
Elastin-like polypeptides	Vancomycin	Hydrogel		173
Hydroxyapatite	Gentamicin	Cement	Rabbit (tibia)	178
Magnetic liposome	rhBMP-2	Nanoparticle	Rat (femur)	193
mPEG-PLGA	Teicoplanin	Nanoparticle	Rabbit (femur)	179
N-isopropylacrylamide		Hydrogel	Rat (intramuscular)	212
Nanobone putty	hBMP-2	Putty	Mouse (intramuscular)	213
Nanohydroxyapatite	Amoxicillin	Microsphere	MG63 cells	214
Oligo(poly(ethylene glycol) fumarate)	TGF-beta 1	Hydrogel	Mesenchymal stem cells	54
Peptide amphiphile	BMP-2	Scaffold	Rat (subcutaneous)	79
PLA-PEG	rhBMP-2	Hydrogel		191
PLGA	Dexamethasone	Nanoparticle	Rat (cranial)	215
	rhGDF-5	Composite		216
	Vancomycin	Microparticle		217
PLGA/CaP	TGF-beta 1	Microsphere	Rat (skull)	218
PLGA/hydroxyapatite	Alendronate	Microsphere	hFOB	219
PLGA-mPEG	Teicoplanin	Hydrogel	Rabbit	179
Poly(NiPAAm-co-AAc)/HA	BMP-2	Composite	MSC	158
Poly(propylene fumarate)/calcium phosphate	Ginsenoside Rg1	Cement	HUVEC	220
Poly(sebacic-co-ricinoleic-ester-anhydride)	Gentamicin	Hydrogel	Rat (tibia)	174,180
Polyurethane	PDGF	Scaffold	MC3T3-E1 Cells	221
	Tobramycin	Scaffold	*S. aureus*	222
Starch-poly-ε-caprolactone	BMP-2	Microparticle	C2C12 Cells	86,196
Tricalcium phosphate	Platelet-rich plasma	Composite	Goat (tibia)	86
Tricalcium phosphate/alginate	IGF-1	Composite	MG-63 and Saos-2	223
Tricalcium phosphate/collagen	rh-PDGF	Microparticle	Rat (tibia)	224

14.7 Citric Acid-based Systems

In 2004, Yang et al. synthesized the first citric acid-based biomaterial through a convenient polycondensation reaction between citric acid and 1,8-octanediol to create poly(octamethylene citrate) (POC).[225] The resulting biodegradable, soft, and elastic material was shown to cover a wide range of mechanical properties, degradation profiles, and surface energies, which are all important in controlling the biological response to an implanted material. The excellent biocompatibility, hemocompatible nature, and tunable mechanical properties of POC drove Yang et al. to utilize the material primarily for small diameter vascular grafts[226] and medical device coatings.[227] Qiu et al. later proposed to combine POC and HA to create a composite (POC-HA) that would have the desired characteristics of a bioceramic suitable for orthopedic tissue engineering.[228] Bone screws fabricated from POC-HA composites displayed improved processability, mechanical properties, and degradation kinetics over previous biodegradable composites. However, the previous design required harsh processing conditions (>120 °C) for polymer network formation rendering them unable to be used in injectable strategies.

To overcome this limitation, our lab has recently developed a new family of *in situ* cross-linkable citric acid-based polymers, which can be cross-linked through free radical polymerization methods to avoid the use of harsh processing conditions required by the previous design. In this system, citric acid, maleic anhydride or maleic acid, and 1,8-octanediol were reacted together in a convenient polycondensation reaction to produce a biodegradable elastomer, poly(alkylene maleate citrates) (PAMC), which could be cross-linked using UV irradiation or redox systems to form a cross-linked network.[229] Maleic anhydride[230] and maleic acid[110] were both used to introduce a vinyl moiety in order to allow for network formation under mild conditions. Unlike the previous citric acid-based designs, this additional cross-linking method allowed for the preservation of valuable citric acid carboxylic acid and hydroxyl chemistries, which could be later used to conjugate bioactive molecules into the bulk material to control cell behavior.[231] To ensure that cells and sensitive drugs/factors could be incorporated and delivered to the injury site, poly(ethylene glycol) and acrylic acid were introduced into the system to create poly(ethylene glycol) maleate citrate (PEGMC), which allowed for water solubility and faster network formation kinetics.[13] The encapsulation of NIH 3T3 fibroblasts and human dermal fibroblasts showed the cytocompatibility of PEGMC and the controlled drug release using bovine serum albumin demonstrated PEGMC potential as a suitable cell and drug delivery vehicle.

To widen the application of PEGMC, our lab set out to develop an injectable, porous, and strong citric acid based-composite, which could be used as a delivery vehicle for cells and drugs in bone tissue engineering applications. PEGMC was combined with various wt.-% of HA to create PEGMC/HA composites.

The rationale behind this biomaterial design are:

(1) citric acid was chosen as a multi-functional monomer, which could participate in pre-polymer formation using a convenient polycondensation reaction while preserving valuable pendant functionalities
(2) to create a completely water-soluble material, which was injectable and provided a suitable environment for the delivery of sensitive cells/molecules, PEG was chosen a di-functional diol
(3) maleic anhydride introduced a vinyl moiety into the polymer backbone, which allowed for network formation using free radical polymerization to avoid the harsh processing conditions of previous designs
(4) to improve the osteointegration capacity and mechanical properties, HA was incorporated as a composite blend
(5) the pendant carboxylic acid chemistries can react with bicarbonates to induce gas foaming and create an injectable porous material.

This new generation of biodegradable citric acid-based elastomer composite offers many advantages over other injectable cell and drug delivery systems in that the valuable pendant chemistries are preserved during network formation, mechanical strength and osteoconductivity are improved using HA, mild conditions are utilized for network formation to enable the delivery of sensitive cells/biomolecules, and a porous construct can be created after delivery using minimally invasive procedures.

The degradation profiles for PEGMC/HA networks showed increasing mass loss with lower concentration of HA. Mechanical compressive tests showed that the PEGMC/HA networks were elastic and achieved complete recovery without any permanent deformation for hydrated and non-hydrated conditions. Human fetal osteoblast (hFOB 1.19) encapsulated in PEGMC/HA hydrogel composites showed that the cells were viable and functional at the end of 21 days of subculture (Figure 14.4A). ECM production was measured for alkaline phosphatase and calcium content, and both were shown to increase after 3 weeks of culture. SEM/EDX analysis of the constructs showed that the PEGMC/HA films were covered with small cauliflower shaped structures after 7 days of incubation in simulated body fluid (Figure 14.4B). The presence of pendant groups in the PEGMC polymer allows for easy modification through the bioconjugation of biological molecules such as type I bovine collagen, and resulted in enhanced cellular attachment and proliferation at the end of day 7 of subculture. Unlike many injectable systems, PEGMC/HA composites could also be fabricated into highly porous architectures from gas foaming techniques *in situ* (Figure 14.4C). Thus, unlike previous injectable materials, PEGMC/HA composites show great potential as an injectable, porous, and strong cell/drug delivery system for orthopedic applications.

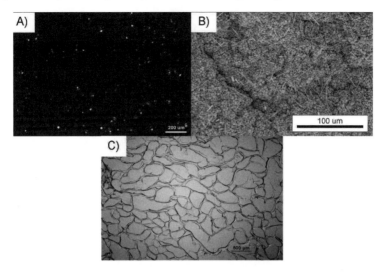

Figure 14.4 (A) Live/dead stain of hFOB 1.19 osteoblasts encapsulated in PEGMC/HA hydrogel after 7 days. (B) Mineralization in SBF for PEGMC/HA composite with 40 wt.% HA at 7 days. (C) 10 μm section of PEGMC scaffold showing the porosity created from a gas foaming technique.

14.8 Future Directions

The use of injectable systems in orthopedic tissue engineering is still in its infancy, and continued advances in biomaterial development and design are required to realize the goal of applying injectable strategies to bone regeneration. Although recent success has been demonstrated in delivering cell and therapeutic agents using injectable-based designs, more studies using synthetic polymer composites to improve the construct mechanical properties while maintaining proper degradation kinetics to match the stringent requirements for bone tissue engineering will be the focus of future studies. In addition to mechanical compliance, research focused on the use of sacrificial porogens to deliver drugs and introduce porosity to promote cell infiltration and the establishment of a vascular network will continue to dominate the future investigations. Thus, as new materials are continually introduced to the field, the growth of knowledge in designing constructs with improved mechanical properties, porosities, and angiogenesis will bring the field closer, developing clinically relevant orthopedic tissues using biodegradable injectable systems.

14.9 Conclusions

The context of this chapter aims to discuss the most recent advances in the use of injectable biodegradable materials for bone tissue engineering. The current clinical need, design criteria, and material property requirements were illustrated followed by an overview of the latest material, cellular, and drug delivery technologies through the use of injectable systems. Finally, the major roadblocks pertaining to the field and the future perspectives to address the current challenges were described. The ability to design injectable systems shows huge potential for the regeneration of damaged orthopedic tissues through minimally invasive procedures. While the initial studies are encouraging and many injectable materials have shown great promise, the regeneration of mechanically compliant and porous constructs with a vascular supply remains a challenge. The precisely controlled and cooperative interaction between the scaffold material, architecture, therapeutics, and cells is imperative to fully regenerate biologically functional engineered bone. The continued advancement in material chemistry and a greater understanding of cell–matrix interactions, metabolic transport, and the cellular events involved in the body's natural healing response will be significant steps in the translation of tissue engineering research into clinical reality.

Acknowledgments

This work was supported in part by an award R21EB009795 from the National Institute of Biomedical Imaging and Bioengineering (NIBIB), and a National Science Foundation (NSF) CAREER award 0954109.

References

1. G. Bran, J. Stern-Straeter, K. Hörmann, F. Riedel and U. Goessler, *Arch. Med. Res.*, 2008, **39**, 467.
2. M. Kofron and C. Laurencin, *Adv. Drug Deliv. Rev.*, 2006, **58**, 555.
3. E. Arrington, W. Smith, H. Chambers, A. Bucknell and N. Davino, *Clin. Orthopaedics Rel. Res.*, 1996, **329**, 300.
4. J. Kretlow and A. Mikos, *Tissue Engineering*, 2007, **13**, 927.
5. M. Ngiam, S. Liao, A. J. Patil, Z. Cheng, C. K. Chan and S. Ramakrishna, *Bone*, 2009, **45**, 4.
6. R. Langer and J. Vacanti, *Science*, 1993, **260**, 920.
7. M. Hiles and J. Hodde, *Int. Urogynecol. J. Pelvic Floor Dysfunct.*, 2006, **17** (Suppl 1), S39.
8. E. Rabkin and F. J. Schoen, *Cardiovasc. Pathol.*, 2002, **11**, 305.
9. J. P. Vacanti and R. Langer, *Lancet*, 1999, **354** (Suppl 1), SI32.
10. J. Temenoff and A. Mikos, *Biomaterials*, 2000, **21**, 2405.
11. J. Kretlow, L. Klouda and A. Mikos, *Adv. Drug Deliv. Rev.*, 2007, **59**, 263.
12. J. L. Ifkovits and J. A. Burdick, *Tissue Eng.*, 2007, **13**, 2369.
13. D. Gyawali, P. Nair, Y. Zhang, R. Tran, M. Samchukov, M. Makarov, H. Kim and J. Yang, *Biomaterials*, 2010, **31**, 9092.
14. Y. Wang, G. A. Ameer, B. J. Sheppard and R. Langer, *Nat. Biotechnol.*, 2002, **20**, 602.
15. K. Athanasiou, C. Zhu, D. Lanctot, C. Agrawal and X. Wang, *Tissue Engineering*, 2000, **6**, 361.

16. B. G. Amsden, G. Misra, F. Gu and H. M. Younes, *Biomacromolecules*, 2004, **5**, 2479.
17. J. R. Porter, T. T. Ruckh and K. C. Popat, *Biotechnol. Prog.*, 2009, **25**, 1539.
18. J. Dey, H. Xu, J. Shen, P. Thevenot, S. R. Gondi, K. T. Nguyen, B. S. Sumerlin, L. Tang and J. Yang, *Biomaterials*, 2008, **29**, 4637.
19. R. Tran, P. Thevenot, Y. Zhang, L. Tang and J. Yang, *Materials*, 2010, **3**, 1375.
20. C. J. Bettinger, J. P. Bruggeman, J. T. Borenstein and R. S. Langer, *Biomaterials*, 2008, **29**, 2315.
21. S. C. Cowin, *J. Biomech.*, 1999, **32**, 217.
22. N. X. Chen, D. J. Geist, D. C. Genetos, F. M. Pavalko and R. L. Duncan, *Bone*, 2003, **33**, 399.
23. C. D. Toma, S. Ashkar, M. L. Gray, J. L. Schaffer and L. C. Gerstenfeld, *J. Bone Miner. Res.*, 1997, **12**, 1626.
24. K. Rezwan, Q. Chen, J. Blaker and A. Boccaccini, *Biomaterials*, 2006, **27**, 3413.
25. M. Dadsetan, T. Hefferan, J. Szatkowski, P. Mishra, S. Macura, L. Lu and M. Yaszemski, *Biomaterials*, 2008, **29**, 2193.
26. D. Logeart-Avramoglou, F. Anagnostou, R. Bizios and H. Petite, *J. Cell. Mol. Med.*, 2005, **9**, 72.
27. J. R. Woodard, A. J. Hilldore, S. K. Lan, C. J. Park, A. W. Morgan, J. A. Eurell, S. G. Clark, M. B. Wheeler, R. D. Jamison and A. J. Wagoner Johnson, *Biomaterials*, 2007, **28**, 45.
28. J. Holmbom, A. Sodergard, E. Ekholm, M. Martson, A. Kuusilehto, P. Saukko and R. Penttinen, *J. Biomed. Mater. Res. A*, 2005, **75**, 308.
29. E. Behravesh, S. Jo, K. Zygourakis and A. G. Mikos, *Biomacromolecules*, 2002, **3**, 374.
30. M. D. Krebs, K. A. Sutter, A. S. Lin, R. E. Guldberg and E. Alsberg, *Acta Biomater.*, 2009, **5**, 2847.
31. P. A. Gunatillake and R. Adhikari, *Eur. Cell. Mater*, 2003, **5**, 1.
32. S. Sakai, K. Hirose, K. Taguchi, Y. Ogushi and K. Kawakami, *Biomaterials*, 2009, **30**, 3371.
33. Y. Tokiwa and B. P. Calabia, *Appl. Microbiol. Biotechnol.*, 2006, **72**, 244.
34. G. C. Steffens, L. Nothdurft, G. Buse, H. Thissen, H. Hocker and D. Klee, *Biomaterials*, 2002, **23**, 3523.
35. E. M. Christenson, K. S. Anseth, J. J. van den Beucken, C. K. Chan, B. Ercan, J. A. Jansen, C. T. Laurencin, W. J. Li, R. Murugan, L. S. Nair, S. Ramakrishna, R. S. Tuan, T. J. Webster and A. G. Mikos, *J. Orthop. Res.*, 2007, **25**, 11.
36. C. D. Reyes, T. A. Petrie, K. L. Burns, Z. Schwartz and A. J. Garcia, *Biomaterials*, 2007, **28**, 3228.
37. R. A. Stile and K. E. Healy, *Biomacromolecules*, 2001, **2**, 185.
38. E. Alsberg, K. W. Anderson, A. Albeiruti, R. T. Franceschi and D. J. Mooney, *J. Dent. Res.*, 2001, **80**, 2025.

39. E. Behravesh, K. Zygourakis and A. G. Mikos, *J. Biomed. Mater. Res. A*, 2003, **65**, 260.
40. J. Burdick and K. Anseth, *Biomaterials*, 2002, **23**, 4315.
41. A. Talal, N. Waheed, M. Al-Masri, I. J. McKay, K. E. Tanner and F. J. Hughes, *J. Dent.*, 2009, **37**, 820.
42. D. Xue, Q. Zheng, C. Zong, Q. Li, H. Li, S. Qian, B. Zhang, L. Yu and Z. Pan, *J. Biomed. Mater. Res. A*, 2010, **94**, 259.
43. H. Nie, B. W. Soh, Y. C. Fu and C. H. Wang, *Biotechnol. Bioeng.*, 2008, **99**, 223.
44. Y. C. Fu, H. Nie, M. L. Ho, C. K. Wang and C. H. Wang, *Biotechnol. Bioeng.*, 2008, **99**, 996.
45. M. Kato, H. Toyoda, T. Namikawa, M. Hoshino, H. Terai, S. Miyamoto and K. Takaoka, *Biomaterials*, 2006, **27**, 2035.
46. L. C. Yeh and J. C. Lee, *Biochim. Biophys. Acta*, 2006, **1763**, 57.
47. R. R. Chen, E. A. Silva, W. W. Yuen and D. J. Mooney, *Pharm. Res.*, 2007, **24**, 258.
48. G. Schmidmaier, B. Wildemann, T. Gabelein, J. Heeger, F. Kandziora, N. P. Haas and M. Raschke, *Acta. Orthop. Scand.*, 2003, **74**, 604.
49. J. M. Kanczler and R. O. Oreffo, *Eur. Cell. Mater.*, 2008, **15**, 100.
50. M. P. Lutolf and J. A. Hubbell, *Biomacromolecules*, 2003, **4**, 713.
51. H. Hall, T. Baechi and J. A. Hubbell, *Microvasc. Res.*, 2001, **62**, 315.
52. L. J. Suggs, M. S. Shive, C. A. Garcia, J. M. Anderson and A. G. Mikos, *J. Biomed. Mater Res.*, 1999, **46**, 22.
53. K. Nguyen and J. West, *Biomaterials*, 2002, **23**, 4307.
54. X. Guo, H. Park, G. Liu, W. Liu, Y. Cao, Y. Tabata, F. K. Kasper and A. G. Mikos, *Biomaterials*, 2009, **30**, 2741.
55. M. D. Timmer, C. G. Ambrose and A. G. Mikos, *J. Biomed. Mater. Res. A*, 2003, **66**, 811.
56. K. A. Davis, J. A. Burdick and K. S. Anseth, *Biomaterials*, 2003, **24**, 2485.
57. I. Strehin, Z. Nahas, K. Arora, T. Nguyen and J. Elisseeff, *Biomaterials*, 2010, **31**, 2788.
58. M. Heggli, N. Tirelli, A. Zisch and J. A. Hubbell, *Bioconjug. Chem.*, 2003, **14**, 967.
59. S. Q. Liu, P. L. Ee, C. Y. Ke, J. L. Hedrick and Y. Y. Yang, *Biomaterials*, 2009, **30**, 1453.
60. R. Jin, L. S. Moreira Teixeira, P. J. Dijkstra, M. Karperien, C. A. van Blitterswijk, Z. Y. Zhong and J. Feijen, *Biomaterials*, 2009, **30**, 2544.
61. L. Wang and J. P. Stegemann, *Biomaterials*, 2010, **31**, 3976.
62. F. Lee, J. E. Chung and M. Kurisawa, *J. Control Release*, 2009, **134**, 186.
63. R. Jin, C. Hiemstra, Z. Zhong and J. Feijen, *Biomaterials*, 2007, **28**, 2791.
64. J. Sperinde and L. Griffith, *Macromolecules*, 1997, **30**, 5255.
65. M. Kim, K. Seo, G. Khang, S. Cho and H. Lee, *J. Polym. Sci. Part A: Polym. Chem.*, 2004, **42**, 5784.
66. M. Kim, H. Hyun, K. Seo, Y. Cho, J. Lee, C. Lee, G. Khang and B. Lee, *J. Polym. Sci. Part A: Polym. Chem.*, 2006, **44**, 5413.

67. B. Jeong, Y. H. Bae, D. S. Lee and S. W. Kim, *Nature*, 1997, **388**, 860.
68. B. Jeong, K. M. Lee, A. Gutowska and Y. H. An, *Biomacromolecules*, 2002, **3**, 865.
69. Y. M. Chung, K. L. Simmons, A. Gutowska and B. Jeong, *Biomacromolecules*, 2002, **3**, 511.
70. Y. M. Kang, S. H. Lee, J. Y. Lee, J. S. Son, B. S. Kim, B. Lee, H. J. Chun, B. H. Min, J. H. Kim and M. S. Kim, *Biomaterials*, 2010, **31**, 2453.
71. J. Cortiella, J. E. Nichols, K. Kojima, L. J. Bonassar, P. Dargon, A. K. Roy, M. P. Vacant, J. A. Niles and C. A. Vacanti, *Tissue Eng.*, 2006, **12**, 1213.
72. S. E. Stabenfeldt, A. J. Garcia and M. C. LaPlaca, *J. Biomed. Mater. Res. A*, 2006, **77**, 718.
73. K. E. Crompton, J. D. Goud, R. V. Bellamkonda, T. R. Gengenbach, D. I. Finkelstein, M. K. Horne and J. S. Forsythe, *Biomaterials*, 2007, **28**, 441.
74. S. Ohya and T. Matsuda, *J. Biomater. Sci. Polym. Ed.*, 2005, **16**, 809.
75. M. C. Branco and J. P. Schneider, *Acta. Biomater.*, 2009, **5**, 817.
76. R. S. Tu and M. Tirrell, *Adv. Drug Deliv. Rev.*, 2004, **56**, 1537.
77. C. T. Wong Po Foo, J. S. Lee, W. Mulyasasmita, A. Parisi-Amon and S. C. Heilshorn, *Proc. Natl. Acad. Sci. U S A*, 2009, **106**, 22067.
78. H. Hosseinkhani, M. Hosseinkhani, A. Khademhosseini and H. Kobayashi, *J. Control Release*, 2007, **117**, 380.
79. H. Hosseinkhani, M. Hosseinkhani, A. Khademhosseini, H. Kobayashi and Y. Tabata, *Biomaterials*, 2006, **27**, 5836.
80. M. A. Royals, S. M. Fujita, G. L. Yewey, J. Rodriguez, P. C. Schultheiss and R. L. Dunn, *J. Biomed. Mater. Res.*, 1999, **45**, 231.
81. A. Hatefi and B. Amsden, *J. Control Release*, 2002, **80**, 9.
82. C. K. Kuo and P. X. Ma, *Biomaterials*, 2001, **22**, 511.
83. L. Chan, Y. Jin and P. Heng, *Int. J. Pharm.*, 2002, **242**, 255.
84. P. Aslani and R. A. Kennedy, *J. Microencapsul.*, 1996, **13**, 601.
85. E. Westhaus and P. B. Messersmith, *Biomaterials*, 2001, **22**, 453.
86. L. Bi, W. Cheng, H. Fan and G. Pei, *Biomaterials*, 2010, **31**, 3201.
87. K. L. Low, S. H. Tan, S. H. Zein, J. A. Roether, V. Mourino and A. R. Boccaccini, *J. Biomed. Mater. Res. B Appl. Biomater.*, 2010, **94**, 273.
88. R. H. Li, M. L. Bouxsein, C. A. Blake, D. D'Augusta, H. Kim, X. J. Li, J. M. Wozney and H. J. Seeherman, *J. Orthop. Res.*, 2003, **21**, 997.
89. H. Xu, M. Weir, E. Burguera and A. Fraser, *Biomaterials*, 2006, **27**, 4279.
90. F. Shen, A. A. Li, R. M. Cornelius, P. Cirone, R. F. Childs, J. L. Brash and P. L. Chang, *J. Biomed. Mater. Res. B Appl. Biomater.*, 2005, **75**, 425.
91. O. Jeon, K. H. Bouhadir, J. M. Mansour and E. Alsberg, *Biomaterials*, 2009, **30**, 2724.
92. A. I. Chou and S. B. Nicoll, *J. Biomed. Mater. Res. A*, 2009, **91**, 187.
93. F. Rask, S. M. Dallabrida, N. S. Ismail, Z. Amoozgar, Y. Yeo, M. A. Rupnick and M. Radisic, *J. Biomed. Mater. Res. A*, 2010, **95**, 105.
94. T. Horio, M. Ishihara, M. Fujita, S. Kishimoto, Y. Kanatani, T. Ishizuka, Y. Nogami, S. Nakamura, Y. Tanaka, Y. Morimoto and T. Maehara, *Artif. Organs*, 2010, **34**, 342.

95. J. W. Hayami, D. C. Surrao, S. D. Waldman and B. G. Amsden, *J. Biomed. Mater. Res. A*, 2010, **92**, 1407.
96. I. E. Erickson, A. H. Huang, S. Sengupta, S. Kestle, J. A. Burdick and R. L. Mauck, *Osteoarthritis Cartilage*, 2009, **17**, 1639.
97. C. Chung, J. Mesa, G. J. Miller, M. A. Randolph, T. J. Gill and J. A. Burdick, *Tissue Eng.*, 2006, **12**, 2665.
98. Y. Sakai, Y. Matsuyama, K. Takahashi, T. Sato, T. Hattori, S. Nakashima and N. Ishiguro, *Biomed. Mater. Eng.*, 2007, **17**, 191.
99. D. J. Quick, K. K. Macdonald and K. S. Anseth, *J. Control Release*, 2004, **97**, 333.
100. J. Burdick, M. Mason, A. Hinman, K. Thorne and K. Anseth, *J. Control. Release*, 2002, **83**, 53.
101. S. Bryant, K. Durand and K. Anseth, *J. Biomed. Mater. Res. A*, 2003, **67**, 1430.
102. F. P. Melchels, D. W. Grijpma and J. Feijen, *J. Control. Release*, 2006, **116**, e98.
103. B. G. Ilagan and B. G. Amsden, *J. Biomed. Mater. Res. A*, 2010, **93**, 211.
104. S. L. Bourke, M. Al-Khalili, T. Briggs, B. B. Michniak, J. Kohn and L. A. Poole-Warren, *AAPS Pharm. Sci.*, 2003, **5**, E33.
105. R. H. Schmedlen, K. S. Masters and J. L. West, *Biomaterials*, 2002, **23**, 4325.
106. E. Behravesh and A. G. Mikos, *J. Biomed. Mater. Res. A*, 2003, **66**, 698.
107. M. C. Hacker, A. Haesslein, H. Ueda, W. J. Foster, C. A. Garcia, D. M. Ammon, R. N. Borazjani, J. F. Kunzler, J. C. Salamone and A. G. Mikos, *J. Biomed. Mater. Res. A*, 2009, **88**, 976.
108. A. K. Shung, E. Behravesh, S. Jo and A. G. Mikos, *Tissue Eng.*, 2003, **9**, 243.
109. S. Wang, L. Lu, J. A. Gruetzmacher, B. L. Currier and M. J. Yaszemski, *Biomaterials*, 2006, **27**, 832.
110. D. Gyawali, R. Tran, K. Guleserian, L. Tang and J. Yang, *J. Biomater. Sci. Polym. Ed.*, 2010, **21**, 1761.
111. M. Weir, H. Xu and C. Simon Jr, *J. Biomed. Mater. Res. A*, 2006, **77**, 487.
112. R. K. Birla, D. E. Dow, Y. C. Huang, F. Migneco, L. Khait, G. H. Borschel, V. Dhawan and D. L. Brown, *In vitro Cell. Dev. Biol. Anim.*, 2008, **44**, 340.
113. J. M. Brunet-Maheu, J. C. Fernandes, C. A. de Lacerda, Q. Shi, M. Benderdour and P. Lavigne, *J. Biomater. Appl.*, 2009, **24**, 275.
114. S. Fu, G. Guo, X. Wang, L. Zhou, T. Liu, P. Dong, F. Luo, Y. Gu, X. Shi, X. Zhao, Y. Wei and Z. Qian, *J. Nanosci. Nanotechnol.*, 2010, **10**, 711.
115. H. Hosseinkhani, M. Hosseinkhani, F. Tian, H. Kobayashi and Y. Tabata, *Biomaterials*, 2006, **27**, 4079.
116. J. Y. Lee, J. E. Choo, Y. S. Choi, J. S. Suh, S. J. Lee, C. P. Chung and Y. J. Park, *Biomaterials*, 2009, **30**, 3532.
117. M. Gungormus, M. Branco, H. Fong, J. P. Schneider, C. Tamerler and M. Sarikaya, *Biomaterials*, 2010, **31**, 7266.

118. G. Turco, E. Marsich, F. Bellomo, S. Semeraro, I. Donati, F. Brun, M. Grandolfo, A. Accardo and S. Paoletti, *Biomacromolecules*, 2009, **10**, 1575.
119. K. Nanno, K. Sugiyasu, T. Daimon, H. Yoshikawa and A. Myoui, *Clin. Orthop. Relat. Res.*, 2009, **467**, 3149.
120. D. Hutmacher, *Biomaterials*, 2000, **21**, 2529.
121. M. Wang, *Biomaterials*, 2003, **24**, 2133.
122. W. Habraken, J. Wolke and J. Jansen, *Adv. Drug Del. Rev.*, 2007, **59**, 234.
123. R. Del Real, J. Wolke, M. Vallet-Regi and J. Jansen, *Biomaterials*, 2002, **23**, 3673.
124. R. Del Real, E. Ooms, J. Wolke, M. Vallet-Regi and J. Jansen, *J. Biomed. Mater. Res. A*, 2003, **65**, 30.
125. M. Ginebra, J. Delgado, I. Harr, A. Almirall, S. Del Valle and J. Planell, *J. Biomed. Mater. Res. A*, 2007, **80**, 351.
126. I. D. Xynos, A. J. Edgar, L. D. Buttery, L. L. Hench and J. M. Polak, *Biochem. Biophys. Res. Commun.*, 2000, **276**, 461.
127. K. Fujimura, K. Bessho, Y. Okubo, N. Segami and T. Iizuka, *Clin. Oral Implants Res.*, 2003, **14**, 659.
128. T. Kokubo, *Biomaterials*, 1991, **12**, 155.
129. R. Z. LeGeros and J. P. LeGeros, *Dense Hydroxyapatite*, in *An Introduction to Bioceramics*, Eds. L.L. Hench and J. Wilson, World Scientific Publishing Co. Pte. Ltd., Singapore, 1993,p. 139.
130. T. R. Blattert, G. Delling and A. Weckbach, *Eur. Spine J.*, 2003, **12**, 216.
131. E. P. Frankenburg, S. A. Goldstein, T. W. Bauer, S. A. Harris and R. D. Poser, *J. Bone Joint Surg. Am.*, 1998, **80**, 1112.
132. E. F. Morgan, D. N. Yetkinler, B. R. Constantz and R. H. Dauskardt, *J. Mater. Sci. Mater. Med.*, 1997, **8**, 559.
133. L. L. Hench, I. D. Xynos and J. M. Polak, *J. Biomater. Sci. Polym. Ed.*, 2004, **15**, 543.
134. Q. Hou, P. Bank and K. Shakesheff, *J. Mater. Chem.*, 2004, **14**, 1915.
135. G. Nicodemus and S. Bryant, *Tissue Eng. Part B: Rev.*, 2008, **14**, 149.
136. J. Kim, I. Kim, T. Cho, K. Lee, S. Hwang, G. Tae, I. Noh, S. Lee, Y. Park and K. Sun, *Biomaterials*, 2007, **28**, 1830.
137. Y. S. Hwang, J. Cho, F. Tay, J. Y. Heng, R. Ho, S. G. Kazarian, D. R. Williams, A. R. Boccaccini, J. M. Polak and A. Mantalaris, *Biomaterials*, 2009, **30**, 499.
138. D. Green, I. Leveque, D. Walsh, D. Howard, X. Yang, K. Partridge, S. Mann and R. Oreffo, *Adv. Funct. Mater.*, 2005, **15**, 917.
139. J. M. Dang, D. D. Sun, Y. Shin-Ya, A. N. Sieber, J. P. Kostuik and K. W. Leong, *Biomaterials*, 2006, **27**, 406.
140. A. Sawhney, C. Pathak and J. Hubbell, *Macromolecules*, 1993, **26**, 581.
141. D. Benoit, A. Durney and K. Anseth, *Tissue Eng.*, 2006, **12**, 1663.
142. C. Nuttelman, M. Tripodi and K. Anseth, *J. Biomed. Mater. Res. A*, 2004, **68**, 773.
143. S. Bruder, N. Jaiswal and S. Haynesworth, *J. Cell. Biochem.*, 1997, **64**, 278.

144. K. Jackson, S. Majka, G. Wulf and M. Goodell, *J. Cell. Biochem.*, 2002, **85**, 1.
145. F. Yang, C. Williams, D. Wang, H. Lee, P. Manson and J. Elisseeff, *Biomaterials*, 2005, **26**, 5991.
146. C. Nuttelman, D. Benoit, M. Tripodi and K. Anseth, *Biomaterials*, 2006, **27**, 1377.
147. S. Jo, H. Shin, A. Shung, J. Fisher and A. Mikos, *Macromolecules*, 2001, **34**, 2839.
148. L. Suggs, E. Kao, L. Palombo, R. Krishnan, M. Widmer and A. Mikos, *J. Biomaterials Sci. Polym. Ed.*, 1998, **9**, 653.
149. S. He, M. Timmer, M. Yaszemski, A. Yasko, P. Engel and A. Mikos, *Polymer*, 2001, **42**, 1251.
150. J. Temenoff, H. Park, E. Jabbari, D. Conway, T. Sheffield, C. Ambrose and A. Mikos, *Biomacromolecules*, 2004, **5**, 5.
151. J. Temenoff, H. Park, E. Jabbari, T. Sheffield, R. LeBaron, C. Ambrose and A. Mikos, *J. Biomed. Mater. Res.*, 2004, **70**, 235.
152. H. Shin, J. Temenoff, G. Bowden, K. Zygourakis, M. Farach-Carson, M. Yaszemski and A. Mikos, *Biomaterials*, 2005, **26**, 3645.
153. R. Payne, J. McGonigle, M. Yaszemski, A. Yasko and A. Mikos, *Biomaterials*, 2002, **23**, 4381.
154. R. Payne, M. Yaszemski, A. Yasko and A. Mikos, *Biomaterials*, 2002, **23**, 4359.
155. D. Wang, C. Williams, Q. Li, B. Sharma and J. Elisseeff, *Biomaterials*, 2003, **24**, 3969.
156. D. Wang, C. Williams, F. Yang, N. Cher, H. Lee and J. Elisseeff, *Tissue Eng.*, 2005, **11**, 201.
157. Q. Li, J. Wang, S. Shahani, D. Sun, B. Sharma, J. Elisseeff and K. Leong, *Biomaterials*, 2006, **27**, 1027.
158. K. Na, S. Kim, B. Sun, D. Woo, H. Yang, H. Chung and K. Park, *Biomaterials*, 2007, **28**, 2631.
159. A. Martins, S. Chung, A. J. Pedro, R. A. Sousa, A. P. Marques, R. L. Reis and N. M. Neves, *J. Tissue Eng. Regen. Med.*, 2009, **3**, 37.
160. L. S. Wang, J. E. Chung, P. P. Chan and M. Kurisawa, *Biomaterials*, 2010, **31**, 1148.
161. G. Wei and P. X. Ma, *Biomaterials*, 2009, **30**, 6426.
162. K. Lee, E. A. Silva and D. J. Mooney, *J. R. Soc. Interface*, 2010, **8**, 153.
163. D. H. Kempen, L. Lu, T. E. Hefferan, L. B. Creemers, A. Maran, K. L. Classic, W. J. Dhert and M. J. Yaszemski, *Biomaterials*, 2008, **29**, 3245.
164. P. Yilgor, N. Hasirci and V. Hasirci, *J. Biomed. Mater. Res. A*, 2010, **93**, 528.
165. C. He, S. W. Kim and D. S. Lee, *J. Control. Release*, 2008, **127**, 189.
166. L. Yu and J. Ding, *Chem. Soc. Rev.*, 2008, **37**, 1473.
167. J. P. Jain, D. Chitkara and N. Kumar, *Expert Opin. Drug. Deliv.*, 2008, **5**, 889.
168. M. Colilla, M. Manzano and M. Vallet-Regi, *Int. J. Nanomedicine*, 2008, **3**, 403.

169. M. Biondi, F. Ungaro, F. Quaglia and P. A. Netti, *Adv. Drug. Deliv. Rev.*, 2008, **60**, 229.
170. A. S. Mistry and A. G. Mikos, *Adv. Biochem. Eng. Biotechnol.*, 2005, **94**, 1.
171. S. H. Lee and H. Shin, *Adv. Drug. Deliv. Rev.*, 2007, **59**, 339.
172. J. D. Kretlow, S. Young, L. Klouda, M. Wong and A. G. Mikos, *Adv. Mater. Deerfield*, 2009, **21**, 3368.
173. S. B. Adams Jr., M.F. Shamji, D.L. Nettles, P. Hwang and L.A. Setton, *J. Biomed. Mater. Res. B Appl. Biomater.*, 2009, **90**, 67.
174. M. Y. Krasko, J. Golenser, A. Nyska, M. Nyska, Y. S. Brin and A. J. Domb, *J. Control. Release*, 2007, **117**, 90.
175. S. L. Henry and K. P. Galloway, *Clin. Pharmacokinet.*, 1995, **29**, 36.
176. S. Hesaraki and R. Nemati, *J. Biomed. Mater. Res. B Appl. Biomater.*, 2009, **89B**, 342.
177. Z. Gou, J. Chang, W. Zhai and J. Wang, *J. Biomed. Mater. Res. B Appl. Biomater.*, 2005, **73**, 244.
178. U. Joosten, A. Joist, T. Frebel, B. Brandt, S. Diederichs and C. von Eiff, *Biomaterials*, 2004, **25**, 4287.
179. K. T. Peng, C. F. Chen, I. M. Chu, Y. M. Li, W. H. Hsu, R. W. Hsu and P. J. Chang, *Biomaterials*, 2010, **31**, 5227.
180. Y. S. Brin, J. Golenser, B. Mizrahi, G. Maoz, A. J. Domb, S. Peddada, S. Tuvia, A. Nyska and M. Nyska, *J. Control. Release*, 2008, **131**, 121.
181. J. M. Kanczler, P. J. Ginty, L. White, N. M. Clarke, S. M. Howdle, K. M. Shakesheff and R. O. Oreffo, *Biomaterials*, 2010, **31**, 1242.
182. Z. S. Patel, S. Young, Y. Tabata, J. A. Jansen, M. E. Wong and A. G. Mikos, *Bone*, 2008, **43**, 931.
183. S. D. Boden, *Clin. Orthop. Relat. Res.*, 1999, S84.
184. C. M. Agrawal and R. B. Ray, *J. Biomed Mater Res*, 2001, **55**, 141.
185. R. T. Kao, S. Murakami and O. R. Beirne, *Periodontol. 2000*, 2009, **50**, 127.
186. D. H. Kempen, M. J. Yaszemski, A. Heijink, T. E. Hefferan, L. B. Creemers, J. Britson, A. Maran, K. L. Classic, W. J. Dhert and L. Lu, *J. Control. Release*, 2009, **134**, 169.
187. P. Yilgor, K. Tuzlakoglu, R. L. Reis, N. Hasirci and V. Hasirci, *Biomaterials*, 2009, **30**, 3551.
188. J. Sohier, G. Daculsi, S. Sourice, K. de Groot and P. Layrolle, *J. Biomed. Mater. Res. A*, 2010, **92**, 1105.
189. M. D. Weir and H. H. Xu, *J. Biomed. Mater. Res. A*, 2010, **94**, 223.
190. H. C. Kroese-Deutman, P. Q. Ruhe, P. H. Spauwen and J. A. Jansen, *Biomaterials*, 2005, **26**, 1131.
191. N. Saito, T. Okada, H. Horiuchi, H. Ota, J. Takahashi, N. Murakami, M. Nawata, S. Kojima, K. Nozaki and K. Takaoka, *Bone*, 2003, **32**, 381.
192. H. Hosseinkhani, M. Hosseinkhani, F. Tian, H. Kobayashi and Y. Tabata, *Tissue Eng.*, 2007, **13**, 11.
193. T. Matsuo, T. Sugita, T. Kubo, Y. Yasunaga, M. Ochi and T. Murakami, *J. Biomed. Mater. Res. A*, 2003, **66**, 747.

194. H. Maeda, A. Sano and K. Fujioka, *Int. J. Pharm.*, 2004, **275**, 109.
195. Z. S. Haidar, R. C. Hamdy and M. Tabrizian, *Biomaterials*, 2009, **31**, 2746.
196. E. R. Balmayor, G. A. Feichtinger, H. S. Azevedo, M. van Griensven and R. L. Reis, *Clin. Orthop. Relat. Res.*, 2009, **467**, 3138.
197. E. L. Hedberg, A. Tang, R. S. Crowther, D. H. Carney and A. G. Mikos, *J. Control. Release*, 2002, **84**, 137.
198. Y. Wang, C. Wan, G. Szoke, J. T. Ryaby and G. Li, *J. Orthop. Res.*, 2008, **26**, 539.
199. H. Zhang, P. Leng and J. Zhang, *Clin. Orthop. Relat. Res.*, 2009, **467**, 3165.
200. E. Verron, O. Gauthier, P. Janvier, P. Pilet, J. Lesoeur, B. Bujoli, J. Guicheux and J. M. Bouler, *Biomaterials*, 2010, **31**, 7776.
201. W. J. Habraken, O. C. Boerman, J. G. Wolke, A. G. Mikos and J. A. Jansen, *J. Biomed. Mater. Res. A*, 2009, **91**, 614.
202. H. P. Stallmann, C. Faber, A. L. Bronckers, A. V. Nieuw Amerongen and P. I. Wuisman, *BMC Musculoskelet. Disord.*, 2006, **7**, 18.
203. H. J. Seeherman, M. Bouxsein, H. Kim, R. Li, X. J. Li, M. Aiolova and J. M. Wozney, *J. Bone Joint Surg. Am.*, 2004, **86-A**, 1961.
204. B. Bai, Z. Yin, Q. Xu, M. Lew, Y. Chen, J. Ye, J. Wu, D. Chen and Y. Zeng, *Spine (Phila. Pa. 1976)*, 2009, **34**, 1887.
205. H. J. Seeherman, K. Azari, S. Bidic, L. Rogers, X. J. Li, J. O. Hollinger and J. M. Wozney, *J. Bone Joint Surg. Am.*, 2006, **88**, 1553.
206. D. X. Li, H. S. Fan, X. D. Zhu, Y. F. Tan, W. Q. Xiao, J. Lu, Y. M. Xiao, J. Y. Chen and X. D. Zhang, *J. Mater. Sci. Mater. Med.*, 2007, **18**, 2225.
207. D. P. Link, J. van den Dolder, J. J. van den Beucken, J. G. Wolke, A. G. Mikos and J. A. Jansen, *Biomaterials*, 2008, **29**, 675.
208. D. J. Park, B. H. Choi, S. J. Zhu, J. Y. Huh, B. Y. Kim and S. H. Lee, *J. Craniomaxillofac. Surg.*, 2005, **33**, 50.
209. S. J. Stephan, S. S. Tholpady, B. Gross, C. E. Petrie-Aronin, E. A. Botchway, L. S. Nair, R. C. Ogle and S. S. Park, *Laryngoscope*, 2010, **120**, 895.
210. N. Alles, N. S. Soysa, M. D. Hussain, N. Tomomatsu, H. Saito, R. Baron, N. Morimoto, K. Aoki, K. Akiyoshi and K. Ohya, *Eur. J. Pharm. Sci.*, 2009, **37**, 83.
211. M. Miyazaki, Y. Morishita, W. He, M. Hu, C. Sintuu, H. J. Hymanson, J. Falakassa, H. Tsumura and J. C. Wang, *Spine J.*, 2009, **9**, 22.
212. H. Uludag, B. Norrie, N. Kousinioris and T. Gao, *Biotechnol. Bioeng.*, 2001, **73**, 510.
213. X. B. Tian, L. Sun, S. H. Yang, Y. K. Zhang, R. Y. Hu and D. H. Fu, *Chin. Med. J. (Engl.)*, 2008, **121**, 745.
214. M. P. Ferraz, A. Y. Mateus, J. C. Sousa and F. J. Monteiro, *J. Biomed. Mater. Res. A*, 2007, **81**, 994.
215. Q. Wang, J. Wang, Q. Lu, M. S. Detamore and C. Berkland, *Biomaterials*, 2010, **31**, 4980.

216. S. Herberg, M. Siedler, S. Pippig, A. Schuetz, C. Dony, C. K. Kim and U. M. Wikesjo, *J. Clin. Periodontol.*, 2008, **35**, 976.
217. A. Billon, L. Chabaud, A. Gouyette, J. M. Bouler and C. Merle, *J. Microencapsul.*, 2005, **22**, 841.
218. A. Plachokova, D. Link, J. van den Dolder, J. van den Beucken and J. Jansen, *J. Tissue Eng. Regen. Med.*, 2007, **1**, 457.
219. X. Shi, Y. Wang, L. Ren, Y. Gong and D. A. Wang, *Pharm. Res.*, 2009, **26**, 422.
220. C. H. Chang, T. C. Liao, Y. M. Hsu, H. W. Fang, C. C. Chen and F. H. Lin, *Biomaterials*, 2010, **31**, 4048.
221. A. E. Hafeman, B. Li, T. Yoshii, K. Zienkiewicz, J. M. Davidson and S. A. Guelcher, *Pharm. Res.*, 2008, **25**, 2387.
222. A. E. Hafeman, K. J. Zienkiewicz, E. Carney, B. Litzner, C. Stratton, J. C. Wenke and S. A. Guelcher, *J. Biomater. Sci. Polym. Ed.*, 2010, **21**, 95.
223. V. Luginbuehl, E. Wenk, A. Koch, B. Gander, H. P. Merkle and L. Meinel, *Pharm. Res.*, 2005, **22**, 940.
224. J. O. Hollinger, A. O. Onikepe, J. MacKrell, T. Einhorn, G. Bradica, S. Lynch and C. E. Hart, *J. Orthop. Res.*, 2008, **26**, 83.
225. J. Yang, A. Webb and A. Guillermo, *Adv. Mater.*, 2004, **16**, 511.
226. J. Yang, D. Motlagh, A. R. Webb and G. A. Ameer, *Tissue Eng.*, 2005, **11**, 1876.
227. J. Yang, D. Motlagh, J. Allen, A. Webb, M. Kibbe, O. Aalami, M. Kapadia, T. Carroll and G. Ameer, *Adv. Mater.*, 2006, **28**, 1493.
228. H. Qiu, J. Yang, P. Kodali, J. Koh and G. A. Ameer, *Biomaterials*, 2006, **27**, 5845.
229. R. Tran, Y. Zhang, D. Gyawali and J. Yang, *Recent Patents Biomed. Eng.*, 2009, **2**, 216.
230. R. Tran, P. Thevenot, D. Gyawali, J. Chiao, L. Tang and J. Yang, *Soft Matter*, 2010, **6**, 2449.
231. J. Yang, Y. Zhang, S. Gautam, L. Liu, J. Dey, W. Chen, R. P. Mason, C. A. Serrano, K. A. Schug and L. Tang, *Proc. Natl. Acad. Sci. U S A*, 2009, **106**, 10086.

CHAPTER 15
Production of Polyhydroxybutyrate (PHB) from Activated Sludge

M. SURESH KUMAR[1] AND TAPAN CHAKRABARTI[2]

[1] Solid and Hazardous Waste Management Division, National Environmental Engineering Research Institute (NEERI), Nehru Marg, Nagpur 440 020, India; [2] Former Acting Director, National Environmental Engineering Research Institute (NEERI), Nehru Marg, Nagpur 440 020, India

15.1 Introduction

Plastics are utilized in almost every manufacturing industry ranging from automobiles to medicine. The synthetic compounds polyethylene, polyvinylchloride and polystyrene are largely used in the synthesis of plastics. These plastics can be easily moulded into any desired products. Because of these reasons enormous quantities of plastics have been used and are still in use all over the world. However, the major problem with plastic is its disposal as plastics are recalcitrant to microbial degradation. In the recent years, there has been increasing public concern over the harmful effects of petrochemical-derived plastic materials in the environment. This has prompted many countries to start developing biodegradable plastics. Polyhydroxyalkanoates (PHA) are the polyester of hydroxyalkanoates synthesized by numerous bacteria as intracellular carbon-storage and energy-storage compounds, and have mechanical properties similar to polypropylene or polyethylene. Short-chain-length PHAs such as poly-3-hydroxybutyrate (PHB) have been studied in depth

RSC Green Chemistry No. 12
A Handbook of Applied Biopolymer Technology: Synthesis, Degradation and Applications
Edited by Sanjay K. Sharma and Ackmez Mudhoo
© Royal Society of Chemistry 2011
Published by the Royal Society of Chemistry, www.rsc.org

and have been produced on a commercial scale.[1] Additionally, polyhydroxyalkanoates have the attractive features of being completely biodegradable, biocompatible and are produced from renewable sources. For the economical production of PHAs, various bacterial strains, either wild type or recombinant, and new fermentation strategies were developed with high content and productivity of PHAs.[2] However, the PHAs production cost is still 8 to 10 times more than that of conventional plastics. The major cost is attributed to the cost of pure culture fermentation and expensive pure substrates. Recently, the production of poly-β-hydroxybutyrate (PHB) by activated sludge from wastewater treatment plants has attracted attention. The activated sludge can use cheap substrates (such as organic waste) and avoid the need of sterilization. Thus, the costs may be substantially reduced.

15.2 Polymers

Polymers are made of two major groups: aliphatic (linear) polymers and aromatic (aromatic rings) polymers. Polyhydroxyalkanoates (PHAs) are aliphatic polymers naturally produced *via* a microbial process using carbohydrate-based medium, which acts as carbon-storage and energy-storage materials in bacteria. They were the first biodegradable polymers to be utilized in plastics. The two main members of the PHA family are polyhydroxybutyrate (PHB) and polyhydroxyvalerate (PHV). Aliphatic polymers such as PHAs, and homopolymers and copolymers of hydroxybutyric acid and hydroxyvaleric acid, have been proven to be readily biodegradable. Such polymers are actually synthesized by microbes, with the polymer accumulating in the microbes' cells during growth. The PHB homopolymer is a stiff and rather brittle polymer of high crystallinity, whose mechanical properties are more or less akin to those of polystyrene, which, however, is less brittle. PHB copolymers are preferred for general purposes as the degradation rate of PHB homopolymer is high at its normal melt processing temperature. PHB and its copolymers with PHV are melt-processable semi-crystalline thermoplastics made from renewable carbohydrate feedstocks through fermentation. They represent the first example of a true biodegradeable thermoplastic produced through the biotechnological route. No toxic by-products are known to result from PHB or PHV production.

Polyhydroxybutyrate-co-polyhydroxyhexanoates (PHBHs) resins are one of the newest types of naturally produced biodegradable polymers. The PHBH resin is derived from carbon sources such as sucrose, fatty acids or molasses *via* a fermentation process. These are 'aliphatic-aliphatic' copolyesters, as distinct from 'aliphatic-aromatic' copolyesters. Besides being completely biodegradable, they also exhibit barrier properties similar to those exhibited by ethylene vinyl alcohol. Procter & Gamble Co. researched the blending of these polymers to obtain the appropriate stiffness or flexibility.

Beside the homopolyester poly-R-3-hydroxybutanoate, consisting of 3-hydroxybutanoate (3HB) only, two main types of copolyesters can be formed by different microorganisms.[3] The first type of PHAs always contain C3 units in the polymer backbone; however, the side chains can contain H, and methyl (or ethyl) groups if prepared with microorganisms like *Ralstonia eutropha*. Propyl to nonyl groups are found in the side chains if the copolyester is prepared with *Pseudomonas oleovorans*. In the latter case, branchings,[4] double bonds,[5] epoxides[6] and aromatic structures[7] can be introduced into the side chain. Furthermore, copolyesters containing *o*-chloroalkanoates (F, Cl, Br) can be produced.[8–10] In the case of *P. oleovorans* and other strains from the group of fluorescent pseudomonads, PHA formation only occurs when the organisms are grown either with fatty acids (butanoate to hexadecanoate) or with alkanes (hexane to dodecane). Doi[11] recently reported a synthesis of a copolyester consisting 3-hydroxybutyrate and 3-hydroxyhexanoate by *Aeromonas cavei*, and Chen[12] isolated a bacterium from oil-contaminated soil able to synthesize the same polyester when fed with glucose and laurylic acid. This copolyester shows an extremely high extension needed to break.

The second type of PHA is a short side chain polyester, containing hydrogen, methyl or ethyl groups in the side chains, and having C3, C4 and C5 units in the backbone of the polymer.[13,14] Carbohydrates, alcohols and short-chain fatty acids are typical substrates for growth and PHA formation for these microorganisms. In most cases, cosubstrates have to be fed to the producing cultures as precursors for copolyester formation.[14,15] Typical precursors that have been used are propionate, valerate or 1,4-butanediol, leading to analogues of 3HB such as 4- and 5-hydroxyalkanoates.

15.3 Storage Polymers in Microorganisms

15.3.1 PHB Biosynthesis

The bacteria synthesize and store PHAs when they lack the complete range of nutrients required for cell development but receive a generous supply of carbon. A deficiency of magnesium, sulfur, nitrogen, phosphate and/or oxygen can initiate PHA biosynthesis. Approximately 300 different bacteria have been reported to accumulate various PHAs.[16] For many of them the polymer, once accumulated, serves both as a carbon source and an energy source during starvation. It constitutes ideal carbon energy storage due to its low solubility, high molecular weight and inert nature, thus exerting negligible osmotic pressure on the bacterial cell.[17] The presence of PHA in a cell frequently, but not universally, retards the degradation of cellular components such as RNA and proteins during nutrient starvation.[18]

PHB synthesis from glucose using *Azotobacter beijerinkii* revealed substantial amounts of polymer accumulation under oxygen limitation conditions.[19–21] The key feature of control in *A. beijerinkii* is the pool size of acetyl-CoA, which may either be oxidized *via* the tricarboxylic acid (TCA) cycle or can serve as a substrate for PHB synthesis; the diversion depends on environmental conditions, especially oxygen limitation, when the NADH/NAD ratio increases. Citrate synthase and isocitrate dehydrogenase are inhibited by NADH, and as a consequence, acetyl-CoA no longer enters the TCA cycle at the same rate. Instead acetyl-CoA is converted to acetoacetyl-CoA by β-ketothiolase, the first enzyme of the PHB biosynthetic pathway, which is inhibited by CoA. When the oxygen supply is adequate, the CoA concentration is high and the β-ketothiolase is accordingly inhibited, thereby preventing PHB synthesis. Senior and Dawes[21] proposed that PHB served not only as a reserve of carbon and energy but also as a sink of reducing power and could be regarded as a redox regulator within the cell.[18]

There are four different pathways for the synthesis of PHAs found to date.

- In *A. eutrophus*, β-ketothiolase carries out the condensation of two molecules of acetyl-CoA to acetoacetyl-CoA. An NADPH-dependent acetoacetyl-CoA reductase then carries out its conversion to 3-hydroxybutyryl-CoA. The third and the final step is the polymerization reaction catalysed by PHB synthase.[18]
- In *Rhodopsuedomonas rubrum*, the pathway differs after the second step where the acetoacetyl-CoA formed by β-ketothiolase is reduced by a NADH-dependent reductase to L-(+)-3-hydroxybutyryl-CoA which is then converted to D-(−)-3-hydroxybutyryl-CoA by two enoyl-CoA hydratases.
- A third type of PHA biosynthetic pathway is found in most *Psuedomonas* species belonging to rRNA homology group I. *P. oleovorans* and other *Psuedomonas* species accumulate PHA consisting of 3-hydroxyalkanoic acid of medium-chain length (MCL) if cells are cultivated on alkanes, alkanols or alkanoic acids.[22,23]
- The fourth type of PHA biosynthetic pathway is present in almost all *Psuedomonas* species belonging to rRNA homology group II. This pathway involves the synthesis of copolyesters consisting of MCL 3HAs from acetyl-CoA. This pathway has not been studied in detail.[2]

15.3.2 Enzymes Involved in Biosynthesis and Degradation

Since 1987, the extensive body of information on poly(3-hydroxybutyrate) [P(3HB)] metabolism, biochemistry and physiology has been enriched by molecular genetic studies. Numerous genes encoding enzymes involved in PHA formation and degradation have been studied and characterized from a variety of microorganisms. Of all the PHAs, P(3HB) is the most extensively characterized polymer, mainly because it was the first to be discovered, in 1962 by Lemoigne at the Institute Pasteur.[24] The P(3HB) biosynthetic pathway consists of three enzymatic reactions catalysed by three distinct enzymes. The first reaction consists of the condensation of two acetyl coenzyme A (acetyl-CoA) molecules into acetoacetyl-CoA by β-ketoacyl-CoA thiolase (encoded by *phbA*). The second reaction is the reduction of acetoacetyl-CoA to (R)-3-hydroxybutyryl-CoA dehydrogenase (encoded by *phbB*). Lastly, the (R)-3-hydroxybutyryl-CoA monomers are polymerized into poly(3-hydroxybutyrate) by P(3HB) polymerase (encoded by *phbC*). Although P(3HB) accumulation is a widely distributed prokaryotic phenotype, the biochemical investigations into the enzymatic mechanisms of β-ketoacyl-CoA thiolase, acetoacetyl-CoA reductase, and P(3HB) polymerase have focused on only two of the natural producers, *Zoogloea ramigera* and *Ralstonia eutropha* (formerly known as *Alcaligenes eutrophus*).[25]

Numerous bacteria synthesize and accumulate PHA as carbon- and energy-storage materials or as a sink for redundant reducing power under the condition of limiting nutrients in the presence of excess carbon source. When the supply of the limiting nutrient is restored, the PHA can be degraded by intracellular depolymerases and subsequently metabolized as a carbon and energy source.[26,27] Two different types of depolymerase systems had been recognized, in *Rhodospirillum rubrum* and *B. megaterium*. Native granules from *R. rubrum* are self hydrolysing, whereas those from *B. megaterium* are quite stable, although a soluble extract from *R. rubrum* was active in the degradation of native granules from *B. megaterium*. Purified polymer or denatured granules did not serve as substrate.[28] A soluble PHB depolymerase isolated from *B. megaterium* yielded a mixture of dimers and monomer as hydrolysis products,[29] whereas the soluble depolymerase from *Alcaligenes* sp. gave D-(–)-3HB as the sole product.[30] A more detailed study of the *B. megaterium* system disclosed that depolymerization required a heat-labile factor associated with the granules together with three soluble components, namely, a heat-stable protein activator, PHB depolymerase and a hydrolase.[31]

15.3.3 Properties of PHB

PHB has some properties similar to polypropylene with three unique features: thermoplastic processability, 100% resistance to water and 100% biodegradability.[31] Booma *et al.*[32] stated that PHB is an aliphatic homopolymer with a melting point of 179 °C and is highly crystalline (80%). It can be degraded at a temperature above its melting point. de Koning[33] reported that the molecular weight of PHB was decreased to approximately half of its original value when it was held at 190 °C for 1 hour. PHAs can have physical properties that range from brittle and thermally unstable to soft and tough, depending upon their composition, *i.e.* PHV/PHB ratios. The physical properties of PHB, *e.g.* crystallization and tensile strength, depend on molecular weight, which is determined by the strain of microorganism employed, growth conditions and the purity of the sample obtained. To improve the physical properties of microbially produced PHB, various attempts were made to synthesize various copolymers of (*R*)-3-HB with better properties. Poly[(*R*)-3-hydroxybutyrate-co-(*R*)-3-hydroxy valerate] (P[3HB-co-3HV]) for instance, was successfully produced on a commercial basis under the trade name of BIOPOL. P[3HB-co-3HV] is characterized by increased elongation to break and low melting temperature, which does not degrade during thermal processing.[34] However, P[3HB-co-3HHx] has low tensile strength. In an attempt to improve the brittleness of PHB films, Barham and Keller[35] applied cold-drawing techniques. Recently, it was demonstrated that both cold-drawing and annealing procedures could be used in improving the mechanical properties of PHB films.[36]

In spite of all these efforts to improve the mechanical properties of PHB and its copolyesters, their commercial application for production of plastic materials are still hindered, due to downstream processing cost and poor yields.

15.3.4 Potential Applications

The possible applications of bacterial PHA are directly connected to their properties pertaining to biological degradability, thermoplastics characteristics and piezoelectric properties. The applications of bacterial PHAs have concentrated on three principal areas: medical and pharmaceutical, agricultural and commodity packaging.[37–39] According to Laferrty et al.[37] the most advanced development of bacterial PHAs is in the medical field, especially pharmaceutical applications, although they have a considerable potential as consumer goods products.

The degradation product of P(3HB), D-(–)-3-hydroxybutyric acid, is a common intermediate metabolic compound in all higher organisms.[37,39] Therefore, it is plausible that it is biocompatible to animal tissues and P(3HB) can be implanted in animal tissues without any toxic manifestation.

Some possible use of bacterial PHAs in the medical and pharmaceutical applications include: biodegradable carrier for long-term dosage of drugs inside the body, surgical pins, sutures and swabs, wound dressing, bone replacements and plates, blood vessel replacements, and stimulation of bone growth and healing by piezoelectric properties. The advantage of using biodegradable plastics during implantation is that it will be biodegraded, i.e. the need for its surgical removal is not necessary.

PHAs are biodegraded in soil. Therefore, the use of PHAs in agriculture is very promising. They can be used as biodegradable carriers for long-term application of insecticides, herbicides or fertilizers, seedling containers and plastic sheaths protecting saplings, biodegradable matrix for drug release in veterinary medicine, and tubing for crop irrigation. Here again, it is not necessary to remove biodegradable items at the end of the harvesting season.

According to Lafferty et al.[37] PHB homopolymer and PHB-PHV copolymer have some properties, i.e. tensile strength and flexibility, similar to polyethylene and polystyrene. Holmes[38] reported that PHAs can be used in extrusion and moulding processes and blended with synthetic polymer, e.g. chlorinated polyethylene, to make heteropolymers. Also, small additions of PHA improves the property of some conventional polymers, e.g. addition of a small amount of PHA reduces the melt viscosity of acrylonitrile. Tsuchikura[40] reported that 'BIOPOL' with high PHV content is more suitable for extrusion blow moulding and extrusion processes, e.g. made into films, sheets and fibres, while 'BIOPOL' with low PHV content is more suitable for general injection moulding processes. According to Lafferty et al.[37] one particular property of PHB films that make it possible to be used for food packaging is the relatively low oxygen diffusivity. Plastics produced from PHAs have been reported to be biodegraded both in aerobic and anaerobic environments.[41] In summary, possible applications of PHAs for commodity goods include packaging films, bags and containers, and disposable items such as razors, utensils, diapers and feminine products.

15.3.5 Biodegradation of PHB

PHAs are biodegradable *via* composting. Optimum conditions for the commercially available Biopol™ (PHA) degradation during a 10-week composting period were 60 °C, 55% moisture, and C:N ratio of 18:1. Biopol™ reached close to a 100% degradation rate under these composting conditions. These aliphatic polymers are suited to applications with short usage and high degradation rate requirements. Shin *et al*.[42] found that bacterial PHB/PHV (92/8 w/w) degraded nearly to completion within 20 days of cultivation by anaerobic digested sludge, while synthetic aliphatic polyesters such as PLA, PBS and PBSA did not degrade at all in 100 days. Cellophane, which was used as a control material, exhibited a similar degradation behaviour to PHB/PHV. Under simulated landfill conditions, PHB/PHV degraded within 6 months. Synthetic aliphatic polyesters also showed significant weight losses through 1 year of cultivation. The acidic environment generated by the degradation of biodegradable food wastes which comprises approximately 34% of municipal solid waste seems to cause the weight loss of synthetic aliphatic polyesters. PHBH resins biodegrade under aerobic as well as anaerobic conditions, and are digestible in hot water under alkaline conditions.

15.4 PHB Production

15.4.1 PHB Production with Pure Substrates

Conventional PHA production by wild-type strains and recombinants is usually performed in two-stage fed-batch cultures, which consists of a cell growth phase and a PHA production phase. In the cell growth phase, nutritionally enriched medium is used to obtain high cell mass during early cultivation. In the sequential PHA production phase, the cell growth is limited by depletion of some nutrients such as nitrogen, phosphorous, oxygen or magnesium.[25] This depletion acts as a trigger for the metabolic shift to PHA biosynthesis.

Sugars such as glucose and sucrose are the most common main carbon source for PHA production because they can be obtained at a relatively low price. The highest P(3HB) productivity from glucose was obtained by a recombinant *E. coli* harbouring *Alcaligenes latus*-derived PHA biosynthesis genes.[43] In this culture, P(3HB) productivity of 4.63 g/L h was obtained, and the P(3HB) concentration and content in the dry cells reached 142 g L^{-1} and 73%, respectively, in 30 h.[44] PHB and its copolymers had been produced from glucose and propionate on a semi-commercial scale first by Zeneca Bioproducts (Billingham, UK) and later by Monsanto (St. Louis, MO, USA).[45] ICI evaluated three organisms as principal contenders for the industrial production of PHB, namely, an *Azotobacter* sp., a *Methylobacterium* sp. and *A. eutrophus*, which collectively embraced a wide range of substrates.[46] The *Azotobacter* sp., which could grow on glucose or sucrose with high polymer yields, proved unstable and also synthesized carbohydrate, thereby diverting substrate from PHB production. The methylotroph, seemingly attractive on account of ICI's wide experience in methanol fermentation technology, gave only moderate yields of polymer of low molecular weight, which was extracted with difficulty. *A. eutrophus*, grown heterotrophically, became the organism of choice which can utilize various economically acceptable substrates.[18,47]

15.4.2 PHB Production with Wastes

PHAs can substitute petroleum-derived polymers, can be produced from renewable resources and are harmless to the environment due to their biodegradability. However, the major hurdle facing commercial production and application of PHA in consumer products is the high cost of bacterial fermentation. It makes bacterial PHA production 5–10 times more expensive than the petroleum-derived polymers such polyethylene and polypropylene.[2] The significant factor of the production cost of PHA is the cost of substrate (mainly carbon source). In order to decrease this cost, the use of cheap carbon sources as substrates have been developed. The researches have been carried out to develop recombinant strains utilizing a cheap carbon source, while corresponding fermentation strategies have been developed and optimized.

40% of total operating expense of PHA production is related to the raw materials, and more than 70% of this cost is attributed to the carbon source. By using cheap substrate sources such as agroindustrial wastes (*e.g.* whey, molasses and palm oil mill effluents), PHA production may be made economic.[43,48] Many waste streams from agriculture are potentially useful substrates. These include cane and beet molasses, cheese whey, plant oils and hydrolysates of starch (*e.g.* corn, tapioca), cellulose and hemicellulose. Because an open-culture system would be used, not all substrates would be equally suitable. For example, starch and cellulose hydrolysates could lead to the growth of glycogen-accumulating organisms.[49] This problem can be easily overcome by acidification of sugars, starch and cellulose hydrolysates with a mixture of volatile fatty acids (VFAs) such as acetic, propionic, butyric and others. Such a mixture is readily converted to PHAs. Oil refinery waste such as cracker condensate and effluent of a partial wet oxidation unit are available as potential sources of volatile fatty acids.[48]

The percolate from the organic wet fraction (OWF) of household waste is considered the most suitable substrate for PHA production. It has a very high volatile fatty acid concentration, is available in large quantities and can be transported easily. The heavy metal content can be removed by precipitation. There are also several reports on the production of volatile fatty acids from anaerobically treated palm oil mill effluent (POME) and utilization of these organic acids for the production of PHA.[50–53]

The uses of alternative carbon sources and corresponding strain developments have been reported. These include glycerol as co-product of many industrial processes using *P. oleovorans*[54] wastewater using activated sludge,[55–62] glutamic acid in the wastewater,[63] olive oil mill effluents,[64,65] palm oil mill effluents,[66] soyabean oil[67] and agricultural waste.[68] However, the polymer concentration and content obtained were considerably lower than those obtained using purified carbon substrates. Therefore, there is a need for development of more efficient fermentation strategies for production of these polymers from a cheap carbon source.

15.4.3 PHB Production by Mixed Culture

There are a few studies on PHA production by mixed cultures using propionate, butyrate, lactate, succinate, pyruvate, malate, ethanol,[69–71] glutamate and aspartate[72] and glucose,[73,74] but no studies on the optimization of nutrient removal and PHA production by mixed cultures are available.

The effect of acetate, acetate and glucose and glucose alone on PHA production under anaerobic–aerobic conditions in a sequential batch reactor (SBR) was evaluated by Hollender et al.[74] A rapid and complete consumption of glucose occurred in the anaerobic period, whereas acetate consumption was slow and incomplete. The produced PHBV included 17% and 82% HV for acetate and glucose substrates, respectively. The maximum amount of phosphate release and PHA storage during anaerobic phase as well as the highest phosphate (poly-P) and glycogen storage during aerobic phase were obtained using acetate. Lemos et al.[75] demonstrated that acetate uptake by polyphosphate-accumulating organisms (PAOs) leads to the production of a copolymer of hydroxybutyrate (HB) and hydroxyvalerate (HV), with the HB units being dominant (69–100% HB; 0–31% HV). With propionate, HV units are mainly incorporated in the polymer. The yield of polymer (YP/S) was found to diminish from acetate (0.97) to propionate (0.61) to butyrate (0.21). Using a mixture of acetate, propionate and butyrate, the PHA synthesized was enriched in HV units.[75,76] In this case, the polymer was composed of 2–28% HB and 45–72% HV.

15.4.3.1 Activated Sludge

It is well known that production and storage of PHAs are integral and essential parts of the excess biological phosphorus removal (EBPR) mechanisms. It is less well known by the wastewater treatment community that bacterial PHAs are used for the manufacture of biodegradable plastics. The primary differences in the two approaches are: (1) EBPR is accomplished using mixed cultures of bacteria, i.e. activated sludge, with the intent of minimizing the amount of phosphorus remaining in solution at the end of the aerobic phase; but (2) plastics production is typically accomplished using pure cultures with the goal of maximizing the amount of stored PHAs at the end of a nutrient-limited or oxygen-limited phase. Thus, for EBPR, the microbial cells are harvested when phosphorus storage is at a maximum while for plastics production, the cells are harvested when PHA storage is at a maximum. Simplistically, the two processes exploit different parts of the same basic biochemical cycle, and this raises the possibility that perhaps wastewaters can be used to produce biodegradable plastics, a marketable by-product. Satoh et al.[72] demonstrated the basic feasibility by showing that microaerophilic conditions could be used to increase the percent PHA in activated sludge to as much as 62%TSS, using sodium acetate as the primary source of volatile fatty acids (VFAs) in the feed. There are industrial wastewaters such as those from cellulose acetate manufacturing that average as much as 1200 mg L^{-1} acetic acid, and would seem to be good substrates for plastics production. Also, fermentation can be used to convert much of municipal sewage to VFAs, and potentially enable sewage to provide inexpensive substrate for plastics production.

According to Dawes[77] there are four principal classes of storage polymers in microorganisms: lipids, carbohydrates, polyphosphates and nitrogen reserve compounds. For lipid storage, the discussion will be focused mainly on polyhydroxyalkanoates (PHAs). Some organisms can accumulate more than one kind of storage polymer. For example, PAOs that are responsible for phosphorus removal in BPR systems store glycogen, PHA and polyphosphate. Dawes and Senior[78] stated that environmental conditions and regulatory mechanisms influence the content of storage polymers accumulated. In addition, the nature of the carbon substrate will determine the type of storage polymers synthesized. For example, when the source of carbon was changed from glucose to acetate, the carbohydrate content of *E. coli* decreased, while the lipid content increased. According to Sasikala and Ramana[79] when a carbon substrate is metabolized *via* acetyl-CoA, *i.e.* no pyruvate is formed as an intermediate, the flow of carbon is mainly to PHA synthesis. However, if a carbon substrate is metabolized *via* pyruvate, glycogen storage is predominant. Dawes[77] stated that the important role of the possession of storage polymers by microorganisms is that such storage materials provide carbon and energy sources for organisms and permit them to maintain viability longer under nutrient limitation conditions. Organisms that do not possess storage polymers tend to utilize cellular contents, *e.g.* RNA and protein, during starvation periods.

15.4.3.2 EBPR

Recently, interest has developed in the use of biological, rather than chemical, processes for phosphorus removal from wastewater. Sequential anaerobic–aerobic operation of activated sludge process is applied to achieve enhanced biological phosphorus removal (EBPR). The simplest system configuration for EBPR consists of two stages in series, the first one being anaerobic and the second aerobic. The activated sludge (biomass) is cycled between anaerobic and aerobic phases, and the influent is supplied under anaerobic conditions. Operation in this model promotes the accumulation of PAOs.[80]

During the EBPR anaerobic phase, carbon substrates such as acetic and propionic acids are taken up to biosynthesize PHV. The energy and reduction equivalent required for PHA biosynthesis are provided by the degradation of intracellular saved polyphosphate (poly-P) and glycogen as well as substrate degradation in the tricarboxylic acid cycle.[81–83] The bulk phosphorus concentration in the anaerobic period is, therefore, increased with time. In the subsequent aerobic phase, where no external carbon source is present, the internally stored PHA are oxidized and used for cell growth, phosphate uptake and poly-P accumulation and glycogen synthesis. In the successful EBPR process, the aerobic phosphate uptake is much higher than the anaerobic phosphate release, which results in net phosphate removal. The carbon substrates obviously play an important role in the EBPR process since they are the raw materials of PHA biosynthesis.

Enhanced biological phosphorus removal (EBPR) is accepted as one of the most economical and environmentally sustainable processes to remove phosphorus (P) from wastewater. It is widely applied for the treatment of domestic wastes, which have typical P concentrations of between 4 and 12 mg PO_4-P L^{-1}.[84] However, adoption of EBPR for the treatment of industrial and agricultural wastewaters is less common. These high-strength wastewaters can be rich in phosphorus, reaching, for example, 125 mg PO_4-P L^{-1} in New Zealand dairy wastewaters.[85] There is limited knowledge about the ability of the EBPR process to deal with such high strength waste streams. The EBPR process relies on PAOs, which can be encouraged to take up significantly more phosphorus than is required for cell growth. In order to achieve a population high in PAOs, an anaerobic contact phase followed by an aerobic contact phase is required. During the anaerobic phase PAOs convert volatile fatty acids (VFA) into poly-β-hydroxylalkanoates (PHA). Energy for this process comes mostly from the use of stored polyphosphates, while the reducing equivalents are provided by the glycolysis of glycogen[86] or alternatively as a product of the TCA cycle.[87] As the polyphosphates are hydrolysed, orthophosphate is released resulting in an increased phosphate concentration in the bulk liquid. During the aerobic phase the PAOs use the internally stored PHAs as an energy source to take up orthophosphate and replenish their polyphosphate reserves. PHA is also used to drive cell growth and glycogen replenishment. During the aerobic phase the polyphosphates are accumulated and result in a net reduction of orthophosphate from the bulk solution. Phosphorus can then be removed from the system *via* wastage of the polyphosphate-rich biomass.

15.5 Factors Affecting PHB Production

15.5.1 Feast/Famine Conditions

Recently, much research emphasis has been laid on the production of PHAs by mixed cultures when exposed to a transient carbon supply. Activated sludge processes are highly dynamic with respect to the feed regime. The biomass subjected to successive periods of external substrate availability (feast period) and no external substrate availability (famine period) experiences what in the literature is often called an unbalanced growth. Under dynamic conditions, growth of biomass and storage of polymer occur simultaneously when there is an excess of external substrate. When all the external substrate is consumed, stored polymer can be used as carbon and energy source. In these cases, storage polymers are formed under conditions that are not limiting for growth. The

Production of Polyhydroxybutyrate (PHB) from Activated Sludge 463

storage phenomena usually dominates over growth, but under conditions in which substrate is present continuously for a long time, physiological adaptation occurs, and growth becomes more important. The ability to store internal reserves gives to these microorganisms a competitive advantage over those without this ability, when facing transient substrate supply. Among the mentioned systems for industrial production of PHAs, the feast and famine approach is the most promising because of high PHA accumulation. This approach promotes the conversion of the carbon substrate to PHA and not to glycogen or other intracellular materials.

Mixed cultures or co-culture systems have been recognized to be important for several fermentation processes. Several studies have claimed the integrity and effectiveness of the system using mixed cultures. This is due to its complex nature of dynamics and the difficulty in analysing the dynamics and control of the system having multiple microorganisms. Unfortunately, the previous researches only focused on the mixed cultures using two or three well-known bacteria. The mixed cultures using unknown bacteria have not been paid much attention due to inconsistent results.

In recent years many studies have been focused on the production of PHAs by mixed culture when exposed to a transient carbon supply.[88,89] Under these dynamic conditions, the biomass subjected to consecutive periods of external substrate accessibility (feast period) and unavailability (famine period) generates a so-called unbalanced growth. During the excess of external carbon substrate, the growth of biomass and storage of polymer occur simultaneously. The uptake is mainly driven to PHA storage with low consideration for biomass growth. The microorganisms are able to accumulate substrate as internal storage products in their cells. Usually these storage products are glycogen, lipids or PHAs. The final PHA content in the cells will be slightly increased during transient condition (shifting from feast to famine period).[90] After substrate is exhausted, the stored polymer can be used as an energy and carbon source to enable them to survive the famine period. Hence, the accumulated PHA will be degraded. Beccari et al.[91] reported that the activated sludge is able to accumulate PHAs up to 50% of cell dry weight.

15.5.2 Microaerophilic Conditions

Although activated sludge acclimatized under anaerobic–aerobic conditions accumulates PHA, there is no guarantee that anaerobic–aerobic operation of the activated sludge process is best for enrichment of PHA-accumulating microorganisms. Ueno et al.[92] and Saito et al.[93] found that sludge accumulated more PHB under aerobic conditions than under anaerobic conditions. Satoh et al.[94] introduced the microaerophilic–aerobic process where a limited amount of oxygen is supplied to the anaerobic zone of anaerobic–aerobic operation. In such conditions, microorganisms can take up organic substrates by obtaining energy through partial oxidative degradation. If the supply of oxygen is sufficient, the microorganism may be able to get enough energy for assimilative activities such as the production of protein, glycogen and other cellular components, *etc*. However, if the supply of oxygen is adequately controlled, the assimilative activity will be suppressed while letting the microorganism accumulate PHA. By using these conditions, PHA accumulators are selected regardless of the ability of microorganisms to accumulate poly-P or glycogen, and the selected PHA accumulators will have a lower tendency to accumulate glycogen. The metabolic model was formulated by Van Aalst-van Leeuwen *et al*.[95] and Beun *et al*.[89] The model includes the following reactions: synthesis of acetyl-CoA from acetate, anabolism reactions, catabolism reactions, electron transport phosphorylation, synthesis of PHB from acetyl-CoA, and the synthesis of acetyl-CoA from PHB.

15.5.3 Carbon/Nitrogen Limitation Conditions

When bacteria, able to store PHAs, are grown in imbalanced media, certain media components like the assimilable nitrogen source (*e.g*. $(NH_4)_2SO_4$) or others are depleted during microbial growth. In such cases, synthesis of important cell constituents like proteins, DNA, RNA, *etc*., cannot take place, and acetyl-CoA derived from incomplete oxidation of carbon compounds will now be fed into the tricarbonic acid cycle at a lower rate than during the normal growth phase. This key component will now enter the metabolic routes of PHA synthesis, and depending on the availability of other acyl-CoA derivatives, acetoacetyl-CoA (or other homologues) will be formed by a condensation reaction. Acetoacetyl-CoA is reduced to (*R*)-3-hydroxybutyryl-CoA, and further used for a polymerization to form the homopolyester poly(*R*)-3-hydroxybutyrate or copolyesters that will be stored by the bacteria in form of globular granules.[96] When growth limitation for PHA-accumulating cells is abolished (*e.g*. by addition of assimilable nitrogen to the production medium), PHA stored in the cells is depolymerized to the monomers *via* oligomers as intermediates. These can be further metabolized to CO_2 and water producing adenosine triphosphate, energy for the cell, in the same time. The same is true for pure PHA extracted from bacterial cells. The polyester can be degraded by extracellular depolymerases, which can be produced by a high number of prokaryotic and eukaryotic microorganisms.

15.5.4 Phosphate Limitation Conditions

Phosphorus can be efficiently removed from wastewaters by the activated sludge process which incorporates alternating anaerobic and aerobic periods. In this process, known as enhanced biological phosphorus removal (EBPR), the influent wastewater flows into an anaerobic zone where a considerable amount of phosphorus (P) is released by the sludge and external carbon substrates (mainly short-chain fatty acids) are taken up and stored as polyhydroxyalkanoates (PHAs). This mixed liquor then flows to an aerobic zone where the stored PHAs are used for growth and polyphosphate (poly-P) accumulation.[97,98] The stored PHAs are used for microbial growth and for glycogen and poly-P synthesis in the aerobic zone. The PHA, poly-P and glycogen storage capacities provide PAOs with a selective advantage in their competition for substrates in EBPR systems.

15.6 PHB Yields and Recovery Processes

Microorganisms which are able to quickly store available substrate and consume the storage to achieve a more balanced growth have a strong competitive advantage over organisms without the capacity of substrate storage.[99] Activated sludge accumulates PHA to around 20% of dry weight under anaerobic conditions. The PHA content of activated sludge can be increased to 62% in a microaerophilic–aerobic sludge process.[94,100] When compared with a pure culture (more than 88% of cell dry weight),[39] the merits of PHA production in open mixed culture would be an enhanced economy, a simpler process control, no requirement of monoseptic processing, and an improved use of wastes. A considerable effort has gone in producing PHA using mixed culture.

Several methods have been used as a recovery process for PHA. These methods include solvent extraction, sodium hypochlorite digestion and enzymatic digestion. Details of each method as well as their advantages and disadvantages will be discussed and summarized here. In most cases, bacterial biomass is separated from substrate medium by centrifugation, filtration or flocculation. Then, the biomass is freeze dried (lyophilized). Basically, mild polar compounds, e.g. acetone and alcohols, solubilize non-PHA cellular materials whereas PHA granules remain intact. Non-PHA cellular materials are nucleic acids, lipids, phospholipids, peptidoglycan and proteinaceous materials. On the other hands, chloroform and other chlorinated hydrocarbons solubilize all PHAs. Therefore, both types of solvents are usually applied during recovery process. Finally, evaporation or precipitation with acetone or alcohol can be used to separate the dissolved polymer from the solvent. PHAs are soluble in solvents, such as chloroform, methylene chloride or 1,2-dicholoroethane. These three solvents can be used to extract PHA from bacterial biomass. In addition, other solvents were also reported to be used to extract PHA, e.g. ethylene carbonate, 1,2-propylene carbonate, mixtures of 1,1,2-trichloroethane with water and mixtures of chloroform with methanol, ethanol, acetone or hexane.

Doi[101] described a chloroform extraction method. PHA is extracted with hot chloroform in a Soxhlet apparatus for over 1 hour. Then, PHA extracted is separated from lipids by precipitating with diethyl ether, hexane, methanol or ethanol. Finally, PHA is redissolved in chloroform and further purified by precipitation with hexane. Ramsay et al.[102] examined the recovery of PHA from three different chlorinated solvents (chloroform, methylene chloride and 1,2-dichloroethane). They obtained the best recovery and purity when biomass was pre-treated with acetone. The optimum digestion time for all three solvents were 15 minutes. Further digestion resulted in a reduction in the molecular weight (MW) of PHA. The degree of recovery when the biomass was pre-treated with acetone were 70, 24 and 66% when refluxed for 15 minutes with chloroform, methylene chloride and 1,2-dichloroethane, respectively. The level of purity of these three solvents under these optimum conditions were 96, 95 and 93%, respectively. Temperatures used with these three solvents were 61, 40 and 83 °C, respectively. The authors emphasized that extraction conditions have a great impact on the degradation of PHA during the recovery process.

Sodium hypochorite solubilizes non-PHA cellular materials and leaves PHA intact. Then, PHA can be separated from the solution by centrifugation. A severe degradation of polymers during sodium hypochorite digestion is frequently reported. Berger et al.[103] observed a 50% reduction in the MW of the polymers when the biomass was digested with sodium hypochorite.[39] Because sodium hypochlorite is a strong oxidant, care has to be taken to select suitable digestion conditions in order to maintain a high molecular weight of the polymers. Ramsay et al.[104] examined the PHA recovery process from *R. eutropha* using hypochlorite digestion with surfactant pre-treatment. Two different surfactants, namely Triton X-100 and sodium dodecyl sulfate (SDS), were investigated. Improvements in purity and molecular weight can be obtained by pre-treating with surfactant prior to the extraction with sodium hypochlorite. They reported that surfactant removed approximately 85% of the total protein and additional protein (10%) was further removed by sodium hypochlorite digestion. They also stated that this method resulted in a high MW of extracted PHA and the recovery time was reduced when compared to surfactant-enzymatic treatment or solvent extraction. In addition, a native PHA granule could be maintained during this treatment, which allows PHA to be used for more diverse applications in comparison to the solvent extraction method. Ramsay et al.[104] stated that the roles of surfactant are well understood. Generally, surfactant disrupts the phospholipid bilayer of the cell membrane and separates PHA granules from other cell materials. In addition, surfactant denatures or solubilizes proteins, *i.e.* facilitating the cell disruption (anionic surfactants, *e.g.* SDS, denature protein; non-ionic surfactants, *e.g.* Tritox100, solubilize protein).

Due to the high cost of solvent extraction, the enzymatic digestion method was developed by ICI. Steps in this process include thermal treatment (100–150 °C) to lyse cells and denature nucleic acids, enzymatic digestion and washing with anionic surfactant to solubilize non-PHA cellular materials. Finally, concentrated PHA from centrifugation is bleached with hydrogen peroxide. According to Steinbuchel,[108] ICI used a mixture of various enzymes during the enzymatic digestion step. These enzymes are lysozyme, phospholipase, lecithinase, proteinase, alcalase and others. These enzymes hydrolyse most of the non-PHA cellular materials but PHA remains intact. Braunegg et al.[109] reported that ICI used proteolytic enzymes, *e.g.* trypsin, pepsin, and papain, and mixtures of those enzymes. Fidler and Dennis[110] investigated a system to recover PHB granules from *E. coli* by expressing T7 bacteriophage lysozyme gene. The lysozyme penetrates and disrupts the cells, and causes PHB granules to be released. The system developed by Fidler and Dennis[110] used a separate plasmid and expressed it at a low level throughout the cell cycle. At the end of the accumulation phase, the cells were harvested and resuspended in the chelating agent, ethylenediaminetetraacetate. This activated the lysozyme to disrupt the cell structure and release PHA granules at a time when PHA accumulation reached the maximum. Triton X-100 was also added to assist the cell disruption. They reported the efficiency of lysis was greater than 99%.

PHB and other PHAs are readily extracted from microorganisms by chlorinated hydrocarbons. Refluxing with chloroform has been extensively used; the resulting solution is filtered to remove debris and concentrated, and the polymer is precipitated with methanol or ethanol, leaving low-molecular-weight lipids in solution. Longer-side-chain PHAs show a less restricted solubility than PHB and are, for example, soluble in acetone. However, the large-scale use of solvents is not economic commercially, and other strategies have been adopted.

15.7 Techno-economic Feasibility

A major drawback to the world-wide commercialization of PHBs is their high price compared to conventional petrochemical-based plastic materials. The current cost of PHB is still around 8–10 times higher than that of conventional plastics.[73] The reasons for high cost of PHB have been due to utilization of pure cultures and substrates; need to maintain sterile conditions, and complex solvent extraction and recovery of PHB from microbial cells. Most of the researches on PHB production have been concentrating on pure culture of microorganisms.[105,106] Activated sludge or mixed culture enriched under suitable/ideal culture conditions is a promising alternative for PHB production. In wastewater treatment, disposal of excess sludge poses disposal problems. Use of activated sludge instead of pure culture for PHB production has several advantages, which include: reduced cost of production, generation of wealth from waste and a viable environmentally friendly process. Ueno et al.[92] applied acetate to laboratory acclimatized anaerobic/aerobic activated sludge and reported that sludge accumulated more PHB under aerobic conditions than they did under anaerobic conditions. Satoh et al.[107] further introduced a new activated sludge process tentatively named as 'microaerophilic/aerobic' activated sludge process and demonstrated 62% PHB accumulation by the mixed microbial population of activated sludge using acetate as growth substrate. This yield was obtained at an incubation period of 30 h. Several methods have been developed for the recovery of PHBs (mostly PHB) from cells. Solvents such as chloroform, methylene chloride, dichloroethane, etc., have been used for extraction of PHB.[103,55] The authors have reported that by using the activated sludge, it is possible to augment the PHB yield up to 64%.[55]

Wider use of PHAs is prevented mainly by their high production cost compared with the oil-derived plastics.[47,43] With the aim of commercializing PHA, a substantial effort has been devoted to reducing the production cost through the development of bacterial strains and more efficient fermentation/recovery processes. From the literature, the major cost in the PHA production is the cost of the substrate.[25,39] Productivity also has an effect on the production costs. However, this is relative to the substrate, and downstream processing apparently has a weak effect on the final cost. A lower PHB content clearly results in a high recovery cost. This is mainly due to the use of large amounts of digesting agents for breaking the cell walls and to the increased cost of waste disposal.

15.8 Conclusions

Waste activated sludge was an effective source of PHB-accumulating microorganisms and though the PHB content using industrial wastewaters was lower than that reported earlier for other wastes, the process provided the benefit of converting otherwise un-utilizable and toxic waste into value-added product – PHB. This would not only utilize the excess sludge generated in a wastewater treatment plant and reduce the load on landfill sites, but would also contribute to a reduction in the cost of PHB production by avoiding sterile conditions and pure carbon sources for growth and maintenance of pure cultures. The added advantage of PHB production using activated sludge was the effective treatment of this wastewater with up to 60% COD removal.

References

1. S. Zhiyong, A. Ramsay, G. Martin and B. A. Ramsay, *Appl. Microbiol. Biotechnol.*, 2007, **75**, 475.
2. S. Khanna and A. K. Srivastava, *Process Biochem.*, 2005, **40**, 607.
3. G. Braunegg, G. Lefebvre, G. Renner, A. Zeiser, G. Haage and K. Loidl-Lanthaler, *Can. J. Microbiol.*, 1995, **41**, 239.
4. K. Fritsche, R. W. Lenz and R. C. Fuller, *Int. Biol. Macromol.*, 1990, **12**, 92.
5. K. Fritsche, R. W. Lenz and R. C. Fuller, *Int. Biol. Macromol.*, 1990, **12**, 85.
6. M. M. Bear, M. A. Leboucherdurand, V. Longlois, R. W. Lenz, S. Goodwin and P. Guerin, *React. Funct. Polymers*, 1997, **34**, 65.
7. Y. B. Kim, R. W. Lenz and R. C. Fuller, *Macromolecules*, 1991, **24**, 5256.
8. C. Abe, Y. Taima, Y. Nakamura and Y. Doi, *Polym. Commun.*, 1990, **31**, 404.
9. Y. Doi and C. Abe, *Macromolecules*, 1990, **23**, 3705.
10. Y. B. Kim, R. W. Lenz and R. C. Fuller, *Macromolecules*, 1992, **25**, 1852.
11. Y. Doi, in *Proceedings of International Workshop on Environmentally Degradable Plastics: Materials Based on Natural Resources*, 1999, p. 15.
12. G. Chen, *Proceedings of International Workshop on Environmentally Degradable Plastics: Materials Based on Natural Resources*, 1999, p. 33.
13. Y. Saito, S. Nakamura, M. Hiramitsu and Y. Doi, *Polym. Int.*, 1996, **39**, 169.
14. Y. Doi, A. Tamaki, M. Kunioka and K. Soga, *Makromol. Chem. Rapid Commun.*, 1987, **8**, 631.
15. M. Kunioka, Y. Nakamura and Y. Doi, *Polym. Commun.*, 1988, **29**, 174.
16. A. Steinbuchel, In *Biomaterials: Novel Materials From Biological Sources*, ed. D. Byrem, Macmillan, Basingstoke, 1991, p. 123.
17. E. A. Dawes and P. J. Senior, *Adv. Micro. Physiol.*, 1973, **10**, 135.

18. A. J. Anderson and E. A. Dawes, *Microbiol. Rev.*, 1990, **54**, 450.
19. F. A. Jackson and E. A. Dawes, *J. Gen. Microbiol.*, 1976, **97**, 303.
20. P. J. Senior and E. A. Dawes, *Biochem. J.*, 1971, **125**, 55.
21. P. J. Senior and E. A. Dawes, *Biochem. J.*, 1973, **134**, 225.
22. R. G. Lageween, G. W. Huisman, H. Prevstig, P. Ketelaar, G. Eggink and B. Witholt, *Appl. Environ. Microbiol.*, 1988, **54**, 2924.
23. M. J. De Smet, G. Eggink, B. Witholt, J. Kingma and H. Wynberg, *J. Bacteriol.*, 1983, **154**, 870.
24. M. Lemoigne, *Bull. Soc. Chim. Biol.*, 1926, **8**, 770.
25. L. L. Madison and G. W. Huisman, *Microbiol. Mol. Biol. Rev.*, 1999, **63**, 21.
26. J. M. Merrick and M. Doudoroff, *J. Bacteriol.*, 1964, **88**, 66.
27. R. A. Gavard, B. Dahinger Hauttecoever and C. Reymand, *C.R. Acad. Sci.*, 1967, **265**, 1557.
28. H. Hippe and H. G. Schlegel, *Arch. Mikrobiol.*, 1967, **56**, 278.
29. R. Griebel, Z. Smith and J. M. Merrick, *Biochemistry*, 1968, **7**, 3676.
30. R. Griebel and J. M. Merrick, *J. Bacteriol.*, 1971, **108**, 782.
31. O. Hrabak, *FEMS Microbiol. Rev.*, 1992, **103**, 251.
32. M. Booma, S. E. Selke and J. R. Giacin, *J. Elastomers Plastics*, 1994, **26**, 104.
33. G. De Koning, *Can. J. Microbiol.*, 1995, **41**, 303.
34. T. Yutaka and C. U. Ugwu, *J. Biotechnol.*, 2007, **132**, 264.
35. P. J. Barham and A. Keller, *J. Polym. Sci. B: Polym. Phys. Ed.*, 1986, **24**, 69.
36. T. Iwata, K. Tsunoda, Y. Aoyagi, S. Kusaka, N. Yonezawa and Y. Doi, *Polym. Degrad. Stabil.*, 2003, **79**, 217.
37. R. M. Lafferty, B. Koratko and W. Korsatco, Microbial production of poly-β-hydroxybutyric acid. In: H.-J. RehnG. Reed (eds.), *Biotechnology*. VCHWeinheim, Vol. 6b, Ch 6, 1988. p. 135.
38. P. A. Holmes, *Phys. Technol.*, 1985, **16**, 32.
39. S. Y. Lee, *Biotechnol. Bioeng.*, 1996, **49**, 1.
40. K. Tsuchikura, 'BIOPOL' Properties and Processing, In: *Biodegradable Plastics and Polymers*, Y. Doi and K. Fukuda (eds.), Elsevier Science, New York, 1994, p. 362.
41. W. J. Page, *Can. J. Microbiol.*, 1995, **41**, 1.
42. P. K. Shin, M. H. Kim and J. M. Kim, *J. Env. Polym. Degrad.*, 1997, **5**, 33.
43. J. I. Choi, S. Y. Lee and K. Han, *Appl. Environ. Microbiol.*, 1998, **64**, 4897.
44. J. I. Choi and S. Y. Lee, *J. Microbiol. Biotechnol.*, 1999, **9**, 722.
45. S. Y. Lee, J. Choi and H. H. Wong, *Int. J. Biol. Macromol.*, 1999, **25**, 31.
46. S. Collins, *Spec. Publ. Soc. General Microbiol.*, 1987, **21**, 161.
47. D. Byrom, *Trends Biotechnol.*, 1987, **5**, 246.
48. K. H. P. Meesters, *Report of Technical University of Delft*, Delft, 1998.
49. T. Mino, W. T. Liu, H. Satoh and T. Matsuo, *Proceedings 10th Forum Appl. Biotechnol. Brugge,* Belgium, 1996, p. 1769.

50. M. A. Hassan, N. Shirai, N. Kusubayashi, M. I. Abdul Karim, K. Nakanishi and K. Hashimoto, *J. Ferment. Bioeng.*, 1996, **82**, 151.
51. M. A. Hassan, N. Shirai, N. Kusubayashi, M. I. Abdul Karim, K. Nakanishi and K. Hashimoto, *J. Ferment. Bioeng.*, 1997, **83**, 485.
52. M. A. Hassan, N. Shirai, H. Umeki, M. I. Abdul Karim, K. Nakanishi and K. Hashimoto, *Biosci. Biotechnol. Biochem.*, 1997, **61**, 1465.
53. A. R. Nor Aini, M. A. Hassan, Y. Shiria, M. I. Abdul Karim and A. B. Ariff, *Mol. Biol. Biotechnol.*, 1999, **7**, 179.
54. R. D. Ashby, D. K. Y. Solaiman and T. A. Foglia, *Biomacromolecules*, 2005, **6**, 2106.
55. M. S. Kumar, S. N. Mudliar, K. M. K. Reddy and T. Chakraborti, *Bioresource Technol.*, 2004, **95**, 327.
56. A. A. Khardenavis, M. S. Kumar, S. N. Mudliar and T. Chakrabarti, *Bioresource Technol.*, 2007, **98**, 3579.
57. A. A. Kardenavis, P. K. Guha, M. S. Kumar, S. N. Mudliar and T. Chakrabarti, *Environ. Technol.*, 2005, **26**, 545.
58. S. N. Mudliar, A. N. Vaidya, M. S. Kumar, S. Dahikar and T. Chakrabarti, *Clean Tech. Environ. Policy*, 2008, **10**, 255.
59. A. A. Kardenavis, A. N. Vaidya, M. S. Kumar and T. Chakrabarti, *Waste Management*, 2009, **29**, 2558.
60. M. Kalyani, M. S. Kumar, A. N. Vaidya and T. Chakrabarti, *J. Env. Sci. Eng.*, 2007, **49**, 164.
61. S. Godbole, S. Gote, M. T. Latkar and T. Chakrabarti, *Bioresource Technol.*, 2003, **86**, 33.
62. P. Chakrabarti, V. Mhaisalkar and T. Charkrabarti, *Bioresource Technol.*, 2010, **101**, 2896.
63. D. Diomisi, M. Majone, A. Miccheli, C. Puccetti and G. Sinisi, *Biotech. Bioeng.*, 2004, **86**, 842.
64. C. Pozo, M. V. Martinez-Toledo, B. Rodelas and J. Gonzalez-Lopez, *J. Biotechnol.*, 2002, **97**, 125.
65. D. Dionisi, G. Carucci, M. P. Papini, C. Riccardi, M. Majone and F. Carrasco, *Water Res.*, 2005, **39**, 2076.
66. M. A. Hassan, O. Nawata, Y. Shirai, N. P. A. Rahman, P. L. Yee, A. Bin Ariff, M. Ismail and A. Karim, *J. Chem. Eng. Jpn.*, 2002, **35**, 9.
67. P. Kahar, T. Tsuge, K. Taguchi and Y. Doi, *Polym. Degradation Stability*, 2004, **83**, 79.
68. M. Koller, R. Bona, G. Braunegg, C. Hermann, P. Horvat, M. Krontil, J. Martinz, J. Neto, L. Pereira and P. Varila, *Biomacromolecules*, 2005, **6**, 561.
69. M. Majone, M. Beccari, D. Dionisi, C. Levantesi and V. Renzi, *Water Res.*, 2001, **42**, 151.
70. M. Beccari, D. Dionisi, A. Giuliani, M. Majone and R. Ramadori, *Water Sci. Technol.*, 2002, **45**, 157.
71. H. Satoh, T. Mino and T. Matsuo, *Water Sci. Technol.*, 1992, **26**, 933.
72. H. Satoh, T. Mino and T. Matsuo, *Water Sci. Technol.*, 1998a, **37**, 579.
73. J. Yu, *Biotechnology*, 2001, **36**, 105.

74. J. Hollender, D. Van Derkrol, L. Kornborger, E. Grerden and W. Dott, *Microbiol. Biotechnol.*, 2002, **18**, 355.
75. C. Lemos, C. Viana, E. N. Sagueiro, A. M. Rmas, S. G. Crespo and M. A. M. Reis, *Enzyme Microb. Technol.*, 1998, **22**, 662.
76. H. Satoh, T. Mino and T. Matsuo, *Water Sci. Technol.*, 1994, **3**, 203.
77. E. A. Dawes, Starvation, Survival and Energy Reserves, In: *Bacteria in Their Natural Environment*, M. Fletcher and G. D. Floodgate (eds.), Academic Press, Florida, 1985, pp. 43–79.
78. E. A. Dawes and P. J. Senior, *Adv. Microb. Physiol.*, 1973, **10**, 135.
79. C. H. Sasikala and C. H. Ramana, *Adv. Appl. Microbiol.*, 1996, **42**, 97.
80. C. D. M. Filipe and G. T. Daigger, *Water Environ. Res.*, 1998, **70**, 67.
81. M. C. Wentzel, L. H. Lotter, G. A. Ekarna, R. E. Loewenthal and G. V. R. Marais, *Water Sci. Technol.*, 1991, **23**, 567.
82. H. Pereira, P. C. Lemos, M. A. M. Reis, J. P. S. G. Crespo, M. J. T. Carrondo and H. Santos, *Water Res.*, 1996, **30**, 2128.
83. T. M. Louie, T. J. Mah, W. Oldham and W. D. Ramsay, *Water Res.*, 2000, **34**, 1507.
84. Metcalf and Eddy, Inc, 2003, *Wastewater Engineering: Treatment, Disposal, Reuse*, McGraw-Hill Book Company, New York, USA, p. 186.
85. P. O. Bickers, R. Bhamidimarri, J. Shepherd and J. Russell, *Water Sci. Technol.*, 2003, **48**, 43.
86. T. Mino, M. C. M. Van Loosdrecht and J. J. Heijnen, *Water Res*, 1998, **32**, 3193.
87. Y. Comeau, K. J. Hall, R. E. W. Hancock and W. K. Oldham, *Water Res*, 1986, **20**, 1511.
88. M. Majone, P. Massanisso, A. Carucci, K. Lindrea and V. Tandoi, *Water Sci. Technol.*, 1996, **34**, 223.
89. J. J. Beun, F. Paletta, M. C. M. Van Loosdrecht and J. J. Heijnen, *Biotechnol. Bioeng.*, 2000, **67**, 379.
90. M. F. M. Din, Z. Ujang, M. C. M. van Loosdrecht, R. Razak, A. Wee and S. M. Yunus, 2004, Accumulation of Sunflower Oil (SO) under Slowly Biosynthesis for Better Enhancement the Poly-β-hydroxybutyrate (PHB) Production Using Mixed Cultures Approach. Faculty of Civil Engineering (FKA), Universiti Teknologi Malaysia (UTM). Submitted to Journal FKA.
91. M. Beccari, M. Majone, P. Massanisso and R. Ramadori, *Water Res.*, 1998, **32**, 3403.
92. T. Ueno, H. Satoh, T. Mino and T. Matsuo, *Polym. Preprint*, 1993, **42**, 981.
93. Y. Saito, T. Soejima, T. Tomozawa, Y. Doi and F. Kiya, *Environ. Systems Eng.*, 1995, **52**, 145.
94. H. Satoh, Y. Iwamoto, T. Mino and T. Matsuo, *Water Sci. Technol.*, 1998, **38**, 103.
95. M. C. M. Van Loosdrecht, M. A. Pot and J. J. Heijnen, *Water Sci. Technol.*, 1997, **35**, 41.
96. K. Sudesh, H. Abe and Y. Doi, *Prog. Polym. Sci.*, 2000, **25**, 1503.

97. W. T. Liu, T. Mino, K. Nakamura and T. Matsuo, *Water Res.*, 1996, **30**, 75.
98. H. Satoh, T. Mino and T. Matsuo, *Water Sci. Technol.*, 1992, **26**, 933.
99. M. C. M. Van Loosdrecht, M. A. Pot and J. J. Heijnen, *Water Sci. Technol.*, 1997, **35**, 41.
100. H. Takabatake, H. Satoh, T. Mino and T. Matsuo, *Water Sci Technol.*, 2002, **45**, 119.
101. Y. Doi, *Microbial Polyesters*, New York, VCH, 1990.
102. J. A. Ramsay, E. Berger, R. Voyer, C. Chavarie and B. A. Ramsay, *Biotechnol. Tech.*, 1994, **8**, 589.
103. E. Berger, B. A. Ramsay, J. A. Ramsay and C. Chavarie, *Biotechnol. Tech.*, 1989, **3**, 227.
104. J. A. Ramsay, E. Berger, B. A. Ramsay and C. Chavarie, *Biotechnol. Tech.*, 1990, **4**, 221.
105. S. Godbole, Studies on the production of microbial polyesters from wastes using biotechnological route. Ph.D. thesis, Nagpur University, Nagpur, 1996.
106. H. Shimizu, S. Sonoo, S. Shioya and K. Suga, Production of poly-3-hydroxybuteric acid (PHB) by *Alcaligenes eutrophus* H16 in fed batch culture, In: *Biochemical Engineering for 2001*, Proceedings of Asia-Pacific Conference, Furusaki *et al.* (eds), Springer, Tokyo, 1992, p. 195.
107. H. Satoh, M. Takashi and T. Massew, *Int. J. Biol. Macromol.*, 1999, **25**, 105.
108. A. Steinbüchel, PHB and other polyhydroxyalkanoic acids, In: *Biotechnology*, Vol. 6. H.-J. Rehm, G. Reed, A. Pühler and P. Stadler (eds), Wiley, Weinheim, 1996, p. 403.
109. G. Braunegg, G. Lefebre and K. F. Genser, *J. Biotechnol.*, 1998, **65**, 127.
110. S. Fidler and D. Dennis, *FEMS Microbiol. Rev.*, 1992, **103**, 231.

Subject Index

Page references to *figures* and *tables* are shown in *italics*.

acrylate groups, polymerization techniques 31
Actinobacillus succinogenes 57
activated sludge, poly(hydroxybutyrate) (PHB) from 452–68
ADME (adsorption, distribution, metabolism and elimination) processes 388 90
agro-polymers *see* thermoplastic agro-polymers
Alcaligenes eutrophus see Ralstonia metallidurans
Alcaligenes latus 83, *169*
alginic acid, applications of 371–2
allylic substitution reaction, palladium-catalyzed *38*
alternating polymerization 38–9
amber, as a moulding material 2
American Chemical Society, Rubber Division 8
α-amino acid *N*-carboxyanhydride (NCA) 295–7
amylopectin 107–8, *109,* 198–9, 368
amylose 107–8, *109,* 198–9, 222, 368
Anaerobiospirillum succiniciproducens 57
anticancer drugs, controlled release 301–2
Art Deco plastics 2
artificial oxygen carriers 302–4
Aspergillus terreus fermentation 57

ASTM Standard Test Methods 336
autooxidation of oils 27–8
Azotobacter beijerinkii 325, 454–5
Azotobacter vinelandii 169

Bacillus megaterium 18, 80, 312, 456
Bacillus polymyxa 57
bacteria, PHA-degrading 317–18, 321
bacterial cellulose (BC) networks 114–15
Bakelite 8
bandages, adhesive 7
'bean soup,' fire-fighting 111
Bell, E 15–16
bile acids as polymer sources 280–2
billiard balls 8
bio-based polymer *see* green polymers
biocompatible polymers, synthetic 6
biodegradable polyesters
 braided polyester as a suturing material 6
 ecological applications 181–2
 future trends 185, 374–8
 medical applications 184–5, 391–408
 properties of commercial products *183*
 sources and classification of 149–51
 structural formulae and acronyms *152*

biodegradable polyesters (*continued*)
  synthesis, properties and degradation of 153–81
  see also thermoplastic agro-polymers
biodegradable polymers *see* green polymers
bioelastomers, biodegradable
  applications 244–5
  bile acids as sources 280–2
  biodegradability 246
  citric acid-related 271–6
  cross-linking 247
  definition and characteristics 243–5
  photo-curing of 254–6
  polycarbonate-related 264–7
  poly(Σ-caprolactone) (PCL) related 254–9
  poly(ester amide) 278–80
  poly(ether ester) 276–8
  poly(glycerol sebacate) 267–70, *271*
  polylactide-related 259–64
  poly(octamethylene maleate (anhydride) citrate) (POMaC) based 282
  poly(triol α-ketoglutarate) 283
  safety 245
  segmented polyurethane derived 248–53
  synthesis 247–8
biomass polymers *see* green polymers
biomedical applications
  biodegradable polyesters 184–5, 391–408
  of catgut 5
  cellular toxicity 408
  of cobweb 4–5
  genocompatibility and toxicogenomics of polymers 409–13
  green polymers 367–8, 388–414
  PLGA based copolymers 393–408
  poly(glycolic acid) (PGA) 391–3

poly(hydroxyalkanoate) (PHA) 91–4, 378
poly(lactic acid) (PLA) 375–6, 391–3
  of rubber, natural (NR) 5
  of silk 4
bionanocomposites
  applications of 121–2, 394–9
  cellulose fibres 103–5, 114–16
  chitin whiskers and chitosan 105–6, 116–17
  definition and general considerations 102–3
  natural rubber (NR) 114
  PLGA-based copolymers 394–9
  poly(lactic acid) (PLA) 111–14, 118–21
  soy protein particles 110–11, 117–18
  starch crystals 107–8, *109*, 117
bioplastics
  abiotic degradation 314
  biodegradation studies 315
  biotic degradation 314–15
  definitions 311–12
  global production capacity 95, *96*
biopolymer *see* green polymers
biopolymer-based nanocomposites
  as biodegradable materials 129–45
  dynamic-mechanical thermal analysis (DMTA) 138
  experimental assessments of properties 131–3
  morphological changes 133–8
Bois Durci, Kölsch Collection 1–2
bone tissue engineering, injectable systems for biocompatibility 425
  cellular behaviour 423, 424–5
  citric acid-based systems 439–41
  drug delivery systems 434–8
  future developments 441–2
  grafting systems 419
  implant biodegradation 423–4
  injectable cell vehicles 431–4

Subject Index

injectable ceramics 429–31
mechanical properties of bone constructs 422–3
network formation 425–9
porosity of the scaffold architecture 423
in situ cross-linking 419
system properties *430, 434*
traditional *vs.* injectable paradigm 420–1
Braconnot, Henri 16
braided polyester as a suturing material 6
Brown, A. J. 17
BS series assessments of biodegradability 336–7

caprolactone (CL) 299
carbohydrates and their derivatives 51–5
carbon dioxide emission 23
carbon neutrality 23–4
carboxyanhydrides *see* α-amino acid *N*-carboxyanhydride (NCA)
cardanol 49–50
Carothers, Wallace Hume 5–6
casein
  invention and history 9–11
  Kölsch Collection 1–3
  plastic artefacts and ornaments 11, *12*
  Spitteler's cat *10*
cashew nut shell liquid (CNSL) 49–50
castor oil
  composition 29
  epoxidized (ECO) 32
  properties and derivatives 27, 33–4
catechol (*o*-hydroxyphenol) 49
catgut, medical uses of 5
cationic polymerization 30
cellular toxicity of polymers 408
Celluloid
  dentures 8
  jewellery box *9*
  uses 7–9

cellulose
  applications 369–70
  bacterial cellulose (BC) networks 114–15
  bacterial production 17–18
  as a bio-based plastic 7–9, 198
  crystalline polymorphs 104–5
  fibres in bionanocomposites 103–5, 114–16
  Kölsch Collection 1
  lignin digests 46–8
  molecular structure *104*
  as a sugar source 51–4
cellulose acetate 370
cellulose nitrate 7–8
chemical cross-linking systems (CCS), bone tissue engineering 427
chemurgy movement 13
chicken feather fibre (CFF) reinforced PLA 120
chitin
  discovery and uses 16, 199, 370
  molecular structure 105, *106*
  whiskers 105–6, 116–17
chitosan
  history of 16–17
  sources and uses 105–6, 116–17, 199, 370–1
Christensen, casein plastics 11
cinnamoyl group, photodimerization of 40–3
citric acid
  bioelastomers 271–6
  bone tissue engineering, injectable systems for 439–41
cobweb, medical uses of 4–5
coconut oil 27
collagen, lattices and scaffolds 14–17, 372–3
composting
  see also green polymers
composting, degradation of green polymers
  assessment of biodegradability 336–8

composting, degradation of green
    polymers (*continued*)
  background and principles 332–4
  biodegradation mechanisms
    334–6
  chemistry of 339–44
  composting systems 344–5
  physical parameters of 344–5
  poly(Σ-caprolactone) (PCL)
    355–7
  poly(ethylene) 346–55
  poly(hydroxyalkanoate)
    (PHA) 346–8
  poly(hydroxybutyrate) (PHB) 458
  poly(lactic acid) (PLA) *351–2*
  polymer blends 338–9, *340–3*
  vermicomposting 345
copolymerization 34–40, 41
Corynebacterium glutamicum 57
crab shell, as a source of chitin 105
Cupriavidus necator 80, 81, 169, *312*,
    326
cyclic carbonates 297–9
cyclic diesters 299

dentures, celluloid 8
dextran 374
Diels–Alder reaction *25–6, 29–30*
differential scanning calorimetry
    (DSC) 219–21
differential temperature analysis
    (DTA) 224
drug delivery systems
  collagen lattices and
    scaffolds 14–17, 372–3
  functionalized poly(L-lactide)
    in 300–2
  injectable systems for bone tissue
    engineering 434–8
  PHAs as drug carrier scaffolds 94
  poly(ethylene glycol) 94
  poly(lactic acid) (PLA) 375–6,
    393–4
  poly(lactide-co-glycolide)
    (PLGA) 393–4
  polyurethanes 379–80

dynamic mechanical (thermal)
    analysis (DMA) 221–3
  of biopolymer-based
    nanocomposites 138

earthworms 345
EBENA 2
ebonite, Kölsch Collection 1
electrospun fibre mats 300–1
emulsification diffusion method,
    nanoparticle fabrication
    397–8
ene reaction *25–6, 29–30*
enhanced biological phosphorus
    removal (EBPR) 461–2
enhanced permeability and retention
    (EPR) effect 301–2
epoxidized oils
  castor oil (ECO) 32
  linseed oil (ELO) 32–3
  soybean oil (ESBO) 30–3
  triglycerides *25, 28,* 31–3
epoxy resin synthesis 39–40
Escherichia coli
  K12 strain 57–9, 80
  thermophilic hydrolase in 335
essential oils
  classification 34, *35*
  extraction 34
  structure of derivatives *35*
eugenol *25,* 34, 44–6, *47*
European Standards, packaging
    requirements 337–8
extracellular degradation of
    PHAs 316–25

fatty acids 24–9
fermentation as a source of
    monomers 55–9, *60*
fire-fighting 'bean soup' 111
Ford, Henry, "growing a car like a
    crop" 12–14
free radical polymerization (FRP),
    bone tissue engineering
    426–7
functionalized poly(L-lactide)

α-amino acid N-carboxyanhydride
(NCA) 295–7
  applications of 299–305
  cyclic carbonates 297–9
  cyclic diesters 299
  derivation from PLA 291–3
  lactones 299
  morpholine dione cyclic
    monomers 293–5
functionalized poly(L-lactides)
  see also poly(lactic acid) (PLA)
fungi, PHA degrading 317–18,
  319, 321
furan derivatives 54–5, *56*

Galen of Pergamon, medical uses of
  silk 4
gelatin
  applications 373
  biodegradation of
    polyhydroxyalkanoates 90
gelatinization of
  thermoplastic 203–4, 225–7
gene delivery polymers *409*
gene expression changes induced by
  PLGA *412*
genes and genetic loci
  phaB (acetoacetyl-CoA
    reductase) 85
  phaC (PHA synthase) 84
  phaG (acyl-CoA-ACP
    transferase) 85
genocompatibility of
  polymers 409–13
glass transition 31, 86, 115,
  138, 201–5, 210, *219*, *220*, 221–9
global production capacity,
  bioplastics 95, *96*
glucaric acid 58
glutathione derivatives 305
glycerol 29
green bionanocomposites *see*
  bionanocomposites
Green Chemistry concept 333
green nanocomposites *see*
  bionanocomposites

green polymers
  carbohydrates and their
    derivatives 51–60
  classification of *390*
  classification of associated
    monomers *25–6*
  environmental considerations 366
  essential oils 34–51
  fermentation as a source of
    monomers 55–9, *60*
  medical applications 367–8, 388–414
  natural biodegradable
    polymers 368–74
  philosophy of 22–4, 333, 388–90
  research initiatives 365–6
  synthetic polymers 374–83
  triglycerides and their
    derivatives 24–34
  see also composting; specific
    polymers
greenhouse gas GHG emissions,
  petroleum *vs.* bio-based
  polymers 95, *96*
growth factors as indicators of
  neovascularization 425
guaiacol (*o*-methoxyphenol) 49
guncotton *see* cellulose nitrate
gutta-percha, mass production of 9

haemoglobin polymers 303–4
Halsted, William Stewart 5
Hardened Wood, Kölsch
  Collection 1–2
Hevea brasiliensis (rubber tree)
  87, 114
Hooke, Robert 15–16
human soft tissues, mechanical
  properties of *245*
hyaluronic acid 373
Hyatt, John Wesley 7–8
hydrogels
  in biomedical polymers 370–8
  in bone tissue engineering 431–4
  for protein-immobilization 304
m-hydroxyphenol *see* resorcinol
o-hydroxyphenol *see* catechol

'in-vessel' composting systems 345
injectable ceramics in bone tissue engineering 429–31
interfacial deposition, nanoparticle fabrication 398–9
ion-mediated gelation systems (IGS) in bone tissue engineering 429
ISO series assessments of biodegradability 336–7

Julian, Percy Lavon 111

Klebsiella pneumoniae 57
Kocher, Theodor 5
kollodium 7
Kölsch Collection, plastic artefacts 1–4, 9, 19
kopal resin
  jewellery box *3*
  sources 2
Krische, Ernst 10
Kyoto Protocol 23

laccases 46–50
lactic acid *see* poly(lactic acid) (PLA)
lactones 299
Lemoigne, Maurice 18, 80, 312
Leuconostoc mesenteriode 17
ligatures, antiseptic 5–6
lignin
  digests *25,* 46–8
  printed circuit board from derivatives *48*
  sources and synthesis 34–5, 40, 46, 198
limonene *25,* 34–40
linseed oil
  epoxidized (ELO) 32–3
  polymerization of 30
  properties 27
liposome-encapsulated haemoglobin (LEH) 303
Lister, Joseph 5

macromolecule delivery systems, PLGA based 400–5
magnetite-PLGA composite nanoparticles 399–400
Manheimia succiniciproducens 57
medical applications *see* biomedical applications
metathesis polymerization 30
3-*O*-methacryloyl-1,2:5,6-diisopropylidene-D-glucose (MDG) 52
o-methoxyphenol *see* guaiacol
5-methyl-5-benzyloxycarbonyl-1,3-dioxan-2-one (MBC) 299
micro-organisms, storage polymers in 454–8
microparticles (MPs) 394–6
microspheres, PLA and PLGA for drug delivery 300
microstructures in thermoplastic agro-polymers *208*
monomers
  renewable (*see* green polymers from renewable monomers)
  thermoplastic agro-polymers *208*
morpholine dione cyclic monomers 293–5
Murray, William S. 11
myrcene *25,* 36–7

N-carboxyanhydride (NCA) of α-amino acids 295–7
nanocomposites *see* bionanocomposites; biopolymer-based nanocomposites
nanoparticles (NPs) 394–6
  fabrication of 396–9
  magnetite-PLGA composite 399–400
nanoprecipitation technique, nanoparticle fabrication 398
natural rubber (NR) *see* rubber, natural (NR)
neovascularization, growth factors as indicators of 425

Subject Index

network formation in injectable bone tissue engineering
  chemical cross-linking systems (CCS) 427
  free radical polymerization (FRP) 426–7
  ion-mediated gelation systems (IGS) 429
  self-assembly systems (SAS) 428–9
nitrocellulose *see* cellulose nitrate
nylon, uses of 6
Nylon 11 34

Oken, Lorenz 15–16
'open to air' composting systems 345
orthopaedic therapeutic carrier systems *437–8*
oxygen carriers, artificial 302–4

palm oil 27
Paracoccus denitrificans 326
Parkes, Alexander 7–8
Parkesine 7–8
Pasteur, Louis 17
Payen, Anselme 103
Penicillium funiculosum 319, 323–4
PET bottles, biodegradable 335–6
petroleum-based polymers 22, 24
  petroleum *vs.* bio-based polymers 95, *96*, 197–8
  synthetic polymers 22
PHA depolymerases
  biochemical properties 318–21
  hydrolysis mechanism 321
  producing bacteria and fungi 317–18, *319*, 321–4
  structure 321–2
pha genes 84–5
phenolic compounds (natural) and their derivatives 34–51
phenylpropanoids 34, *35*, 40–6
photo-curing bioelastomers 254–6
  polycarbonate-related 267
photodimerization of cinnamoyl group 40–3
Physick, Philip Syng 5

pinene *25,* 34–40
plastic artefacts, Kölsch Collection 1–4
plasticizers
  ethylene glycol in soy products 14–15
  in thermoplastic agro-polymers 201–2, 216–17, 232–3
plastics, incineration of 3
poly(1,8-octanediol co-citrate) (POC) 271–6
poly(α-hydroxy acids) 153–61
  see also specific acids
poly(alkyl cyanoacrylates) 382
poly(alkylene dicarboxylate)
  degradation of 177–81
  properties 177–81
  synthesis of 174–6
poly(alkylene maleate citrates) (PAMC) 439–41
polyamides 55, 58, *59,* 380
polyanhydrides 380–1
polycarbonate
  bioelastomers 264–7
  cyclic 297–9
polycondensation reactions 34, 150–1, 156–9
poly(Σ-caprolactone) (PCL)
  biodegradability 130–45, 164–6
  bioelastomers 254–9
  composting degradation of 355–7
  in nanocomposites 129–30, 377–8
  properties 164–6, 377
  synthesis of 161–4, 377
poly(ester amide)
  bioelastomers 278–80
polyesters *see* biodegradable polyesters
poly(ether ester) bioelastomers 276–8
poly(ethylene), composting degradation of 346–55
poly(ethylene glycol)
  in bioelastomers 277
  drug delivery systems 94, 375–6
  as a plasticizer 14–15, 119, 227
poly(ethylene glycol) maleate citrate (PEGMC) 439–41

poly(ethylene oxide) (PEO)
  brushes for protein separation and purification 304–5
  in drug delivery systems 301
poly(glycerol sebacate) bioelastomer 267–70, *271*
poly(glycolic acid) (PGA), biomedical applications 391–3
poly(hydroxyalkanoate) (PHA)
  biocompatibility 88–9
  as biodegradable products 3–4
  biodegradation of 89–90, 171–4, 311–27, 346–8
  bioelastomers 252
  classification 313
  composting degradation 346–8
  degrading bacteria and fungi 317–18, *319*, 321
  depolymerases 318–22
  economic considerations 95–7
  extracellular degradation of 316–25
  granule structure 312, *313*
  as green biofuels 94–5
  history of production 80, 312
  industrial applications 90–1, 378
  intracellular degradation of 325–6
  medical applications 91–4, 378
  monomer derivatives *82, 83*
  occurrence and biosynthesis 18, 79–86, 166–70
  properties 86–8, 171–4, 313–14, 378
  structure 80–1, 83, *167, 313,* 378
  substrates 86, *87*
poly(hydroxybutyrate) (PHB) -co-poly(hydroxyhexanoate) (PHBH) resins 453
  applications 457–8
  biodegradation 458
  biosynthesis 454–6
  limiting factors 462–5
  production systems from wastes 458–62
  properties 456–7
  as a short chain PHA 452–4
  techno-economic feasibility 467–8
  yields 465–7
poly(hydroxyvalerate) (PHV) 453
poly(lactic acid) (PLA)
  as a biodegradable product 3–4, 130–45, 159–61
  biomedical applications 375–6, 391–3
  in bionanocomposites 111–14, 118–21, 129–30, 375–6
  chicken feather fibre (CFF) reinforcing 120
  composting degradation of *351–2*
  drug delivery systems 375–6
  plasticization 119
  properties 159–61
  related bioelastomers 259–64
  size exclusion chromatography (SEC) of 142–5
  structure and classification *26, 112, 154*
  synthesis of 55–6, 154–9, 291–3
  see also functionalized poly(L-lactides)
poly(lactide-co-glycolide) (PLGA)
  biomedical applications 393–9
  current research 376–7
  drug delivery systems 393–4
  gene expression changes induced by *412*
  macromolecule delivery systems 400–5
  magnetite-PLGA composite nanoparticles 399–400
  structure *376*
  tissue engineering systems 405–8
polymersome-encapsulated haemoglobin (PEH) 303–4
poly(octamethylene maleate (anhydride) citrate) (POMaC) bioelastomers 282
polyolefins 129
polyols 28, 30–4
poly(phenylene oxide) 48–9
polyphosphoesters (PPEs) 381–2
polypropylene, as a suturing material 6
polysaccharides 198–9

poly(trimethylene terephthalate) (PTT) 56–7
poly(triol α-ketoglutarate) bioelastomers 283
polyurethane
 current research 379–80
 degradable segmented bioelastomers 248–53
 drug delivery systems 379–80
 monomer sources 32, 34
printed circuit board from lignin derivatives 48
proteins
 denaturing of 226–7
 secondary structure of 229–31
 separation and purification 304–5
 as thermoplastic agro-polymers 200, 204–6
Pseudomonas spp. 85, 169, 455
 P. lemoignei 322–3
 P. oleovorans 454
 P. putida 326

Ralstonia metallidurans 80, 81, 169, 455, 456
Ralstonia pickettii 335
renewable monomers, classification of 25–6
renewable polymer *see* green polymers
resorcinol (*m*-hydroxyphenol) 49
Rhizopus oryzae 335
Rhodopsuedomonas rubrum 455
Rhodospirillum rubrum 326, 456
ring-opening polymerization (ROP) reactions 30, 150–1, 155–9, 292–3
rubber, natural (NR) 114
 as a bio-based plastic 7
 in bionanocomposites 114
 dentures 8
 mass production of 9
 medical uses of 5

salting out method, nanoparticle fabrication 397–8
Satow, Sadakichi 13

Schleiden, Matthias 15–16
Schönbein, Christian Friedrich 7–8
Schwann, Theodor 15–16
self-assembly systems (SAS) in bone tissue engineering 428–9
sericin 6
shape-memory polymers, light-induced 41–2, *43*
shellac
 Kölsch Collection 1
 mass production of 9
silk
 history of 4–6
 medical uses of 4
silk fibroin
 chemical structure *4*
 sericin from 6
size exclusion chromatography (SEC) of PLA 142–5
Society of Plastics Engineers (SPE) 8
Society of the Plastics Industry (SPI) 8
solvent diffusion method, nanoparticle fabrication 396–7
solvent evaporation method, nanoparticle fabrication 396
soy protein
 as a biodegradable product 3
 concentrates (SPC) 110
 denaturing of 226–7
 extract (SPE) 110
 extraction and uses 110–11, 117–18
 isolates (SPI) 110
 isolation and history 12–15
 polymer mechanical properties *209*, 210–18
soybean oil 27
 epoxidized (ESBO) 30–3
Spitteler, Adolf 10
 Spitteler's cat inventing casein *10*
spontaneous emulsification method, nanoparticle fabrication 396–7
starch
 amylopectin 107–8, *109*, 198–9, 368
 amylose 107–8, *109*, 198–9, 222, 368

starch (*continued*)
  applications 368–9
  crystals 107–8, 117, 233–6, 368–9
  as a source of PHAs 87
  as a source of sugars 51
  thermoplastic starch (TPS) 198–9, 202–4, *209*, 210–18, 232–3
Streptomyces albus 335
Streptomyces lividans 88
structure–property–processing relationship *207*
styrenated oils 30–1
styrene 30–1
sugar diols 53–4
sugars as monomer sources 51–4
sunflower oil 27
synthetic polymers
  petroleum-based 22
  from renewable monomers 22–4

terpenoids 34–40
thermal analysis techniques *219*
thermal gravimetric analysis (TGA) 223
thermally induced gelation systems (TGS) in bone tissue engineering 428
Thermobifida fusca 335
thermophilic hydrolase 335
thermoplastic agro-polymers
  additives 201–2
  cellulose nitrate 24
  characterization considerations 206–10
  classification of *198*
  degradation processes 225–9
  mechanical properties *209*, 210–18
  monomers and microstructures *208*
  versus petroleum-based polymers 95, *96*, 197–8
  sources of 198–200
  synthesis of 200–6

thermal properties *209*, 218–36
  see also biodegradable polyesters
tissue engineering
  bone, biodegradable injectable systems for 419–42
  lattices and scaffolds 14–17, 19
  PHA scaffolds 91–4
  PLGA based systems 405–8
toxicity of polymers 408
toxicogenomics of polymers 409–13
trehalose 57–8
triglycerides and their derivatives
  composition and chemical structures 27
  monomers from 24–9
  polymers synthesized from 29–34
  synthesis of multifunctional reactive monomers from *28*
trimethylene carbonate (TMC) 299
tulipalin 51
tung oil 27
tyramine adducts 251–2

United Nations Framework Convention on Climate Change (UNFCCC) 23
urushi lacquer 50
urushiol *25*, 49–50

Van Tieghem, Philippe Édouard Léon 17
vanilla oil 50–1
vanillin 50–1
vermicomposting 345

Xylonite 8

yeasts and yeast-like organisms 57–8

Zoogloea ramigera 456